INTELLIGENT MANUFACTURING SYSTEMS 1997

SYSTEMS 1997

(IMS'97)

T0264378

A Proceedings volume from the 4th IFAC Workshop,
Seoul, Korea, 21 - 23 July 1997

Edited by

JONGWON KIM

School of Mechanical and Aerospace Engineering,
Seoul National University, Seoul, Korea

Published for the

INTERNATIONAL FEDERATION OF AUTOMATIC CONTROL

by

PERGAMON

An Imprint of Elsevier Science

UK	Elsevier Science Ltd, The Boulevard, Langford Lane, Kidlington, Oxford, OX5 1GB, UK
USA	Elsevier Science Inc., 660 White Plains Road, Tarrytown, New York 10591-5153, USA
JAPAN	Elsevier Science Japan, Tsunashima Building Annex, 3-20-12 Yushima, Bunkyo-ku, Tokyo 113, Japan

First edition 1997

Library of Congress Cataloging in Publication Data

A catalogue record for this book is available from the Library of Congress

British Library Cataloguing in Publication Data

A catalogue record for this book is available from the British Library

ISBN: 9780080430256

This volume was reproduced by means of the photo-offset process using the manuscripts supplied by the authors of the different papers. The manuscripts have been typed using different typewriters and typefaces. The lay-out, figures and tables of some papers did not agree completely with the standard requirements: consequently the reproduction does not display complete uniformity. To ensure rapid publication this discrepancy could not be changed: nor could the English be checked completely. Therefore, the readers are asked to excuse any deficiencies of this publication which may be due to the above mentioned reasons.

The Editor

Transferred to digital print 2009

Printed and bound in Great Britain by CPI Antony Rowe, Chippenham and Eastbourne

IFAC WORKSHOP ON INTELLIGENT MANUFACTURING SYSTEMS 1997

Sponsored by
International Federation of Automatic Control (IFAC)
 Technical Committee on Advanced Manufacturing Technology (MIT)

Co-sponsored by
IFAC Technical Committees on
- Robotics (MIR)
- Manufacturing, Modeling, Management and Control (MIN)
- Intelligent Autonomous Vehicles (TVI)
- AI in Real-Time Control (CCA)
- Real-Time Software Engineering (CCR)
International Federation for Information Processing (IFIP-TC5)
Korea Science and Engineering Foundation (KOSEF)
Automation and Systems Research Institute (ASRI) at SNU
Korea Advanced Institute of Technology (KAITECH)

Organized by
Engineering Research Centre for Advanced Control and Instrumentation (ERC-ACI)
Seoul National University (SNU), Seoul, Korea
on behalf of
Institute of Control, Automation and Systems Engineers (ICASE), Korea

PREFACE

The IMS'97, the fourth in the series of IFAC workshops on Intelligent Manufacturing Systems (IMS), was held in Seoul, Korea, on July 21–23, 1997. It was sponsored by the IFAC Technical Committee on Advanced Manufacturing Technology (IFAC-TC MIT), and organized by the Engineering Research Center for Advanced Control and Instrumentation (ERC-ACI) at Seoul National University, Seoul, Korea, on behalf of the Institute of Control, Automation and Systems Engineers (ICASE) in Korea.

Rapid progress that we are experiencing in the area of modern manufacturing is probably most evident through the developments in intelligent manufacturing systems (IMS). The same fast advancements, that make the field of IMS so exciting, have also made the objective of achieving a balanced technical program a challenging task. The International Program Committee (IPC) wanted the Workshop to include the most notable and recent results, but still to reflect the versatility of maturing IMS technologies

In the Workshop, the importance of intelligence in modern manufacturing has gained considerable recognition from engineers and researchers due to today's unforeseen manufacturing environment change. To further signify the importance, this Workshop focused on the issue "intelligent manufacturing," especially, with two intriguing keynote speeches, a special invited session on the worldwide IMS Project and two tutorial programs as well as the 64 papers from 16 countries worldwide. We do hope that this event has provided the excellent opportunity to identify the future trends as well as exchange and learn ideas and experiences in intelligent manufacturing.

We were most thankful to all members of the IPC of the Workshop and the IFAC-TC MIT for their guidance and contribution in enhancing the scientific content of IMS'97 by means of careful refereeing. We have to also acknowledge the financial support from the Korea Advanced Institute of Technology (KAITECH) and many Korean manufacturing industries. Finally, the special appreciation is due to the authors who enthusiastically participated in the Workshop.

Professor Marek B. Zaremba
University of Quebec in Hull, Canada
IPC Chairman of IMS'97

Professor Hyung Suck Cho
KAIST, Korea
NOC Chairman of IMS'97

Professor Jongwon Kim
Seoul National University, Korea
Editor for IMS'97

CONTENTS

CONCURRENT ENGINEERING

SCHEDULING II

MANUFACTURING MANAGEMENT SYSTEMS II

MANUFACTURING CELL DESIGN

NEURAL NETWORK APPLICATIONS

CAD/CAM/CAE I

ROBOTICS I

MANUFACTURING MANAGEMENT SYSTEMS III

PRODUCTION/PROCESS PLANNING

FUZZY THEORY AND APPLICATIONS

ROBOTICS II

MONITORING

CAD/CAM/CAE II

MANUFACTURING CELL CONTROL

REMOTE RENEWAL BY AGGREGATING REAL AND VIRTUAL MODELS IN MANUFACTURING SYSTEMS

Hiroyuki Goto
Toshiaki Kimura

*Japan Society for the promotion
of Machine Industry (JSPMI)
1-1-12, Hachiman-cho, Higashikurume-shi
Tokyo203, Japan
goto@tri.jspmi.or.jp*

Kenichiro Mori
Norio Yoshikawa
Tadaaki Hotta

*OMRON Corporation
2-5-48, Tadao, Machida-shi,
Tokyo 194, Japan
kenichiro_Mori@omron.co.jp*

Motoo Asamori
Youichi Kamio

*Toyo Engineering Corporation
8-1, Akanehama 2-chome, Narashino-shi,
Chiba 275, Japan
asamori@rd.toyo-eng.co.jp*

Yoshiro Fukuda

*Hosei University
3-7-2 Kagino-cho, Koganei-shi,
Tokyo 184, Japan
fukuda@is.hosei.ac.jp*

Abstract: In this paper, a new remote renewal method in manufacturing systems is proposed. In order to support renewal effectively the method includes not only the feed-forward structure of planning, design, construction, operation, and maintenance, but also the feed-back information about know-how and experiences of field operations. The method assumes a virtual enterprise environment which accounts for and unifies both manufacturing and design even though they typically operate remotely. The method supports gathering and categorizing data such as signals from equipment, trouble information, quality management information, operators' know-how and other information. A demonstration, based on actual field-gathered data and its associated renewal activities, verified the feasibility of the renewal-centered method.

Keywords: Renewal processes, Production systems, Manufacturing systems, Modelling, Computer Simulation

1. INTRODUCTION

The environment surrounding the manufacturing industries are changing substantially. These changes include structural changes in supply and demand, development and applications of new technology, and various new measures for ecological conservation.

Under these changes, the life cycle of manufacturing systems tend to become relatively longer compared with that of products. As the industries' policy for manufacturing system changes from scrap and build to renewal or improvement, renewal activity in the production life cycle is becoming more important than ever.

Although facility life cycle (Iwata, et al.,1995) and virtual manufacturing system (Onosato and Iwata,1993) were proposed, the industries are not sufficiently equipped with tools, methods and techniques that should consistently support all phases from design to renewal of manufacturing systems, and also should design, implement and operate manufacturing systems and maintain them for many years.

As production activities of enterprises become more international, it is often observed that a manufacturing system is operated remotely from the designing department. In such cases, remote operation (Mitsuishi,1995) and virtual enterprise (NIIIP,1995) were applied. For realizing the remote operation, it is important to ensure smooth information exchange between manufacturing departments, and to modify or improve the system efficiently at a low cost in accordance with changes in manufacturing environment.

Advances in information technology proceed rapidly, and new tools and techniques have possibility to apply to every phase of the life cycle of the manufacturing system.

In view of these trends, we propose a new method which aggregates real models and virtual models without a consciousness of geographical position. This method can implement, operate and evaluate the system in all phases from design to renewal on shop floor.

In this paper, Chapter 2 defines renewal activity. Chapter 3 describes methods of remote renewal. Chapter 4 discusses an experimental result using a demonstration system, and Chapter 5 describes concluding remarks.

This study has been conducted under "The Globeman 21 Project", a part of the IMS International R&D Program.

2. RENEWAL ACTIVITY AND KAIZEN

Nowadays the products life is becoming shorter than ever. If a manufacturing facility is composed of module units, then the unit can be replaced frequently. Manufacturers will be able to improve the performance of their manufacturing systems by conducting such renewal activity. This continuous improvement activity is called KAIZEN.

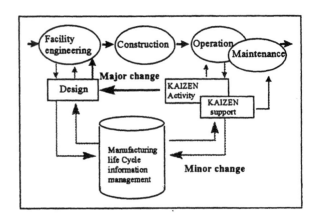

Fig.1. Life cycle of manufacturing system and renewals

However, KAIZEN activity can not always be applied to the global manufacturing environment, because it strongly depends on operators' know-how and experiences of field operations of the system, and will be very difficult to fully support with an information system.

This research is intended to realise the environment which remotely supports KAIZEN activity using up-to-date information technology. It is supposed that if the model of manufacturing system is made in the design phase, then the model information can be converted into electronic data. Information of field operations is monitored from the operating facilities. If any difference between the model information and the monitored information is observed, the monitored information is transferred to the model. Using the monitored information, a simulation is carried out to identify the cause of the difference. It is investigated and also counter measure is considered. After the consideration, the manufacturing system is re-configured to improve the performance of the system. This is defined as major change. In a another case of solution, the system is not re-configured. For example, only system parameters are changed to fit the model. This is defined as minor change. Finally all these activities are defined as the remote renewal. Relationship between life cycle of manufacturing system and renewal is shown in Fig.1.

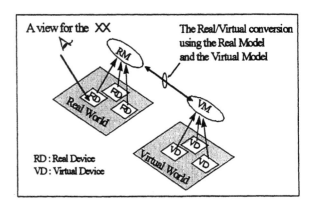

Fig.2. Conceptual scheme of virtual and real models

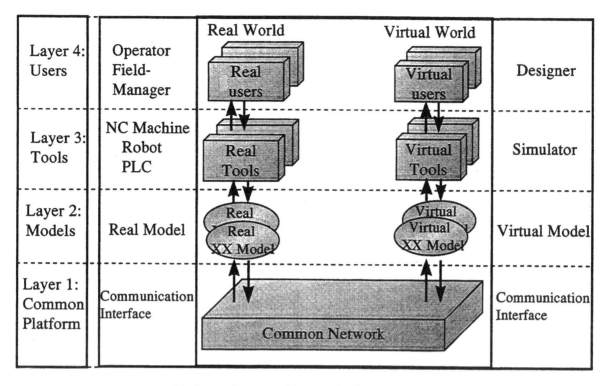

Fig.3. A reference architecture for function models

3. A METHOD OF REMOTE RENEWAL

3.1 Requirements for remote renewal activity

A manufacturing system consists of various devices which have their own information items and structures. Information of a device is independently defined and is not compatible with that of others, so it is difficult to exchange information among devices.

To realise information exchange, interface program is needed, which is not general but particular. Therefore a common communication platform is needed which shows the generic meaning of information independent of the structure dedicated to the device. Information will be exchanged at different places in the global manufacturing environment rather than in the same factory. The communication platform is required in the global manufacturing environment independent of the spatial distance.

3.2 Aggregation of virtual model and real model

An actually existing manufacturing system is composed of manufacturing devices, control information, management information, material flows, and transaction processes among them. Design, implementation and operation of a manufacturing system is very complicated because all the elements above are handled together, and

therefore problems are difficult to identify if happen. In order to identify problems, models viewed from a specific points must be created. For example, relating to the problem, a model will be made from many viewpoints such as productivity of manufacturing system, operability of devices, quality of products and other information.

Modeling discussed in this paper is limited to manufacturing devices. The devices are defined as a real model and a virtual model (Mori, et al.,1996). A real model is modeled from the viewpoint of control information on the basis of manufacturing devices in the real world. It defines information in the control system and transition of its states. If the modeling is limited to control systems, the behavior in the real world can be described using the real model. A virtual model represents manufacturing devices in several types of simulators or other information processing worlds (virtual world) . The virtual model is created from the viewpoint of control information. It expresses an ideal state of transition and control under given conditions in the virtual world. The behavior in the virtual world can be described using the virtual model . Fig.2 shows a conceptual scheme of aggregation of virtual model and real model.

In order to support the remote renewal method effectively, a new reference architecture for function model is proposed. In this concept, a hierarchically arranged group of services and functions are completed from a conceptual view point of control

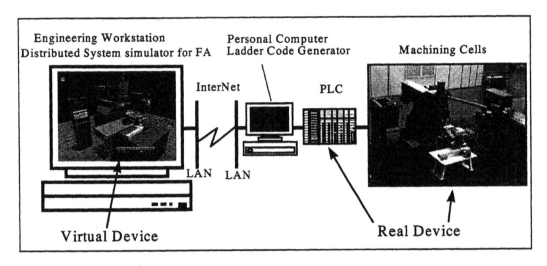

Fig. 4. Configuration of the demonstration system

Table 1 Definition of the real model

---

i) geometrical data
  · device
  · material
  · layout of devices
ii) state definition chart
  · motion sequence for devices
  · motion sequence for cells
iii) I/O control sequence
  · timing chart
iv) cycle time data
  · measured motion cycle time
v) control program
  · PLC
  · NC
  · Robot

Table 2 Definition of the virtual model

---

i) 3D computer graphics model
  · device
  · material
  · layout of devices
ii) discrete event definition program
  · control flow logic
  · device configuration
iii) I/O control sequence
  (not designed)
iv) cycle time data
  · ideal motion cycle time
v) control program
  (not designed)

---

information. Fig.3 shows the reference architecture for function models. The architecture has four layers. The common platform layer provides communication interface to the models layer; that is, this layer issupposed to aggregate virtual and real models. The models layer consists of virtual and real models. Applications such as Robot, NC,PLC and simulator are include in the tools layer. The users layer takes request and gives output to the users such as operator, field manager and designer. This reference architecture includes not only the feed-forward structure of planning, design, implementation, operation, and maintenance, but also the feed-back information about know-how and experiences of field operations. The architecture assumes the method aggregating the real and virtual environments which accounts for and unifies both manufacturing and design even though they typically operate remotely. The architecture supports remote gathering and categorizing of data such as signals from equipment, troubleshooting information, quality management information, operators' know-how, and other necessary information. Information will be exchange by the common communication platform at different place in the global manufacturing environment.

4. DEMONSTRATION

From the view point of control information, a demonstration, based on real field-gathered operation data and its associated renewal activities, was conducted to verifies the feasibility of the renewal-centered architecture. The demonstration used a discrete-event simulator, robot, conveyor, buffer storage and milling machine. The virtual environment and device were designed by OMRON Corporation. The virtual model was defined in Table 1. The real environment and device were independently built by JSPMI. The real model was defined in Table 2. Therefore, the virtual environment was located at OMRON in Machida City, and the real environment were located at JSPMI in Higashikurume City. All experiment data was exchanged through the Internet. The configuration of the demonstration system is shown in Fig.4.

4.1 Design and implementation of the virtual and real manufacturing system

Using a discrete-event simulator, device layout, machining time, transportation method, and other parameters in the virtual environment were modeled. The simulation in the virtual environment was performed to compare the system throughput of the configuration with buffer storage and without buffer storage.

As a result of the simulation, it was predicted that the system throughput satisfied the requirements of the system configuration without a buffer storage for machining Job 1. To operate the real machining-cell, a sequence of control was defined. By using the control information on the simulator, a control program was implemented, and downloaded onto the controller directly. Then the real machining-cell was operated by the downloaded program. The operation was consistent with the simulation. The implementation of the machining cells was completed without trial and error.

4.2 Improvement of the simulation by data feed-back from the real environment

Operating information, such as machining time and other information recorded in the machining cells were fed back to the simulator through the Internet. As to fluctuations and errors which were not predicted by the simulation at the design stage, their main cause was found to be from the insufficient expressing capability of the virtual device model. As the model on the basis of our analysis was modified, the discrepancy between the real field operation data and simulation became smaller and agreed.

4.3 Renewal of the manufacturing system

Using the modified virtual model, the system productivity with buffer storage, and without buffer storage was simulated for Job 2. In Job 2, the buffer storage condition is more effective than in the case of Job 1. By repeated simulation, it was concluded that the model with buffer storage was most suitable for both machining Job 1 and Job 2. Then the device layout in the real environment was renewed. As a

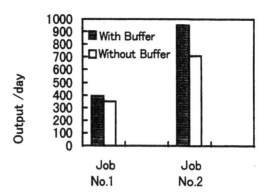

Fig.5. Result of renewal activity

result, the simulation using the modified virtual model enabled us to improve the system productivity. The result of renewal activity is shown in Fig.5.

5. CONCLUSIONS AND REMARKS

A new method for the remote renewal of the manufacturing system is proposed. It has been verified that the proposed new method is feasible and can efficiently realize the continuous improvement of manufacturing systems.

Some additional work will be necessary for the remote renewal of a practical system .

- Investigation of other information archiving for the remote renewal.
- Design for a multimedia data base architecture for renewal.
- Development of detailed models and simulations.

REFERENCES

Iwata, K. et al.(1995). Proceedings of EI'95, pp.154-167
Mori,K.et al.(1996).Proceedings of APMS'96, pp.309-314.IFIP

Mitsuishi,M. et al.(1995). Proceedings of IROS'95, pp.13-20.IEEE
NIIIP.(1995).NIIIP cycle0,Revision 6, The NIIP Consortium.
Onosato,M and Iwata,K.(1993).Annals of CIRP Vol.42,No1,pp.475-478.CIRP

Integrated Network for Decentral Intelligent Manufacturing Control and Automation

Prof. Dr.-Ing. Jürgen Gausemeier, Dipl.-Ing. Gerrit Gehnen

Heinz Nixdorf Institut, Universität-GH Paderborn
Fürstenalle 11, D-33102 Paderborn, Germany
Tel. +49 5251 606267, Fax +49 5251 606268
email: gausemeier@hni.uni-paderborn.de, gehnen@hni.uni-paderborn.de

Abstract: Information processing in manufacturing is today structured hierarchically. There is no close interaction between the seperate layers. To solve real world problems a more flexible approach is required. The presented approach is based on intelligent objects. These objects represent the manufacturing objects like robots, material flow systems, NC-machines or human workplaces. They communicate over a peer to peer network and fieldbusses. That approach integrates the layers of automation and manufacturing control, enables control over the Internet using JAVA-components and reduces the effort for configuration and operation of manufacturing control.

Keywords: Automation, Intelligent manufacturing systems, decentralized control systems, Networks, Objects

1. MANUFACTURING CONTROL AND AUTO-MATION - TWO SEPERATE WORLDS

The overall requirements like speed, reduction of costs and flexibility of manufacturing are increasing in a very fast manner. It is not necessary to go in details with these requirements in this paper. The most important requirement to fulfill is the flexibility of manufacturing processes. Lotsize one is common to many production processes today.

Information technology is used in the manufacturing in two layers: The automation layer and the control layer. There is a strong separation between these layers. The reason for this separation are the different structure of the processed information. While in the automation layer a large number of binary information with high real-time requirements has to be processed the data of the control layers are more complex like order data or quality data, but there is no requirement for hard real-time transmission.

It is very common to visualize the hierarchical structure of the control systems in manufacturing as a pyramid with the factorywide MRP-system at the top and the field layer at the bottom of the pyramid.

Information flow in the layers today is often networked. Cell controllers communicate over fieldbusses and coordinate the inter-cell material flow, applications on the manufacturing control layer use client-server architecture and access company wide databases through product data management systems.

Between the layers today a network is not state of art. Orders and process parameters are sent down from one layer to the next with increasing detailism, responses about the results are give up to the upper layers with increasing abstraction. Data exchange is usually done in batch runs once a day.

The separation of the layers, especially between the two lower layers causes problems regarding the flexibility of the manufacturing process. While the common view is, that the separation hides the structure and the functionalities between the layers and reduces complexity the separation disables the possibility of flexible control of manufacturing processes. With modern agent software architectures the required complexity handling is done.

2. FUNCTIONALITIES OF THE LAYERS

The functionalities of the two layers looks very different at the first view. The automation layer is characterized by hard realtime requirements and a large number of binary datasets. The basic functionalities of information systems at these level are the distribution of NC-data, the control of discrete binary and analogue actors and the monitoring with discrete sensors. There is no planning functionality at this level. Common for all automated systems are the large number of sensors and actors and as result the large number of binary information. Interaction with the user is done on binary level with lamps and pushbuttons or with small displays, showing one or two process parameters and the possibility to enter new values for these parameters. The environment requirements are strong. So most components are encapsulated in rugged cases which protect them from dust, liquid and electromagnetic interference.

The manufacturing control layer has different functionalities. The main task is the scheduling of manufacturing jobs and the distribution of these jobs to the workplaces, either automatic or manual. The manufacturing control today uses DNC-systems for the administration of instructions for manufacturing for each part. Quality control is done by integrated systems. Actions are taken by the manufacturing control system, when quality flaws occurs. The data processed by the manufacturing control system are more complex than in the automation system. These complex data are manufacturing orders, NC-programs, quality reports and may other. There is not the hard real-time requirement like in the automation layer. Manufacturing control systems interact with the user via graphical user interfaces like PC-based systems.

3. DECENTRAL APPROACHES

In the last few years different approaches are taken to decentralize the information technology in manufacturing as well as in other applications.

In the manufacturing control level there are different client-server-approaches and distributed systems, which are restricted in most cases to the order scheduling. With the large number of orders, which must be handled by intelligent manufacturing systems today, parallel scheduling is important. In the case of disturbances, which have impact to the production schedule, fast reschedule is necessary. But with the increasing complexity of production processes and

the amount of information technology for process control and monitoring it is necessary to go further. More and more central systems get bottlenecks of manufacturing control and automation. So decentral approaches are getting more and more popular in all types of information processing, not only in manufacturing automation. Complete decentral manufacturing control with scheduling, control and monitoring is topic of research at some places today (Gausemeier *et al.*, 1995; Kieß 1995; Schwall, 1996).

The approach of distributed manufacturing control is, that every object of the real word manufacturing like robots, NC-machines, material flow systems or human workplaces is modeled in holons or Intelligent Objects. Each Intelligent Object consists of functions for scheduling, control and monitoring of the real object (Figure 1). The scheduling of manufactuing jobs is done by negotiation between the scheduling parts of the Intelligent Objects. The control and monitoring parts are connected via fieldbus or point-to-point connection to the real device.

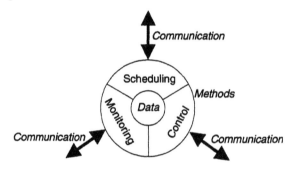

Fig. 1: The Intelligent Object

The Intelligent Object is designed object-oriented with a encapsulation of data. Access to the object is only possible over defined communication interfaces. The Intelligent Objects inherit mehtods and data from genericObject types.

The Intelligent Objects communicates over a software network with other Objects, the User via a graphical user interface or with external systems. The communication system has interfaces to the manufacturing process via fieldbus gateways. The communication system is desinged like a software bus. The entire system is called Intelligent Object Network or short ION. (Figure 2). The Intelligent Objects in the architecture represent the real world objects in the manufacturing process.

The Intelligent Objects are placed in PC's or workstations and communicate with each other. A PC is capable to manage a large number of these Intelligent Objects. Over networks like Ethernet more computers can be added to the system to scale up the entire performance. With implementation of the Intelligent Objects in standard programming languages and a communication system supporting heterogeneous systems it is possible to integrate high-end workstations or mainframes for scheduling while other parts of the manufacturing control systems run on cheap standard PC's.

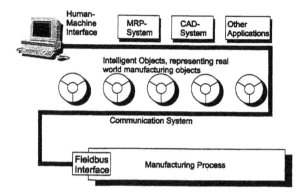

Fig. 2: ION-Architecture

In automation level the requirements are different as shown above. There is no scheduling but the requirement for direct process interface in this level. So the Decentral Intelligent Automation DIA was developed. At the moment two DIA-architectures are available at the market: the Controller Area Network CAN from Bosch (Etschberger 1994) and the Local Operating Network LON (LONWORKS 1993) from Echelon. While they are different in their capabilities and the network structure, common to both is, that they integrate cheap controllers (up to 60 with CAN, up to 32k with LON) over a network. Each controller is programmed with an individual application program. The controllers communicate over the network. The controllers have hardware interfaces for the field interface. Sensors and actors can be interfaced to the nodes as well as complex subsystems like robots (Figure 3).

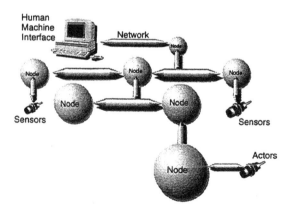

Fig. 3: Decentral Intelligent Automation Network

The applications on the nodes use the communication protocol of the network. LON for example is programmed with a special C-dialect called Neuron-C. This language bases on the network-protocol LONWORKS and defines network-wide variables, which are automatically transmitted over the network from one controller to each other. On the targets the change of the network-variable generates an event, which is processed by the application. The software on the nodes can be handled as functional objects with communication interfaces and properties.

These objects have three significant attributes: The software, the network communication interface and the required hardware interface. The software split up in the application code and configuration properties. Application code often is installed during manufacturing of a hardware component, the configuration parameters are set up during installation of the network.

In result there are six aspects of interfaces for a software object in the DIA-architecture, which the can be visualized as a cube:

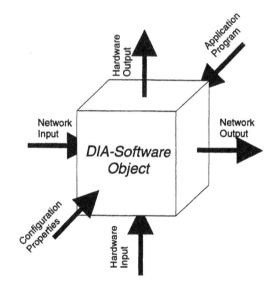

Fig. 4: DIA-Software Objects

Because the controllers, called nodes, are powerful enough, it is possible to integrate often more than one software object in a hardware node. The placement of the software objects in the hardware nodes depends on the required hardware interfaces, the computing power and the communication rate. If two software objects communicates with a high data rate it is wise to place them into the same node to reduce the network load (Figure 5). In larger DIA-networks it is not uncommon, that the network capacity is the limiting factor. So intelligent distribution of objects over the network is the key factor for performance of an DIA-solution.

In addition to these, for this special puropose developed approaches it is posible to use JAVA-technology for visualization and remote control of manufacturing processes. JAVA is the new object oriented hardware independent programming language developed by SUN Microsystems. Today in most cases JAVA is used only as tool for desining pages for the WWW or for lightweight office-applications. But the basic technology is well suited for the design of graphical human machine interfaces for distributed processes. JAVA components can be distributed over a TCP/IP-network with the RMI-specification. The user can so easy interact from every PC in the world connected with the Internet to the manufacturing system. For such applications security issues are very important.

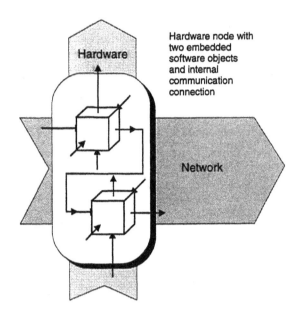

Fig. 5: Relation between software objects and hardware nodes.

4. INTEGRATED NETWORK ARCHITECTURE

Looking in detail at the ION-Architecture it can bee seen, that the relation between the three parts of a Intelligent Object can be loose, sometimes looser than the inter-object relations. So it is necessary for a performant system to distribute parts of a single Intelligent Object over the network. The scheduling part could be implemented on a high performance workstation with access to databases, while the control and monitoring are closer to the DIA-level. In addition the scheduling components of different Intelligent Objects are very similar and differ only in parameters while the control and monitoring are special for every real object in the manufacturing. So it is clear, that for optimal efficiency the three parts of Intelligent Objects should use different technologies. For the integration of manufacturing control and automation the control and monitoring parts are important.

The parts for control and monitoring can be implemented in different ways: either a monolithic object, implemented on a workstation like the scheduling, with a fieldbus interface for controlling a NC-based machine or as a composition of smaller objects - a decentral system within the decentral system. These decentral objects are coordinating subtasks of the Intelligent Object - in fact they are DIA-networks.

So the Intelligent Object is structured as a monolithic scheduling part, implemented on an workstation network and a control and monitoring part, implemented on a DIA-network. The interface between the parts is designed like the inter-object interface.

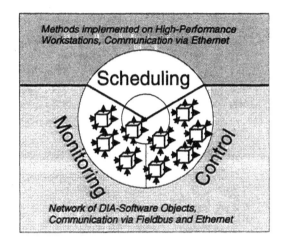

Fig. 6: Internal structure of Intelligent Objects

With the new network technologies the integration between ION and DIA is possible. In the implementation at the Heinz Nixdorf Institut ION is realized on UNIX-workstations and for DIA the LON-system is used. ION uses as communication middleware the PVM software standard and Objective-C as programming language. The LON uses Neuron-C, the LONWORKS network protocol and a component architecture called LONMARK, which implements the DIA-software objects (LONMARK 1995).

The link between these two systems is done via a PC-based interface under the operating system LINUX. LON supports a special node type called Host Application, which is programmed in any high-level-language and implements the necessary parts of the LONTALK protocol. The PC is equipped with a interface board to the network. The host-application converts the LONTALK-protocol into the higher level communication protocol.

The communication is not restricted. Parallel to the PVM communication a interface to JAVA-applications is possible. Based on the JAVA-Remote Method Invocation API and the Beans component architecture the system connects to WWW-based applications (Kuüger and Neubert 1996; SUN Microsystems 1996). The JAVA-Beans architecture is very similar to the DIA-software object architecture shown above. For network communication Beans use an event mechanism with registered listeners. At installation time parameters can be set up by the configuration property interface. Instead of hardware-interfaces to the process JAVA-beans use the possibility of modern GUI-systems. So they are well suited for creating HMI's on graphical screens.

The WWW-based JAVA-beans architecture enables new possibilities connecting manufacturing control with other applications. A new dimension of EDI can be opened with the data exchange between supplier and customer over secure internet-connections. Remote monitoring with access to every sensor and actor from every computer in the world is possible.

The upcoming JAVA-embedded API and JAVA-processors will support this trend.

LONMARK objects and JAVA-beans give the system configurator the possibility to introspect the features of the components. With a sophisticated configuration tool a plug'n'play operation of manufacturing control is possible. The features of each component can be detected and presented to the system. The material flow, the process structure and guidelines for scheduling can be given graphically. An example for graphical configuration tool is the LONTROL tool developed at the Heinz Nixdorf Institut in cooperation with FASTEC, a engineering company in Paderborn. The basic principle is, that the user draws the material flow layout with a PC-based tool. The back-end of the tool derives from this graphical layout the control data and creates the software for the control system. (Gausemeier *et al.* 1996).

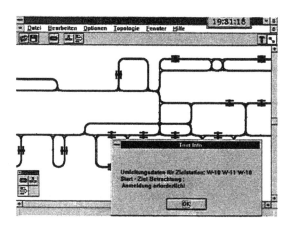

Fig. 7: Graphical material flow control with LON-TROL

5. BENEFITS

Networked objects reduce the complexity of the automation task. More than 50 percent of the costs in automation are driven by engineering efforts. Information technology give the chance to lower these costs by using component architectures. Changes in manufacturing processes are faster to handle by graphical component configuration. The user is able to create his personal view of the process with the Beans-architecture, while the manufacturing control system gathers and analyzes all relevant data.

The JAVA-Beans architecture opens all possibilities of today available Internet-technologies like remote monitoring and gives the possibility to generate hardware independent graphical user interfaces from building blocks.

The network uses the capabilities of the installed hardware in an optimal manner, tries to minimize the impact of disturbances by using the cooperating agent network and gives the today required flexibility. Lot-size one is no problem for the architecture. Together with NC-programs automation jobs like material or tool handling, quality management can be adapted. The configuration of the manufacturing control and automation system is easy even for untrained persons.

6. CONCLUSION

The integrated network for manufacturing control and automation based on component architectures gives more performance by decentral computation, a better user interface and a more flexible architecture for the control system. The architecture is currently successful tested at the Heinz Nixdorf Institut.

REFERENCES

Etschberger, K. (1994). *Controller Area Network: CAN; Grundlagen, Protokolle, Bausteine, Anwendungen,* Carl Hanser Verlag, München

Gausemeier, J.; Gehnen, G.; Gerdes, K.-H.: (1995) Intelligent Object Networks - The Solution for Tomorrows Manufacturing Control Systems. *Proceedings ICCIM '95,* Singapur.

Gausemeier, J.; Buxbaum, H.-J.; Förste, S.; Gehnen, G. (1996) Decentral Control Architecture for Modular Flow Systems. *Proceedings CAD/CAM Robotics and Factories of the Future,* London, England.

Kieß,J.U. (1995) *Objektorientierte Modellierung von Automatisierungssystemen,* Springer Verlag, Berlin, Heidelberg, New York.

Krüger, J.; Neubert, A. (1996) :Maschinendiagnose über das Internet, *ZwF* **91,** 12, Carl Hanser Verlag, München.

LONWORKS Engineering Bulletin (1993) *LONTALK Protocol,* Echelon Corp., Palo Alto

LONMARK (1995) *Application Layer Interoperability Guidelines ,* Lonmark Ass., Palo Alto

Dr. Schwall, E. (1996) Software-Standardisierung und -Flexibilität bei fertigungsleittechnischen Kundenprojekten, *Beitrag auf dem Workshop Informationssysteme für die Fertigung,* iwb TU München,

SUN Microsystems Inc. (1996) *JAVA Beans Specification,* Mountain View, California

FACILITATING MANUFACTURING MESSAGE TRANSFER AND EXECUTION [MMTE] IN A FMC ENVIRONMENT

A.K.Shrivastava
W.Zhao

Mechanical Engineering Department
Monash University, Australia

Abstract: Manufacturing messages play a very critical role in FMC (Flexible Manufacturing Cell) systems to send and initiate control signals to operate various computer controlled participating machines. This paper describes the issues and a solution related to the transfer of data and operational messages necessary to operate a fully automated FMC, comprising of systems from various vendors. It also presents an alternate ISO-OSI based Manufacturing Message Transfer and Execution [MMTE] communication software which was developed with file transfer and manufacturing message process capabilities to facilitate the operations of a FMC comprising of non-standard multi-vendor machines.

Keywords: Flexible Manufacturing Systems, Integrated Manufacturing Systems, Data Handling Systems.

1. INTRODUCTION

FMC is a manufacturing system, used to produce products in small to medium batch sizes, including a one-off job, effectively. This ability of FMC makes it suitable to react quickly to customers' requirements. FMC can be a subset of CIM (Computer Integrated Manufacturing) system, where its elements are interconnected through computer communication network (Shrivastava, *et al.*, 1993; Anuar, *et al.*, 1990). An FMC can be a single vendor turn-key system. The data and manufacturing message transfer in a turn-key system is specifically pre-designed so that the participating machines can be operated smoothly. However, more desirable FMC systems comprise of multi-vendor machines and elements to take advantages of specially designed CNC machines more suitable for certain job-shop situations. Although CNC machines used are capable of working independently as stand alone machines, they may not work together in conjunction with each other because of the compatibility requirements. The solution of these problems on individual basis can be very expensive and lead to a low performance of the system (Rizzardi, 1988; MAP users group, 1986).

In 1980, General Motor (GM) conceived the ideas of Manufacturing Automation Protocol [MAP] (Rizzardi, 1988; MAP users group, 1986) to solve the incompatibility problems and to facilitate the exchanges and processes of manufacturing messages in multi-vendor manufacturing systems. This protocol was developed on the basis of ISO-OSI seven-layer reference model. In this model, File Transfer, Access, and Management [FTAM] (International Standard Organisation, 1991) is used to deal with files, and Manufacturing Message Specification [MMS] is used to exchange and process manufacturing messages. The FTAM has been defined by both FTAM protocol standard ISO - 8571 (International Standard Organisation, 1991) and MMS (Rizzardi, 1988; MAP users group, 1986) protocol standard ISO - 9506 to deal with files. However, the MMS protocol standard recommends to use the FTAM if necessary, because the FTAM can deal with more types of files whereas MMS only supports unstructured binary files as per the requirements (Rizzardi, 1988). Within FMC, file transfer is one of manufacturing message services, and file types are usually ASCII files, or binary files. It would appear that FTAM is more suitable to deal with files in FMC implementations. On the other hand, MMS has been used to facilitate manufacturing message flowing and processing for such events as monitoring, result reporting, program executing, message exchange and so on, amongst programmable devices on job-shop floors. Application of ISO-OSI protocols can provide a solution to incompatibility problems, but the solution are expensive and tend to have lower performance.

In this paper, application of a new cost effective architecture developed on the basis of OSI reference model in a multi-vendor FMC environment has been discussed. This architecture provides low cost and improved performance solution to MMTE. The file transfer codes and protocols were developed using the FTAM protocol standard ISO - 8571, which in turn was integrated by other manufacturing message protocols developed or utilised to suit an FMC comprising of multi-vendor devices interconnected via a star network with RS 232 interfaces. The modified OSI protocols use an altogether distinct three layer architecture for transferring files in the FMC environment. It differs from the three layer Enhanced Performance Architecture (EPA) of the MAP (Rizzardi, 1988) as follows:

- It uses the FTAM at the application layer, while the EPA uses MMS.
- It uses connection mode services at the data link layer, while the EPA uses acknowledged connectionless mode services.
- It uses RS 232 interfaces as its physical layer implementation, while the EPA uses the IEEE 802.4 standard.

Fewer layers used provide improved system performance. RS 232 interfaces are still very popular because of their cost effectiveness and ease of implementation in low cost FMC applications (Zhang, et al., 1993).

FATM was used to create open system implementation environment, and was integrated with other services (not conforming to MMS protocols) provided by the participating software, such as message processing and execution, interlocking and alarm reporting. The purpose of these services was to adequately utilise manufacturing messages to achieve coordinated work between participating machinery to provide flexible manufacturing environment.

Data link layer provides data exchange using the developed MMTE model. The performance tests of this architecture were carried out using the data throughputs and device utilisations at the data link level (Lazowska, et al., 1984). Data throughputs indicate systems' capacities. The test result of data throughput were satisfactory, as they were almost equal to the specified data rate for different buffer sizes defined by verse frame lengths. The utilisation tests were used to reveal the bottleneck of the system. This information can be used for

potential improvements. The test results indicated that the system performance was almost equal to the specified data rate transfer, which is not the case in OSI systems (Dawyer, et al., 1987). It is therefore clear that this model has potential for the implementation of open system architecture for FMC systems at a moderate cost without compromising the performance.

2. DESIGN OF MMTE USING THE MODIFIED ISO-OSI ARCHITECTURE

The MMTE software known as I2-R2 , has been developed to provide file transfer services based on the FTAM protocol standards, and to complement and integrate the services offered by the commercially available Operating system (MS DOS) SCADA (WIZCON), Network (Makedrv) , PLC (STEP-5), CNC (DNC5), and Robot (RTX-ARMP) programs used to operate the FMC. As mentioned the MMTE architecture consists of three layers, namely application layer, data link layer and physical layer. The application layer had been designed using the FTAM standard ISO-8571 (International Standards Organisation, 1991). The data link layer, connection mode services have been defined by ISO-8802 standard (International Standards Organisation, 1989). The physical layer uses non ISO-OSI RS 232 standard, for which naming and addressing techniques as defined by ISO 7498-3 standard (International Standard Organisation, 1988), has been used.

In application layer the FTAM protocol facilitates the file transfer and execution. The FTAM based necessary codes for MMTE have been developed using the coding rules specified in Abstract Syntax Notation One (ASN.1) of ISO 8825. This protocol supports text files and binary files, using the class 0 service as defined by the FTAM protocol standard. The operations of this protocol, such as initialisation, file select, file open, file read or write, file close, file deselect, and termination are based on the achievable agreements codes using the ASN.1. If the agreement conditions are satisfied between the two participating devices, then the file transfer activity goes ahead, otherwise initialisation is aborted.

The Data Link Layer (DLL) comprises of Logical Link Control (LLC) and Medium Access Control (MAC) sub-layers. The design of the LLC protocol of this work follows the ISO 8802-2 standards. The design of MAC frames follows the ISO 8802-1 standard; it also refers to other frame structures currently used. The connection-mode services are used, to provide sequential control, error control and flow control services for reliable data delivery. To enhance the performance at the data link layer, fewer address space is used in a MAC frame because of the limitations imposed by the number of RS 232 interfaces that can be used. Because of reduced address space this data link layer design results in a more efficient and reliable protocol.

The basic design at the physical layer follows RS 232 standard, including cable wiring and parameter setting. The techniques which are common to most applications of this interface have been studied and applied. The details of the techniques are described in the following section. To implement this interface to the modified OSI architecture, an additional address service named directory facility service has been designed. This service uses basic service to link abstract addresses, logical addresses, and physical addresses, and provides its service to link device titles to the abstract addresses. The linkage of titles and abstract addresses are registered by the directory facility service. The service begins with asking system users to input device titles to corresponding abstract addresses at the start of a working session. After the connection is initiated, the service asks the users to choose one of devices that needs to be connected for further communication. It then provides the specific linkage of requested addresses including LLC address, MAC address, abstract address, logical address, and physical address. The advantages of this design are its flexibility and user friendliness. The flexibility is achieved as any required titles can be assigned at the beginning of a working session. In addition, system users need not deal with abstract addresses all the times, because they can be set at the beginning of a working sessions

Experiments of this design have been conducted at the application layer and data link layer. This paper presents the test result at the data link layer, because it reflects basic property of the system and has higher application values. The throughput test result at the data link layer is presented in Figure -1.

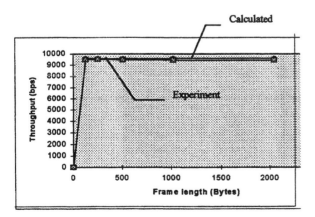

Figure-1:Mean throughput tested at the data link layers

As the rated speeds of PCs' are much faster than the maximum data transfer speed through the media

or link's, the throughput of the modified model is close to the data rate, even if the buffer size is set as 128 bytes. The utilization of each device is also assessed. The assessment indicated that the link is the bottleneck and should be used to determine the system performance. The test results also indicate that the overheads at the data link layer do not have any significant influence on the system performance. The suggested architecture therefore provides an efficient and very cost effective design that can be used to modify many FMCs that are still using RS 232 interfaces.

3. BASIC TECHNIQUES USED TO DESIGN RS 232 INTERFACES

The FMC used as a test bed has been designed as a machining centre (Shrivastava, et al., 1993). The programmable devices of this cell are linked by a star network using two standard RS 232 interfaces and one multipoint RS 232 port (8 ports) through a host computer. The number of nodes to be serviced and the low cost of implementation decided the choice of this network. In this cell, a CNC lathe, a CNC mill, a conveyor, and two robots are integrated by a host computer and a PLC via RS 232 interfaces.

Because of the importance of the RS 232 interfaces in this work, fundamental techniques of RS 232 interfaces have been reviewed as they determine this system's configuration ability or expansion flexibility, which in turn provides a basis for implementing RS 232 interfaces at the physical layer of the modified OSI architecture. The configuration ability provides a system the ability to deal with architecture changes, including hardware architecture changes and software architecture changes. To configure a system, both hardware compatibility and software compatibility aspects of that system should be estimated closely. Sometimes a hardware's configuration ability is limited by the ability of its supporting software. It is not always the case that a communication software can provide a proper support to configure its hardware. The software configuration ability is mainly determined by the ability of the software drivers designed to support the corresponding hardware configuration. Among many aspects of a software driver, the address techniques are important because these influence the configuration ability in implementing RS 232 interfaces.

The abstract addresses method to address various RS 232 ports has been applied. The abstract addresses are linked to corresponding logical addresses, through physical connections with the help of software drivers. In the case of standard port on the host, software driver 'STD.SYS' provides these services. In the case of multipoint adapter on the host software driver 'TH.SYS' provides these services . In the case of robot, software driver 'RBT.SYS' provides these services. The 'STD.SYS', 'TH.SYS', and 'RBT.SYS' are software interrupt drivers generated by 'Makedrv.exe' (Decision-computer international, 1990).

In the I2-R2 software, the functions 'Com-port checking and modification' and 'System configuration' are designed to provide services to check, modify and configure the RS232 settings and the corresponding connected devices. The function 'Com-port checking and modification' is an abstract symbol to a system user. This function is linked to the software 'Modeset.exe' supplied with the multipoint communication adaptor. The system users can use this function to check the setting of each port, and to modify the setting, if necessary. The function 'System configuration' can be used to reconfigure the settings of the multipoint adapter and standard ports, including addresses and interrupt handling, buffer size assignment and so on. For instance, if one robot needs to be added to the system, when this robot is plugged in the system, the information channel for this robot becomes available to the system user. In this way, regardless of whether programmable devices are added to, or dropped from the system, or change positions, as long as the devices exist and users know their abstract addresses in the system, these devices are available to the system users because the software drivers are set to provide full support to their hardware counterpart

I2-R2 provides more flexibility to add or drop devices in a cell or from manufacturing processes. This feature provided an open system architect for future expansion or modification of the FMC.

4. MANUFACTURING MESSAGE TRANSFER AND EXECUTION (MMTE)

In the FMC, distributed and hierarchical software architectures are used by the host and element controllers, to coordinate manufacturing message transfer and execution to drive the entire system. The softwares are independent of each other, but they are interlocked when in the system run mode.

The manufacturing messages are exchanged and processed through the functions provided by the application softwares to coordinate activities among the participating elements of systems. Manufacturing messages can be divided into three group: control messages, status messages, and report messages. They can also be classified into off-line or on-line messages. These messages are exchanged among the cell supervisory software, which are Wizcon and I2 part of MMTE (I2-R2), the element server software R2, PLC program, robot arm movement programs, and CNC machine programs. Among these, off-line manufacturing message

processes include CNC machine code generation and robot arm movement program generation.

CNC machine codes can be generated by using CAD\CAM package in this case CADDSMAN (The ISR group Limited, 1992). Robot arm movement programs can be created using robot program library ARMP.TPU (Universal Machine Intelligence Limited, 1986). These programs can be downloaded to the robots and the CNC machines using on-line services provided by the softwares, which are DNC5 for the mill, COPY command of MS-DOS for the lathe, I2-R2 for the robots.

Wizcon and I2-R2 provide on-line message processing used in this cell. They can be used for operation sequence and cycle control, event log, alarm report, operator communications, file transfer and management, process result record and report, execution of remote programs, and remote variable write.

On the host computer, Wizcon and the I2 serve as cell control softwares. These softwares complement each other to coordinate the cell. Under the control of Wizcon, PLC programs coordinate cell process sequences. The R2 program serves as server software on the robots to coordinate robots' activities. The CNC control programs are used to control machine tools.

The PLC when instructed by Wizcon exchanges the messages with cell elements to achieve the required control. It also allows the PLC to set certain conditions to authorise the activities of the machine tools and robots. Besides, Wizcon can be used to choose process sequence according to the features of parts. In addition, it logs events to keep track of cell activities, and report alarm situation such as broken state of the light guard system. Wizcon can display cell activities dynamically in real-time, and provides an operator communication channel to system users on the cell controller. Thus, this software serves as a cell supervisor to coordinate certain activities in this cell.

To operate this cell properly, the I2 is used to process certain messages that can not be processed by Wizcon, but are essential for the cell operation. The I2 provides 'FTAM' service, to ship robot arm movement files to the robots. These files are essential to operate robots according to different part process sequences. The I2 also uses 'FTAM' function to upload files that record process results to the cell computer. These data are necessary to production management. It can tell the server software R2 on the robots to execute a robot arm movement program based on process sequences, and assign the number of process cycles based on the number of parts required to be processed. Limited remote file system management is provided. For example, it can request robot server software R2 to report its the contents of directories on the robot controller.

As a robot server software, the R2, which is a counterpart of the I2, has been designed to coordinate activities on the robots. It positions itself between the cell control software I2 and robot arm movement programs. A server software on an element controller is necessary (Anuar, and Mohamad, 1990; Zhang, and Mo, 1993). The R2 provides a manufacturing message environment for the robots and makes flexible message flow possible. Without it, the cell can not be used effectively. The advantage of this software architecture on the robot controllers is the independence of the participating softwares. Because of this independence, the R2 resides on the robots permanently, while robot arm movement programs can either reside on the robots for a period if the same or similar programs are to be used for several products, or can be downloaded to the robots to cater for new products. In addition to that, resided robot arm movement programs can also be overwritten to enhance memory utilisation.

The R2 uses its 'FTAM' function to receive files from the cell computer and store them to a filestore according to the instructions from the I2. It also can receive the number of process cycles from the I2, and then deliver it to robot arm movement programs as the number of parts required to be processed. It also receives the execution command from the I2 to enable a specific arm movement program. It also creates a common filestore for arm movement programs to report process messages. It ships the files recording process messages to the cell computer, using the 'FTAM' function.

On the CNC machine controls, different manufacturing message protocols are used which are designed by the suppliers based on their requirements. A similar hierarchy software architecture can be identified on the CNC machines. Their basic control softwares, which are equivalent to the R2, are used by these machine tools to provide a manufacturing message environment to coordinate activities within the machines, while the machine codes necessary to process parts are created off-line, downloaded to the machines, and used to produce parts.

Originally, Wizcon has been chosen as a system supervisory software. Indeed, it has a privilege role to authorise most of activities in the cell through the PLC. Although it can authorise system to start, it can not provide appropriated termination conditions for some activities. For instance, it fails to provide a condition for robot arm movement programs to exit running or robot execution state. To address this disadvantage, the I2-R2 package provides services to set a condition for robot arm movement programs to

exit the robot execution state. The basic interlock here is that Wizcon starts a cell operation sequence, while the I2-R2 finalises this sequence. This interlock is specifically reflected by the operation of the robot arm movement programs and CNC machine programs. Wizcon starts each element of the cell through the PLC. After operation cycles, although all elements in the cell may be in the paused state, they do not really exit current operation cycle for new part operation. Thus, they are not ready for an assignment for next arm movement program. For leading the system into a state for next operation assignment, the I2-R2 provides service through assigning the number of cycles. Through this, after completing process cycles assigned, each element in the cell will be in the system state for next assignment because the number of cycles becomes deterministic. To the robots, this is achieved directly, to the CNC machines and conveyor, this is achieved through the interlock among the PLC, conveyor, CNC machines and robot controllers.

The complementary combination of capabilities of Wizcon and I2-R2 provides adequate conditions for a satisfactory transfer and execution of manufacturing messages in the cell, as and when required.

5. CONCLUDING REMARKS

The proposed architecture which uses a modified 3 Layer ISO-OSI architect can be applied to majority of PC controlled FMC applications in which RS 232 interfaces are most commonly used. It provides an ideal conversion solution to incorporate universal file transfer functions and manufacturing message transfers and executions at a very affordable cost and without compromising performance at the data link layer. Care should be taken to study some fundamental aspects, such as system compatibility, manufacturing message process and exchange, network configuration, and software issues before and during implementation. These issues are especially critical for a heterogeneous FMCs. It should also be noted that proper transfer and execution of appropriate software through open file transfer mechanism to each cell controllers considerably enhances operational flexibility, and expansion and modification possibilities of FMCs.

6. REFERENCES

Anuar, Z. and Mohamad, B. (1990). *A hierarchical distributed micro-computer control system for controlling flexible manufacturing cell,*, University Teknologi Malaysia, Malaysia.

Decision-computer international (1990), *Makedrv*, Decision-computer international, Twain,.

Dwyer, J. and Ioannou, A. (1987), *MAP and TOP: advanced manufacturing communications*, Kogan Page.

International Standards Organisation (1991). *Information processing systems-Open Systems Interconnection-File transfer, access and management, ISO 8571*, International Standards Organisation.

International Standards Organisation (1989). *Information processing systems-Local area network, part 2 logical link control, ISO 8802*, International Standards Organisation.

International Standards Organisation (1988). *Information processing systems-Open Systems Interconnection- Basic reference model-Name and addressing, ISO 7498-3*. International Standards Organisation

Lazowska, E. D., Zahorjan, G. S., and Draham, G.S. (1984). *Quantitative system performance-computer system analysis using queuing network models*, Prentice-Hall.

MAP users' group (1986). *Manufacturing Automation Protocol*, USA, p. 198-199, appendix 26.

Rizzardi, V. A. (1988). *Understanding MAP—Manufacturing Automation Protocol*, Dearborn, Mich, Society of Manufacturing Engineers, Publications Development Department, Reference Publications Division.

Shrivastava, A. K., Barnard, B. W. and Nguyen, H. (1993), *Development & Implementation of An FMS with Future Expansion Possibility*, p. 41-49, JSPE-IFIP WG 5.3 Workshop on The Design of Information Infrastructure Systems for Manufacturing, The University of Tokyo, Japan,.

The ISR Group Limited (1992), *The CADDSMAN software series*, The ISR Group Limited, Australia.

Universal Machine Intelligence Limited (1986), *Inside RTX*, Universal Machine Intelligence Limited.

Zhang, W., and Mo, J. P. T. (1993), *A control system for assembly robots using serial communication interface*, The institution of Engineers, Australia, 1993 International Conference on Assembly.

A WEIGHTED LOAD BALANCING HEURISTIC THAT MINIMIZES MAKESPAN IN ALTERNATIVE ROUTING AND MACHINE ENVIRONMENT

Kidong Kim*, Hyung Sang Hahn*, Chankwon Park,
Namkyu Park***, Young Soo Lee***, and Jinwoo Park******

* Institute for Advanced Engineering
** Dept. of MIS, Youngsan Univ. of International affairs
*** Korea Institute of Industrial Technology
**** Dept. of Industrial Engineering, Seoul National Univ.

Abstract : Traditionally, the problems of manufacturing technology and manufacturing management have been treated independently. The process planning problem and the scheduling problem are a good example. In this research, we propose a scheduling system which solves the process planning problem and the scheduling problem simultaneously. We argue that we can obtain a better and practical scheduling solution by dynamically changing the processing machines and operations as the shop condition changes. The proposed scheduling system takes the initial process plan for alternative machines and operations represented by an AND/OR graph as its input. Other informational inputs to the system are part order and shop status. The system then generates new process plans and schedules iteratively, until we finally fix the process plan after verifying the performance of the schedules by a simulation study. Experimental results show that the proposed scheme provides a viable solution for real world scheduling problems.

keywords : integrated manufacturing system, process planning, scheduling, AND/OR graph, Largrangian relaxation, weighted load balancing.

1. INTRODUCTION

In most manufacturing environments, manufacturing activities can be classified into two main domains; the domain of manufacturing technology and the domain of manufacturing management. Such activities as design, process planning and manufacturing activities belong to the former, while MPS(Master Production Schedule), MRP(Material Requirement Planning) and scheduling belong to the latter. Traditionally, activities of the above two domains have been treated separately. However, they could be and should be integrated to make manufacturing systems more efficient. In this research, we intend to integrate the process planning and scheduling activities in an attempt to integrate the field of production technology with production management.

In usual manufacturing practice, we first generate a process plan by selecting the operations and machine tools among alternatives, taking into account only the technological requirements for manufacturing. Then schedules are generated using the generated process plan. The prevailing assumption has been that the process plan should be used as given and can not be changed during the scheduling stage. This assumption reduces the solution space, and consequently results in longer part delivery time and increased manufacturing cost(Laliberty et al., 1996; Zhang, 1993). In this paper, we introduce a new and better heuristic for minimum makespan scheduling problem, where new process plans are iteratively generated to improve schedules. Experimental results are included to show the viability of the proposed approach.

2. STATE OF THE ART

In this section, a sketch of previous studies concerning the routing and machine selection problems is provided.

The machine selection problem and the scheduling problem are understood as hierarchical problems(Stecke, 1983; Brandimarte et al., 1995; Modi et al., 1995). The process planning problem, including machine selection problem, is solved first, and then the scheduling problem is solved using the process planning problem solution. The most commonly used objective in the process planning problem is balancing the work load. The most common objective in the scheduling problem is minimizing the makespan. This frequently leads to the situation where the solution in the process planning problem is inconsistent with the objective of the scheduling problem. In other words, the solution for the process planning problem in balancing the load can be worse than the solutions based on other objectives(Kim, 1993).

The alternative routing is defined here to include alternative operations and alternative sequences of operations in addition to alternative machines. There are very few studies which deal with alternative routing issues with this expanded concept. Zijm(Zijm, 1995) and Brandimarte (Brandimarte et al., 1995) studied this topic. Zijm developed an algorithm that selects machines and operations which minimizes the processing time of each part. Since his algorithm does not take into account the shop status, he is not able to integrate the scheduling problem with the process planning problem. Brandimarte solved this integrated problems hierarchically. He applies load balancing and makespan minimizing as the objectives of the process planning and scheduling problems, respectively. This scheme can also lead to inconsistent solutions for the above problems.

In section 3, we present a methodology which generates a better solution both for the process planning problem and scheduling problem.

3. PROPOSED SYSTEM FOR ROUTING AND MACHINE SELECTION

The proposed system is designed to yield the process plan and schedule which minimizes the makespan. The inputs to the proposed system include the shop status, the order information, and the AND/OR graph which represents the process plan of the parts to be made. The AND/OR graph could describe both the alternative machines and operations(Catron et al., 1991 ; Mettala et al., 1993; Owen, 1993; Lee et al., 1995). Seven major classes of nodes are defined and used in this research: AND SPLIT, AND JOIN, OR SPLIT, OR JOIN, OPERATION, START, and TERMINAL. Fig. 1 shows an example of AND/OR graph. The notation of AS_ij represents the j-th AND SPLIT

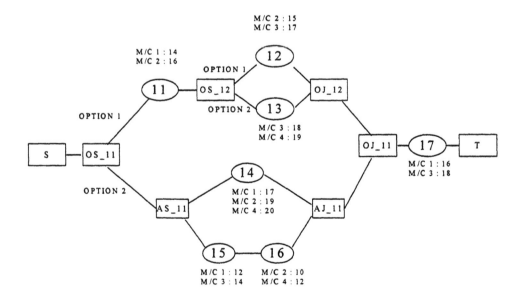

Fig. 1. The AND/OR graph of the example part 1

of part i, and all the nodes are named in a similar manner.

Deriving solutions for the process planning and scheduling problems in real manufacturing environment is not an easy task because available time to solve the problem is limited and the shop status changes frequently. In order to solve this problem within a relatively short period of time, we propose an architecture of hierarchical and iterative solution procedure.

The proposed system generates process plans and schedules sequentially, and fixes the process plan after verifying the performance of the schedule through a simulation study. The process plan is generated by selecting the operations and machines to manufacture the parts with a view to balance the weighted work load. A simulation is followed using dispatching rules and the makespan is determined from the simulation analysis. The test procedure checks the termination condition and updates the weights of machines if needed. These procedures of process planning, simulation, and checking are repeated until the termination condition is met. Fig. 2 shows the flow chart of the proposed system.

3.1 The scheduling module

The main idea of this study is that the process plan can be changed if the performance of its corresponding schedule is not satisfactory, and that the performance of the process plan can be checked only by the performance of its schedule.

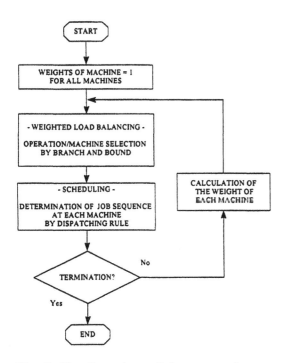

Fig. 2. The flow chart of the proposed system

In this context, actual schedules are generated by such simple dispatching rule as FIFO(First In First Out), and simulation is conducted to check the scheduling performance. The simulator precisely describes the targeted manufacturing system, and the shop status such as previously allocated jobs and machine break downs are reflected in the simulation study. The parts with the highest priority, which is based on due dates, are input to the factory first.

3.2 Calculation of the weight of each machine

The simulation study leads to the calculation of each machine. The bottleneck machine is the one whose makespan is the longest. The operations assigned to the bottleneck machine could be reassigned to others in the next process planning stage by the weights calculated here. The weight of a machine is defined as the makespan of that machine divided by the makespan of the bottleneck machine. So the value of the weight ranges from 0 to 1. The processing time of each operation at a machine is multiplied by the weight of that machine. One caveat though. In the case of zero weight, the processing time of the operation at the zero-weighted machine is zero, which is illogical. So we constrain the minimum weight by some small positive value, namely epsilon. After the simulation study of t-th iteration, the weight of machine k in t+1-th iteration is calculated using equation (1) below, where $MS_{t,k}$ is the makespan of machine k in t-th iteration.

$$LW_{t+1,k} = \max \left\{ \frac{MS_{t,k}}{\max\limits_{k}(MS_{t,k,}\ \epsilon)} \right\} \quad (1)$$

3.3 Termination conditions

There are three conditions for terminating the iteration of the proposed system. One condition is that all jobs to be machined meet their due dates. The second condition is that the improvement of the makespan after adjusting the weights of machines, does not reach the predefined level. The last condition is that the execution time of process planning and scheduling algorithm is completed within the allotted time.

4. WEIGHTED LOAD BALANCING

The operations and machines needed to manufacture the parts are determined at the weighted load balancing module. There may be some loads which are allocated previously, and machine break downs can occur unexpectedly. The machine break down can be viewed as a work load because the machine is not available during that time period. In these circumstances, new operations are assigned to the machines to balance the sum of the old load and new load. The assignment implies the selection of branching option at OR SPLIT node and the selection of machine at OPERATION node in the AND/OR graph.

To solve this assignment problem, we propose a 0-1 mixed integer program, and suggest a branch and bound algorithm to solve the problem. We use the depth first search strategy, and the lower and upper bounds are calculated at each search node. The lower bound is obtained by the Largrangian relaxation method while the upper bound is obtained by the suggested heuristic algorithm.

4.1 The mathematical model for weighted load balancing

In the mathematical model for weighted load balancing, the routings and machines are selected to balance the weighted work load at each machine. The proposed 0-1 mixed integer program model is as follows.

Indices used in the model
i : index for job
j, m : index for node- OPERATION or OR node
p, q : index for option at OR node
k : index for machine

Constants and sets used in the model
P_{ijk} : the processing time of operation j of part i at machine k
OR_i : the set of OR node indices of part i
OP_i : the set of operation node indices of part i
SN_{ijm} : the set of OR node indices belonging to the m-th option of OR SPLIT node j of part i, except for indices of twice or more nested OR nodes
SO_{ijm} : the set of operation node indices belonging to the m-th option of OR SPLIT node j of part i, except for indices of operation nodes belonging to nested OR nodes
LP_{km} : the processing time of pre-assigned operation m at machine k
LW_{tk} : the weight of machine k in t-th iteration

Decision variables
Z_{ijm} : 1, if m-th option of OR node j of part i is selected
0, otherwise

Y_{ijk} : 1, if operation j of part i is assigned to machine k
0, otherwise
L_t : the maximum weighted load in t-th iteration

Problem 1 : the 0-1 mixed integer program for weighted load balancing in t-th iteration

Min L_t (2)
subject to
$$\sum_p Z_{ijp} = 1 \text{ for all } i, j,$$
$$\text{only } j \in OR_i, \ j \not\in \bigcup_{m,q} SN_{imq} \quad (3)$$
$$\sum_p Z_{ijp} = Z_{imq} \text{ for all } i, j,$$
$$\text{only } j \in OR_i, \ j \in SN_{imq} \quad (4)$$
$$\sum_k Y_{ijk} = 1 \text{ for all } i, j,$$
$$\text{only } j \in OP_i, \ j \not\in \bigcup_{m,p} SO_{imp} \quad (5)$$
$$\sum_k Y_{ijk} = Z_{imp} \text{ for all } i, j,$$
$$\text{only } j \in OP_i, \ j \in SO_{imp} \quad (6)$$
$$\sum_m LP_{km} + LW_{tk} \sum_i \sum_j P_{ijk} Y_{ijk} \le L_t \quad \text{for all } k \quad (7)$$
$$Z_{ijm}, Y_{ijk} \in \{0,1\} \quad (8)$$

The equations (3) and (4) are constraints to select one operation at each OR node. The equations (5) and (6) are constraints to select one machine at each operation node selected. Constraint (7) and objective function (2) is designed to balance the weighted load by minimizing the maximum weighted load. Constraint (8) is for integrity of variables.

The generalized assignment problem is known as NP-hard problem(Fisher, 1981). The above problem becomes a sort of generalized assignment problem if only the constraints of (5) and (6) remain. So the above problem is also NP-hard.

4.2 The upper bound of L_t used at B&B

The above problem is solved using B&B(Branch & Bound) method. The upper bound of L_t used in B&B is obtained by a heuristic algorithm. The heuristic algorithm is as follows.

Constants and functions used in the heuristic
M : the number of machines
LP_k : pre-assigned work load of machine k
OS_{ij} : the OR SPLIT node j of part i
NJ_{ij} : the OPERATION node j of part i
$n(NJ_{ij})$: the number of machines which could process the operation j of part i.
$|SO_{ijp}|$: the number of operations which belong to the p-th option of OR SPLIT node j of part i.
$max(\cdot)$: the maximum value among $\cdot$

min($\cdot$) : the minimum value among $\cdot$

The heuristic algorithm for L_t is composed of two main parts. One is the select option at each OR SPLIT node to select operations among alternative ones and the other is to assign a machine to each selected operation.

In order to perform the select option, local weight of each machine(W_k) needs to be calculated by equation (9).

$$W_k = 1, \text{ for all } k, \text{ if } \Delta = 0$$
$$= \frac{LP_k}{\Delta} \text{ for all } k, \text{ if } \Delta \neq 0 \qquad (9)$$

where, $\Delta = \max(LP_1, LP_2, \cdots, LP_M) - \min(LP_1, LP_2, \cdots, LP_M)$

Equation (9) shows that as the load assigned increases, the local weight also increases. And the weighted average processing time of each operation(WP_{ij}) is obtained by equation (10).

$$WP_{ij} = \frac{\sum_k W_k P_{ijk}}{n(NJ_{ij})} \qquad (10)$$

The weighted average processing time is the average processing time of each operation considering the weight caused by the difference of the pre-assigned loads at each machine.

The weighted average processing time of p-th option of OR SPLIT node(WO_{ijp}) is calculated by equation (11).

$$WO_{ij} = \frac{\sum_{j \in SO_{ijp}} WP_{ij}}{|SO_{ijp}|} \qquad (11)$$

An option which has the minimum value can be selected after all these calculations have been done. That is, the operations which may cause lower work load are selected among alternatives.

After having selected the operations among the alternatives, the machine is assigned to each selected operation, one after another, until all the operations are considered. The following criteria are used sequentially to select an operation to be considered.
i) the operation which has the smallest number of alternative machines
ii) the operation which has larger gaps between the minimum processing time and the second minimum processing time
iii) the operation which may be assigned to the bottleneck machine

When a machine is assigned to the considered operation, we choose
i) the machine whose load could be balanced with other machines
ii) the machine which has the smallest processing time for the considered operation

4.3 The lower bound of L_t used in B&B method

The lower bound of L_t is obtained by solving the Largrangian relaxation problem of the problem 1 in section 4.1. The constraint (7), multiplied by Lagrangian multiplier λ_k, is intended to be included in the objective function (2). The Largrangian multipliers are defined as follows.

$$\lambda_k(\geq 0), \quad k=1,..,M, \quad \text{for} \quad \text{constraint} \quad (7) \qquad (12)$$

The Largrangian relaxation problem of problem 1 can then be obtained.

Problem 2 : the Largrangian relaxation problem of problem 1

$$\text{Min } L_t + \sum_k \lambda_k (\sum_m LP_{km} + LW_{tk}\sum_i\sum_j P_{ijk}Y_{ijk} - L_t) \qquad (13)$$

subject to
$$\lambda_k \geq 0 \quad \text{for all } k \qquad (14)$$
(3), (4), (5), (6), and (8) of problem 1

If the λ_k's are determined, the variables which influence the objective function are only L_t and Y_{ijk}. In the constraints, the Y_{ijk} variables appear only once and are restricted by Z_{ijk}. And so, the optimal solution of Problem 2 is obtained by a sort of greedy algorithm after Z_{ijk} are determined.

5. NUMERICAL EXPERIMENT

To check the performance of the proposed system, we conducted several experiments. The process plan of parts used in the experiments are pre-determined. The number of alternative machines for each operation is first determined at random among integers [1,...,5]. Then, the alternative machines are specifically determined in a similar way for the number of machines. The processing time of each operation at each alternative machine is also determined randomly among integers [10,...,20]. Table 1 summarizes the result of the experiment. We apply four different sets of process plans for parts in our experiment. The values under the heading of

Table 1 The experimental results (the values : makespan)

Data Set	pure load balancing				weighted load balancing				reduction rate
	mean	s.d.	min	max	mean	s.d.	min	max	
1	68.34	8.89	53	98	65.92	5.7	58	82	0.96
2	66.3	13.35	44	90	62.4	9.06	44	74	0.94
3	83.3	11.74	66	100	79.5	10.32	66	99	0.93
4	77.3	15.17	58	113	69.2	8.13	58	84	0.89

weighted load balancing are those obtained by our system. The values in table 1 are obtained by 10 replications for each data set. These experiments were performed at SUN Ultra Workstation. The algorithm is programmed in C, and a SPARC C Compiler 4.0 is used.

From the table, we can see that the pure load balancing strategy is inferior to the weighted load balancing strategy.

6. CONCLUSION

We propose a new approach for the scheduling problem which incorporates both process planning and scheduling problems. The experimental data shows that the proposed method can be a very viable alternative for a more effective process planning and scheduling system than the conventional methods which are based only on load balancing objectives or minimizing makespan objectives. We hope that our results induce interests in other competent colleague researchers in the integration approach that considers both manufacturing technology and management simultaneously.

REFERENCE

Brandimarte, P. and M. Calderini (1995). A hierarchical bicriterian approach to integrated process plan selection and job shop scheduling. Int. J. Prod. Res., Vol. 33, No. 1, pp. 161-181.

Catron, B.A. and S.R. Ray (1991). ALPS: A Language for Process Specification. Int. J. of CIM, Vol. 4, No. 2, pp. 105-113.

Fisher, M. (1981). "The Lagrangian relaxation method for solving integer programming problems. Management Science, Vol. 27, No. 1, pp.1-18.

Kim, Y.-D. (1993). A study on surrogate objectives for loading a certain type of flexible manufacturing systems. Int. J. Prod. Res., Vol. 31, No. 2, pp. 381-392.

Laliberty, T.J., D.W. Hildum, N.M. Sadeh, J.

McA'Nulty, D. Kjenstad, R.V.E. Bryant, and S.F. Smith (1996). A blackboard architecture for integrated process planning/ production scheduling. from http://agile.cimds. ri.cmu.edu.

Lee, S., R.A. Wysk and J.S. Smith (1995). Process planning interface for a shop floor control architecture for computer-integrated manufacturing. Int. J. Prod. Res., Vol. 33, No. 9, pp. 2415-2435.

Mettala, E.G. and S. Joshi (1993). A compact representation of alternative process plans/routings for FMS control activities. J. of Design and Manufacturing, Vol.3, pp. 91-104.

Modi, B.K. and K. Shanker (1995). Models and solution approaches for part movement minimization and load balancing in FMS with machine, tool and process plan flexibilities. Int. J. Prod. Res., Vol. 33, No. 7, pp. 1791-1816.

Owen, J. (1993), STEP An Introduction, INFORMATION GEOMETERS.

Sarin, S.C. and C.S. Chen (1987). "The machine loading and tool allocation problem in a flexible manufacturing system. Int. J. Prod. Res., Vol. 25, No. 7, pp. 1081-1094.

Shanker, Kripa and A. Srinivasulu (1989). Some solution methodologies for loading problems in a flexible manufacturing system. Int. J. Prod. Res., Vol. 27, No. 6, pp. 1019-1034.

Stecke, K.E. (1983). Formulation and solution of nonlinear integer production planning problems for flexible manufacturing systems. Management Science, Vol. 29, No. 3, pp. 273-288.

Zhang, H.-C. (1993). IPPM - A prototype to integrate process planning and job shop scheduling functions. Annals of CIRP, Vol. 42, No. 1, pp. 513-518.

Zijm, W.H.M. (1995). The integration of process planning and shop floor scheduling in small batch part manufacturing. Annals of CIRP, Vol. 44, No. 1, pp. 429-432.

ONE-BY-ONE PRODUCTS INPUT METHOD BY OFF-LINE PRODUCTION SIMULATOR WITH GA

Hidehiko Yamamoto

Dept. of Opto-Mechatronics, Faculty of Systems Engineering,
Wakayama University
930, Sakaedani, Wakayama-shi, 640, Japan
tiger@sys.wakayama-u.ac.jp

Abstract: This paper is focused on the problem for one-by-one products input into Flexible Transfer Line (FTL) and describes the research of an off-line production simulator connected Genetic Algorithm (GA) system in order to search a better solution. New individual expression ,crossover operations, mutation operations in GA and their application examples are also described. When a single FTL manufactures a variety of products, there is a problem to keep a production rate called production levels. The developed production simulator connected GA system to have a wide solution search space can solve the problem for production levels. The relation between a production simulator and GA system is that GA system generates some of the input information to operate a production simulator.

Key Words : One-by-one production, Production simulator,
Genetic algorithm, Flexible transfer line, Recurring individual

1. INTRODUCTION

Because of users' taste variety, FMS (Flexible Manufacturing System) and FTL (Flexible Transfer Line) that a single production line manufactures a variety of products have been developing and working (Monden, 1983). In the production line, it is important to decide which parts are input into the production line in order to fit the timing of users' needs and to increase productivity. Currently, the change from batch production to one-by-one production that input each of parts into a production line is seen.

For example, there are the one-by-one input method according to production rate and the method with the constraints of appearance rates such as "Assembly interval of a sun-roof car is every ten" which is from engineers' experience. Although these methods

realize one-by-one production, they include the problem that they don't search better solutions.

This paper is focused on the problem for one-by-one parts input into FTL and describes the research of an off-line production simulator connected Genetic Algorithm (GA) system in order to search a better solution. New individual and crossover in GA and their application examples are also described.

2. ONE-BY-ONE PRODUCTION

FTL that a single production line manufactures variety parts according to a decided production rate is expressed as an automatic production line including some machine tools and some automatic conveyers connected with machine tools, as shown in *Figure 1*. By inputting some variety of parts into the FTL

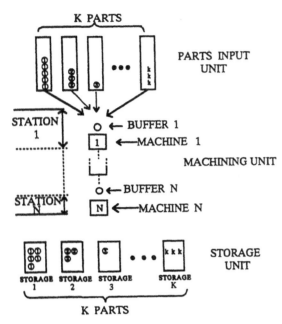

Figure 1 FTL model

entrance, FTL manufactures to keep the decided production rate. One-by-one production means the production style to input each variety of parts one-by-one under the constraint to keep the production rate. In order to satisfy the constraint, it is necessary for the production style to have production levels not to arise the input number difference between each variety of parts in a limited time. The methods of the production levels have been developed (Hitomi, 1978). In the methods, however, the problem that the methods ignore other many feasible solutions for input sequence is remaining.

3. PRODUCTION SIMULATOR CONNECTED GA

In order to solve the problem of conventional production levels methods, this research takes a wide solution search space by connecting a production simulator with GA system. New individual expression for the sake of keeping a production rate of the simulator is proposed in this paper and the search for one-by-one product input sequence to keep a production rate is described.

The relation between production simulator and GA system is that GA system generates one of the input information to operate a production simulator (Yamamoto, 1995). The relation is realized by the following algorithm.

[Algorithm connected Production Simulator and GA]
STEP1: Generate a population that have n pieces of individuals corresponding to randomly selected genes.
STEP2: Carry out simulations with input information corresponding to each individual and calculate each individual's fitness by using the simulation results.

STEP3: Carry out the elitist strategy to send high rank individuals' e pieces to the next generation.
STEP4: Select a pair of individuals by a probability based on fitness and give them crossover operation.
STEP5: Select an individual by a mutation probability (Pm) and the individual participates in mutation operation.
STEP6: If the individuals' number generated in STEP4 and STEP5 are equal to (n-e), continue to STEP7. If not, return to STEP4.
STEP7: Regard n pieces of individuals (sum up e acquired in STEP3 and (n-e) acquired in STEP6) as the next generation's population and finish one cycle of GA operation.
STEP8: If the finish condition is not satisfied, return to STEP2. If satisfied, this algorithm is finished.

4. RECURRING INDIVIDUAL EXPRESSION

4.1 Recurring Individual

The individuals generated in the algorithm of the previous section correspond to input information of a production simulator. For example, in a case where the variety of parts are A, B, and C, the sequence A,B,C,A,B,C,··· or A,A,C,B,A,C,B,··· corresponds to an individual. The conventional method to generate an individual is randomly to select a gene and to assign the gene into a locus (Golderg, 1989a).

However, because one-by-one production has the constraint to keep the goal production rate, the input parts rate must be considered in generating individuals. The conventional method to generate individuals does not consider the rate of each gene in an individual.

If the simulation time is tremendous, input parts number also becomes tremendous. If the conventional method of an individual expression that is a sort of a belt-type chromosome (Holland, 1975), the individual length will become thousands or millions.

In order to solve the problems, because of using the characteristic of production levels, the new individual expression to keep a gene rate and to keep a short individual length is used. If a variety of product is uniform during a limited time, it is regarded that production levels can be performing during the time.

Considering a constant length of an individual length and performing production levels on the individual length, it is regarded that the individual has the individual length whose gene rate is constant. The production levels are realized as the following algorithm.

[Algorithm for Generating Primitive Individuals]
STEP1: Based on the gene rate for v kinds of genes, $r(1) : r(2) : ··· : r(v)$, the possibility, $P(v)$, to assign

on a locus for each of genes is calculated as the following ;

$$P(V) = \frac{r(v)}{\sum\limits_{v=1}^{v} r(v)} \quad \cdots\cdots(1)$$

STEP2: By using the assign possibility P(v), select a gene from among v kinds of genes with roulette strategy. The selected gene is expressed as " i ".

STEP3: Randomly select an integer constant "a", and decide the value q(v) corresponding to a gene quantity in an individual and the individual length δ with the following equations;

$$q(v) = a \times r(v) \quad \cdots\cdots(2)$$

$$\delta = \sum\limits_{v=1}^{v} q(v) \quad \cdots\cdots(3)$$

STEP4: Regard the cumulative value of assigned genes as Sum(v) and give the primitive value the following;

$$Sum(v) = 0 \quad \cdots\cdots(4)$$

STEP5: Carry out the following rule to the accumulative value Sum(i) for gene " i " and this gene quantity q(i);

[*if*] $Sum(i) \neq q(i)$

[*then*]   Continue to STEP6

[*else*] Regard the gene i assignment as bad and go to STEP9 in order to proceed another gene assignment.

STEP6: Judge the gene "i" as the assignment gene and assign the gene to the vacant left locus.

STEP7: Renew the accumulative value of gene "i" with the following equations;

$$_{new}Sum(i) = Sum(i) + 1 \quad \cdots\cdots(5)$$

$$Sum(i) = {_{new}}Sum(i) \quad \cdots\cdots(6)$$

STEP8: By comparing an individual length δ with the current accumulative values Sum(v) for each gene, carry out this algorithm's finish judgment with the following rule.

[*if*] $\delta = \sum\limits_{v=1}^{v} Sum(v)$

[*then*]   Regard acquired chromosome as an individual and finish this algorithm.

[*else*]   Continue to STEP9

STEP9: By using a roulette strategy with the assign probability P, select a gene from among v kinds of genes, regard the gene as "i" and go back to STEP5.

For example, let us consider four kinds of products, A,B,C, and D. When these products production rates are 4 : 3 : 1 : 2, the assignment probability P(A) of

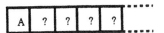

Figure 2 Initial individual generating

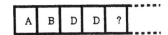

Figure 3 Production rate keeping

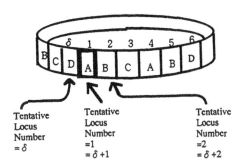

Figure 4 Recurring individual example

gene A in STEP1 is calculated as below.

$$P(A) = \frac{4}{4+3+1+2} = 0.4 \quad \cdots\cdots(7)$$

In the same manner, the probabilities of B, C and D are P(B)=0.3, P(C)=0.1 and P(D)=0.2. By using the probabilities, STEP2 selects a gene. For examples, in a case where i = A is selected, through the processes of STEP3~6, A is assigned to the left vacant locus of a chromosome in *Figure 2*. In STEP7, Sum(A) becomes 1 and carry out the rule judgment in STEP8. When an integer constant "a" is chosen as 1 in STEP3, δ value becomes 4+3+1+2=10. In this case, the *if* part of the rule in STEP8 is not matched with δ value because the left of the rule, δ =10, and the right (Sum(A) + Sum(B) + Sum(C) + Sum(D) = 1) are not same. Then, in order to select another gene, STEP9 is carried out and repeat the same cycle.

This algorithm keeps a containing rate by the rule judgment of STEP5. For example, let us consider the case that the product D is selected again in STEP9 in deciding the fifth locus in *Figure 3*. In this case, the left of the *if* part in STEP5 is 2 and the right is 2. Because this case is not matched with the *if* part, the *else* part is carried out. The *else* part regards the product D as unsuitable and makes a repetitive cycle from STEP9 to select another product. This is because if the product D is selected, the production rate, 4 : 3 : 1 : 2, will not be kept.

In STEP3, a random constant "a" is chosen as a constant individual length. If the constant "a" is regarded as one cycle and the cycle is repeated several times, it is not necessary for GA system to memory a tremendous length of an individual.

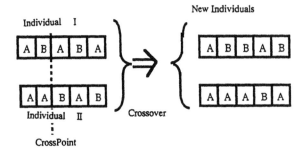

Individual I

Individual II

CrossPoint

New Individuals

Crossover

Figure 5 Conventional crossover

In this way, it is possible for the algorithm not to change the containing rate and to generate an adequate length of an individual.

The individual generated in the algorithm is expressed as a ring-type individual shown in *Figure 4*. The ring-type individual is considered as the first cycle that regards the locus to which the gene selected in STEP2 of the algorithm is assigned as a tentative locus number "1" and the tentative locus number "δ" which is the last result by continuing to add 1 to the right direction of the ring. The tentative locus number 1 is regarded as the tentative locus number "$\delta + 1$" corresponding to the start of the second cycle. In the same manner, by recurring the above operation, a single individual is expressed. This individual is called recurring individual.

4.2 Crossover to Keep Rate

This production simulator repeats GA operations by using recurring individuals. In the GA operations, it is necessary to consider a containing rate. That is, the crossover operation described in STEP4 of Section 3 needs the considerations of a containing rate. This is because if the conventional crossover operations not to include the gene containing rate for an individual are carried out in spite that primitive individuals whose containing rates are satisfied can be generated (Goldberg, 1989b ; Syswerda, 1989), there is a possibility that many individuals whose containing rates are not satisfied are generated after the crossover operations. For example, let us consider the one-point crossover as shown in *Figure 5*. Before crossover, the containing rate of genes A and B for individuals I and II is A : B = 3 : 2. On the contrary, after crossover, two kinds of individuals whose containing rates are A : B = 2 : 3 and A : B = 4 : 1 are generated. In this way, if the conventional crossover method is used, containing rate will be broken and a large amount of lethal genes will also be generated.

The recurring individual has the following characteristic and the above problem can be solved if the characteristic is applied to a crossover. There is the characteristic that gene containing rate in one cycle of the individual length δ keeps constant whichever locus in one cycle is selected as a tentative locus number 1. By using this characteristic of a recurring individual, the crossover operation algorithm not to change a containing rate is realized as below.

[Crossover Algorithm to Keep Containing Rate]
STEP1: Based on fitness, randomly select a pair of recurring individuals as crossover chromosomes and choose $S_{STEP1} = 1$ as loop times.
STEP2: Select one individual from among the individuals selected in STEP1. Call the left end locus of the selected individual a left end number (C_{left}). Randomly select the integer number from 1 to (δ - C_{left} + 1). Call the Lth locus in the selected individual on your right a right end number (C_{right}). The genes from C_{left} to C_{right} corresponds to the left part of a chromosome to be moved because of a crossover and is called a moving chromosome and the L corresponds to a crossover length. Choose $S_{STEP2} = 1$ as loop times.
STEP3: Find a quantity for each variety of genes located in locuses from C_{left} to (L-1)th. Regards the found quantity for v varieties as α (v).
STEP4: Carry out STEP4-1 ~ STEP4-5 to the individual that is the other individual not selected in STEP2 from among a pair of individuals.
   STEP4-1: Regard the individual as a tentative locus number 1 and express it as C'_{left}. That is , $C'_{left} = 1$. Also, express a tentative C_{right} as C'_{right} and calculate the value as the following.

$$C'_{right} = C'_{left} + (L-1) \cdots (8)$$

   STEP4-2: Substitute 1 for primitive search time t.
   STEP4-3: Find each quantity for v varieties of genes located in locuses from C'_{left} to C'_{right} and regard the quantities as β (v).
   STEP4-4: Carry out the following rule.
   [*if*] α (v) = β (v)   for all of v genes.
   [*then*]   Regard the values of C'_{left} and C'_{right} as a pair of list and make list = (C'_{left} , C'_{right}). Add the list as one element of crossover candidate lists and continue to STEP5.
   [*else*]   Renew C'_{left} and C'_{right} by adding 1 to each value of C'_{left} and C'_{right}.
   STEP4-5: Add 1 to t and judge the following rule.
   [*if*]   t = δ
   [*then*]   Continue to STEP5.
   [*else*]   Return to STEP4-3.
STEP5: Carry out the following rules on the crossover candidate lists.
[*if*]   The crossover candidate lists correspond to null and the loop times is $S_{STEP1} = S_1$ (optional integer).
[*then*]   Repeat from STEP1.
[*if*]   The crossover candidate lists correspond to null and loop times is $S_{STEP2} = S_2$ (optional integer).
[*then*]   Randomly select C_{left} and crossover length L

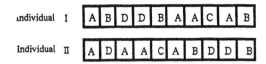

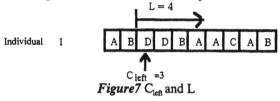

Figure 6 Selected individuals by crossover

Figure7 C_{left} and L

again and repeat from STEP3. Renew loop times S_{STEP1} by adding 1.

[*if*] The crossover candidate lists correspond to null.

[*then*] Randomly select crossover length L again and repeat from STEP3. Renew loop times S_{STEP2} by adding 1.

[*if*] The crossover candidate lists do not correspond to null.

[*then*] Continue to STEP6.

STEP7: Generate new two individuals by exchanging a moving chromosome and the genes between C'_{left} and C'_{right} selected in STEP6 and finish this algorithm.

For example, let us consider the two individuals, individual I and II that have four kinds of genes as shown in *Figure 6*. *Figure 7* shows the individual I selected in STEP2 and its C_{left} and L are 3 and 4. In this case, STEP3 finds v(A) = 1, v(B) = 1, v(C) = 0, v(D) = 2.

In STEP4-1, C'_{left} = 1 and C'_{right} =1+(4-1)= 4 are acquired. In STEP4-3, β (A) = 3, β (B) = 0, β (C) = 0, and β (D) = 1 are acquired.

In this case, as α (v) = β (v) of the *if* part in STEP4-4 is not satisfied, the *else* part is carried out. Search times t=2 by renewing C'_{left} and C'_{right} to 2 and 5 is decided. Till the *if* part of STEP4-4 is satisfied, this repetitive search is carried out. If the *if* part is not satisfied in spit of δ times' searches, to change a pair individual for crossover, to change locuses which are the objects of crossover or to change crossover length is carried out by the three rules of STEP5.

When the *if* part of STEP4-4 is satisfied, an element is randomly selected from among a crossover candidate lists in STEP6. This example is shown in *Figure 8*. Because the developed algorithm chooses the locus sequence whose gene containing rates are same by the rule of STEP4-4, the individuals' gene containing rates never be changed after crossover operations.

In this way, the crossover operations with recurring individuals can select the starting point of an

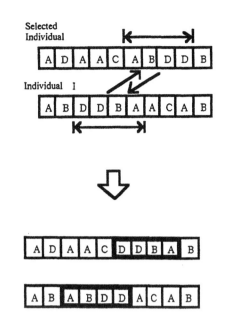

Figure 8 Crossover operation

individual cycle δ times. As a result, it is possible to have a wide search in a candidate set of locus sequences whose gene containing rates are same.

4.3 Mutation Operation

In mutation operation, a gene containing rate must not be changed before and after mutation. The following algorithm keeps a gene containing rate constant.

STEP1: Randomly select an individual from among a population.

STEP2: Randomly select two locus locations of this individual.

STEP3: Carry out the following rule.

[*if*] The contents between the two locuses are not same.

[*then*] Exchange the two contents and finish this algorithm.

[*else*] Return to STEP2.

4.4 Fitness

Fitness in this GA system is calculated, based on a production efficiency. Regarding a goal production outputs for each variety of products in a standard production time and those of a simulation as Q(v) and Q'(v), the production efficiency Ef(v) for each variety of products is expressed with the following equation.

$$Ef(v) = \frac{Q'(v)}{Q(v)} \quad \cdots\cdots(9)$$

As there is few differences between each individual fitness just by using cumulative values of a production efficiency, fitness f is calculated by the following equation.

$$f = (\sum_{v=1}^{v} Ef(v) - \varepsilon)^2 ,$$

(ε : optional constant) ······(10)

5. APPLICATION EXAMPLES FOR PRODUCTION SIMULATIONS

The developed algorithm is applied for FTL example. The FTL has 8 stations (st_1, st_2, ······ , st_8) assembly line and assembles 10 products. Products are input into the station st_1 one-by-one. Assembling times for each variety of products in each station are not same. The goal cycle time of this line is 12 seconds. The production rate for 10 products is P_1 : P_2 : ······ : $P_{10} = 9 : 6 : 7 : 6 : 8 : 7 : 3 : 2 : 4 : 1$. FTL has glitches and they randomly happen every 120 seconds. Their fix time is constant, 20 seconds.

The conditions of GA system are the following; population size = 20, mutation probability = 1%, elite individual = 2, integer constant (a) = 1, individual length (δ) = 53, optional constant (ε) = 865, optional integers $S_1 = 5$ and $S_2 = 5$. As the finishing condition of STEP8 in Section3 algorithm, the time when generations = 60 is adopted.

Several simulations by changing a series of a random number were carried out. *Figure 9* shows one of the results. In the figure, a fine curve shows an average fitness among 20 individuals and a bold curve shows a maximum fitness among the 20. Fitness converge appears from about generation 40. This kind of results is shown in other examples of a series of a random number. The resulted production rate was P_1 : P_2 : ···

··· : $P_{10} = 8.505 : 5.902 : 6.854 : 5.902 : 7.829 : 6.854 : 2.927 : 1.951 : 3.902 : 1$ which is very similar to a goal production rate.

6. CONCLUSIONS

This paper describes the development of an off-line production simulator connected GA system in order to realize one-by-one production for a variety of products. The production simulator includes recurring individuals, their crossover operations and mutation operations not to change the results of a production rate for each variety of products.

The developed production simulator was applied for a FTL model that assembles a variety of products. As a result, acquired production rate was very similar to a goal production rate. The developed production simulator can be used in starting a production plan of FTL that manufactures a variety of products.

REFERENCES

Goldberg(1989a), Genetic Algorithm in Search, Optimization and Machine Learning, Addison-Wesley.

Goldberg(1989b), Alles, Loci, and Travelling Salesman Problem, Proc. Of ICGA-89.

Hitomi, K.(1983), Production Management, Corona Pub. (in Japanese).

Holland(1975), Adaption in Natural and Artificial Systems, Univ. Michigan Press.

Monden, Y.(1983), Toyota Production System, Institute of Industrial Engineers, Atlanta, GA.

Syswerda(1989), Uniform Crossover in Genetic Algorithm, Proc. Of ICGA89.

Yamamoto, H.(1995), Simulator of Flexible Transfer Line including Genetic Algorithm and its Applications to Buffer Capacity Decision, Proc. of the 3rd IFAC Workshop on Intelligent Manufacturing Systems, pp127-132.

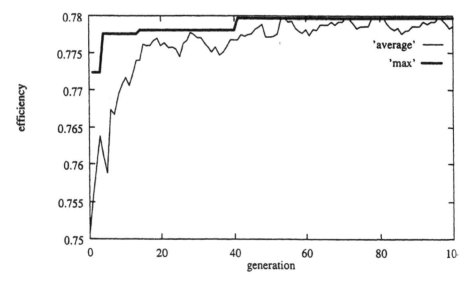

Figure 9 Fitness results

BRANCH AND BOUND LOWER BOUNDS FOR THE HYBRID FLOWSHOP

Omar Moursli

Institut d'Administration et de Gestion, Université Catholique de Louvain
1, place des doyens 1348, Louvain-La-Neuve, BELGIUM
Tel. 32.10.47.83.76, Fax. 32.10.47.83.24, moursli@qant.ucl.ac.be

Abstract : The Hybrid FlowShop HFS consists in the classical multistage flowshop, each stage being composed of identical parallel machines. Minimizing the makespan is NP-Hard in general. To schedule the HFS, Brah and Hunsucker (91), have designed a branch and bound algorithm. In order to solve larger instances, tight lower bounds are required to reduce the enumeration space and time. This paper introduces three improvements of Brah's algorithm and three new lower bounds. The latter are based upon the relaxation of the problem into a single stage sub-problem. Although large problems remain very hard, numerical tests have shown that the new algorithm has drastically reduced the number of the explored nodes as well as the running time needed to find an optimal solution.

Keywords: hybrid flowshop, branch and bound, lower bound, scheduling with release dates and tails, non-preemptive scheduling.

1. INTRODUCTION

The Hybrid FlowShop, HFS, i.e. multiprocessor flowshop, is found in such many different industries as cosmetic, pharmaceutical, textile, and food industries. It consists in the classical multistage flowshop, each stage $j=1,...,S$ being composed of M_j parallel machines, i.e. identical in this case (Figure 1). Our task lies in scheduling a set of jobs I, where a job consists of one operation for each stage. Those operations have to be realized sequentially. At each stage j, each job i has a given processing time p_{ij}, $i \in I=\{1,...,N\}$, $j \in \{1,...,S\}$. The objective is build a schedule with a minimum length -makespan- since the set of jobs usually corresponds to a well-defined set of lots which have to be produced within a short period of time, i.e. one period. This set of lots has been defined by the planning system. At each stage the machines are identical and able to process one job at a time. Job preemption is not allowed. A schedule consists in allocating a specific machine to each operation of each job as well as fixing starting times. This problem is NP-Hard in general

In order to schedule the HFS, Brah and Hunsucker (1991), have proposed a branch and bound algorithm based on the branching scheme introduced by Bratley *et al.*, (1975), for scheduling the parallel machines. Their branching scheme

consists in the enumeration of all possible sequences of all the jobs over all the machines and all the stages. The sequences enumeration complexity is very high. Tighter lower bounds and efficient implementation are then required to reduce the enumeration space and time. This paper introduces three improvements of Brah's algorithm as well as three new lower bounds.

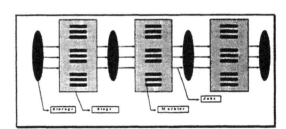

Fig. 1. A three-stage hybrid flowhop

The three new lower bounds are based upon the relaxation of the main problem into a single stage sub-problem with release dates and tails. Section 2 introduces this relaxation and recalls some known results about the single stage sub-problem (Jackson's heuristic and some lower bounds). Section 3 reminds Brah's branching scheme as well as their lower bounds. Section 4 presents the improvements of Brah's algorithm. In section 5 the three new lower bounds are presented. In order to compute a root upper bound, two sets of heuristics are considered, i.e. Johnson's algorithm based heuristics and Jackson's heuristic based ones. Section 6 deals with these upper bounds. The paper ends with some comments on the implementation of the algorithm as well as on numerical tests.

2. THE SINGLE STAGE SUB-PROBLEM

2.1 The Relaxation

If the capacity at all stages is relaxed (i.e. there are as many machines as jobs) except at one stage s, the original problem becomes a problem of scheduling a set of jobs I = {1,...,N}, with release dates r_{is} and tails q_{is} on M_s identical parallel machines. In order to define the corresponding single stage sub-problem, the release dates as well as the tails of the jobs need to be defined. The tails are the amount of time (i.e. the processing and the waiting time) a job will spend at the following stages r (r>s) after it has been processed at stage s. The jobs release dates and tails, in the single stage relaxed sub-problem, can be defined for each stage s as follows :

$$r_{is} = \sum_{l=1}^{s-1} Pil \; ; \; q_{is} = \sum_{l=s+1}^{S} Pil \; ; \; for \; s = 1,...,S$$

As a matter of fact, this relaxation has come up with S (i.e. S stages) parallel-machines problems with release dates and tails. This problem, in the non-preemption case, is NP-Hard (Carlier 1987). In order to schedule the *one stage* sub-problem (i.e. r_i, p_i, q_i, i∈I and M parallel machines) Jackson's heuristic, see (Carlier 87), can be used.

2.2 Jackson's heuristic

This heuristic starts from time 1 and schedules the machines by moving forward in time. Each time a machine becomes available, the job with the largest tail among the available ones starts. The Gantt chart (Figure 2) illustrates the way Jackson's heuristic operates.

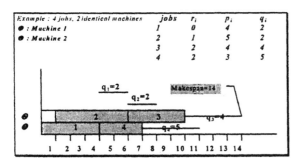

Fig. 2 : Schedule using Jackson's heuristic

2.3 Lower Bounds for the single stage sub-problem

Many lower bounds, of various quality and complexity, with respect to the problem of minimizing the makespan have been proposed in the literature. In (Vandevelde *et al.,* 1995) the interested reader can find an extensive development of these lower bounds. Hereafter a list some of these lower bounds is presented: the Job Bound (JB *O(n)*), the Set Bound (SB *O(n)*) and the Subset Bound (SSB *O(n²)*).
Let

- I = {1,...,N} be a set of jobs
- $R_M(I)$ ={[1] ,[2], ... ,[M]} be a set of M jobs with the smallest release dates
- $Q_M(I)$ ={(1), (2),...,(M)} be a set of M jobs with the smallest tails

$$JB(I) = \max_{i \in I} (r_i + p_i + q_i) \qquad (1)$$
$$SB(I) = \lceil (\sum_{i \in R(I)} r_i + \sum_{i \in Q(I)} q_i + \sum_{i \in I} p_i) / M \rceil \qquad (2)$$
$$SSB(I) = \max_{V \subset I, |V| > M} SB(V) \qquad (3)$$

3. THE ALGORITHM PROPOSED BY Brah and Hunsucker (1991)

3.1 The branching scheme

The branching scheme is based upon the enumeration of all possible sequences of all the jobs at all stages. It starts at stage one and enumerates all the jobs sequences over all the parallel machines. It then moves forward, in the same way to the following stages, until the last stage has been reached. The enumeration is realized by holding a partial schedule at each node and deciding to add a job from the non-scheduled ones either on the current machine or on a new one. The computation of lower bounds and a better feasible solution allow to discard paths (i.e. nodes) leading to solutions with worse makespan.

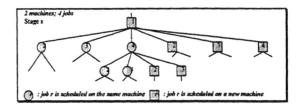

Fig. 3. possible sequences at stage s starting by job 1)

The sequences enumeration is accomplished by generating a tree which contains two types of nodes, i.e. the square nodes and circle nodes (Figure 3). If the path passes through a square node, the candidate job i is scheduled on the current machine. However, if the path passes through a square node, job i is scheduled on a new machine which becomes the current machine. For each stage there are N levels, one for each job. Hence there are N*S levels to explore so as to find a complete schedule.

3. 2 The Lower Bounds

Notation
- A : a set of jobs such that $A \subset I$
- $A' = A \cup \{i\}$ where job $i \in I$, $i \notin A$
- I-A' : a set of the jobs not yet scheduled
- $S_j(A)$: a partial schedule involving all the jobs on all machines through stage j-1 along with a sequence of job set A at stage j
- $S_j(A')$: represents the schedule formed by appending job i to $S_j(A)$
- $C[S_j(A'),k]$: the completion time of the partial sequence on machine k, where k is one of the M_j parallel machines at stage j
- $ACT[S_j(A')]=$

$$\sum_k C[Sj(A'),k] \; /M_j + \sum_{i \in I-A'} p_{ij} /M_j \quad (4)$$

i.e. the average completion time and processing time of jobs in set I-A' at stage j

- $MCT[S_j(A')] = \max_k \; C[S_j(A'),k] \quad (5)$

i.e. the maximum completion for the scheduled workload at stage j

- q_{ij} : tail of job i at stage j ; $q_{ij} = \sum_{j'=j+1}^{S} p_{ij'}$

Brah and Hunsucker (1991) have introduced two lower bounds (i.e. LBM and LBJ):

Let $\quad \alpha = ACT[S_j(A')] + \min_{i \in I-A'} q_{ij}$

$\quad \beta = MCT[S_j(A')] + \min_{i \in A'} q_{ij}$

$\quad \gamma = \min_k C[S_j(A'),k] + \max_{i \in I-A'} (p_{ij} + q_{ij})$

$LBM[S_j(A')] = \alpha \quad$ if $ACT[S_j(A')] \geq MCT[S_j(A')]$
$\qquad\qquad\quad = \beta \quad$ otherwise $\quad (6)$
$LBJ[S_j(A')] = \gamma \quad (7)$

$\quad LB = \max \{ LBM[S_j(A')] , LBJ[S_j(A')] \} \quad (8)$

4. IMPROVEMENTS

Further notation
For any partial schedule $S_j(A')$ let
- ω be the current machine
- $k = 1,..., \omega-1$ be one of the machines already scheduled in $S_j(A')$
- $m_j = \{ \omega+1,...,M_j \}$ be the set of machines not yet scheduled
- $V = \{\omega\} \cup m_j$ be the set of the leftover machines
- c_{ij} be the completion time of job i at stage j
- r_{ij} be release date of job i at stage j
- $r_{ij} = \max (r_{ij}, c_{ij-1})$ and $r_{i1} \geq 0$
- $c_{ij} = \max (C[S_j(A),k], r_{ij}) + p_{ij}$ where job i is assigned to machine k and $S_j(A)$ is the partial schedule augmented by job i at stage j.

Since the remaining jobs will be scheduled only on the leftover machines, Brah & Hunsucker (91),

have improved LBM. They suggest that the lower bound is more effective when dividing the sum of the processing times of the jobs in I-A' by the number of the remaining machines instead of M_j

4.1 lower bound on the completion time of the machines not yet scheduled at stage j

In Brah's lower bounds the completion times of the machines are as follows :

$C[S_j(A'),k] \geq 0$ for $k = 1,..., \omega$
$C[S_j(A'),k] = 0$ for $k \in m_j = \{\omega+1,..., M_j \} \quad (9)$

The improvement of the completion time of the machines in set m_j will eventually lead to the improvement of the lower bounds (i.e. these values are used in the computation of LBM and LBJ).

For any partial schedule, the completion times (which are also the availability times or the earliest starting times) of the machines $k \in m_j$ are set to the minimum release date of the not yet scheduled jobs at stage j.
$\quad C[S_j(A'),k] = \min_{i \in I-A'} r_{ij} \; ; \; k \in m_j$

Furthermore, let $r_{[tj]}$ be the tth smallest release date in set I-A' at stage j
$C[S_j(A'),k] = r_{[tj]} \; k = \omega+1,...,M_j \; ; \; t = k-\omega,...,|m_j| \quad (10)$

For any partial schedule $S_j(A')$ the release dates of the not yet scheduled jobs (i.e. $i \in$ I-A') are known. Actually no machine at any stage will be kept idle unless N<M_j. In order to start $|m_j|$ not yet scheduled jobs at stage j on the $|m_j|$ not yet scheduled machines, they have at least to wait the $|m_j|$ smallest release dates.

4.2 At any stage the lower bound of any node is at least equal to the father's node lower bound

If at some node a lower bound becomes active, sometimes it produces a value smaller than the father's node lower bound. Hereafter two cases where a node $(S_j(A'))$ lower bound is smaller than its father's one $(S_j(A))$ are presented.

Case 1.
(i) $LBM[S_j(A')]$ is active (i.e. > $LBJ[S_j(A')]$)
(ii) $ACT[S_j(A')]$ becomes smaller than $MCT[S_j(A')]$
(iii) $\min_{i \in I-A} q_{ij} - \min_{i \in A'} q_{ij} >$ MCT $S_j(A')$ - ACT $S_j(A')$.

Case 2.
(i) $LBJ[S_j(A')]$ is active (i.e. > $LBM[S_j(A')]$)
(ii) $\min_k C[S_j(A'),k] = \min_k C[S_j(A),k]$

i.e. the added job did not change the minimum completion time

(iii) $\max\limits_{i \in I-A} p_{ij} + q_{ij} < \max\limits_{i \in I-A} p_{ij} + q_{ij}$

i.e. i^* has been assigned $A'=A \cup \{i^*\}$

with $i^* = \arg \max_{i \in I-A'} p_{ij} + q_{ij}$

The LB is set: $LB[S_j(A')] \geq LB[S_j(A)]$ (11)

4.3 Improvement of LBJ [Sj(A')] and LBM[Sj(A')]

Since jobs in set I-A' are going to be scheduled only on the machines in set V, LBJ [$S_j(A')$ can be improved by computing the minimum completion time only over the leftover machines i.e. $k \in V$.

Since α and β are both lower bound on the HFS problem, LBM is set : $LBM[S_j(A')] = \max\{\alpha, \beta\}$

5. NEW LOWER BOUNDS

The next lower bounds are based upon the relaxation of the problem into a single stage sub-problem with release dates and tails. Given a partial schedule $S_j(A')$ the following are known:

1. All the jobs are scheduled from stage 1 through stage j-1, i.e. $\forall i \in I$, c_{ij-1} is known
2. All the jobs in set A' are scheduled at stage j :
   - $\forall i \in A'$, c_{ij} is known
   - $\forall i \in I-A'$, r_{ij} is known
   - for $k = 1,...,\omega$, $C[S_j(A'),k]$ are known
   - machines $k \in m_j = \{\omega+1,...,M_j\}$ are idle and their earliest starting times are known
3. The remaining jobs in set I-A' will be assigned to the machines in set V.

5.1 Job bound (JB)

Since each job i will pass through stage j, spend there some time p_{ij}, and at least its tail q_{ij} at the following stages $j'=j+1...S$ (j'>j). A lower bound on their completion times at the last stage S can be computed :

(i) For the jobs in set A' :
$$JB1[S_j(A')] = \max_{i \in A'} (c_{ij} + q_{ij}) \quad (12)$$

(ii) For the jobs in set N-A', first lets compute their earliest starting times :
$$EST_{ij} = \max\{r_{ij}, \min_{k \in V} (C[S_j(A'),k])\} \quad \forall i \in I-A'$$

$$JB2[S_j(A')] = \max_{i \in I-A'} (EST_{ij} + p_{ij} + q_{ij}) \quad (13)$$

(iii) For stage j : $JB[S_j(A')] = \max (JB1, JB2)$ (14)

5.2 Set bound (SB)

Set bound on set I (SB1). According to the relaxation mentioned in section 2, SB(I) is a lower bound for the single stage sub-problem with release dates and tails. For a partial schedule $S_j(A')$, the HFS problem can be relaxed to a single stage j sub-problem and the SB at stage j, can be computed.

Let $\delta = \sum\limits_k C[S_j(A'),k] + \sum\limits_{i \in I-A'} p_{ij}$

$Q_{Mj}(I)=\{(1j),(2j),...,(M_j)\}$ be set of M_j jobs with smallest tails at stage j

$$SB1[S_j(A')] = \lceil \delta + \sum_{i \in Q_{Mj}(I)} q_{ij}) / M_j \rceil \quad (15)$$

δ is the equivalent part of the release dates and processing times of the set bound SB(I). The completion times of the machines k=1,..., ω, are known and can be regarded as the first ω minimum release dates. For the machines k=ω+1,...,M_j, and according to above introduced improvements, their completion times are computed as in equation (10).

Set bound on set I-A' (SB2). Since jobs in set I-A' can only be scheduled on the machines in set V, the HFS can be regarded as only made up of this set of machines. Actually, for the other machines, k=1,...,ω-1, JB1[$S_j(A')$] is a lower bound on their completion times.

Let $R_{|mj|}(I-A') = \{ [1j], [2j], ..., [|m_j|j] \}$ be the set of $|m_j|$ jobs in set I-A' with smallest release dates.
$J=I-A' \cup \{i^*\}$; $i^* = \arg \min_i q_{ij}$ where the minimum is over all the jobs assigned to the machine ω
$Q_{|mj|+1}(J) = \{ (1j),(2j), ... ,(|m_j|+1j)\}$ be the set of $|m_j|+1$ jobs in set J with smallest tails at stage j

$$SB2[S_j(A')]=\lceil C[S_j(A'),\omega] + \sum_{i \in R_{|mj|}(I-A')} r_{ij} + \sum_{i \in I-A'} p_{ij} + \sum_{i \in Q_{|mj|+1}(J)} q_{ij}) / |m_j|+1 \rceil$$
(16)

C[$S_j(A')$, ω] is the minimum release dates the machine ω has to wait before starting any other job. The minimum tail accounted for the machine ω is the minimum tail of the jobs already assigned to it, which is taken into account in the set J.

$$SB[S_j(A')] = \max \{SB1[S_j(A')], SB2[S_j(A')] \} \quad (17)$$

5.3 Hybrid set bound (HSB)

Each time the partial schedule $S_j(A)$ is augmented by assigning a job i to machine ω, the release date of that job i, at the following stages s>j, is updated :

$$r_{ij+1} = \max \{C[S_j(A'),\omega], r_{ij+1}\}$$

$$r_{is} = r_{is-1} + p_{is-1} ; \quad s = j+2,...,S \quad (18)$$

34

Each time, these release dates are updated, a more effective $SB_s(I)$ can be computed for each of the following stages $s > j$, and a global lower bound for the original problem can be obtained.

$$HSB[S_j(A')] = \max_{s = j+1,...,S} SB_s(I) \qquad (19)$$

HSB is time consuming but has the big advantage of detecting a possible bottleneck stage among the non-scheduled ones, and in which case it would give a tighter lower bound.

5.4 sub set bound (SSB)

In the above mentioned lower bounds, the set bound (SB) is used for the relax single stage sub-problem. Actually the sub set bound (see equation 3) can also be used for the relaxed problem.

6. UPPER BOUND

Many heuristics do exist for the HFS. The best known makespan can be taken as a root note upper bound. Hereafter two sets of heuristics are presented.

6.1 Heuristics based upon Johnson's algorithm (1954)

CDS_F. Guinet and Solomon (1996), have turned the HFS into a classical flow shop. To do so they have divided the processing time of each job by the number of machines of the corresponding stage and applied the CDS heuristic (Campbell et al., 1970). CDS schedules the classical flowshop by a permutation list. Actually, by aggregating stages, CDS creates S-1 fictitious two-machine flowshop problems and uses Johnson's algorithm so as to schedule each of them. As a result Johnson's algorithm delivers a permutation list for each two-machine problem. The list of the problem (among the S-1 fictitious two-machine flowshop problems) with the best (least) makespan is chosen to schedule the flowshop. Like in CDS, Guinet and Solomon (1996), use this list to schedule the HFS. The processing times of the S-1 two-machine (A & B) problems are computed as follows :

$$p_A = \sum_{s=1}^{t} p_{is} \; ; \; p_B = \sum_{s=K-t+1}^{S} p_{is} \; ; t = 1,...,S-1$$

CDS_H. A new heuristic is introduced. The K-1 Johnson's lists are built (using the original processing times). Each of them is used to schedule the whole HFS, instead of scheduling only the K-1 two-machine fictitious problems. In other words the permutation lists coming from each two-machine fictitious problem are transformed it into an HFS feasible schedules. Thus the best schedule among

the k-1 ones is chosen. To schedule the HFS with a permutation list the list is started from the beginning at each stage. Every time a machine is available the next job of the list is assigned.

6.2 Heuristics based upon Jackson's heuristic

Moursli and Pochet (1996), have designed heuristics for scheduling the HFS. These heuristics are based on scheduling *one stage at a time.* In order to schedule each single stage sub-problem, they have used Jackson's heuristic and Carlier's branch and bound (Carlier 1987). The scheduling of one stage -starting and completion time of the jobs- imposes constraints on the other stages. The already scheduled stages may then need rescheduling. These heuristics differ in the order the stages are sequenced. Hereafter lets recall three of these heuristics using Jackson's heuristic to schedule the single stage sub-problem.

Sequentially Forward (SEQ_F). This heuristic schedules the shop stage by stage, starting from the first stage and going sequentially forward until the last stage has been reached. The completion time of each job at a scheduled stage is taken as a release date for that job at the following stage.

Sequentially Forward and rescheduling the critical stages(SEQ_F_C). This heuristic does behave in the same way as SEQ_F but every time a new stage is scheduled, some of the already scheduled stages may need rescheduling.

Shifting Bottleneck Procedure (SBP_H). This heuristic applies the Shifting Bottleneck Procedure (SBP) to schedule the HFS. The SBP has been proposed by Adams & al (1988), for scheduling a job shop problem. For the HFS, SBP_H schedules the shop stage by stage. At each iteration it schedules the bottleneck stage among the non-scheduled ones. The bottleneck stage (i.e. the one to be scheduled next), among the non-scheduled ones, is the stage, when scheduled, giving the highest makespan. Every time a new stage is scheduled, some of the already scheduled stages need to be rescheduled.

Rescheduling the critical stages. For SEQ_F_C as well as for SBP_H the new added schedule (of a single stage sub-problem) may alter the data (release dates and tails) of the already scheduled stages and hence they are no longer feasible or their solutions (makespan) are no longer effective according to these new data. A stage is critical if its data has been altered. In which case it needs rescheduling so as to come up with a feasible or a better solution, see (Moursli and Pochet 1996).

7. IMPLEMENTATION AND NUMERICAL TESTS

7.1 Lower bound computation

As far as the lower bounds are concerned, the five 5 lower bounds (the improved LBM and LBJ, JB, SB and HSB) can be used at each node. In order to reduce the computation complexity the lower bounds are computed in the non-decreasing order of their complexities (first LBM and LBJ followed by JB, SB1, SB2, HSB). A lower bound is computed only if the preceding one fails to discard the evaluated node. Hence, by doing so, a computation effort is consented only if it is really needed.

7.2 Numerical tests

The branch and bound algorithm has been programmed using an object oriented language (C++) and the tests was carried out on a Pentium 120. The processing times were generated uniformly for all the jobs at all stages with $p_{ij}=$ [1..100]. Different configurations have been designed and 20 instances have been generated for each them. Brah's original algorithm have been compared to the new one on the same sets of problems. Both algorithms have started with the same best known upper bound. All the instances have been stopped after 10 minutes-running time.

Table 1 gives the results of the carried out tests. Column N gives the number of jobs. Columns M_j (j=1,...,5) give the number of parallel machines at stage j. Columns 'Time' give the average running time in seconds of the corresponding algorithm. Columns '%' give the deviation percentage of the corresponding algorithm to the best known lower bound (0 means the algorithm is optimal). Columns 'Brah', 'BaB' and 'FirstUB' respectively give Brah's algorithm, the new one and the root node upper bound results ('%' and 'Time'). The root node upper bound is giving by the best of the five heuristics (see section 6).

CONCLUSION

Although large problems remain very hard to solve, numerical tests show that the new algorithm has drastically reduced the number of the explored nodes as well as the running time needed to reach an optimal solution. The extra running time needed for the new lower bounds has been largely compensated by the reduction of the explored nodes.

REFERENCES

Adams, J., E. Balas, D. Zawack (1988). The Shifting Bottleneck Procedure. Management Science vol. 34,391-401, No. 3, March.

Brah, S. A., Hunsucker J. L. (1991). Branch and bound algorithm for the flow shop with multiprocessors, European Journal of Operational Research 51, 88-89.

Bratley, P., Florian M., Robillard P. (1975). Scheduling with earliest start and due date constraints on multiple machines. Naval Research Logistics Quaterly, vol 22, n°1, 165-173.

Campbell, H. G., Dudek R. A., Smith M. L. (1970). A heuristic algorithm for the n job, m machine sequencing problem. Management Science, Vol. 16, n°10, 630-637.

Carlier, J. (1987). Scheduling jobs with release dates and tails on identical machines to minimize the Makespan. European Journal of Operations Research 29 298-306.

Guinet, A., Solomon M. (1996). Scheduling hybrid flowshops to minimize maximum tardiness or maximum completion time. Int. J. Prod. Res., 34, 1643-1654.

Jackson, J. R., (1955). Scheduling a production line to minimize maximum tardiness, Research 43, Management Science Research Project, University of California Los Angeles.

Johnson, S. M. (1954). Optimal two and three stage production schedules with setup time included. Naval Res. Logist. Q.1, 61-68.

Moursli, O., Pochet, Y. (1996). Heuristics for scheduling a Multiprocessor flowshop. Advances in Industrial Engineering Applications and Practice I. Houston Texas. Conference proceeding 456-463.

Vandevelde, A., H. Hoogeveen, C. Hurkens, J. K. Lenstra (1995). Lower bounds for the multiprocessor flow shop. Working paper. Eindhoven University of Technology, The Netherlands.

Table 1 : Comparing Brah's algorithm with new the one

N	M1	M2	M3	M4	M5	BaB %	BaB Time	Brah %	Brah Time	FirstUB %
4	2	2				0	0.02	0	0.02	9.85
4	2	2	2			0	0.04	0	0.07	5.93
4	2	2	3	2	2	0	1.30	0	2.22	6.60
6	2	2				0	0.18	0	1.13	7.30
6	2	2	2			0	4.2	0.4	87	10.30
8	2	2				0	7.7	0.8	230	4.80
10	2	2				0	57	1.9	550	3.07
15	2	2				1	430	1.5	481	1.56
8	2	2	2			0	120	8.3	581	10.90
6	2	2	3	2	2	1.7	190	6.2	444	8.38
10	2	2	2			5.1	490	12.0	600	12.50
20	3	3				2.3	600	3.5	600	3.5

DYNAMIC DISTRIBUTION OF PROCESSES FOR INTELLIGENT SYSTEM DESIGN

M. B. Zaremba, W. Fraczak, and M. Iglewski

Département d' informatique, Université du Québec à Hull
101 St-Jean-Bosco, Hull, Quebec, Canada

Abstract: This paper focuses on software implementation issues related to rapid design and prototyping of distributed systems that integrate the methods and techniques of computational intelligence, such as genetic search, neural networks, and fuzzy logic, with symbolic and algorithmic processing. In order to dynamically assign computational processes to the resources of a distributed computing system, a process algebraic approach is proposed that is based on MAPA (Multi-Action Process Algebra), with special emphasis placed on process synchronization problems.

Keywords: Distributed artificial intelligence, concurrent programming, system design, process algebra, decision support systems.

1. INTRODUCTION

Intelligent computing architectures, often inspired by the functionality of biological structures, which are able to rapidly process complex information in the context of changing requirements and constraints are increasingly finding their place at the heart of modern adaptive manufacturing systems. These architecture combine a variety of techniques, as well as a large body of dedicated subsystems, to form efficient, hybrid information processing machines. The design of such systems becomes a non-trivial task and requires a well-structured approach that takes implementation issues into account.

A general model for configuring tasks in an intelligent control system was developed by Albus (1994). It offers a canonical form of functional architecture for real-time control systems, and was initially applied in a space station Flight Telerobotic Servicer project (Albus *et al.*, 1989). Being a reference model, it disregards the implementation issues. A detailed description of a toolkit for hybrid intelligent system design can be found in (Khebbal and Sharpington, 1996). Problems related to the

integration of different computing paradigms are addressed among others by Medskar (1994), Harris and Brown (1994), Lee and Takagi (1993). The need for integrated computing-aided environments in solving complex engineering problems was formulated by MacFarlane *et al.* (1989). Software systems using hybrid technologies to assist in the design process were reported by Zeigler and Kim (1996) and Kohn *et al.* (1995).

A vast majority of published reports concerning intelligent systems design pertains to classical von Neumann computer architectures, whereas the research on parallel and distributed computing systems is still in its initial phase. A framework for automatic synthesis of large-scale parallel applications in the area of embedded signal processing and instrumentation was presented by Ledeczi (1995). Heitbreder *et al.* (1997) discuss a hybrid Predicate/Transition Petri Nets and neural network approach for specification of distributed design processes. The development of systems for distributed computing platforms is inherently less general and more dependent on the target computing architecture, since system performance and a proper selection of system architecture depend on the level of granularity of elementary tasks and their communication scheme. In the case of intelligent system design, the task granularity level is usually high, and therefore best suited for distributed, rather than massively parallel, architectures of computing systems.

A principal motivation of this paper is to provide a generic, expressive and effectively implementable approach for dynamic distribution and synchronization of processes running in a distributed environment. The approach is based on the use of a simple, though powerful, process algebra called MAPA (Multi-Action Process Algebra) and proposed by Fraczak (1995). Although process algebras have been used for a couple of decades to specify, verify and analyze behavioral aspects of computer systems (Milner, 1983; van Eijk, 1989; Best *et al.*, 1992), their use in distributed applications has been limited largely due to the lack of efficient methods and tools for construction of the distributed program from the formal specification. In this paper, a distributed implementation of MAPA using a simple message passing paradigm is proposed. In the first part of the paper, a proposed functional architecture of the distributed system for intelligent design is presented. The next section describes the architecture of the parallel computer that serves as an implementation platform for the intelligent design system. The last part of the paper discusses a distributed implementation of the Multi-Action Process Algebra used for dynamic distribution of computational processes.

2. INTELLIGENT DISTRIBUTED SYSTEM DESIGN

2.1 Functional architecture

A functional diagram of the software part of our intelligent design system is shown in Fig. 1. It is composed of two principal parts. The first relates to problem definition and evaluation. The user defines the design problem in terms of generic tasks, such as classification and nonlinear function approximation, and provides the structure of functional relationships between the tasks. The evaluation of the final system architecture and parameters is performed after the evaluation of partial results offered by individual component methods and their combinations. This part of the software architecture, executed in close interaction with the designer, operates principally on a local processor, with no need for distribution of computing processes.

The second part of the software system relates to the solution of the generic component tasks. Its main functional elements are a set of methods and

techniques from the area of computational intelligence, complemented with selected statistical techniques and data processing programs. This part, in order to run at a high performance level, has to make an extensive use of parallel and distributed processing. The modules and processes which are most suitable for distributed execution are indicated by a dotted line in Fig. 1.

2.2 Component methods

The principal methods that will eventually be available to the user are:
a) Computational intelligence methods:
- neural networks
- genetic algorithms
- fuzzy logic
b) Rule-based reasoning methods
c) Polynomial networks:
- GMDH
- AIM
d) Decision trees.
e) Statistical techniques:
- linear regression
- multivariate analysis
- nearest neighbours
These methods are accompanied by general tools for scientific visualization, data preprocessing and database management.

An important feature of the integration process is its ability to handle and create hybrid systems, i.e. systems where different techniques are combined in a manner that enhances system performance or overcomes the limitations of individual techniques. Goonatilake and Khebbal (1992) distinguish three classes of hybrid systems: function-replacing, intercommunicating, and polymorphic hybrids. In function-replacing hybrid systems, a specific function of one technique is replaced by another technique. For example, a neural network or a genetic algorithm can be used to replace the manual operation of defining fuzzy membership functions. In interconnecting hybrids, the solution is obtained through the integration of separate, independently executing intelligent modules. Achieving the functionality of different intelligent techniques by applying a particular technique characterizes polymorphic hybrids. In the distributed system presented in this paper, the most representative and pertinent class of hybrid systems that can be efficiently used for solving a vast majority of problems is interconnecting hybrids. There are two reasons for that. First, the design problem can usually be subdivided into distinct sub-tasks, and specific intelligent processing techniques can be assigned to best solve each sub-task. Second, the architecture of the computing platform best supports medium and high granularity level problems,

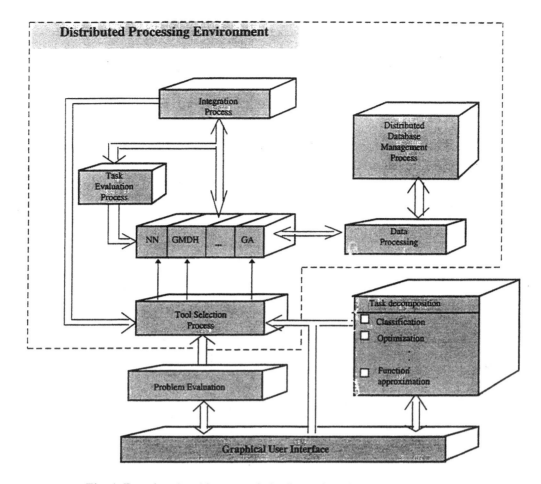

Fig. 1. Functional architecture of distributed intelligent system design.

defined in terms of self-contained, independent modules with limited communications requirements.

3. COMPUTER ARCHITECTURE

Development of the intelligent design system is performed on a 16-node parallel computer AVX3. Each node is equipped with a 604 PowerPC microprocessor, 512 MB RAM memory, and a 9 GB hard disk drive. The processors communicate through a Fast Ethernet bus or through a 40 Mb/s point-to-point communication switch. The architecture of the system is shown in Fig. 2. Each processor has its own IP number; the extension of the internal parallel architecture into an entirely distributed system is therefore simple and straightforward.

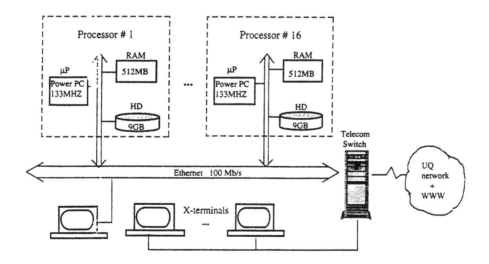

Fig. 2. Parallel computer architecture.

4. TASK DISTRIBUTION

Fast operation of the intelligent design system in a distributed environment of an *a priori* unknown configuration requires a theoretical approach that will provide generic, expressive, and effectively implementable methods and tools for dynamic distribution and synchronization of processes. The existing toolsets, based predominantly on a process algebraic approach, are mainly used for specification, verification, and analysis of distributed algorithms and systems. Those functionalities are useful during the design phase but of little help in the implementation phase, since they do not implement the passage from the specification into the program. In this paper we propose a distributed interpreter of Multi-Action Process Algebra - a simple yet expressive algebraic representation that allows for implementation of essential operations for distributed processing, such as general choice composition or multi-way synchronizations.

4.1 MAPA formalism

The syntax of MAPA includes four operators - prefixing, composition, restriction, and recursion - and can be resumed by the following grammar:

$$P ::= m;P \mid P \, ch \, P \mid P : ch \mid g$$

where m is a multi-action, i.e., a finite multi-set over a set of actions, ch is a channel, and g is a process variable. It is required that for each process variable g there exist a unique defining equation $g \doteq P$.

The intuitive meaning of the syntactic constructs of MAPA is very similar to those in CCS. The term $(m;P)$, called *prefixing*, represents a process able to perform a multi-action m, and after that behaves like P. In MAPA, a multi-action represents a simultaneous execution of all contained actions. A single action can be sending or receiving a signal. Conjugated action would represent sending or receiving of the same signal. Synchronization on the level of actions is assumed to be binary (a signal can be received only once) and synchronous (sending of a signal can occur only if someone receives it). Thus, a process $(m;P)$ can perform multi-action m if its environment performs the conjugate multi-action. The term $(P \, ch \, Q)$ describes a *composition* process that behaves like two processes, P and Q, running in parallel and communicating via an oriented channel ch (Fig. 3a). The construct $(P : ch)$ is called *restriction* (or *interfacing*). It behaves like P but its communication capabilities are limited to channel ch, i.e. only the multi-actions of P allowed by channel ch will be visible from the outside (Fig. 3b). Using process variable g, one can build processes with infinite behaviours.

a)
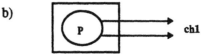

b)
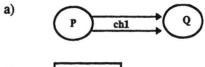

Fig. 3. Syntactic constructs:
a) composition operator,
b) restriction operator.

The operational semantics of MAPA is a mapping of the process terms into the states of a labeled transition system, where states are all elements of process states, labels are multi-actions, and transitions are defined by a set of rules. Different sets of rules result in different operational semantics that have their advantages and disadvantages. Proving that labeled transition system graphs generated by the corresponding sets of rules are weakly bisimular (van Glabbeek, 1993) indicates that the equivalent MAPA process terms are bisimular. The bisimularity is of practical importance, since it allows the designer to switch from one operational semantics to another in order to facilitate a proof or an implementation.

4.2 Distributed implementation

In order to develop a dynamically distributed implementation of MAPA, the semantics introduced in the previous section has to be turned into an interpreter. This section discusses the basic requirements and general structure of a distributed MAPA interpreter. A point-to-point message passing paradigm for process communication is assumed.

General idea. Let us present the general idea of process distribution using the example of a term $((P \, ch_1 \, Q) \, ch_2 \, R) \, ch_3$. The term describes a system of three processes, P, Q, and R, running in parallel and connected through channels as shown in Fig. 4.

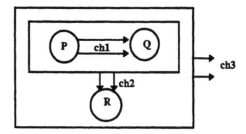

Fig. 4. Distribution of term $((P \, ch_1 \, Q) \, ch_2 \, R) \, ch_3$.

The system communicates with the environment via channel ch_3. Adequate behaviour of the system with reference to the MAPA semantics requires the specification of a communication protocol between the system and the environment. The environment can accept one of the multi-actions generated by the system by sending an acknowledgment message stating which multi-action has been chosen. In general, a chosen multi-action can be in conflict with some other multi-actions sent to the environment, thus automatically disabling all those multi-actions. In order to inform the environment of a conflict situation between proposed multi-actions, an approach has been adopted (Fanchon, 1995; Corradini and De Nicola, 1997) in which the conflict resolution is based on the notion of *location*. By definition, two multi-actions are in conflict iff they share a location, i.e. the sets of all processes involved with the multi-actions are not disjoint. In this way, the environment, by knowing the locations of all proposed multi-actions, can easily determine the conflict situation. It then updates the set of gathered multi-actions by deleting all that are in conflict with the chosen one. The environment is then ready to restart, i.e., to choose the next multi-action from the remaining ones or to listen to the new multi-actions proposed by the system.

The distribution of processes follows the form of the MAPA term, yielding a binary tree of two types of nodes: *process agents* and *channel agents*. Process agents are always leaves so they have no children, while channel agents are the inner nodes. A distribution of the term illustrated in Fig. 4 is shown in Fig. 5. Multi-actions proposed by the environment are initialized by process agents, i.e. the leaves of the graph. A channel agent, which plays the role of the environment to its children, gathers the multi-actions proposed by its children and, depending on the values of local attributes, generates new multi-actions and sends them to the parent and/or accepts some of the multi-actions by sending acknowledgment messages to the children.

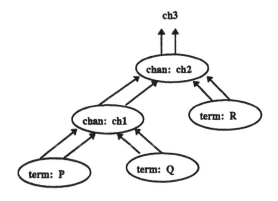

Fig. 5. Distribution graph.

Behaviour of a process agent. A process agent is a sequential agent with the following attributes:
- ident - a unique identifier of the agent;

- term - an executable MAPA process term;
- filter - a description of the communication channel between the process agent and the environment.

The behaviour of the process agent depends mainly on the value of the attribute term. In particular, the case for term = Q ch' R involves dynamic distribution of processes. The process agent terminates by splitting into three agents, a new channel agent and two new process agents. The attribute chan of the channel agent is initialized to ch', and the attributes term of the new process agents are initialized to Q and R. The attribute filter of the channel agent inherits its value from the terminated process agent.

Behaviour of a channel agent. A more complex behaviour is performed by a channel agent which has to communicate with its three neighbors: the environment (parent), and two children (right and left). The role of the channel agent is to gather multi-actions sent by children, to synchronize them with respect to the value of its attribute chan, to choose one of the multi-actions - possibly after negotiation with its environment - and to send the acknowledgments to the children. Apart from chan, a channel agent has four other attributes: filter, O_L, O_R, and sent. The attributes O_L, O_R collect messages coming up from the children. The attribute sent gathers as its value all process agent identifiers related to multi-actions sent to the environment. This information is used when the channel agent receives a message with an empty multi-action. If the empty multi-action is in conflict with a multi-action already sent to the environment, the empty multi-action can be chosen locally by the channel agent, without any interaction with the environment.

5. CONCLUSIONS

In this paper we have proposed an architecture of a hybrid intelligent design system operating on a high performance parallel computing platform. The type of design problems handled by this system calls for hardware and software architectures appropriate for tasks with medium and high levels of granularity. The issues related to dynamic task distribution and synchronization motivated the development of a formal methodology based on a process algebra approach (Multi-Action Process Algebra) and a message passing paradigm. Two types of agents, process agents and channel agents, were proposed for distributed implementation of MAPA. Conflicts between competing multi-actions are resolved through the use of specially defined sets called locations. The proposed methodology can at the same time be useful not only for dynamic process distribution and synchronization, but also for other purposes, such as verification of distributed algorithms in a way similar to the functionality of dedicated theorem provers. This results from the compatibility of the proposed semantics with the Structured Operational Semantics (Plotkin, 1981) specifications.

REFERENCES

Albus, J.S., H.G. McCain and R. Lumia (1989). *NASA/NBS Standard Reference Model for Telerobot Control System Architecture (NASREM)*. National Institute of Standards and Technology, Technical Report 1235, Gaithersburg, MD.

Albus, J.S. (1994). *A Reference Model Architecture for Intelligent Systems Design*. National Institute of Standards and Technology, Technical Report 5502, Gaithersburg, MD.

Best, E., R. Devilliers and J. Hall (1992). The Box Calculus; A new causal algebra with multi-label communication. In *Advances in Petri Nets*, **LNCS 609**, Springer-Verlag, Berlin.

Corradini, F. and R. De Nicola (1997). Locality based semantics for process algebras. *Acta Informatica*, **34**, pp. 291-324.

Fanchon, J. (1995). Algebras of located and directed processes. Laboratoire de Recherche en Informatique, Rapport de Recherche 987, Orsay, France.

Fraczak, W. (1995). Multi-action process algebra. In *Algorithms, Concurrency and Knowledge*. **1023**, Springer- Verlag, Berlin.

Goonatilake, S. and S. Khebbal (1992). Intelligent Hybrid Systems. *Proc. First Int. Conf. On Intelligent Systems*, Singapore, pp. 207-212.

Harris, C. J. and M. Brown (1994). Advances in Neurofuzzy Algorithms for Real Time Modelling, Control and Estimation. *Postprints of IFAC Workshop on Safety, Reliability and Applications of Emerging Intelligent Control Technologies*, Hong Kong, pp. 145-155.

Heitbreder, O., B. Kleinjohann, L. Kleinjohann and J. Tacken (1997). Intelligent Design Assistance with SEA. *Proc. IEEE Int. Conf. on Engineering of Computer-Based Systems*, Monterey, CA, pp. 279-286.

Khebbal, S. and C. Sharpington (1996). *Rapid Application Generation of Business and Finance Software*, Kluwer Academic Publishers, Dordrecht.

Kohn, W., J. James, A. Nerode, K. Harbison and A. Agrawala (1995). A Hybrid Systems Approach to Computer-Aided Control Engineering. *IEEE Control Systems*, **4**, pp.14-25.

Ledeczi, A. (1995). *Parallel Systems with Flexible Topology*. Ph.D. Thesis, Dept. of Electrical and Computer Engineering, Vanderbilt University.

Lee, M.A. and H. Takagi (1993). Integrating Design Stages of Fuzzy Systems Using Genetic Algorithms. *Proc. 2nd IEEE Int. Conf. on Fuzzy Systems*, San Francisco, CA, pp. 612-617.

MacFarlane, A.G.J., G. Gruebel and J. Ackermann (1989). Future design environments for control engineering. *Automatica*, **25**, pp. 165-176.

Medskar, L. (1994). *Hybrid Neural Network and Expert Systems*. Kluwer Academic Publishers. Boston.

Milner, R. (1983). Calculi for Synchrony and Asynchrony. *Theoretical Computer Science*, **25**, pp. 267-310.

Plotkin, G.D. (1981). A structural approach to operational semantics. Report DAIMI FN-19, Computer Science Dept., Aarhus University.

van Eijk, P.H.J., C.A. Vissers and M. Diaz (1989). *The Formal Description Technique LOTOS*. Elsevier Science Publishers, Amsterdam.

van Glabbeek, R.J. (1993). The linear time-branching time spectrum II; the semantics of sequential systems with silent moves. In: *CONCUR'93* (E. Best, Ed.), Springer-Verlag, Berlin.

Zeigler, B.P. and J.W. Kim (1996). A High Performance Modelling and Simulation Environment for Intelligent Systems Design. *Int. J. of Intelligent Control and Systems*, **1**, pp. 83-100.

AN INTER-ORGANIZATIONAL INFORMATION SYSTEM INFRASTRUCTURE FOR EXTENDED (OR VIRTUAL) ENTERPRISES

PARK Kyung Hye and FAVREL Joël

Lab. PRISMa / INSA de Lyon
Bat.502, 69621 Villeurbanne cedex, France

Abstract : The terms "extended enterprise" and "virtual enterprise" have been used in articulating the 21st century global manufacturing enterprise strategy. One of the key requirements is to develop an Information System (IS) infrastructure to control the interoperability of the distributed, heterogeneous and concurrent systems in the participating organizations. This article presents an innovative IS infrastructure to help describing global information technology support for these inter-organizational enterprises with the background of modern information and communication technology. Three types of technology are focused on: Data warehouse, Process warehouse and Intranet/Extranet.

Keywords : Enterprise modeling, Integration, Information systems, Communication systems, Information technology, Networks, Process models

1. INTRODUCTION

Manufacturing is a complex application domain, and it becomes more and more dependent on information through the use of computers and computer-controlled machines. According to Chen (1995), advanced manufacturing enterprises are evolving towards a more agile framework, which must meet demands for quick response to changing technology innovations, production specifications, market conditions, and business constraints. These changes will require significant advances in information system (IS) infrastructures, which consist of a variety of distributed and intelligent information and communication systems.

The environmental changes for advanced global manufacturing enterprises would be summarized as :
- Globalization. Global economy and markets without border like a world of Cyberspace.
- « Coopetition ». Enterprises are competing and also cooperating at the same time.

- Customer-driven production. Customers will be deeply integrated into all aspects of the product cycle. They demand high quality, low cost and fast delivery of increasingly customized products - mass customization.
- Peer-to-peer relationships. Suppliers will be integrated into the product cycle.
- Technological innovation. Technologies for telecollaboration require and enable organizational units to be distributed globally.

In the arena of advanced manufacturing, so-called Next Generation Manufacturing Enterprises (NGME), a number of concepts such as "agile manufacturing enterprise", "virtual enterprise" and "extended enterprise" have emerged, requiring the manufacturing to be highly information-intensive.

The Extended Enterprise or the Virtual Enterprise are concepts which have been used to characterize the global supply chain of a single product in an environment of dynamic networks of companies engaged in many different complex relationships

(Møller, 1996). These concepts are supported by extensive use of information and communication technologies. In this article, these two kinds of inter-organizational manufacturing systems are focused on.

The objectives of this article are :
- to describe organizational aspects of an extended or a virtual enterprise
- to list Information and Communication Technology (ICT) supports for these inter-organizational manufacturing systems
- to design an innovative IS infrastructure to help describing global ICT support
- to focus on some essential technologies : Data warehousing, Process warehousing, Intranet/ Extranet
- and to present a few considerations on the design and implementation of such IS.

2. ORGANIZATIONAL ASPECTS OF AN EXTENDED OR VIRTUAL ENTERPRISE

While CIM mostly concerns intra-enterprise integration (i.e. integration of the business processes of a given enterprise), the Extended Enterprise is concerned with inter-enterprise integration (i.e. inter-networking of enterprise interacting along a common supply chain). Extended enterprises (E.E.) span company boundaries, include complex relationships between a company, its partners, customers, suppliers and its markets.

The organizational aspects of an extended enterprise can be summarized as : 1) Globalization of exchanges, 2) Subcontracting, 3) Partnership. The companies in an Extended Enterprise must coordinate its internal systems (intra-organizational activities) with other systems in the supply chain and further must be flexible and prepared for adapting to change.

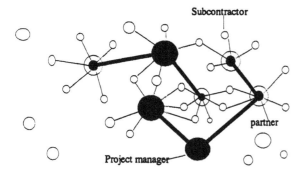

Figure 1. Extended Enterprise (Jagou, 1993)

Virtual enterprises (V.E.) are created to address a specific market opportunity, made up from parts of two or more different enterprises, and designed to facilitate gathering productive resources quickly, broadly and concurrently. They group together on bases of cost effectiveness and product uniqueness without regard for organization size, geographic location, computing environments, technologies deployed, or processes implemented.

In virtual enterprises, manufacturers operate as nodes in a network of suppliers, customers, engineers, and other specialized service functions (Davidow, 1995). According to Nagel and Dove, the virtual enterprise emerges as a result of the information technology revolution. They materialize by selecting skills and assets from different firms and synthesizing them into a single, electronic business entity (Frankwick, 1995).

The main objective of a virtual enterprise is to allow a number of organizations to rapidly develop a working environment to manage a collection of resources contributed by the organizations towards the attainment of some common goals (Su, 1995). Because each partner brings a strength or core competence to the enterprise, the success of the project depends on all cooperating as a unit.

A virtual enterprise sometimes has a relatively short life. For example, two or more enterprises may join together temporarily to fulfill a specific government contract. When the contract has been achieved, the virtual enterprise has no further use and is broken up. Other virtual enterprise may be more permanent, forging their relative strengths together to face a growing market opportunity (Sims, 1996). In that case, a V.E. becomes like an E.E. which has relatively more stable organizational structure.

Extended or virtual enterprise environment will be able to manufacture and assemble products at lower cost and higher quality with less risk and shorter lead times. The participating organizations share costs, skills, and core competencies which collectively enable them to access global markets with world-class solutions that could not be provided individually.

3. INFORMATION AND COMMUNICATION TECHNOLOGY (ICT) SUPPORTS

In this age of global economy, advanced manufacturing systems require the cooperation and collaboration of multiple organizations throughout a nation or across multiple nations. To quickly exploit fast-changing worldwide product manufacturing opportunities, extended or virtual enterprises need to be formed by these organizations to pull together the best of their resources. Therefore, it is essential that the organizations participating in such enterprises are able to rapidly and flexibly develop a common working environment to manage and use their resources toward the attainment of their business goals. Data and computer resources are key resources which need to be shared (Su, 1995). However, they are generally managed by dissimilar software systems running on heterogeneous computing platforms. Thus, one of the key requirements of an extended or virtual enterprise is to develop an information system to support the interoperability of distributed and heterogeneous systems for the purpose of accessing and sharing the necessary data and resources.

Viable IS places specific technical requirements. These ICT supports can perhaps be summarized as :
- Distributed data processing and application-to-application connectivity
- Groupware technology for the participants' process integration
- Communication between the individuals and computer systems of the organizations involved.

3.1 Data/Application Integration Technologies

The operation of the inter-organizational enterprises requires to take up of database technologies which are near to the current state of the art. The main challenges are distributed data processing and application-to-application connectivity across enterprises (between participating organizations of the extended or virtual enterprise).

Several important categories of these technologies have been identified : 1) DBMS/ PDMS, 2) Client/ Server, 3) Middleware and 4) Data warehouse.

3.2 Process Integration and Groupware

Inter-organizational enterprises will succeed only in an environment of teamwork which includes employees, management, customers, suppliers and government, all of them working to achieve common enterprise goals. As business operations get distributed, geographically or organizationally, corporations are facing increasing pressure to find effective means of integrating their organizational processes. While process integration can be certainly enhances by means of network connections and computer-supported collaboration.

It is important to note that Groupware is a relatively new umbrella term, describing a new set of technologies that support person-to-person

collaboration and it provides tools to solve "collaboration oriented" business problems.

Several categories of process integration technologies have been identified: 1) Groupware, 2) Workflow management, 3) Document Management System (DMS) and 4) Process warehouse.

3.3 Communication Technology

This is a key component to successful implementation of an inter-organizational enterprise because this enterprise is an expression of the market driven requirement to embrace external resources in the enterprise without owning them. Enterprise networks are very complex environments that involve the interconnection of a wide variety of computer systems such as portable PCs and personal digital assistants (PDA), desktop PCs and workstations, servers, and mainframes, with a wide variety of communication channels such as dial-in and mobile access via modems, local area networks (LANs), wide area networks (WANs), and the Internet.

Several categories of these technologies have been identified: 1) Inter-enterprise networking technologies such as ISDN or ATM, 2) EDI, and 3) Internet/Intranet/Extranet.

4. AN IS INFRASTRUCTURE FOR EXTENDED OR VIRTUAL ENTERPRISE

Extended or virtual enterprises require an information system (IS) infrastructure for sharing inter-organization information and actively releasing information to other partner organizations in the enterprise through the infrastructure. Advanced information and communication technologies would be necessary.

An innovative IS infrastructure for E.E. or V.E. is presented in Figure 2. This infrastructure is an integration effort of all these three essential ICT groups described before. It connects ultimately everyone within a workgroup or an inter-organizational enterprise. Each participating organization of the enterprise has access to the others with Extranet passing a fire-wall. And it is also an effective integration framework of information technology which is a systematic collection of enablers for information and process integration.

Introduced into this figure is a kind of a network(Intranet)-centric IS logic supporting the communication, exchange and sharing of data between the partner organizations.

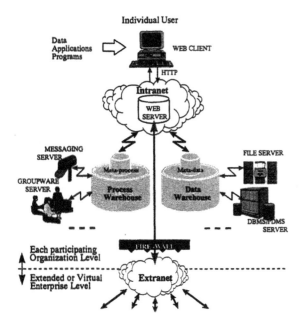

Figure 2. An Information System Infrastructure

The Web can become an extension of the IS for inter-organizational enterprises. It is, therefore, possible to communicate directly with their clients, suppliers and subcontractors in systematic ways. This would be also a general implementation of the EDI concept. In brief, Intranet or Extranet is a platform for accessing information. It plays the role of middleware between the client posts and the global IS of an enterprise. Thanks to Intranet/Extranet, we can improve inter-service communication, information sharing and DB server access.

In summary, this IS infrastructure facilitates communications between and within virtual organizations, allowing development of widely dispersed extended or virtual enterprises.

5. ESSENTIAL TECHNOLOGY FOR THE IS INFRASTRUCTURE

The concepts and characteristics of some essential technologies for this IS infrastructure are described.

5.1 Data warehouse

Dynamic environments, with their ever-changing requirements, are served best by a simple, easily changeable data architecture (e.g., a highly normalized relational structure) rather than an intricate structure that requires rebuilding after every change. Data warehousing is the process of integrating, extracting and staging enterprise-wide operational and external data into a single repository

from which end users can easily run queries, make reports, and perform analysis (BusinessObjects, 1996). A data warehouse is a decision support environment that leverages data stored in different sources, organizing it and delivering it to decision makers across the enterprise, regardless of their platform or technical skill level.

The most widely recognized definition of a data warehouse is a « subject-oriented, integrated, time variant, non-volatile collection of data in support of management's decision making process » (Weldon, 1997). Subject-oriented means the data warehouse focuses on the high-level entities of the business. Integrated means the data is consistently stored. In the data warehouse, there is only one coding scheme. Time variant is for semester, fiscal year, and pay period. Lastly, non-volatile means the data doesn't change once it gets into the warehouse.

There are different levels of data within the data warehouse. Some data is very detailed. Other data is summarized. Other older detailed data is placed in secondary storage. In addition, there is a component of the data warehouse known as « meta data ». Meta data, or information about data, is a directory as where the contents of the data warehouse are and where the contents came from.

The warehouse is the « glue » holding enterprise data stores together until a mature repository comes along. As data warehouses continue to mature in sophistication and usability, the data accumulated within an enterprise will become more organized, more interconnected, more accessible, and more generally available to the participants of the enterprise. By linking the data warehouse to other systems - both internal and external to the organization - we can share information with other business entities with little or no custom development. The definitions and coding standards in the warehouse are necessary.

5.2 Process warehouse

According to Davidow and Malone (1995), virtual enterprises will succeed only in an environment of teamwork which includes employees, management, customers, suppliers, and government all working to achieve common enterprise goals. Groupware or collaborative computing technologies let these organizations share information and enable people to work together in teams.

As business operations get distributed, geographically or organizationally, enterprises are facing increasing pressure to find effective means of integrating their organizational processes. They often communicate, collaborate and coordinate via a set of processes. Hammer and Champy (1993) defined a process as « A collection of activities that takes one or more kinds of input and creates an output that is of value to the customer ». Thus a process integration means the reorganization of structural relationships among process entities for enhanced performance.

Each organization must stock the knowledge in a process reference which can be used dynamically. Scheer (1997) named it « process warehouse ». It is necessary to optimize the process structure for managing and analyzing the evolution of the enterprise.

In the technical point of view, a process warehouse and a data warehouse are the same. But the contents are different because the data represent only a part of process cycle. A process warehouse collects all of the process descriptions and combines them in a reference.

To build a process warehouse, processes are defined by the data descriptions and also by the necessary functions for data treatment, the organizational structure which support the process and the relations between these different components. We must anticipate to open the process warehouse to all of the partner organizations in the enterprise where the processes might be described differently. Therefore, each organization must describe the enterprise's « meta processe » in some standardized manner.

5.3 Intranet/Extranet

The success of the Internet as an enabling technology has everything to do with its near-universal availability and technical homogeneity. Together, these qualities mean that every enterprise can connect using Internet technology, not just physically, but at all layers of network complexity. This is a strategic truth, providing one of the keystones for the so-called extended or virtual enterprise.

There are, of course, several very real problems to be overcome, of which security is the most notorious (though inter-network management may turn out to be more perfidious). Nevertheless, the presence of a common communications infrastructure is already being perceived as a key enabler of the inter-organizational systems (Sims, 1996).

One of the main facilities of Internet, open communications, has arguably the greatest potential to add value for an enterprise. If two systems are

both connected to the Internet, then nothing else has to be done to provide a common communications channel between those two systems. This is one of the wonders of TCP/IP, the protocol used on the Internet.

Deployment of Web servers inside the organization to enhance internal communications is part of an important trend known as Intranet. An Intranet is simply an "internal" version of the Internet, using the same base TCP/IP protocol, and therefore capable of using the same server and client products. Extranet is an "external" version of the Intranet to enhance the communications between the partners in an extended or virtual enterprise.

Intranet/Extranet merges all kinds of client-server technology : DBMS and Groupware. These technologies are totally complementary and would be also dissociable. Therefore, Intranet/Extranet became integrated part of IS. It becomes the privileged media for transmitting shared process information or structured data of a data warehouse. Once the information is captured by the standard format of this media, it is accessible for anyone in cheap price, without technological or organizational limit between the different actors of the inter-organizational enterprise.

6. CONSIDERATIONS

There are several problems to be overcome for the design and implementation of inter-organizational IS. A few considerations obtained during our study are summarized :

- User profile description : who the users (parts of an E.E. or a V.E.) are, where they are, their participation levels, ...
- Compatible semantics (a shared understanding of the meanings of things) : a well-defined set of process
- Standardization efforts: CALS, STEP, CORBA,...
- Secure inter-network connections : authentication, access control, cryptography, ...
- Use of intelligent agents to replace or support human actions, or to initiate actions with little or no human interaction
- Technical supports : videoconferencing systems, BBS (Bulletin Board System), ...
- Specified applications for working together : shared databases, shared applications or models, project management software, tele-engineering.

7. CONCLUSION

As market competition increases and consumer behavior fluctuates, it has become critical to deliver the right product, at the right time, ahead of the competition. To meet this challenge, various forms of organizational alliances have existed. In this article, the terms « extended enterprise » and « virtual enterprise » have been focused on in articulating the 21st century global manufacturing enterprise strategy.

This article examines the organizational aspects of this kind of inter-organizational systems and shows that organizational and technical issues are closely inter-related, especially in the context of integrated systems.

The organizational changes of manufacturing enterprises need to implement newly developed Information Systems infrastructure. It also needs to control the interoperability of the participating organization's distributed, heterogeneous and concurrent systems. In order to meet this requirement, it is necessary to model all things of interest to the IS such as data, process and communication technologies and organizational structures. This IS infrastructure includes different kinds of ICT (Information and Communication Technology). In this article, three groups of these technologies are described : 1) Data/Application Integration Technologies, 2) Process Integration and Groupware, 3) Communication technology.

With an integration effort of these ICT, a new IS infrastructure is presented. For designing this infrastructure, three essential technologies ; Data warehouse, Process warehouse, Intranet/Extranet ; are focused on. Finally, a few considerations for design and implementation of such IS obtained during our study are summarized.

In summary, this article presents an information system infrastructure for an extended or virtual enterprise which describes a global network-centric support for such inter-organizational enterprises with the background of modern information and communication technology. It shows how network-centric information technology would change the art of possible for the so-called extended or virtual enterprises.

REFERENCES

Bloch, M. and Y. Pigneur (1995). The Extended Enterprise : a descriptive framework, some enabling technologies and case studies in the Lotus Notes environment. Paper submitted for publication in the *Journal for Strategic Information Systems*, available on http://haas.berkeley.edu/~bloch/docs/icnom/paper_ee.htm.

Business Objects (1996). *Data warehouse - Delivering Decision Support to the Many*, 24 p. Technical White Paper Series of Business Objects.

Chen, S.S (1995). Role of the Information Infrastructure and Intelligent Agents in Manufacturing Enterprise. *Journal of organizational computing*, Vol.5(1), pp.53-67.

Davidow, W. and M. Malone (1995). *L'Entreprise à l'âge du Virtuel* (French version), 286 p. MAXIMA, Paris.

Frankwick, G.L., K. Elston and L. Laubach K (1995). *Motivators and Barriers to participation in a Virtual Enterprise*. Available on Web http://orcs.bus.ikstate.edu/phase.1/phase1/motivate.html

Hammer, M. and J. Champy (1993). *Le reengineering*, 247 p. Dunod, Paris.

Hardwick, M. and D. Spooner (1995). An Information Infrastructure for a Virtual Manufacturing Enterprise. *Proceedings of "A Global Perspective CE'95" Conference*, Mclean (VA, USA), pp.417-429.

Jagou, P (1993). *Concurrent Engineering* (French version), 140 p. Hermès, Paris.

Møller, C (1996). Interorganizational communication systems. *Proceedings of the international conference of APMS*, Kyoto, pp. 59-64.

Sandoval, V (1996). *Intranet - le réseau d'entreprise*, 152 p. Hermès, Paris.

Scheer, A.W (1997). Interview - Modéliser l'organisation de l'entreprise. *Informatique magazine*, mars 97, pp.68-70.

Sims, O (1996). Enabling the Virtual Enterprise. *Object Currents*, Oct.96.

Smith, R.L. and Wolfe, P.M (1995). Implications of client/server systems on the virtual corporation. *Computers Ind. Engng.*, Vol.29, No.1-4, pp.99-102.

Su, S.Y.W., H. Lam and J. Arroyo-Figueroa (1995). An extensible knowledge base management system for supporting rule-based interoperability among heterogeneous systems. *Proceedings of the '95 ACM International Conference on Information and Knowledge Management*, pp.1-10.

Tapscott, D. and A. Caston (1995). *L'entreprise de la deuxième ère - La révolution des technologies de l'information* (French version), 395 p. Dunod, Paris.

Weldon, J-L. and A. Joch (1997). Data Warehouse Building Blocks. *Byte*, Jan. 97, pp.82-88.

Williamson, M.D.T. and R.L Storch (1996). The collaborative engineering process within the framework of the virtual enterprise. *Proceedings of the '96 APMS International Conference*, Kyoto, pp.49-54.

ENGINEERING PROCESS MODELLING OF AN INTELLIGENT ACTUATION AND MEASUREMENT SYSTEM: FROM THE USERS' NEEDS DEFINITION TO THE IMPLEMENTATION

E. Neunreuther, B. Iung, G. Morel, J.B. Leger

Centre de Recherche en Automatique de Nancy (CNRS URA DO 821)
Equipe Génie des Systèmes Intégrés de Production
Faculté des sciences - BP 239
54506 Vandoeuvre Cedex - France
Phone : (33) 3.83.91.24.08- Fax (33) 3.83.91.23.90 - e-mail : iung@cran.u-nancy.fr

Abstract : This paper aims at describing the engineering process modelling of a particular Intelligent Actuation and Measurement System (**IAMS**) : an experimental laboratory mock-up whose finality is to control a level. Thanks to mechanisms, the modelling process allows to map the functional description of the users' needs, to a distributed operational architecture composed of field devices, a fieldbus and computerised systems. The users' needs are expressed through the Control, Maintenance and Technical Management (**CMM**) domains in keeping with the European IAMS[1] ESPRIT projects.

Key-words: Intelligent instrumentation, Hierarchically intelligent control, Distribution systems, Integration, Shop-floor oriented systems, Systems engineering.

1. INTRODUCTION : THE INTEGRATION AND DISTRIBUTION PARADIGMS

The world of instrumentation is undergoing a major evolution shift as embedded intelligence and fieldbus communication allow a migration of functionality into all parts of distributed automation system complying with integrated requirements related to integrated Control, Maintenance and Technical Management domains (CMMS concept). So, the engineering process modelling of an Intelligent Actuation and Measurement System implementing the CMMS concept, supports a paradox based on two paradigms :

- the **integration paradigm** which consists in integrating the CMM domains to make up a coherent whole to gain a real business advantage,
- and the **intelligence distribution paradigm** at the shop-floor level which consists in distributing somewhat of intelligence into all parts of the CMMS to gain **interoperability** and expect **interchangeability**.

---

[1] Intelligent Actuation & Measurement System projects :
DIAS : ESPRIT II 2172 Distributed Intelligent Actuators and Sensors
PRIAM : ESPRIT III 6188 Prenormative Requirements for Intelligent Actuation and Measurement
EIAMUG : ESPRIT III 8244 European Intelligent Actuation and Measurement User Group

1.1. From the hierarchical model to the distributed one :

In relation to our particular IAM process, the paradigms means first that the Integration requires to consider the MMMS (Manufacturing, Maintenance and technical Management System) and then its CMMS informational view as an architecture at the shop-floor level integrating activities of (Galara, *et al.*, 1993) :
- Manufacturing whose finality is to modify the shape of the market products,
- Maintenance whose finality is to retain or to restore through time manufacturing resources,
- and technical Management whose finality is to connect by information and through space the Manufacturing and the Maintenance domains.

Second, the Interoperability is reached by the functions composing the IAM when they have the ability to operate with each other in order to respect the IAM user requirements. The interoperability can be extended to the Interchangeability if an IAM device can take the place of another (substitution), by keeping all its characteristics (operational, environmental, mechanical, etc.) and its interoperability (Lorentz, *et al.*, 1993). Finally, somewhat of **Intelligence**, in fact information processing for interoperability and interchangeability, is embedded into distributed field devices which have not only their own processing but also their own information (information world) (Morel, *et al*, 1994). An Intelligent Actuation and Measurement system is characterised in the European IAMS ESPRIT projects *as the complete set of the intelligent actuation functions and of the intelligent measurement functions needed to support the rationalised automation modelled with the CMMS[2]*. Therefore, in relation to the known global manufacturing contexts such as :

- the C.I.M.E. : the **CMMS** modelling framework can be considered as a subset of enterprise modelling frameworks (CIM-OSA, GERAM (Bernus, *et al*, 1996) and the **interoperability** can be seen as a means of integrating the field devices: a hierarchical (top-down, co-ordinating) organisation.
- the I.M.S. (Intelligent Manufacturing System) : the **IAMS** can be considered as a means of reaching an autonomy level such at those quoted in the intelligence definition of the IEEE Control Systems Society (Antsaklis, 1994) and the **interchangeability** as an answer for the adaptation of the control structure or architecture related to the environment changes : a distributed (bottom-up, co-operating) organisation.

---

[2] ESPRIT III-PRIAM n°6188 project, PRIAM dictionary

1.2. The IAMS modelling process:

To solve the integration and distribution paradox, the IAMS engineering process has to be based on a modelling framework which allows to define and use reference models and static mechanisms, and expected dynamic ones for the future, for each engineering life cycle step in order to ensure the system consistency. This modelling framework which is common to the CIM and the IMS approaches, is materialised by a three-axis referential (Mayer, *et al.*, 1996) :

- an axis which takes into account the system modelling levels, all along its life cycle through the functioning (user's needs definition), the technological (physical system), and the engineering (mapping from the functioning modelling to the technological one) levels,
- an axis which allows to define at each modelling level, the functional, the behavioural and the informational views,
- and an axis which imposes through the life cycle, to integrate the Control, the Maintenance and the technical Management domains.

1.3. The IAMS experimental platform :

The practical result of the modelling process of our IAMS, is a laboratory platform that is composed of two integrated valves, a pump, a level and a flow transmitter, a control system, a maintenance system, a technical management system and a FIP field-bus (Fig. 1). This IAM system finality is to stock a water volume, between an input water flow representing the disruption, and an output water flow.

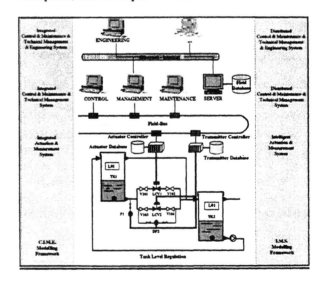

Fig. 1 The IAM methodological and technological laboratory platform.

At the technological modelling step, the IAM users' functions of the shop-floor level, are distributed between the field devices, such as the level control loop in the controller, the flow control loop in the redundant valves; and the Control, Maintenance and Technical Management systems. Due to the function distribution, the field devices have, firstly, in order to guarantee the integration objective, to be co-ordinated to meet the needs expressed by the users at the functioning level, and secondly, in order to contribute to the intelligence objective, to co-operate to meet the architecture adaptation needs.

To face up to these objectives, this paper aims at describing some mechanisms in the modelling framework which allow to ensure the interoperability between the devices and the CMM systems, and to extend these mechanisms to provide the device interchangeability.

2. FROM THE INTEROPERABILITY ...

As the device co-ordination is related to the interoperability degree of the components, the IAM device interoperability depends :

- on the **operational interoperability** which has to check among others, the communication interoperability, that is to say the device services compatibility, and the database interoperability,
- on the **functional interoperability** which has to verify the semantics and the type of information exchanged between IAM applications.

2.1. The functional interoperability :

To contribute to the functional interoperability mainly related to the finality of application and of each application component, the IAM system has to be structured to identify the standard control, maintenance and technical management process functions. This structuring is based on system approach which organises each process activity (physical world), through the product flow in relation to its stocking, transport and processing.

From an autonomous module ... In the "symbolic" world dual to the physical one, a Control activity which processes the informational flow, can be associated to each process activity and physically connected to its, through the action and observation nature activities. According to this basic circular loop (to control, to act, to operate and to observe) linking the two worlds, the variable emergence mechanism application (Petin, *et al*, 1996), lead to a first functional Control solution, describing the totality of the user needs in terms of control functions (Fig. 3). Indeed the engineering process, keeping the product finality flow, allows by using automation and mechanical know-how, an emergence of the necessary control loops to reach a control loop

directly controlled. In the end, the procedure leads to an IAM system composed of a level, a flow, a valve opening and a valve position control loops (from the input flow to the output one). The identified control activities can be compared to control autonomous modules which transform event and data flow (Fig. 2a) :

-the event flow process function aims at transforming a set point into action in accordance with the control system finality,
-the data flow process function aims at transforming the data to inform the event flow process function.

These two major transformations, kernel of the control modules, can be modelled according to the behavioural filter concept (Lhoste and Morel, 1996) by a command function for the event flow processing, a monitoring function for the data flow processing and a mode management function to connect the command and the monitoring functions (Fig. 2b). A generalisation of the module activities leads to the formalisation of a Automation Functional Module, receiving and generating events, and structured around four generic functions implementing the objective validation, the observation validation, the command execution, and the report elaboration. This module owns a first autonomy degree because on the one hand the command objective (the set point) can be accepted or not according to its internal state and to the associated technological operator status (the observations) and on the other hand the module is able to continuously monitor (by forecasting, comparison and analysis) its behaviour in relation to the consistency of the objectives, the commands and the observations.

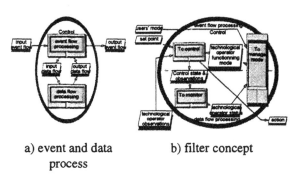

a) event and data   b) filter concept
   process

Fig. 2 Event and data flow processing

The re-use of the same autonomous module principle for the domains other than the Control, allows to structure the whole of the IAM system. Indeed from a dual point of view to the Control, a Maintenance activity which processes the resource flow, is associated to each controlled process activity and also linked to the resource activity through the action and observation nature activities.

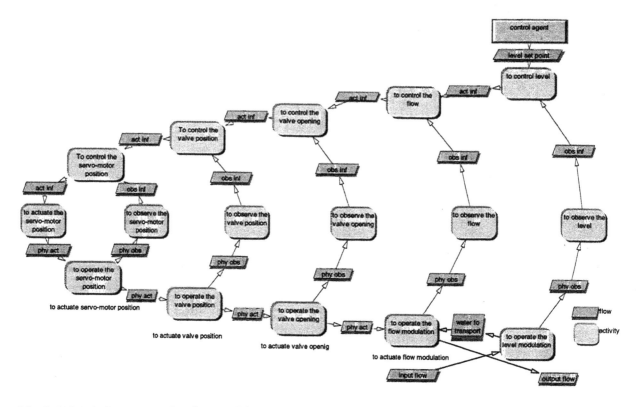

Fig. 3. IAM platform processing flow model.

The identification of all the maintenance variables lead to a first Maintenance functional solution, describing the totality of user needs in terms of maintenance modules. The engineering process, keeping the resource finality flow, allows by using maintenance and mechanical know-how, an emergence of the necessary maintenance loops, to observe resource degradation, to decide and to make a corrective or preventive maintenance (to retain or to restore manufacturing resources). An autonomous and reusable control and maintenance module is therefore defined for each structuring level (level, flow, valve opening, valve position and servo-motor position control loops) leading to an IAM system composed of modules assembling.

A such architecture is a first step towards the functional interoperability because one module is independent of the others modules (a modification of one module does not imply a modification of the environmental modules). For example the level control module is independent of the flow control module which can be a valve or a pump module. For the interoperability issue, the IAMS and REMAFEX ESPRIT projects have proposed, for each module, to standardise a **FCS** (Functional Companion Standard) including the functional and the informational parts of its supported control and maintenance functions. The FCS are defined in relation to a set of {F, O, DF} where :

- F is the Function supplying a service to the component environment,

- O represents all the Objects produced or consumed by the function,
- DF characterises the Data Flow which associates the objects to the function.

The FCS concept has initiated the user's function block standardisation for continuous process of the IEC SC 65C - WG7.

... To an open integrated intelligent module. A first up-grading of the autonomous module towards an intelligent one is to really "open" it towards its environment knowledge. That means the module have to exchange information (and not only events) to co-operate with the others and to integrate all the domains such as the Control, the Maintenance and the technical Management ones (CMMS concept). The informational exchange concerns, among others, between the Control and Maintenance domains, the information on the device failure and availability, and between the Control and the Technical Management domains, the device statistic information. The exchange is realised by transactional process (report transfer function of the Automation Functional Module) through the information communication (transactional flow) and its storage (information management system) (Fig. 4). The interoperability of the information system is solved by an IAM reference data model which allows with the help of specialisation and explosion / globalisation mechanisms (Ducateau and Picavet, 1995) to define a data model for each structuration level (generic model) in accordance with the STEP/EXPRESS standardisation principle normalising the objects life cycle data model.

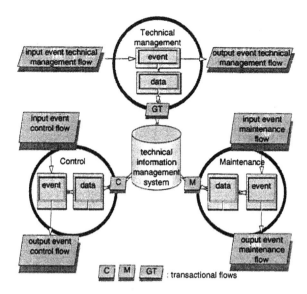

Fig. 4. CMMS integration.

2.2.Operational interoperability :

At the engineering level, the FCS and data models are distributed between the external systems, composed of the control, maintenance and technical management systems and the internal ones composed of the field devices. The FCS distribution implies that a same module can be shared into different systems generating consequently new transactional flows. From the distribution can result a redundant system (several resources for the same finality fulfilment) and/or a resources sharing (a same resource with several accesses such as Control, Maintenance,...) requiring to manage conflict and co-operation : the decision-making. So, the distribution has to guarantee an operational interoperability necessary for the devices co-ordination in order to fulfil the users' requirements and a decision-making processing in order to select the FCS and data base allocation.

That means for the interoperability, firstly, the informational part of the FCS which composes each external and internal system interface, has to be broadcast on a communication network to be exchanged. This translation is made from a "communication reference model" by defining **CCS** (Communication Companion Standard), expressing the information and the services used by the communication bus. The CCS guarantee the mapping of the FCS with an application service of particular field-bus OSI model, by a vectorisation of the information according to the communication services. Secondly, the data models have to be shared between the field devices (short term information and device technical information) and the technical management system (long term information and platform technical information) in co-operative databases by guaranteeing the information consistency.
In our IAMS, the decision-making is underlined between the CMM domains through only a resource

sharing which the processing is centralised in the external systems (Fig. 5) and not distributed into the field devices (hierarchical architecture).

For example, the choice of the "flow function" allocation expressed by the "mode" flow, is calculated by the controller according to the valve status. So if the devices characteristics change, it is necessary to re-configure or to re-parametrise the controller that is opposite to the interchangeability principle.

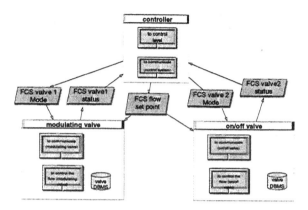

Fig. 5 IAM platform hierarchical architecture.

3. ... TO THE INTERCHANGEABILITY PROSPECT

To reach the interchangeability, meaning the system react/respond to changes in the environment, the architecture have to be more flexible and bring responsiveness (agility). This objective can be fulfilled increasing the device autonomous degree by a decision task allocation in all field-devices, such as in the self-organised systems as the holonic systems (Valckenaers, *et al*, 1994).

In this case, a concrete distributed architecture is build dynamically on the basis of the device aggregation from an organisational architecture instanciation. Indeed the architecture design is managed by an arbitrator and can implemented multi-agent algorithms as contract net algorithms (David and Smith, 1983).
For example, the Fig. 6 shows the IAM architecture built from the "control flow" functions allocation with the contract net algorithm use.
This resulting structure is a first step towards the autonomous and co-operating components by increasing the interchangeability degree of the IAM system because a substitution of an equipment by another informs the arbitrator which realises a new announcement to design again the concrete architecture.

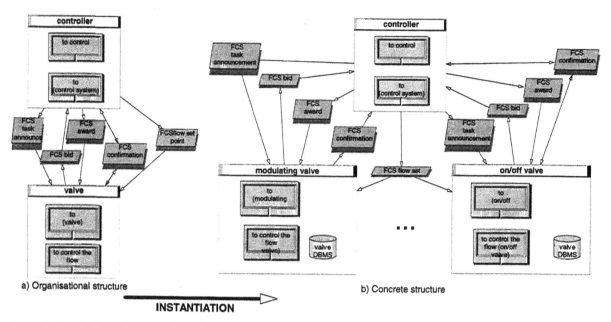

a) Organisational structure

INSTANTIATION

b) Concrete structure

Fig. 6 IAM platform distributed architecture

4. CONCLUSION

The main result of our contribution to the IAMS engineering process modelling is the formalisation of mechanisms such as the FCS, the CCS, ... which have to ensure a functional and operational interoperability asked to meet the users' needs. These mechanisms lead to the distribution of the transactional and transformational processings directly into the field devices complying with the CMMS requirements. However the degree of autonomy of these equipment is very low because the decision-making is more centralised in the external intelligent system such as the control one. To face up to the necessary responsiveness of the manufacturing systems, our prospect is therefore to evolve from an IAM hierarchical structure to an IAM distributed one based on autonomy and co-operation concepts. This new R&D orientation is in progress through our participation to the IMS-WG n°21955 long term research.

REFERENCES

Antsaklis P. (1994). Defining Intelligent Control. In : *proceedings of the 1994 IEEE International Symposium on Intelligent Control.* Columbus Ohio USA.

Bernus P., L. Nemes and T.J. Williams (1996). Architectures for Enterprise Integration, Chapmann & Hall editor

Davis R. and R.J. Smith (1983). Negociation as a Metaphor for Distributed Problem Solving. In : *Artificial Intelligence.* **Volume 20 No 1**, pp 63-109.

Ducateau C.F. and M. Picavet (1995). Progressive Adjusting Process for Data Modelling. In : *International Conference on Industrial Engineering and Production Management.* pp 235-244, Marrakech Maroc.

Galara D., J.M. Favennec, B. Iung and G. Morel (1993). Distributed Intelligent Actuators and Sensors. In : *7th European Computer Conference COMPEURO'93.* Paris Evry.

Lorentz P., F. Beaudoin and M. Castello (1993). Intelligent Transmitters : Need for Interoperability and Interchangeability. In : *6th International Exhibition and Congress for Sensor & Systems.* Nuremberg, Germany.

Lhoste P. and G. Morel (1996). From discrete event behavioural modelling to intelligent actuation and measurement modelling. In : *proceedings of ICIM-NOE conference (ASI'96) of Life Cycle Approaches to Production Systems,* Toulouse France.

Mayer, F., G. Morel, B. Iung and J.B Leger (1996). Integrated manufacturing system meta-modelling at the shop-floor level. In : *proceedings of ICIM-NOE conference (ASI'96) of Life Cycle Approaches to Production Systems,* Toulouse France.

Morel G., P. Lhoste and B. Iung (1994). Towards intelligent actuation and measurement system. In : *proceedings of ICIM-NOE conference (ASI'94) on Intelligent control and integrated manufacturing systems.* pp 121-126, Rion Patras Greece.

Petin J.F., B. Iung, E. Neunreuther and G. Morel (1996). Contribution méthodologique à l'Actionnement et la Mesure Intelligents. In : *European Journal of Automation* (Hermes Ed, in French), **Volume 30 No 6,** pp 897-918.

Valckenaers P., H. Van Brussel, F. Bonneville, L. Bonagaerts and J. Wyns (1994). IMS Test Case 5 : Holonic Manufacturing Systems. In : *Preprints of IMS'94, IFAC workshop.* Vienna.

AUTOMOTIVE BODY ASSEMBLY MODELING
FOR DIMENSIONAL CONTROL USING STATE SPACE MODELS

Jianjun Shi and Jionghua Jin

In-process Quality Improvement Research Laboratory
S.M. Wu Manufacturing Research Center
Department of Industrial and Operations Engineering
The University of Michigan
Ann Arbor, MI 48109-2117
Phone: 313-763-5321,   Fax: 313-764-3451,   e-mail: shihang@engin.umich.edu

Abstract: In this paper, a state space modeling approach is developed for the dimensional control of autobody assembly processes. In this study, a 3-2-1 scheme is assumed for the sheet metal assembly. Several key concepts, such as tooling locating error, part accumulative error, re-orientation error, are defined. The inherent relationships among these error components are developed. Those relationships finally lead to a state space model which describes the variation propagation throughout the assembly processes. An observation equation is also developed to represent the relationship between observation vector (the in-line OCMM measurement information) and the state vector (the part accumulative error). Potential usage of the developed model is discussed in the paper.

Keywords: manufacturing system modeling fault diagnosis and control, automotive body manufacturing.

1. INTRODUCTION

Dimensional control is one of the most important challenges in automotive body assembly. Due to the complexity of autobody assembly process, it normally requires dozens of fixtures to assemble, on average, 150-250 parts. The complexity of the assembly line places high demands on the tooling design, manufacture, and diagnosis for improving autobody quality.

In recent years, the implementation of in-line Optical Coordinate Measurement Machines (OCMM) in the automotive industry has provided new opportunities for assembly fault diagnosis. OCMM gages are installed at the end of major assembly processes, such as framing, side frames, underbody, etc. The OCMM measures 100 to 150 points on each major assembly with a 100% sample rate. These inspected points are located on many of the individual parts of the autobody. As a result, the OCMM provides tremendous amounts of dimensional information, which can be used for assembly process control. However, effective utilization of this measurement information, especially for assembly fault diagnosis, is still a challenge.

Recent research exploring fault isolation issues in autobody assembly has focused on a statistical descriptions of variation patterns (Hu and Wu, 1992) and the detection of failing assembly stations (Ceglarek et al., 1994). Hu and Wu (1992) investigated the description of the dimensional faults by in-line measurement data using Principal Component Analysis (PCA). Ceglarek et al. (1994) described a systematic method of identifying failing stations and faulty parts in the assembly line. Additionally, they described a rule-based approach to identify root causes of dimensional faults in the fixture. Their rule-based approach is based on heuristic knowledge which specifies a fixed level of detail about the position and control directions of the fixture locators. More recently, the diagnosis of a *single* fixture has been studied using principal component analysis (Ceglarek and Shi, 1996) and Least Square estimation with hypothesis testing (Apley and Shi, 1995).

The aforementioned research activities have significantly advanced process monitoring and diagnosis for dimensional control of body assembly processes. However, the lack of models describing the overall assembly process has imposed a large constraint on developing advanced diagnosis techniques for body assembly processes.

This paper attempts to resolve the above mentioned challenges by developing a state space modeling approach which defines 1) the part accumulative dimensional error as a state vector, 2) the fixture tooling locator error as a control vector, 3) the geometric relationship of assembled part variation stack-up as the dynamic matrix, 4) The geometric relationship between the tooling locators and part orientation as the control matrix, and, 5) the assembly station number which serves as the time index in the state space model.

In this paper, assumptions and definitions will be given first in the section 2.2 and 2.3 to describe the part dimensional variations and fixture failures. Then, several theorems will be developed in the subsection 2.3.1 and 2.3.2 to describe the inherent relationship between the tooling locator error and part orientation error in the defined body coordinates. Based on those results, a state space model is developed in the section 3.1 and 3.2 to represent the relationship between the tooling error and part dimensions for the overall assembly processes, which involves all assembly stations in the processes. Some potential usage of the developed state space model is discussed in the section 4. Finally, a summary and conclusion is given in the section 5.

2. DESCRIPTION AND HYPOTHESIS OF AUTOMOTIVE BODY ASSEMBLY PROCESS

2.1 Automotive Body Assembly Process

The automotive body without doors, hoods, fenders and truck lid is called the "Body in White (BIW)". In a BIW assembly line, depending on the complexity of the product, there are typically 50 to 80 assembly stations which assemble 150 to 250 sheet metal parts. An assembly station normally has two or more assembly fixtures, and each fixture holds one part to be assembled with the other parts. In this paper, it is assumed that one assembly station contains only two fixtures. Based on their functions, the components of a BIW are usually divided into structural and non-structural parts. Structural parts, such as rails, plenum and door hinge reinforcements are much more rigid than non-structural parts, such as the door outer, cowl-side, roof, etc. Past research indicates that a structural part usually has much larger impact on the automotive body dimensional accuracy [ABC,1993 and Takezawa,1980]. Thus, only structural parts will be considered in the later modeling procedure.

In order to conduct the state space modeling of the assembly process, a body coordinate system shown in Fig. 1 is used, which is defined on a normal (perfect) BIW without any assembly error. The origin of the body coordinate system is defined in the front center

of the vehicle and below the underbelly. The X-Y-Z axes are shown in the figure. This definition of the body coordinate system has been widely used in auto industry in product and process design.

2.2 Fixture Layout and Fixture Error

In sheet metal assembly, locating pins and NC blocks are widely used in fixtures to determine the part location and orientation in an assembly process. For a rigid part, a 3-2-1 principle is the most common layout method. As shown in Fig.2, a typical 3-2-1 fixture contains several key tooling locators: (1) a four-way pin P_1 to precisely locate the hole in X and Z directions; (2) a two-way pin P_2 to locate a slot in Z direction; together, the two pins constrain the part rotation and translation in the X-Z plane; and (3) three NC Blocks to locate the part in the Y direction. In this paper, a general modeling procedure is presented which focuses on the X-Z plane as shown in Fig.2.

Fixture errors (also called as tooling locator errors, or tooling faults) result from many different factors, such as a worn locator, missing block, or broken pin, etc. In this paper, the following definitions are given to describe a fixture error.

<u>Definition 1:</u> *Fixture Error Vector:* For station i, the tooling locating error in the X-Z plane for a 3-2-1 fixture is represented by

$$\Delta \mathbf{P}(i) = \left(\Delta x_{P_1}(i), \quad \Delta z_{P_1}(i), \quad \Delta z_{P_2}(i) \right)^T \quad (1)$$

where $\Delta \mathbf{P}(i)$ is the fixture error vector for locator points P_1 and P_2 of station i; $\Delta x_{P_1}(i)$, $\Delta z_{P_1}(i)$ and $\Delta z_{P_2}(i)$ represent the locating error for the 4-way pin P_1 and 2-way pin P_2 in the X and Z direction respectively; and the superscript T means matrix transpose.

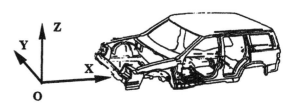

Fig. 1. Body coordinate system

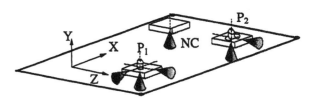

Fig. 2. 3-2-1 fixture layout

2.3 Part Variation

In order to study the assembly process variation, the part orientation in the body coordinate needs to be described. Based on the rigid part assumption, any single part orientation can be represented with a point on the part and an orientation angle in the body coordinate. Since variation is the focus of the research, only the deviation, or error, is included in the model. Thus, we have the following definition on part point and part error vector.

<u>Definition 2</u>: *Part point and part error vector:* A "part point A" is defined to represent the part orientation in the body coordinate. The part error vector represented by the part point A is described as:

$$\mathbf{X}_A(i) = \begin{pmatrix} \Delta x_A(i) & \Delta z_A(i) & \Delta\alpha(i) \end{pmatrix}^T \qquad (2)$$

where $\Delta x_A(i), \Delta z_A(i)$ are the deviation errors at point A in the X and Z directions in the body coordinates at station i; $\Delta\alpha(i)$ is the part orientation angle error of this part at station i, which is defined as positive when it is a counterclockwise.

A part point A can be any point on a part. However, the part point A should be the same throughout the assembly process for a given part. The part error vector defined in Eq. (2) can specifies the part orientation error in the X-Z plane. It should be emphasized that the part error vector reflects the "accumulative error", which means the total assembly error occurs in the previous assembly processes up to the current assembly station.

The part error vector comes from two independent root causes in the assembly process.

(i) The first one is due to the fixture error in the current assembly station, which could be due to worn out locating pins, missing or loose tooling locators, etc. This part error is call part locating error and is defined as follows.

<u>Definition 3</u>: *Part locating error:* The part error, which is represented by the part point A due to the fixture error vector $\Delta\mathbf{P}(i)$ at the current assembly station i, is represented as $\mathbf{F}_A(i)$.

(ii) The other one is the part error due to the part reorientation around the locator points at the current station i. The part has error after assembled in the last station, which can be represented by the locating points on the part. The tooling locating points are used in the current assembly station. As a result, the deviation of those points will be re-set to zero (or design intent). Meanwhile, the part error is changed simultaneously. This part movement, which resets the error of the tooling locating point to zero at the current station, is called the "reorientation" of the part. Due to this reorientation, the accumulative error up to the last assembly station will be "reoriented" following the tooling locating points on the part. This

part reorientation error is defined in Definition 4.

<u>Definition 4</u>: *Part reorientation error:* The part error represented by the part point A due to the part reorientation movement to compensate the stack-up variations up to the previous assembly station i-1 at the locator points of the current station i is called the part reorientation error and represented as $\mathbf{T}_A(i-1)$.

A simple example is shown in Fig.3 to illustrate the variations resulted from the two sources in an assembly process. In the example, three parts are assembled together by two stations. The original assembly process without fixture errors are shown in Figs. 3(a) and 3(b), respectively. The real assembly process with a fixture error $\Delta Z_{P_1'}(1)$ at station 1 is shown in Figs.3(c) and 3(d). The point P_s (s=1,2; i=1,2) represents the locator point P_s of the first subassembly part at the station i; and the point $P_s'(i)$ (s=1,2; i=1,2) represents the locator point P_s' of the second subassembly part at the station i. In this example, the locator points $P_1(2)$ and $P_2(2)$ are designed as the same points $P_1'(1)$ and $P_2'(1)$. In this case, only the part 2 at station 1 has a deviation due to $\Delta Z_{P_2'}(1)$. The assembly error due to the fixture error at station 1 is the part locating error, as defined in Definition 3.

At the second assembly station, the assembly error due to the part reorientation movements is shown in Fig.3(d). In this station, the points $P_1(2)$ and $P_2(2)$ are used as locating points to hold the first subassembly part (part 1 & part 2). Thus, the assembly error (or deviation) accumulated in previous assembly stations in those two points will be reset as zero. This result is based on the fact that all tooling locators (pins) are calibrated as the design-intent. Due to this "reorientation nature", the part 1, which has been assembled with part 2 in the first station, will be moved with part 2. As the assembly error in part 2 reset to zero, the assembly

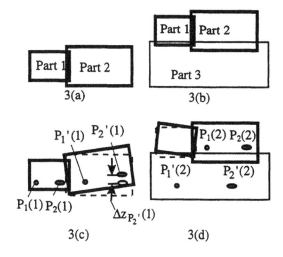

Fig. 3. One example of variation root causes

error in part 1 has been created simultaneously. The assembly error due to such a part reorientation is defined in definition 4.

Based on the definitions 3 and 4, the part variation represented by the point A can be described as follows:

$$\mathbf{X}_A(i) = \mathbf{X}_A(i-1) + \mathbf{F}_A(i) + \mathbf{T}_A(i-1) \qquad (3)$$

where i = 1, 2, 3, ... N is the number of assembly stations, with N as the last assembly station in the process. The following two sections develop expressions for $F_A(i)$ and $T_A(i-1)$, respectively.

2.3.1 Part Variations via Locator Errors at the Current Assembly Station

In the modeling of an assembly process, it is essential to understand the relationship between the part variation and tooling locating errors. More specifically, how does a fixture error leads to a part assembly error? The following theorem states the relationship.

<u>Theorem 1</u>: The part locating error represented by the point A, $F_A(i)$, can be calculated from the fixture error vector, $\Delta\mathbf{P}(i)$, using the following equation:

$$\mathbf{F}_A(i) = \mathbf{Q}_{A,P_1}(i) \quad \Delta\mathbf{P}(i) \qquad (4)$$

where $\mathbf{Q}_{A,P_1}(i)$ is the coordinate transformation matrix from the fixture error vector to the part locating error represented by the point A at station i. $\mathbf{Q}_{A,P_1}(i)$ is given by

$$\mathbf{Q}_{A,P_1}(i) = \begin{pmatrix} 1 & \dfrac{L_Z(A,P_1)}{L_X(P_1,P_2)} & -\dfrac{L_Z(A,P_1)}{L_X(P_1,P_2)} \\ 0 & 1-\dfrac{L_X(A,P_1)}{L_X(P_1,P_2)} & \dfrac{L_X(A,P_1)}{L_X(P_1,P_2)} \\ 0 & -\dfrac{1}{L_X(P_1,P_2)} & \dfrac{1}{L_X(P_1,P_2)} \end{pmatrix} \quad (5)$$

Proof: At station i, the part orientation angle error due to the tooling locator error is noted as $F_\alpha(i)$, and can be calculated by

$$F_\alpha(i) = \frac{1}{L_X(P_1,P_2)}\left[\Delta z_{P_2}(i) - \Delta z_{P_1}(i)\right] \qquad (6)$$

Where $L_X(P_1,P_2) = x_{P_2} - x_{P_1}$, (x_{P_1}, z_{P_1}) and (x_{P_2}, z_{P_2}) are the coordinates of points P_1 and P_2 in the body coordinate. Based on Eq.(6), a matrix expression is obtained by

$$\mathbf{F}_{P_1}(i) = \mathbf{Q}_{P_1,P_1}(i) \quad \Delta\mathbf{P}(i) \qquad (7)$$

that is :

$$\begin{pmatrix} \Delta x_{P_1}(i) \\ \Delta z_{P_1}(i) \\ F_\alpha(i) \end{pmatrix} = \begin{pmatrix} 1 & 0 & 0 \\ 0 & 1 & 0 \\ 0 & -\dfrac{1}{L_X(P_1,P_2)} & \dfrac{1}{L_X(P_1,P_2)} \end{pmatrix} \begin{pmatrix} \Delta x_{P_1}(i) \\ \Delta z_{P_1}(i) \\ \Delta z_{P_2}(i) \end{pmatrix} \quad (8)$$

For any two points A and B on a rigid body as shown in Fig.4, when a tooling locator error occurs, the points A and B move to A* and B*. Based on the homogeneous transform, the deviation on these two points can be simply expressed by Eq. (9)

$$\begin{pmatrix} \Delta x_B \\ \Delta z_B \\ \Delta\beta \end{pmatrix} = \begin{pmatrix} 1 & 0 & -L_Z(A,B) \\ 0 & 1 & L_X(A,B) \\ 0 & 0 & 1 \end{pmatrix} \begin{pmatrix} \Delta x_A \\ \Delta z_A \\ \Delta\beta \end{pmatrix} \qquad (9)$$

Where $\Delta\beta$ is an orientation angle error as shown in Fig. 4.

This transform relationship can be simply noted as a $M_{A,B}$ and will be used in the following sections describing the deviation relationship between any two points A and B on a rigid body, i.e.,

$$\mathbf{M}_{A,B} = \begin{pmatrix} 1 & 0 & -L_Z(A,B) \\ 0 & 1 & L_X(A,B) \\ 0 & 0 & 1 \end{pmatrix} \qquad (10)$$

Based on Eqs. (7) and (10), the part locating error represented by point A can be further obtained as follows:

$$\begin{aligned} \mathbf{F}_A(i) &= \mathbf{M}_{A,P_1}(i) \quad \mathbf{F}_{P_1}(i) \\ &= \mathbf{M}_{A,P_1}(i) \quad \mathbf{Q}_{P_1,P_1}(i) \quad \Delta\mathbf{P}(i) \quad (11) \\ &= \mathbf{Q}_{A,P_1}(i) \quad \Delta\mathbf{P}(i) \end{aligned}$$

where $\quad \mathbf{Q}_{A,P_1}(i) = \mathbf{M}_{A,P_1}(i) \quad \mathbf{Q}_{P_1,P_1}(i) \qquad (12)$

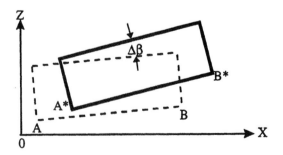

Fig. 4. Part error referenced by point A or B

58

2.3.2 Part Variations via Part Reorientation Errors

In the last subsection, the relationship between the part locating error and the fixture error vector has been studied. The study are focused on the part locating error due to current fixture faults. In this subsection, it will focus on the effect of stack-up variations resulting from the previous station and its reorientation error due to the reorientation movements. As the first step, the accumulative locating point error is defined in definition 5.

Definition 5: *Accumulative locating point error.* The part accumulative error up to station i-1 represented by the locating point P_s (s=1,2; P_s is used as the locators in station i) is called as an *accumulative locating point error* and represented by:

$$X_{P_s}(i-1) = \left(x_{P_s}(i-1) \quad z_{P_s}(i-1) \quad \Delta\alpha(i-1)\right)^T \quad (13)$$

As discussed before, the part error can be represented by any selected "part point A" and part error vector as defined in Definition 2. If we select the part point as the locating point P_s, the part error vector under this representation is called as "accumulative locating point error" as defined in Definition 5. This definition is important when we further study the reorientation error in the following theorem.

Theorem 2: The part reorientation error represented by the part point P_1 at station i can be obtained from the part error vector represented by the part point P_1 at station i-1 and the part error vector represented by the part point P_2 at station i-1 using the following equations

$$T_{P_1}(i-1) = \left(D(i) \quad G(i)\right)\begin{pmatrix} X_{P_1}(i-1) \\ X_{P_2}(i-1) \end{pmatrix} \quad (14)$$

where

$$D(i) = \begin{pmatrix} -1 & 0 & 0 \\ 0 & -1 & 0 \\ 0 & \dfrac{1}{L_x(P_1,P_2)} & 0 \end{pmatrix} \quad (15)$$

and

$$G(i) = \begin{pmatrix} 0 & 0 & 0 \\ 0 & 0 & 0 \\ 0 & -\dfrac{1}{L_x(P_1,P_2)} & 0 \end{pmatrix} \quad (16)$$

The P_1 and P_2 are the two tooling locators in the current station i, and can be either on the same part or different parts at station i-1.

Proof: As shown in Fig.5, the locator points P_1 and P_2 on the station i before reorientation of the part is plotted with a solid line and their positions after reorientation are plotted with a dashed line. The

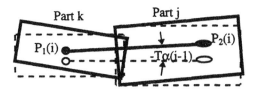

Fig. 5. Locators on the different parts k and j

dimensional variations $x_{P_1}(i-1)$, $z_{P_1}(i-1)$, and $z_{P_2}(i-1)$ of the accumulative locating point errors will be adjusted to zero. The part orientation error represented by point P_1 is:

$$T_{P_1}(i-1) = \left(T_X(i-1) \quad T_Z(i-1) \quad T_\alpha(i-1)\right)^T \quad (17)$$

which can be determined by:

$$T_X(i-1) = -x_{P_1}(i-1), \quad (18)$$

$$T_Z(i-1) = -z_{P_1}(i-1), \text{ and} \quad (19)$$

$$T_\alpha(i-1) = -\frac{1}{L_X(P_1,P_2)}(z_{P_2}(i-1) - z_{P_1}(i-1)) \quad (20)$$

Rewriting (18) to (20) as a matrix expression gives Eqs. (14) to (16).

Theorem 2 indicates the relationship between the part locating error represented by a locating point and the reorientation error when only two parts are involved in station i. If there are three parts involved in station, the two locating points, P_s (s=1,2) may be located in either two separated parts or the same part. If this situation, the part reorientation error of the part, on which the locating points are located, can be obtained from theorem 3 and 4 respectively.

Theorem 3: If points A_k, A_j, and A_r are selected as the part points of part k, j and r at station i respectively, and the locator points P_1 and P_2 are located on the different parts k and j, respectively, then the part reorientation error of part r represented by the part point A_r can be obtained by:

$$T_{A_r}(i-1) = \left(H_{rk}(i-1) \quad H_{rj}(i-1)\right)\begin{pmatrix} X_{A_k}(i-1) \\ X_{A_j}(i-1) \end{pmatrix} \quad (21)$$

where $H_{rk}(i) = M_{A_r,P_1}(i) \quad D(i) \quad M_{P_1,A_k}(i) \quad (22)$

$$H_{rj}(i) = M_{A_r,P_2}(i) \quad G(i) \quad M_{P_2,A_j}(i) \quad (23)$$

Proof: Based on Eq.(9), it can be obtained:

$$T_{A_r}(i-1) = M_{A_r,P_1}(i) \quad T_{P_1}(i-1) \quad (24)$$

59

$$\mathbf{X}_{P_1}(i-1) = \mathbf{M}_{P_1,A_k}(i) \ \ \mathbf{X}_{A_k}(i-1) \qquad (25)$$

$$\mathbf{X}_{P_2}(i-1) = \mathbf{M}_{P_2,A_j}(i) \ \ \mathbf{X}_{A_j}(i-1) \qquad (26)$$

So, substituting Eqs.(14), (25) and (26) into Eq.(24), gives Eq.(27).

$$\mathbf{T}_{A_r}(i-1) = \mathbf{M}_{A_r,P_1}(i)\big(\mathbf{D}(i) \ \ \mathbf{G}(i)\big)$$
$$\begin{pmatrix} \mathbf{M}_{P_1,A_k}(i) & \Theta \\ \Theta & \mathbf{M}_{P_2,A_j}(i) \end{pmatrix}\begin{pmatrix} \mathbf{X}_{A_k}(i-1) \\ \mathbf{X}_{A_j}(i-1) \end{pmatrix} \qquad (27)$$

Theorem 4: If the locator points P_1 and P_2 on station i are located on the same part k, then the part reorientation error of part r represented by the part point A_r, $\mathbf{T}_{A_r}(i-1)$, can be calculated by

$$\mathbf{T}_{A_r}(i-1) = -\mathbf{M}_{A_r,A_k}(i) \ \ \mathbf{X}_{A_k}(i-1) \qquad (28)$$

Proof: Based on Eq.(9), the deviation vectors $\mathbf{X}_{A_j}(i-1)$ can be obtained by:

$$\mathbf{X}_{A_j}(i-1) = \mathbf{M}_{A_j,A_k}(i) \ \ \mathbf{X}_{A_k}(i-1) \qquad (29)$$

Substituting Eq.(29) into Eq.(21), gives:

$$\mathbf{T}_{A_r}(i-1) = \big(\mathbf{H}_{rk}(i-1) \ \ \mathbf{H}_{rj}(i-1)\big)$$
$$\begin{pmatrix} \mathbf{I} & \Theta \\ \Theta & \mathbf{M}_{A_j,A_k}(i) \end{pmatrix} \begin{pmatrix} \mathbf{X}_{A_k}(i-1) \\ \mathbf{X}_{A_k}(i-1) \end{pmatrix} \qquad (30)$$

Substituting Eqs.(9), (15), (16), (22) and (23) into Eq.(30), Eq. (30) reduced to

$$\mathbf{T}_{A_r}(i-1) = -\mathbf{M}_{A_r,A_k}(i) \ \ \mathbf{X}_{A_k}(i-1) \qquad (31)$$

As a special case of theorem 4, where r=k in Eq.(31),

$$\mathbf{T}_{A_k}(i-1) = -\mathbf{X}_{A_k}(i-1) \qquad (32)$$

The interpretation of Eq. (32) is that when the locator points P_1 and P_2 are located at the same part k, the part error vector of part k at station i, $\mathbf{X}_{A_k}(i)$, is only determined by the part locating error at the current station i, i.e., substituting Eq. (32) into Eq.(3) gives

$$\mathbf{X}_{A_k}(i) = \mathbf{F}_{A_k}(i). \qquad (33)$$

3. ASSEMBLY PROCESEE MODELING BASED ON STATE SPACE MODEL

3.1 State Equation

In this section, a state space model will be developed

to describe the part dimensional variations during an assembly process. A state variable vector $\mathbf{X}(i)$ is defined by including all the assembly part error vectors and represented as

$$\mathbf{X}(i) = \begin{pmatrix} \mathbf{X}_{A_1}(i) \\ \vdots \\ \mathbf{X}_{A_n}(i) \end{pmatrix} \qquad (34)$$

where, i (i=1,2,...,N) is the assembly station, N is the total number of assembly stations in the assembly process; n is the total number of parts to be assembled in the whole assembly process; $\mathbf{X}_{A_i}(i)$ is the part error vector of part i represented by the part point A_i on the same part.

Based on the two types of part dimensional variation sources represented in Eq. (3), the state equation at station i can be expressed by:

$$\mathbf{X}(i) = \mathbf{H}(i-1) \ \ \mathbf{X}(i-1) \ + \ \mathbf{B}(i) \ \ \mathbf{U}(i) \qquad (35)$$

$\mathbf{X}(i)$ and $\mathbf{X}(i-1)$ include all part error vectors at station i and station i-1 respectively, and $\mathbf{X}(0)$ is equal to a zero vector, which means the part variation caused by stamping process is ignored. $\mathbf{U}(i)$ is the control vector at station i, which is defined as the fixture error vector for both subassembly parts at station i. P_s and P_s' (s=1,2) are the locator points of subassembly parts 1 and 2 at station i respectively. $\mathbf{U}(i)$ can be expressed by

$$\mathbf{U}(i) = \begin{pmatrix} \Delta\mathbf{P}(i) \\ \Delta\mathbf{P}'(i) \end{pmatrix}$$
$$= \big(\Delta x_{P_1} \ \ \Delta z_{P_1} \ \ \Delta z_{P_2} \ \ \Delta x_{P_1'} \ \ \Delta z_{P_1'} \ \ \Delta z_{P_2'}\big)^T \qquad (36)$$

Thus, the control matrix $\mathbf{B}(i)$ has dimension $3n \times 6$ and is given by

$$\mathbf{B}(i) = \begin{pmatrix} \mathbf{Q}_{A_1,P_1}(i) & \Theta^{3\times3} \\ \vdots & \vdots \\ \mathbf{Q}_{A_{i-1},P_1}(i) & \Theta^{3\times3} \\ \Theta^{3\times3} & \mathbf{Q}_{A_i,P_1'}(i) \\ \Theta^{3(n-i)\times3} & \Theta^{3(n-i)\times3} \end{pmatrix} \qquad (37)$$

where, $\Theta^{J\times K}$ is a zero matrix with dimension $J \times K$.

The matrix $\mathbf{H}(i-1)^{3n\times3n}$ in Eq.(35) is the dynamic matrix:

$$\mathbf{H}(i-1) = \begin{pmatrix} \mathbf{H}_{i-1}^{3(i-1)\times3(i-1)} & \Theta^{3(i-1)\times3(n-i+1)} \\ \Theta^{3(n-i+1)\times3(i-1)} & \Theta^{3(n-i+1)\times3(n-i+1)} \end{pmatrix}$$
$$(38)$$

$$H_{i-1} = \begin{pmatrix} I & \Theta & \cdots & \Theta & H_{1k} \\ \vdots & \vdots & \vdots & \vdots & \vdots \\ \Theta^{3\times3} & \Theta & \cdots & \Theta & H_{(i-1)k} \\ \Theta & \cdots & \Theta & \Theta & H_{1j} & \Theta \\ \vdots & \vdots & \vdots & \vdots & \vdots & \vdots & \vdots \\ \Theta & \cdots & \Theta & \Theta & \cdots & I \end{pmatrix} \quad (39)$$

the dimension of all the zero matrix Θ and unit matrix I in Eq. (39) are 3×3.

3.2 Observation Equations

The measurement points are usually different from the part points. The observation equation can be expressed by

$$Y(i) = C(i) \ X(i) + W(i), \quad (40)$$

(i) $Y(i)$ is an observed vector related to all the measurement points at station i, which is expressed by:

$$Y(i) = \begin{pmatrix} Y_1(i) & \cdots & Y_r(i) & \cdots & Y_n(i) \end{pmatrix}^T \quad (41)$$

$Y_r(i)$ represents the deviations at the measurement points $R_{j,r}$ (j=1,2,... m_r) on part r at station i, that is:

$$Y_r(i) = \begin{pmatrix} x_{1,r} & z_{1,r} & \cdots & x_{m_r,r} & z_{m_r,r} \end{pmatrix}^T \quad (42)$$

Where m_r is the total number of measurement points on part r. Therefore the dimension of the vector $Y(i)$ is equal to $2 \sum_{k=1}^{n} m_k \times 1$.

(ii) $C(i)$ is an observation matrix, which can be expressed by:

$$C(i) = \begin{pmatrix} C_1(i) & & & & \\ & \ddots & & \Theta & \\ & & C_r(i) & & \\ & \Theta & & \ddots & \\ & & & & C_n(i) \end{pmatrix} \quad (43)$$

where $C_r(i)$ dimension is $2 \sum_{k=1}^{n} m_k \times 3$, and based on Eq.(9), the matrix $C_r(i)$ related to part r can be calculated by:

$$C_r(i) = \begin{pmatrix} 1 & 0 & -L_Z(R_{1,r}, A_r) \\ 0 & 1 & L_X(R_{1,r}, A_r) \\ \vdots & \vdots & \vdots \\ 1 & 0 & -L_Z(R_{m_r,r}, A_r) \\ 0 & 1 & L_X(R_{m_r,r}, A_r) \end{pmatrix} \quad (44)$$

(iii) $W(i)$ is an white noise representing measuring

noise, and $Cov(W_{Q_{s,r}}(i), W_{Q_{t,r}}(i)) = \sigma^2 \delta_{Q_{s,r}, Q_{t,r}}$, which means measuring error at any measurement points s and t on part r are not correlated, where $\delta_{Q_{s,r}, Q_{t,r}}$ is the Kronecker delta.

4. DISCUSSION ON THE USAGE OF THE DEVELOPD MODEL

The proposed modeling approach and results will have great impact on dimensional control for automotive body assembly. This research lays the foundation for implementing advanced system identification and control theory in process design, monitoring, and diagnosis for body assembly. Detailed results of this methodology will be developed on future research. Here, a brief summary of the main problems and how these problems may be approached is provided.

1) In-line OCMM sensor placement strategy:

In order to perform process monitoring and diagnosis, in-line dimensional measurement are essential. How to place sensors in the assembly line so as to minimize the cost and maximize the amount of information has been a challenging issue for years. Using the developed state space model and observation model, concepts similar to classical "observability" can be applied for sensor placement evaluation. In addition, the observation matrix expresses the relationship between the observation variable (in-line sensor measurement information) and state variable (part accumulative error). Thus, a sensitivity study and optimization can be performed to select the sensor location which provides maximum diagnosability.

2) Variation simulation for the assembly processes

The developed state space model describes the mechanisms of variation propagation for the whole assembly processes. Thus, a variation simulation can be conducted by solving the state space equation with given initial conditions.

3) Fixture tooling locator design and optimization:

The state space model provides a quantitative relationship between the tooling locator error and the part error vector. The concepts similar to classical "controllability" in control theory can be used to evaluate the impact and effectiveness of the tooling locators on part dimensional control. Some mechanical joint design philosophies, such as design slip plane and design gaps (Ceglarek and Shi, 1997), can be easily incorporated in the state space model as a constraint conditions.

4) Monitoring and diagnosis for assembly processes

One of the major contributions of this research is

dimensional control of body assembly. Based on the state space model, many well-developed algorithms in control and system science can be directly applied to process monitoring and control of the body assembly processes. Examples include 1) Kalman filtering for state estimation, which provides the part accumulative error and identification of large variation parts; 2) Abrupt change detection techniques for fixture tooling failure (e.g. broken or missing tooling pins); 3) System identification technique to model the assembly process and compare the identified model with the design intent, which provides the evaluation of process and tooling design.

5. SUMMARY AND CONCLUSION

The complexity of the assembly line due to the number of parts and stations and its high production rate, places high demand on the tooling equipment. Tooling failure diagnosis based on in-line measurements is an important issue in autobody dimensional integrity, and various efforts have been made in the past. However, there were no models available to describe the overall body assembly processes for the purposes of dimensional control in manufacturing.

This paper develops a modeling technique for autobody assembly using state space models. A 3-2-1 fixturing mechanism is assumed, and emphasis on the dimension control for the X-Z plane are considered. Various variation error components (part, tooling, etc.) and their representations are defined. Furthermore, the inherent relationship among those error components are studied, leading to a state space model.

The state space model is developed solely based on the assembly product design, process configuration and tooling/fixture design. Thus, it can be obtained in the early design stage for dimensional control analysis and process diagnosis.

The major contribution of this research is to provide a state space model to describe the overall body assembly variation propagation. As a result, many well developed algorithms in control and system science can be used in dimensional control of body assembly.

It should be pointed out that further research on the modeling techniques are indeed needed. In the paper, a 3-2-1 fixture and rigid part assumption are made in the derivation, which covers 68% of total part in a typical autobody (Shiu et al., 1996). Relaxing this assumption to non-rigid parts should be studied further, and the concept of beam modeling (Shiu et al, 1996; 1997) has great potential in this respect. The second area of the research is to extend the assumptions to allow multiple parts to be assembled in one assembly station. The authors believe that this extension is straightforward with improved notation.

REFERENCE

ABC, (1993). Variation Reduction for Automotive Body Assembly. *Annual Report for Advanced Technology Program (NIST)*. Autobody Consortium (ABC) and University of Michigan, Ann Arbor.

Apley, D. and Shi, J. (1995). Diagnosis of Multiple Fixture Faults for Panel Assembly. *Proceedings of the 95' ASME Winter Annual Meeting*.

Ceglarek, D., Shi, J. (1996). Fixture Failure Diagnosis for Auto Body Assembly Using Pattern Recognition. *ASME Transactions, Journal of Engineering for Industry*, Vol. 118, pp55-65.

Ceglarek, D., Shi, J. (1997). Design Evaluation of Sheet Metal Joints for Dimensional Integrity. to appear in *ASME Transactions, Journal of Manufacturing Science and Technology*.

Ceglarek, D., Shi, J. and Wu, S.M. (1994). A Knowledge-based Diagnosis Approach for the Launch of the Auto-body Assembly Process. *Trans. of ASME, Journal of Engineering for Industry* vol. 116, no. 3, pp491-499.

Hu, S., Wu, S.M., (1992). Identifying Root Causes of Variation in Automobile Body Assembly Using Principal Component Analysis. *Trans. of NAMRI, vol. XX*, 311-316.

Shiu, B., Ceglarek, D., and Shi, J. (1996). Multi-Station Sheet Metal Assembly Modeling and Diagnostics. *NAMRI/SME Transactions*, Vol. 23, pp199-204.

Shiu, B., Ceglarek, D. and Shi, J. (1997). Flexible Beam-based Modeling of Sheet Metal Assembly for Dimensional Control. *NAMRI/SME Transactions*, Vol. 24, pp49-54.

Takezawa, N. (1980). An Improved Method for Establishing the Process-Wise Quality Standard. *Rep. Stat. Appl. Res., JUSE*, vol. 27, No. 3, pp63-75.

FAILURE ANALYSIS AND DIAGNOSIS USING DISCRETE EVENT MODELS: THEORY AND CASE STUDY*

K.-H. Cho* J.-T. Lim*,1

*Dept. of Electrical Engineering
Korea Advanced Institute of Science and Technology
Taejon, 305-701, Korea*

Abstract: In this paper, a discrete event model is introduced to improve the reliability of a discrete event dynamical system (DEDS). An analytical framework for fault-tolerant supervisory control systems including a systematic way for analyzing DEDSs to classify faults and failures quantitatively, a search of tolerable fault event sequences (TFESs) embedded in the system, an automated failure diagnosis scheme with respect to the nominal normal operating event sequences, and the supervisory control for TFESs, is presented. A case study of an ion implantation system is described.

Keywords: discrete event dynamical system, failure analysis, failure diagnosis, supervisory control, fault-tolerant system

1. INTRODUCTION

As the demands on reliability and safety of modern complicated systems are increasing, the analytical and numerical work on failure diagnosis is in progress today and many techniques have been developed (Lapp and Powers, 1977), (Gertler and Anderson, 1992), and (Handelman and Stengel, 1989). Recently, this problem has also been studied in the framework of discrete event dynamical systems (DEDSs) (Sampath *et al.*, 1994). In the literature (Patton and Chen, 1994), the unexpected changes in the system, such as component faults and variations in operating conditions, are classified qualitatively. A "fault" is understood as an unexpected change in the system that tends to degrade the overall system performance, although it may not represent the "failure" of physical components. The term fault rather than failure is used to denote a malfunction rather than a catastrophe. The term failure suggests a complete breakdown of a system component or function, whereas the term fault may be used to indicate that a malfunction is present but it may be tolerable. However there exists no clear and quantitative classification at this time.

We propose in this paper a DEDS approach to the problems of failure analysis and diagnosis, and fault-tolerant supervisory control. We adopt the framework proposed by Ramadge and Wonham (Ramadge and Wonham, 1987) for the study of fault-tolerant supervisory control systems (FTSCSs). The overall model is thus a state model of the open loop system dynamics with external control. This approach is applicable to systems that fall naturally in the class of DEDSs; moreover, for the purpose of diagnosis and supervision, continuous-variable dynamical systems can often be viewed as DEDSs at a higher level of abstraction. One of the major advantages of the proposed method is that it does not require detailed indepth modeling of the system to be analyzed and hence is ideally suited for the diagnosis and the fault-tolerant supervisory control of large complex systems like semiconductor

---

* This work was supported by the Korean Science and Engineering Foundation (96-0102-05-01-3).
¹ To whom all correspondence should be addressed.

processes, power plants, and communication networks. See (Ramadge and Wonham, 1987) and (Ramadge and Wonham, 1989) for a synopsis of the framework and some of the principal results.

Our approach to failure diagnosis and fault-tolerant supervisory control involves two major steps: developing failure diagnosis scheme through failure analysis on the discrete event model (DEM) of the system followed by construction of the fault-tolerant supervisory controller if possible. The DEM that we develop captures both the normal and the failed behavior of the system. The failures are modeled as abnormal events with their associated states and the objective is to infer about past occurrences of the source failures on the basis of the observed events. Throughout this paper, we assume that a supervisor can record all the state transitions and events generated by the plant.

In this paper, a systematic way for analyzing DEDSs is proposed to classify faults and failures quantitatively and to find tolerable fault event sequences embedded in the system. An automated failure diagnosis scheme w.r.t. the nominal normal operating event sequences and the supervisory control for tolerable fault event sequences are presented. Finally, we present an analytical framework for FTSCSs.

This paper is organized as follows. In the remainder of this section we give some notation. In Sect. 2, we present the quantitative definitions of faults and failures, and their classification algorithm. In Sect. 3, the problem of failure diagnosis, tolerable fault event sequences, and supervised fault-tolerant system with supervisor failure diagnosis is studied. The case study of an ion implantation system is considered in Sect. 4. Finally, the conclusions are formulated in Section 5.

Notation

$A(q)$	set of events possible to occur after state q
$B(q)$	set of events possible to occur before state q
$D(e)$ $(D(s))$	starting state of event e (string s) (domain of e (s))
$R(e)$ $(R(s))$	ending state of event e (string s) (range of e (s))
aug-event	event augmented with its originating state
$Ac(q_0, q)$	set of events in the driving sequences from initial state q_0 to q

2. FAILURE ANALYSIS

In this section, the quantitative definitions of faults and failures with their classification algorithm are presented.

The unexpected changes in the system can be described as the set of abnormal events which are uncontrollable and unexpected to occur during the normal operating mode. Let Σ_{an} be the set of abnormal events and Σ_n be the set of normal events. The set Σ_{an} is a subset of Σ_{uc}, the set of uncontrollable events, and the total set of events Σ can be partitioned as $\Sigma = \Sigma_c \dot{\cup} \Sigma_{uc} = \Sigma_n \dot{\cup} \Sigma_{an}$ where Σ_c is the set of controllable events. Then Σ_{an} can be further partitioned by the set of fault and failure events defined in the following with their originating states.

Definition 1. : The aug-event (σ_f, q_f) is a "fault w.r.t. Q_m" and σ_f is called a "fault event" if

(i) $\sigma_f \in \Sigma_{an}$,
(ii) there exists at least one event string $s \in L/\sigma_f := \{t \in \Sigma^* | \sigma_f t \in L\}$ such that $\sigma_f s$ leads to the marker states set Q_m and for each aug-event (s_i, q_i) of $s = s_1 s_2 \cdots s_n$, any $\sigma \in A(q_i) - \{s_i\}$ for all s_i corresponding to q_t belongs to Σ_c or is an another fault event.

Definition 2. : The aug-event (σ_f, q_f) is a "failure w.r.t. Q_m" and σ_f is called a "failure event" if $\sigma_f \in \Sigma_{an}$ and (σ_f, q_f) is not a fault.

To classify faults and failures in a quantitative manner, we define an index function as follows.

Definition 3. : The "index function" of each state q, $I_m(q)$, is defined according to the corresponding set of marker states Q_m as;

$$I_m(q) := \begin{cases} 1, & \text{if } q \in Q_m \text{ or if there exists } \sigma \in \\ & A(q) \text{ such that } I_m(R(\sigma)) = 1 \\ & \text{and any } \sigma' \in A(q) - \{\sigma\} \text{ is in } \Sigma_c \\ & \text{or } I_m(R(\sigma')) = 1, \\ 0, & \text{otherwise.} \end{cases}$$

For each state of a given DEDS, we can assign systematically the value of index function through the following algorithm.

Algorithm (Assigning index functions)

Step 1. Assign $I_m(q_m) = 1$ where $q_m \in Q_m$.
Step 2. For each $q_i \in D(B(q_j))$ where $I_m(q_j) = 1$, assign 1 if $I_m(R(A(q_i) - B(q_j))) = 1$ or $A(q_i) - B(q_j) \subset \Sigma_c$.
Step 3. Assign 0's for all the remained states.

For a given DEDS, we can classify faults and failures quantitatively through the following proposition.

Proposition 4. : The aug-event (σ, q) with $\sigma \in \Sigma_{an}$ is a fault w.r.t. Q_m if and only if $I_m(q) = 1$ and it is a failure w.r.t. Q_m if and only if $I_m(q) = 0$.

3. FAILURE DIAGNOSIS AND FAULT-TOLERANT SYSTEM

The nominal work procedure in the system can be represented by some normal event sequences whether an unexpected change may lead them to certain failures or not. We can formally define such event sequences as follows.

Definition 5. : The normal event sequence which derives the initial state to the marker one is called a "nominal normal operating event sequence (NNOES)" since it represents a nominal operating sequence when a given system is in its normal status.

During the work process along the NNOES, an unexpected change may deviate it to a certain failure. In this case, the origin of deviation is called a source failure and is defined formally as follows.

Definition 6. : For a given failure, the failure that makes the given failure event be deviated from the NNOES for the first time is called a "source failure" of the given failure.

To maintain a high level of performance for complex systems, e.g., power plants and semiconductor processes, etc., it is crucial that failures are detected promptly and diagnosed so that a corrective action can be taken to reconfigure the system. For a systematic failure diagnosis, we define a set of node functions in the following.

Definition 7. : A set of "node functions" of q w.r.t. NNOES, $S_J(q) = \{J(q)\}$, is defined as;

$$J(q) := \begin{cases} (2n+1)2^m, & \text{if } q \text{ is reachable from states in the NNOES, which is again reachable from the initial state by } n \text{ transitions, and there are } m \text{ abnormal events in the transitions from NNOES and the first of them is abnormal,} \\ 0, & \text{otherwise.} \end{cases}$$

Then we know the source failure and the number of abnormal events occurred from NNOES for a given failure by the following proposition.

Proposition 8. : The set of source failures of (σ_f, q) w.r.t. NNOES is
$$S_f(q) := \{(\sigma_{fs}, q_s) | J(q_s) = J(q) \bmod 2 \text{ with } J(q) \neq 0\}.$$

When some abnormal events occurred during the work process, if we can still find an another event sequence reachable to the marker state or if we can eliminate the path to the abnormal events, then the work procedure is called "fault-tolerable".

Definition 9. : The event sequence which consists of normal events or fault events and which derives the initial state to the marker one is called a "tolerable fault event sequence (TFES)" if, for each normal event, all the possible events following the corresponding states belong to Σ_c or another fault events.

Definition 10. : The given DEDS G is "fault-tolerable w.r.t. Q_m" if there exists at least one TFES w.r.t. Q_m.

We can easily determine whether a given DEDS is fault-tolerable or not using the index function in the following.

Proposition 11. : The given DEDS is fault-tolerable w.r.t. Q_m if and only if $I_m(q_0) = 1$.

Once TFESs were found, we can make the system evolve along the TFESs through supervisor. The supervisor is designed on the basis of recognizer for L_g which is interpreted as a legal language (Ramadge and Wonham, 1987). In this case, let $L_c(S/G) = L_g = K := \{\text{TFESs}\} \subset L_m(G)$. Thus if we find TFESs, the supervised fault-tolerant system can be easily constructed.

Definition 12. : The overall supervised system S/G of the fault-tolerable plant G w.r.t. Q_m with supervisor S is "fault-tolerant w.r.t. Q_m" if $L_c(S/G) = \text{TFESs}$ w.r.t. Q_m.

As it follows from Proposition 4–11, we propose the analytical framework for FTSCS as follows:

First, for a given DEDS G do failure analysis by off-line to classify faults and failures, and to find any TFES embedded in the system if it exists. Once we found TFESs, let them be a legal language K. If there is no TFES then let NNOESs be a legal language K. Next, design a supervisor S along the legal language K. During the operation, the control system monitors the supervised behavior of G. If a failure is detected, then the control system starts failure diagnosis to find the source failure and change the status of G into repairing mode. If a fault is detected, then the control system automatically reconfigures along the TFES within K. The overall structure of the proposed FTSCS is shown in Fig. 1.

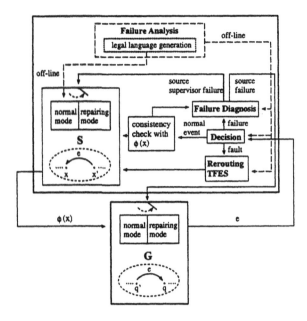

Fig. 1. Overall structure of FTSCS.

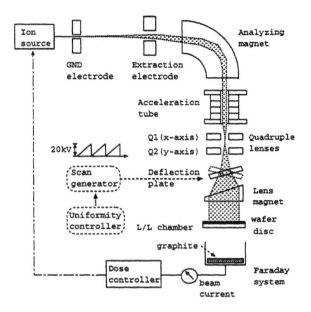

Fig. 2. Overall schematic diagram of IIP system.

4. CASE STUDY OF SEMICONDUCTOR MANUFACTURING SYSTEM : ION IMPLANTATION PROCESS

Ion implantation (IIP) is a process to inject impurities into the surface of a silicon wafer. This is a primary process for modern VLSI circuits fabrication at many steps of the manufacturing processes (Jaeger, 1993). To meet high precision and to ensure high reliability, it is required to minimize any unscheduled down-time of the equipments in the system. Thus, an analytical tool for failure diagnosis scheme is crucial. Despite its widespread use, IIP remains a poorly understood operation; moreover, it is prohibitively difficult to obtain its accurate mathematical model. However IIP system is suited to be analyzed in a DEDS framework and we apply our results to the analysis and improvement of IIP operation. Figure 2 illustrates an overall IIP system. The main control objective is to maintain a uniform injection ratio on the whole wafer surface in the process chamber during the implantation process. The uniform injection part is composed of quadruple lenses to deflect the ion beam in x, y-direction and deflection plate to tune the deflected beam.

Consider the uniform injection part of IIP in Fig. 2. Suppose that the quadruple lens Q_2 of this system is subject to an abnormal event which causes it to be overdeflected and the deflection plate to an abnormal event which causes it to be overtuned. The overdeflection of the quadruple lenses can be complemented by undertuning the deflection plate in the corresponding direction. Figure 3 illustrates the component models. In Fig. 3, DQ_1 (Deflect Q_1) and DQ_2 (Deflect Q_2) are the normal controllable events of quadruple lenses. The quadruple lenses are also subject to an abnormal event,

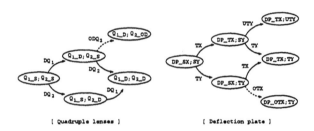

[Quadruple lenses]   [Deflection plate]

Fig. 3. Component DEM for uniform injection part of IIP system.

ODQ_2 (Overdeflect Q_2). Their states are composed of Q_1_S;Q_2_S (Q_1_Set;Q_2_Set), Q_1_D;Q_2_S (Q_1_Deflected;Q_2_Set), Q_1_S;Q_2_D (Q_1_Set;Q_2_Deflected), Q_1_D;Q_2_D (Q_1_Deflected;Q_2_Deflected), and Q_1_D;Q_2_OD (Q_1_Deflected;Q_2_Over Deflected). The deflection plate has similarly the normal controllable events, TX (Tune in X), TY (Tune in Y), and UTY (UnderTune in Y), and an abnormal event, OTX (OverTune in X). Its states comprise DP_SX;SY (Deflection Plate_Set in X;Set in Y), DP_TX;SY (Deflection Plate_Tuned in X;Set in Y), DP_SX;TY (Deflection Plate_Set in X;Tuned in Y), DP_TX;TY (Deflection Plate_Tuned in X;Tuned in Y), DP_TX;UTY (Deflection Plate_Tuned in X;UnderTuned in Y), and DP_OTX;TY (Deflection Plate_OverTuned in X;Tuned in Y). The DEM of the overall system is obtained through parallel composition (Heymann, 1990) and is illustrated in Fig. 4. The composed states and their contents are shown in Table 1. In this case, the initial state is 1 and the marker states are 15 and 22 in Fig. 4. The index function values of each state are also shown in Table 1. From these, we know that the aug-events with $I_m(q) = 1$ such as $(ODQ_2,7)$, $(ODQ_2,9)$, etc are faults and the other aug-events with $I_m(q) = 0$ such as $(OTX,5)$, $(ODQ_2,11)$, etc are failures.

Table 1. States definition of DEM in Fig. 4 with index functions and set of node functions.

state(q)	contents	$I_m(q)$	$S_J(q)$	state(q)	contents	$I_m(q)$	$S_J(q)$
1	$(Q_1\_S;Q_2\_S, DP\_SX;SY)$	1	$\{1\}$	16	$(Q_1\_D;Q_2\_OD, DP\_TX;TY)$	0	$\{0, 3 \times 2^1\}$
2	$(Q_1\_S;Q_2\_S, DP\_TX;SY)$	1	$\{0\}$	17	$(Q_1\_D;Q_2\_OD, DP\_SX;TY)$	0	$\{0, 3 \times 2^1\}$
3	$(Q_1\_S;Q_2\_S, DP\_TX;UTY)$	1	$\{0\}$	18	$(Q_1\_D;Q_2\_OD, DP\_OTX;TY)$	0	$\{0, 3 \times 2^2\}$
4	$(Q_1\_S;Q_2\_S, DP\_TX;TY)$	1	$\{0\}$	19	$(Q_1\_D;Q_2\_D, DP\_SX;SY)$	1	$\{5\}$
5	$(Q_1\_S;Q_2\_S, DP\_SX;TY)$	0	$\{0\}$	20	$(Q_1\_D;Q_2\_D, DP\_TX;SY)$	1	$\{7\}$
6	$(Q_1\_S;Q_2\_S, DP\_OTX;TY)$	0	$\{0\}$	21	$(Q_1\_D;Q_2\_D, DP\_TX;UTY)$	0	$\{0\}$
7	$(Q_1\_D;Q_2\_S, DP\_SX;SY)$	1	$\{3\}$	22	$(Q_1\_D;Q_2\_D, DP\_TX;TY)$	1	$\{9\}$
8	$(Q_1\_D;Q_2\_S, DP\_TX;SY)$	1	$\{0\}$	23	$(Q_1\_D;Q_2\_D, DP\_SX;TY)$	0	$\{7\}$
9	$(Q_1\_D;Q_2\_S, DP\_TX;UTY)$	1	$\{0\}$	24	$(Q_1\_D;Q_2\_D, DP\_OTX;TY)$	0	$\{0, 7 \times 2^1\}$
10	$(Q_1\_D;Q_2\_S, DP\_TX;TY)$	0	$\{0\}$	25	$(Q_1\_S;Q_2\_D, DP\_SX;SY)$	1	$\{3\}$
11	$(Q_1\_D;Q_2\_S, DP\_SX;TY)$	0	$\{0\}$	26	$(Q_1\_S;Q_2\_D, DP\_TX;SY)$	1	$\{0\}$
12	$(Q_1\_D;Q_2\_S, DP\_OTX;TY)$	0	$\{0\}$	27	$(Q_1\_S;Q_2\_D, DP\_TX;UTY)$	0	$\{0\}$
13	$(Q_1\_D;Q_2\_OD, DP\_SX;SY)$	1	$\{3 \times 2^1\}$	28	$(Q_1\_S;Q_2\_D, DP\_TX;TY)$	0	$\{0\}$
14	$(Q_1\_D;Q_2\_OD, DP\_TX;SY)$	1	$\{0, 3 \times 2^1\}$	29	$(Q_1\_S;Q_2\_D, DP\_SX;TY)$	1	$\{0\}$
15	$(Q_1\_D;Q_2\_OD, DP\_TX;UTY)$	1	$\{0, 3 \times 2^1\}$	30	$(Q_1\_S;Q_2\_D, DP\_OTX;TY)$	0	$\{0\}$

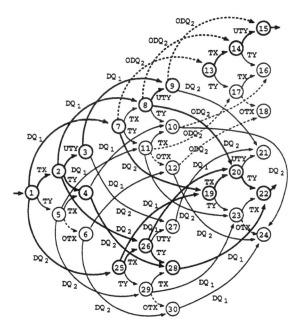

Fig. 4. Overall DEM for uniform injection part of IIP system.

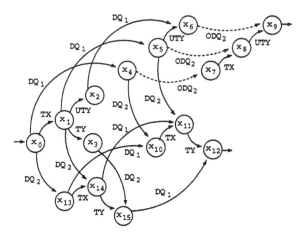

Fig. 5. Recognizer for K.

state transition diagram for the recognizer of K in Fig. 5 can serve to define S; it just remains to identify the state feedback map ϕ. For each state x of S, $\phi(x)$ is a map

$$\phi(x) : \{c_1, c_2, \cdots, c_5\} \mapsto \{0, 1\},$$

i.e., a binary evaluation of each of the controls $c_1 \sim c_6$. Thus it is enough to define

$$\phi(x)(c_i) = \begin{cases} 1, & \text{if an edge corresponding to } c_i \text{ is on } K, \\ 0, & \text{otherwise.} \end{cases}$$

The resulting control patterns are shown in Table 2. The supervisor $\mathcal{S} = (S, \phi)$ then certainly

Table 2. Control data for $\mathcal{S}$.

state	ϕ	state	ϕ
x_0	11100	x_8	00010
x_1	11011	x_9	00000
x_2	10000	x_{10}	00100
x_3	01000	x_{11}	00001
x_4	01000	x_{12}	00000
x_5	01010	x_{13}	10100
x_6	00000	x_{14}	10001
x_7	00100	x_{15}	10000

Consider the NNOESs that lead to the marker state 22, NNOESs := $(DQ_1 DQ_2 + DQ_2 DQ_1)(TX$ $TY + TY TX)$. For failure diagnosis, the sets of node function values w.r.t. the NNOESs are shown in Table 1. If a certain failure is detected then we can find its source failure from these values according to Proposition 8. For example, if $(OTX, 17)$ is detected then the source failure may be $(ODQ_2, 7)$ or $(DQ_2, 7)$.

From Proposition 11, this system turns out to be fault-tolerable. Let the TFESs be the legal language K. Hence, $K = \{$ TFESs $\} = DQ_1(ODQ_2$ $TX\ UTY + DQ_2\ TX\ TY) + DQ_2(DQ_1\ TX\ TY$ $+ TX(TY\ DQ_1 + DQ_1\ TY)) + TX(DQ_1(ODQ_2$ $TX\ UTY + DQ_2\ TX\ TY) + UTY\ DQ_1\ ODQ_2 +$ $TY\ DQ_2\ DQ_1 + DQ_2(DQ_1\ TY + TY\ DQ_1))$. We can construct a supervised fault-tolerant system through a supervisor designed on the basis of K. The design procedure of supervisor is given in the next. Apparently, $\Sigma_c = \{DQ_1 : c_1, DQ_2 :$ $c_2,\ TX : c_3,\ UTY : c_4,\ TY : c_5\}$ where c_i is a control value and $\Sigma_{uc} = \{ODQ_2, OTX\}$. The

determines

$$L(\mathcal{S}/G) = \overline{K}, \quad L_m(\mathcal{S}/G) = K.$$

Furthermore $\mathcal{S}$ is complete w.r.t. G and it is also a quotient supervisor (Ramadge and Wonham, 1987).

5. CONCLUSIONS

We have studied failure analysis and diagnosis issues related to large complicated systems from the point of view of DEDSs. This paper has been focused on analyzing DEDSs to classify faults and failures quantitatively and to find TFESs embedded in the system. An automated failure diagnosis scheme w.r.t. the NNOESs and the supervisory control for TFESs have been provided. Finally, an analytical framework for FTSCSs is proposed.

6. REFERENCES

Gertler, J.J. and K.C. Anderson (1992). An evidential reasoning extension to quantitative model-based failure diagnosis. *IEEE Trans. System, Man, and Cybernatics* **22**, 275–289.

Handelman, D.A. and R.F. Stengel (1989). Combining expert system and analytical redundancy concept for fault-tolerant flight control. *J. of Guidance, Control, and Dynamics* **12**, 39–45.

Heymann, M. (1990). Concurrency and discrete event control. *IEEE Control Systems* **10**, 103–112.

Jaeger, R.C. (1993). *Introduction to microelectronic fabrication*. Addison-Wesley. Massachusetts.

Lapp, S. and G. Powers (1977). Computer aided synthesis of fault trees. *IEEE Trans. Reliability* **26**, 2–13.

Patton, R.J. and J. Chen (1994). Review of parity space approaches to fault diagnosis for aerospace systems. *J. of Guidance, Control, and Dynamics* **17**, 278–285.

Ramadge, P.J. and W.M. Wonham (1987). Supervisory control of a class of discrete event processes. *SIAM J. of Control and Optimization* **25**, 206–230.

Ramadge, P.J. and W.M. Wonham (1989). The control of discrete event systems. *Proc. IEEE, Special Issue on Discrete Event Dynamic Systems* **77**, 81–98.

Sampath, M., R. Sengupta, S. Lafortune, K. Sinnamohideen and D. Teneketzis (1994). Failure diagnosis using discrete event models. In: *Proc. 33rd IEEE Conf. on Decision and Control*. pp. 3110–3116. Lake Buena Vista, Florida.

INCORRECT OBSERVATIONS IN FAILURE DIAGNOSIS OF DISCRETE EVENT SYSTEMS

Jiří Pik

*Institute of Information Theory and Automation, Academy of Sciences,
Pod vodárenskou věží 4, 182 08 Prague 8, Czech Republic, pik@utia.cas.cz*

Abstract: Incorrect observations in the language-based approach to failure diagnosis of discrete event systems are considered. The rho-bar distance based on event-to-event operations is utilized and an extension of the diagnostic procedure is proposed. A simple example illustrating the developed ideas is included.

Keywords: fault detection; diagnosis; discrete event systems; uncertainty.

1 INTRODUCTION

Failure diagnosis of complex systems belongs to the topical research problems. A language-based approach to failure diagnosis of discrete event systems (DES) has been presented in (Sampath, *et al.*, 1995; Sampath, *et al.*, 1996). A methodology for on-line fault detection and isolation as well as for off-line analysis of the diagnosability properties of the system using diagnosers is provided in the framework of formal languages and corresponding state machine representations. The failures are modelled as unobservable events and the normal and failed system behaviour is described by a regular language. The diagnoser is a finite state machine constructed from the model of the system. The diagnosis provided by the diagnoser depends on two factors (Sampath, *et al.*, 1996): (i) the system model from which the diagnoser is synthesized, and (ii) the observation sequences considered by the diagnoser. Unmodelled dynamics and incorrect observations lead to inconsistencies in the diagnoser observations. To determine the nature of those, further studies are necessary (Sampath, *et al.*, 1996).

An event deformation model has been introduced

in (Pik, 1996) to deal with different kind of event uncertainty in discrete event systems. Using transformations of event sequences based on event-to-event operations, an equivalence relation over the set of the event sequences and a corresponding induced partition are defined.

In the paper, incorrect observations in the language-based approach to failure diagnosis of discrete event systems are considered. To evaluate the distortion of event sequences, the rho-bar distance (Gray, 1990) is utilized; some basic concepts are summarized in Section 2. In Section 3, incorrect observations are modelled using event-to-event operations. An extension of the diagnostic procedure is proposed and a simple example illustrating the approach is included in Section 4.

2 RHO-BAR DISTANCE

Different approaches ranging from pure experimental to more theoretically based ones can be considered to introduce a measure of similarity or distortion of the sequences of events. In what follows, the rho-bar distance (Gray, 1990) based on event-to-event operations is considered.

Let E be an alphabet, an operation over E is an ordered pair $s = (a, b)$ such that $a, b \in E \cup \{\epsilon\}$ and $s \neq (\epsilon, \epsilon)$, where ϵ denotes the sequence consisting of no symbols.

The operation $s = (a, b)$ is called

1. deletion if $b = \epsilon$,

2. insertion if $a = \epsilon$,

3. substitution otherwise.

A sequence Y results from the application of the operation $s = (a, b)$ to a sequence X, written $X \overset{s}{\Rightarrow} Y$, if $X = A_1 a A_2$ and $Y = A_1 b A_2$, where $A_1, A_2 \in E^*$. To transform a sequence X into Y, a series of operations $S = s_1, s_2, \ldots, s_q$ is needed, $q > 0$, such that $X = X_0 \overset{s_1}{\Rightarrow} X_1$, $X_1 \overset{s_2}{\Rightarrow} X_2, \ldots, X_{q-1} \overset{s_q}{\Rightarrow} X_q = Y$.

Two modifications of the transformation are considered. While the first modification computes the probability $p(Y/X)$ of the transformation X into Y based on a stochastic mapping of the alphabet of events, the second modification introduces the Levenshtein metric for an optimal representation of the event sequences. A nonnegative real number $w(s)$ called a weight of the operation $s = (a, b)$ is associated with each event operation. The following properties are required for all $a, b, c \in E \cup \{\epsilon\}$

(i) $w(a, b) \geq 0$ and $w(a, b) = 0$ iff $a = b$,

(ii) $w(a, b) = w(b, a)$,

(iii) $w(a, b) + w(b, c) \geq w(a, c)$.

The notion of $w(s)$ is extended to a series of operations $S = s_1, s_2, \ldots, s_m$ using

$$w(S) = \sum_{i=1}^{m} w(s_i) \text{ and } w(S) = 0 \text{ for } m = 0.$$

The weighted distance $d_w(X, Y)$ from $X \in E^*$ to $Y \in E^*$ is defined by

$$d_w(X, Y) = \min_{S} \{w(S) \; : \; S \text{ is a series of}$$
operations which transforms X into Y $\}$.

There exists an illustrative graphical interpretation of the computation procedures for $p(Y/X)$ or $d_w(X, Y)$ using a dynamic programming search in the lattice.

A class of similarity measures instead of a single one is obtained through the specification of the weights of the operations. The proper choice of the weight values usually reflects our knowledge and insights into the considered problem. Another approach based on a weight parametrization and a given sample set of strings is possible. Further, a generalized form of considered operations can be proposed to make possible an application of the context-dependent operations instead of the context-free event-to-event ones.

3 MODELLING OF INCORRECT OBSERVATIONS

Several sources of observation errors may be supposed in the considered approach to failure diagnosis. Uncertainty of events may be a consequence of the possibly ambiguous event recognition, observation and/or of the transmission of the event sequences through a noisy channel. As a result of uncertainties the diagnoser G_d sees an observation record that is different from the event sequence of the system S.

In (Pik, 1996), the following event uncertainties are distinguished: (i) an actual event is not observed by G_d, (ii) an observed event is not generated by S, and (iii) an actual generated event by S is observed by G_d as a different event.

The same sources of observation errors are listed in (Sampath, *et al.*, 1996): the diagnoser may miss seeing an event, it may assume that an event occurred when none did, or it may mistake one event for another.

To model occurrences of the considered errors in the event sequences, the event-to-event operations (i) deletion, (ii) insertion, and (iii) substitution are utilized. As the probabilities or the weights are associated with these operations, the rho-bar distances of the observation record and the corresponding event sequences of the diagnoser can be computed to quantify a consequence of the inconsistencies in the observations.

A partition on the set of event subsequences can be defined by non-numerical clustering and/or by transformation of event sequences. The subsequences belonging to the same partition block are represented by a macro-event symbol and an event deformation model can be built, (Pik, 1996).

From another point of view, a possibility to abstract from details and to work in terms of the simpler modules or abstractions is provided by the representation based on macro-event symbols.

4 INCORRECT OBSERVATIONS IN FAILURE DIAGNOSIS

An event inconsistent with the current state of the diagnoser may be contained in the observation record in the process of diagnosis. The differences between the event sequence generated by the system and the event sequences of the diagnoser are modelled using the event-to-event operations and the corresponding rho-bar distances are determined. Supposing an inconsistency in observations, the diagnoser enters the state following the event belonging to the more likely or optimal event sequence representation.

More precisely, to deal with an event inconsistent with the current state of the diagnoser, a three-step extension of the diagnostic procedure is proposed:

Suppositions:

- (*i*) the current state q_i of a finite state machine model of the diagnoser G_d following the observation record $X_i = \ldots e_{-2}e_{-1}$, $X_i \in E^*$,

- (*ii*) an event e_0 inconsistent with the current state q_i,

- (*iii*) single occurrences of observation errors located in the subsequence X_{i_l} of X_i with the length $\mid X_{i_l} \mid = l$, where $X_{i_l} = e_{-l}e_{-l+1}\ldots e_{-2}e_{-1}$,

- (*iv*) a prefix X_1 of a sequence X, denoted $X_1 \leq X$, if there exists X_2 such that $X_1 X_2 = X$, and $X, X_1, X_2 \in E^*$.

Step 1: Inverse transitions from state q_i of G_d along the trace $e_{-1}e_{-2}\ldots e_{-l+1}e_{-l}$ are performed. After those, state q_j is reached, $\delta_g(q_j, e_{-l}\ldots e_{-1}) = q_i$, where an extended transition function δ_g of G_d is considered, $\delta_g : Q \times E^* \to Q$.

Step 2: The rho-bar distances of the sequence $X_{i_l} e_0 = e_{-l}e_{-l+1}\ldots e_{-2}e_{-1}e_0$ and traces $t \leq t^* \in L(G_d, q_j, m)$ are determined, where $L(G_d, q_j, m)$ denotes the set of all traces t^* that originate from state q_j of G_d with $\mid t^* \mid = m$, $m = 2l + 1$. The trace t' for that

(*i*) $t' \leq t^* \in L(G_d, q_j, m)$ and

(*ii*) $d_w(X_{i_l} e_0, t') = \min\limits_{t \leq t^* \in L(G_d, q_j, m)} \{d_w(X_{i_l} e_0, t)\}$

is denoted by t_d.

Step 3: The new current state q_k of G_d is entered following the trace t_d, $\delta_g(q_j, t_d) = q_k$.

Example 1.

Consider a finite state machine model of the diagnoser G_d (depicted in Figure 1), the alphabet $E = \{a, b\}$, and the length l of subsequences of the observation record, $l = 2$. Suppose that G_d is in the current state q_{11} following the observation record $X_{11} = \ldots baab$, and the event b is seen by the diagnoser now. As this event is inconsistent with state q_{11}, the next current state is obtained using the proposed extension.

Step 1: Starting in state q_{11}, the inverse transitions following ba are performed, the resulting state is q_4.

Step 2: The distances $d_w(abb, t)$, where $t \leq t^* \in L(G_d, q_4, 5)$ are determined, where
$L(G_d, q_4, 5) = \{aaaba, aabba, baaba, babba, abaab, ababb, bbaab, bbabb\}$. Depending on the choice of the probabilities or the weights of the event-to-event operations w_s, the resulting trace t_d may be found to be $t_d = aabb \leq aabba \in L(G_d, q_4, 5)$, $d_w(abb, t_d) = \min\limits_{t \leq t^* \in L(G_d, q_4, 5)} \{d_w(abb, t)\}$.

Step 3: Starting in state q_4, the new current state of the diagnoser G_d following trace $t_d = aabb$ is obtained, this state is q_{10}.

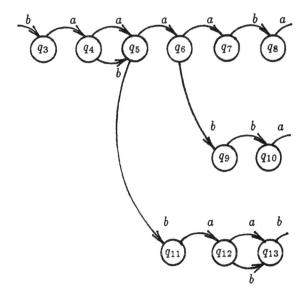

Fig.1. A part of the graph representation of the diagnoser G_d.

ACKNOWLEDGEMENT

This work was partially supported by the grants GA AV ČR #A2075505 and GA ČR #102/96/1671.

REFERENCES

Gray, R.M. (1990). Entropy and Information Theory. Springer-Verlag. New York, Berlin, Heidelberg.

Pik, J. (1996). An event deformation model and its application. *Proc. of the workshop on Discrete Event Systems, WODES'96*, pp.352-355, IEE, London, U.K.

Sampath, M., R. Sengupta, S. Lafortune, K. Sinnamohideen and D. Teneketzis (1995). Diagnosability of discrete-event systems. *IEEE Transactions on Automatic Control*, **40**, pp.1555-1575.

Sampath, M., S. Lafortune and D. Teneketzis (1996). A language-based approach to failure diagnosis of discrete event systems. *Proc. of the workshop on Discrete Event Systems, WODES'96*, pp.261-271, IEE, London, U.K.

A FAULT DETECTION ISOLATION OBSERVER DESIGN
BY HYBRID DISTURBANCE DECOUPLING APPROACH

Kee-Sang Lee · Tae-Geon Park

Department of Electrical Engineering
Dankook University
#8 Hannam-Dong, Yongsan-Gu, Seoul, 140-714, Korea
Phone: 82-2-709-2575, Fax: 82-2-795-8771
e-mail: keesang@soback.kornet.nm.kr

Abstract: A fault detection isolation observer developed so far requires that the number of unknown inputs (including the faults of no interest and disturbances to be algebraically rejected) must be less than that of outputs. The existence condition is hard to meet and restricts the practical use of a fault detection isolation observer. The development of the fault detection isolation observer with some mild existence condition has been an open problem. The purpose of this paper is to propose a fault detection isolation observer which gives a viable solution to the problem. The basic concept of this proposition is the use of the unknown disturbance modelling approach together with the algebraic decoupling approach. A numerical example shows the applicability of the proposed fault detection isolation observer.

Keywords: Fault detection isolation; Fault detection isolation observer; Unknown input observer; Algebraic decoupling approach; Disturbance modelling approach.

1. INTRODUCTION

Increasing complexity and large scale property of today's industrial process not only make the achievement of the desired level of system reliability difficult but also increase the shutdown cost, and various theories and techniques have been developed for enhancing system reliability. Among them, the model-based fault detection and isolation system (FDIS) has been an important research area. The FDIS can be classified into many categories according to the type of model, the method of residual generation and the diagnosis algorithm adopted in the scheme. Here, our attention will be focused on the observer based approach in which several state observers are employed to generate the residuals that will be used for detection and isolation of the faults. In this scheme, the performance of the FDIS largely depends on the quality of the residuals provided by the observers. It is important to design the observer that is sensitive to some preselected faults and invariant to the remaining faults, disturbances and modelling errors so that the faults can be isolated by the simple decision logic with the prescribed confidence level. Hereafter, an unknown input observer (UIO) will be called by a fault detection isolation observer (FDIO) when it is sensitive to some particular faults and invariant to some other faults and unknown disturbances, while an UIO is invariant to all the unknown inputs (disturbances) to the process. No intentional classification of disturbances is made in the design UIO. There have been a number of studies for the robust residual generation based on the UIO theory. Viswanadham and Srichander (1987) developed FDIS by directly utilizing the UIO of Kudva, *et al.* (1980). Frank and Wünnenberg (1989) systematically described a unified approach to the design of robust observer schemes for fault detection isolation (FDI). More recently, an approach for detecting and identifying sensor and actuator faults in uncertain systems was presented by Saif and Guan (1993). A design procedure and the necessary and sufficient condition for the existence of FDIO were given by Hou and Müller (1994) through a new formulation of the FDIO problem (Massoumnia, *et al.*, 1989; Koenig, *et al.*, 1996; Koenig, *et al.*, 1997). The reduced order UIO and its application to the FDI problem in a class of state delayed dynamical systems were addressed by Yang and Saif (1996). It is well known that almost all the existing UIOs and FDIOs take the algebraic decoupling approach to guarantee the invariant property to the faults of no interest and disturbances, and they have the existence condition that is hard to meet. That is: the number of disturbances must be less than that of outputs. This condition restricts the practical application of the UIO to the FDI problem in which multiple faults are concerned (Saif and Guan, 1993). This problem has been well recognized and is still an open problem to be solved.

The purpose of this paper is to suggest the FDIO that yields a viable solution to the problem. The basic concept of this proposition is the use of the unknown disturbance modelling approach (Lee, *et al.*, 1996) together with the algebraic decoupling approach.

The paper is organized as follows. In Section 2, the system including the faults is formulated and the structural constraints imposed upon the previous studies of FDIO (or UIO) is pointed out. Such a crucial limitation is removed through a new design concept of FDIO and the feature of the proposed FDIO is illustrated in Section 3. Section 4 shows the applicability of the FDIO through an example.

2. PROBLEM DESCRIPTION

Assume that our system with the faults can be described by the linear time invariant model (Koenig, *et al.*, 1996):

$$\dot{x} = Ax + Bu + \sum_{i=1}^{k} T_i m_i \qquad (1a)$$

$$y = Cx \qquad (1b)$$

where $x \in \mathbf{R}^n$ is the state vector, $u \in \mathbf{R}^m$ is the input vector, k is the number of possible faults, $m_i \in \mathbf{R}^n$ are the ith fault mode, and $T_i \in \mathbf{R}^n$ are the fault signature matrix. Matrices A, B and C are (n, n), (n, m) and (p, n) matrices, respectively. In multiple observer based FDI schemes, a number of FDIOs are employed to isolate the fault mode(s). Each FDIO is a special kind of UIO which is sensitive to a subset of the fault modes and insensitive to the remaining fault modes. How to partition the fault modes into two subsets depends on the design features: simplicity of isolation logic, ability of multiple fault isolation, and existence condition of each FDIO, etc.. Hereafter we will call the

set of the fault mode to which an FDIO is sensitive by the fault of interest; otherwise, the fault of no interest. Rewriting the system (1) with the two subsets of the fault mode gives

$$\dot{x} = Ax + Bu + Fd + \overline{F}w \qquad (2a)$$

$$y = Cx \qquad (2b)$$

where $d \in \mathbf{R}^\alpha$, $w \in \mathbf{R}^\beta$, rank $(F) = \alpha$, rank $(\overline{F}) = \beta$, rank $(C) = p$ and $\alpha + \beta = k$. In (2a), Fd and $\overline{F}w$ are defined as

$$Fd = \sum_{i \in f,\, i=1}^{\alpha} T_i m_i \qquad (3a)$$

$$\overline{F}w = \sum_{i \in \bar{f},\, i=1}^{\beta} T_i m_i \qquad (3b)$$

where f is the index set of the faults of interest, d, and $\bar{f}$ denotes the index set of the faults of no interest, w. Obviously, $f \wedge \bar{f} = \emptyset$ and $f \vee \bar{f} = \{1, 2, \cdots, k\}$. The matrix $\overline{F}$ consists of the signature matrix of the faults of no interest and therefore may be regarded as the coefficient matrix of some unknown inputs. F denotes the coefficient matrix of the faults of interest. FDIO design problem is to construct an observer which is insensitive to the faults of no interest, w in (2a). As it was noticed before, almost all the existing FDIOs take the algebraic decoupling approach, so they have the existence condition:

$$\text{rank } (C\overline{F}) = \text{rank } (\overline{F})$$

This condition cannot be satisfied whenever the number of outputs is less than that of the faults of no interest. The condition restricts the practical application of the FDIO and must be replaced by some mild condition to achieve reliable fault diagnosis.

3. A NEW DESIGN CONCEPT OF FDIO

In this section a new FDIO design concept is proposed. Remember that the problem is to design an FDIO such that the fault vector w in (2a) should not affect the quality of the state estimates, or equivalently, the performance of the observer. The main idea of our proposition is the combined use of two useful but different techniques: the algebraic decoupling approach and the unknown disturbance modelling approach.

3.1 Partitioning the faults of no interest

The first step of the FDIO design is to partition the faults of no interest, w in (2a), into two sub-vectors:

$$w = [\, w_1^T \: : \: w_2^T \,]^T \qquad (4)$$

where the effect of the fault vector $w_1 \in \mathbb{R}^q$ ($q < p$) on the state estimates can be removed by an algebraic way, and all remaining faults form another fault vector $w_2 \in \mathbb{R}^{\beta-q}$. Then, the system (2) becomes

$$\dot{x} = Ax + Bu + Fd + \left(\overline{F_1} \ \ \overline{F_2} \right) \begin{pmatrix} w_1 \\ w_2 \end{pmatrix} \quad (5a)$$

$$y = Cx \quad (5b)$$

where $\overline{F_1} \in \mathbb{R}^{n \times q}$ and $\overline{F_2} \in \mathbb{R}^{n \times (\beta-q)}$.

Because the effect of the fault vector w_2 cannot be algebraically removed, they should be treated in some other way so that the resultant FDIO provides the correct state estimates in the face of the faults of no interest. The unknown disturbance modelling approach is taken in this study. Hereafter, let us call w_1 by the algebraically rejectable fault vector and w_2 by the modelled fault vector. Theorem 1 gives the explicit partitioning conditions.

Theorem 1. A partitioning of (4) is proper, if
(a) rank $(C\overline{F_1})$ = rank $(\overline{F_1})$ = q; and
(b) the pair $(C : PA)$ is completely observable where

$$P = (I_n - \overline{F_1}(C\overline{F_1})^+ C) \in \mathbb{R}^{n \times n}$$

$$(C\overline{F_1})^+ = ((C\overline{F_1})^T(C\overline{F_1}))^{-1}(C\overline{F_1})^T.$$

Condition (a) is necessary for the algebraic rejection of the fault vector w_1. It is easily proved by showing that there exists a transformation matrix, F, for (5) that makes $P\overline{F_1} = 0$. Since rank $(C\overline{F_1}) = q$, we have $(C\overline{F_1})^+ (C\overline{F_1}) = I_q$. Therefore,

$$P\overline{F_1} = (I_n - \overline{F_1}(C\overline{F_1})^+ C)\overline{F_1}$$
$$= \overline{F_1} - \overline{F_1}(C\overline{F_1})^+ (C\overline{F_1}) = 0.$$

Condition (b) is to guarantee the existence of an FDIO for the transformed system. It should be noticed that the transformation matrix is almost always singular, and the pair $(C : PA)$ may not be observable even if the pair $(C : A)$ is proved to be observable. In such cases the number of the rejectable fault vector w_1 that will be algebraically rejected should be reduced and rearranged so that the observability condition is satisfied.

3.2 Modelling the fault vector

The effect of the fault vector w_2 on the state estimates cannot be algebraically rejected. In many cases, however, the faults that may occur in physical processes have some waveform structures such as jump(step), ramp and exponential function etc.. So, a fault can be modelled by the differential equation:

$$\dot{z_{2i}} = E_{2i} z_{2i}, \quad i = 1, \cdots, \beta-q \quad (6a)$$
$$w_{2i} = H_{2i} z_{2i}. \quad (6b)$$

If the sufficient information of the faults are available, the parameter matrices can be identified. Otherwise, the faults may be modelled as the solution of (6) with the matrices:

$$E_{2i} = \begin{bmatrix} 0_{(\delta_i-1) \times 1} & I_{(\delta_i-1)} \\ 0_{1 \times 1} & 0_{1 \times (\delta_i-1)} \end{bmatrix} \in \mathbb{R}^{\delta(\beta-q) \times \delta(\beta-q)}$$

$$H_{2i} = [I_1 \ \ 0_{1 \times (\delta_i-1)}] \in \mathbb{R}^{(\beta-q) \times \delta(\beta-q)}.$$

where δ_i is the order of the polynomial:

$$w_{2i} = \sum_{i=0}^{\delta_i-1} a_i t^i, \quad \delta_i \geq 1.$$

The dynamical equation (6) provides an effective model not only for unknown disturbances but also for process faults and sensor faults with the assumption that they are of the waveform structures. The $(\beta-q)$-dimensional fault vector is modelled by arranging the model of each fault vector to diagonal form as

$$\dot{z_2} = E_2 z_2 \quad (7a)$$
$$w_2 = H_2 z_2. \quad (7b)$$

3.3 Augmented system and FDIO design

The augmented system is obtained by combining the system (5) with the fault model (7):

$$\dot{x_a} = A_a x_a + B_a u + F_a d + \overline{F_a} w_1 \quad (8a)$$
$$y = C_a x_a \quad (8b)$$

where

$$x_a = \left[x^T \ \ z_2^T \right]^T, \quad A_a = \begin{bmatrix} A & \overline{F_2} H_2 \\ 0 & E_2 \end{bmatrix}, \quad B_a = \begin{bmatrix} B \\ 0 \end{bmatrix},$$

$$F_a = \begin{bmatrix} F \\ 0 \end{bmatrix}, \quad \overline{F_a} = \begin{bmatrix} \overline{F_1} \\ 0 \end{bmatrix}, \quad C_a = [C \ \ 0],$$

$$A_a \in \mathbb{R}^{(n+\delta(\beta-q), n+\delta(\beta-q))}, B_a \in \mathbb{R}^{(n+\delta(\beta-q), m)},$$
$$F_a \in \mathbb{R}^{(n+\delta(\beta-q), a)}, \quad \overline{F_a} \in \mathbb{R}^{(n+\delta(\beta-q), q)},$$
$$C_a \in \mathbb{R}^{(p, n+\delta(\beta-q))}.$$

The problem is to minimize the following criteria with respect to the fault vector w_1 (Koenig, *et al.*, 1997).

$$\xi(w_1) = \| \dot{y} - C_a A_a x_a - C_a B_a u \\ - C_a F_a d - C_a \overline{F_a} w_1 \|_2 \quad (9)$$

where $\dot{y} - C_a A_a x_a - C_a B_a u - C_a F_a \dot{d}$ is a fixed vector and $\| \cdot \|_2$ denotes the Euclidean norm. The solution is given by

$$w_1 = (C_a \overline{F_a})^+ (\dot{y} - C_a A_a x_a - C_a B_a u - C_a F_a d) \\ + (I_q - (C_a \overline{F_a})^+ (C_a \overline{F_a})) w_1 \quad (10)$$

with rank $(C_a \overline{F_a})$ = rank $(\overline{F_a})$ = q, where $\overline{w_1}$ can be considered as new disturbances and

$$(C_a \overline{F_a})^+ = ((C_a \overline{F_a})^T \ (C_a \overline{F_a}))^{-1}$$
$$\cdot \ (C_a \overline{F_a})^T \in R^{\ q \times p}$$

With the rank condition above, $P\overline{F_a} = 0$ where

$$P = (I_{n+\delta(\beta-q)} - \overline{F_a}(C_a \overline{F_a})^+ C_a)$$
$$\in R^{\ (n + \delta(\beta-q)) \times (n + \delta(\beta-q))}. \qquad (11)$$

Therefore, substituting (10) into (8a) gives

$$\dot{x_a} = PA_a x_a + PB_a u + PF_a d \qquad (12a)$$
$$+ \overline{F_a}(C_a \overline{F_a})^+ \dot{y}$$

$$y = C_a x_a. \qquad (12b)$$

where the algebraically rejectable fault vector was disappeared. If the pair $(C_a : PA_a)$ is completely observable, a full-order FDIO is given by

$$\widehat{x_a} = PA_a \widehat{x_a} + PB_a u + \overline{F_a}(C_a \overline{F_a})^+ \dot{y} \qquad (13)$$
$$+ K(y - C_a \widehat{x_a})$$

Defining a new state $z = \widehat{x_a} - \overline{F_a}(C_a \overline{F_a})^+ y$ in (13) removes the time derivative of the measurement output vector and gives the well-known expression:

$$\dot{z} = Nz + Gy + Hu \qquad (14a)$$
$$\widehat{x_a} = z + Ly \qquad (14b)$$

where $z \in R^{\ (n + \delta(\beta-q))}$, $\widehat{x_a} \in R^{\ (n + \delta(\beta-q))}$, and $N \in R^{\ (n + \delta(\beta-q)) \times (n + \delta(\beta-q))}$, $G \in R^{\ (n + \delta(\beta-q)) \times p}$, $H \in R^{\ (n + \delta(\beta-q)) \times m}$, and $L \in R^{\ (n + \delta(\beta-q)) \times p}$ must be determined such that $\widehat{x_a}$ will asymptotically converge to x_a if no fault of interest occurs.

With the new state $z = \widehat{x_a} - \overline{F_a}(C_a \overline{F_a})^+ y = \widehat{x_a} - Ly$, the FDIO (13) becomes

$$\dot{z} = (PA_a - KC_a)z + PB_a u \qquad (15)$$
$$+ [K(I_p - C_a L) + PA_a L]y.$$

Then, the coefficient matrices of the FDIO (14) are obtained from (15):

$$N = PA_a - KC_a \qquad (16a)$$
$$G = K(I_p - C_a L) + PA_a L \qquad (16b)$$
$$H = PB_a \qquad (16c)$$
$$L = \overline{F_a}(C_a \overline{F_a})^+ \qquad (16d)$$

Define the state estimation error vector by

$$e = \widehat{x_a} - x_a \qquad (17)$$

then, the error dynamics become

$$\dot{e} = Ne - PF_a \dot{d}. \qquad (18)$$

The error dynamic equation (18) shows that the full-order FDIO (14) with the coefficient matrices (16) is invariant to the faults of no interest and sensitive to the faults of interest. Theorem 2 summarizes the existence condition of the asymptotic FDIO (14) for the augmented system (8).

Theorem 2: If
(a) The pair $(C_a : PA_a)$ is observable; and
(b) rank $(C_a \overline{F_a})$ = rank $(\overline{F_a})$ = q $(p > q)$,
then there exists an FDIO (14) with the coefficient matrices (16).

It is straightforward to show that the conditions (a) and (b) in Theorem 2 are equivalent to two conditions in Theorem 1 because the model (7) of the fault vector w_2 is an observable canonical form. The proposed new methodology yields the following characteristics, namely

· it resolves one of the major practical difficulties of all the observer based FDISs previously reported that the number of unknown inputs must be less than or equal to that of outputs and is capable of detecting and isolating simultaneous faults more than the number of outputs.

· in conventional FDIS based on UIOs or FDIOs, the eigenspectrum of the UIO (or FDIO) cannot be arbitrarily assigned if the number of outputs is equal to that of unknown inputs (Saif and Guan, 1993). This problem was removed in the proposed scheme.

· the residuals for detecting and isolating faults are independent of the modelled fault vector. Therefore the proposed FDIO has the invariance property to unknown inputs (including the algebraically rejectable fault vector).

· the FDIO offers additional degree of freedom in the selection of the faults to be rejected by the algebraic decoupling approach.

· as a byproduct, the FDIO regenerates the unknown fault modes so that they can be used to accommodate the fault.

Remark: As a special case, (i) if we can partition the fault vector of no interest and corresponding transmission matrix as $w_1 = u$ and $\overline{F_1} = \overline{F}$, then all the faults of no interest can be rejected by the algebraic method and the observer design procedure is therefore reduced to the pure algebraic approach, and (n)th order (lowest) observer may be constructed. (ii) if $w_2 = u$ and $\overline{F_2} = \overline{F}$, then all the faults of no interest must be modelled and the resultant observer with highest order of $(n + \delta\beta)$ may be constructed by the use of the unknown fault vector modelling approach. It is noteworthy that case (ii) is the worst case in view points of dimensionality and estimation errors of the observer.

4. NUMERICAL EXAMPLE

A numerical example is taken to the system with matrices:

$$A = \begin{bmatrix} -1 & 0 & 1 \\ 0 & -2 & 1 \\ 0 & 1 & -3 \end{bmatrix}, \quad B = \begin{bmatrix} 1 \\ 1 \\ 0 \end{bmatrix}, \quad C = \begin{bmatrix} 1 & 0 & 0 \\ 0 & 1 & 1 \end{bmatrix}$$

(19)

Assume the total number of possible process faults is $k = 3$. Possible faults are described by the fault signatures T_i, $i = 1, \cdots, 3$ with the indices corresponding to the fault number:

$$T_1 = [1\ 0\ 0]^T, \quad T_2 = [0\ 1\ 0]^T, \quad T_3 = [0\ 0\ 1]^T.$$

Choose $\beta = 2$ and $\alpha = 1$, and define $f = (1)$, $\overline{f} = (2,3)$ for FDIO 1 and $f = (2)$, $\overline{f} = (3,1)$ for FDIO 2 and $f = (3)$, $\overline{f} = (1,2)$ for FDIO 3. In this case, FDIO i should be sensitive to only fault i and robust to all other two faults. Choose matrices F, $\overline{F} = [\overline{F_1} : \overline{F_2}]$ and vectors d, $w = [w_1 : w_2]$ with $q = 1$ as in Table 1, where $p = 6$. Matrices E_2 and H_2 for modelling the fault vector w_2 are given with $\delta = 2$ in Table 2. In Table 3, the coefficient matrices N, G, H, L of each FDIO with eigenvalues of $N = \{-1, -2, -3, -5, -5\}$ are represented. The residual generation process simply consists of subtracting the measured state vector from the state vector estimated by the observer i:

$$r_i = |\ \hat{x}^i - x\ |, \quad i = 1, \cdots, 3. \quad (20)$$

where $\hat{x}^i \in \mathbf{R}^n$ is the state vector estimated by the observer i. The detection logic is

$$v_i = \begin{cases} 0 & \text{if } r_i < \varepsilon_i \\ 1 & \text{if } r_i > \varepsilon_{i,} \end{cases} \quad i = 1, 2, \cdots, 3. \quad (21)$$

where the thresholds are $\varepsilon_1 = \varepsilon_2 = \varepsilon_3 = 0.7$.

All zero v_is' mean that the faults of interest has not occurred, while if at least one among v_is' is not equal to zero, it indicates the occurrence of the fault(s) of interest. In related to Table 1, the isolation logic of Table 4 is chosen to identify the faults where $s_i = 1$ means that fault i has occurred. There are two zeros in each column, which implies that FDIO i should be sensitive to the fault i and invariant to the other two faults for correct isolation. Two faults that occur simultaneously are taken into account. In order to show the effectiveness of the proposed FDIOs in FDI and unknown fault modelling, simulations are performed for the fault modes,

$m_1 = m_2 = m_3 = 2 + 0.3\sin(0.1t)$ and the system input, $u = 1 + 0.2\sin(t)$. Fig. 1 shows the FDI results for three cases that two simultaneous faults occur: (a) m_1 and m_2, (b) m_2 and m_3, and (c) m_3 and m_1. It is assumed that the faults occurred at 5 second. From those figures it is easily recognizable that non-zero residual values due to the faults of no interest disappear after a short transient. By proper choice of the threshold values for each residual, the detection delay can be minimized.

Table 1 Matrices $F, \overline{F}$ and vectors d, w

FDIO	F	d^T	$[\overline{F_1} : \overline{F_2}]$	$[w_1^T : w_2^T]$
1	$[T_1]$	$[m_1]$	$[T_2 : T_3]$	$[m_2 : m_3]$
2	$[T_2]$	$[m_2]$	$[T_3 : T_1]$	$[m_3 : m_1]$
3	$[T_3]$	$[m_3]$	$[T_1 : T_2]$	$[m_1 : m_2]$

Table 2 Matrices E_2 and H_2

FDIO	E_2	H_2
1	$\begin{bmatrix} 0 & 1 \\ 0 & 0 \end{bmatrix}$	$[1\ 0]$
2	$\begin{bmatrix} 0 & 1 \\ 0 & 0 \end{bmatrix}$	$[1\ 0]$
3	$\begin{bmatrix} 0 & 1 \\ 0 & 0 \end{bmatrix}$	$[1\ 0]$

Table 3 Coefficient matrices N, G, H, L of each FDIO

FDIO	N	G
1	$\begin{bmatrix} -7 & -0.5 & 0.5 & 0 & 0 \\ 13 & -4.5 & -0.5 & -1 & 0 \\ -13 & -0.5 & -4.5 & 1 & 0 \\ -61 & 0 & 0 & 0 & 1 \\ -30 & 0 & 0 & 0 & 0 \end{bmatrix}$	$\begin{bmatrix} 6 & 0 \\ -13 & -1 \\ 13 & 1 \\ 61 & 0 \\ 30 & 0 \end{bmatrix}$
2	$\begin{bmatrix} -8 & -0.5 & 0.5 & 1 & 0 \\ 0 & -4 & -1 & 0 & 0 \\ 0 & -1 & -4 & 0 & 0 \\ -17 & 0 & 0 & 0 & 1 \\ -10 & 0 & 0 & 0 & 0 \end{bmatrix}$	$\begin{bmatrix} 7 & 1 \\ 0 & 1 \\ 0 & -1 \\ 17 & 0 \\ 10 & 0 \end{bmatrix}$
3	$\begin{bmatrix} -5 & 0 & 0 & 0 & 0 \\ 0 & -6.6 & -3.6 & 1 & 0 \\ 0 & -0.4 & -4.4 & 0 & 0 \\ 0 & -13.4 & -13.4 & 0 & 1 \\ 0 & -7.5 & -7.5 & 0 & 0 \end{bmatrix}$	$\begin{bmatrix} 0 & 0 \\ 0 & 4.6 \\ 0 & 1.4 \\ 0 & 13.4 \\ 0 & 7.5 \end{bmatrix}$

FDIO	1	2	3
H^T	$[1\ 0\ 0\ 0\ 0]$	$[1\ 1\ -1\ 0\ 0]$	$[0\ 1\ 0\ 0\ 0]$
L^T	$\begin{bmatrix} 0 & 0 & 0 & 0 & 0 \\ 0 & 1 & 0 & 0 & 0 \end{bmatrix}$	$\begin{bmatrix} 0 & 0 & 0 & 0 & 0 \\ 0 & 0 & 1 & 0 & 0 \end{bmatrix}$	$\begin{bmatrix} 1 & 0 & 0 & 0 & 0 \\ 0 & 0 & 0 & 0 & 0 \end{bmatrix}$

Table 4 Isolation Logic

Fault	v_1	v_2	v_3	Decision Logic
m_1	1	0	0	$s_1 = v_1$
m_2	0	1	0	$s_2 = v_2$
m_3	0	0	1	$s_3 = v_3$

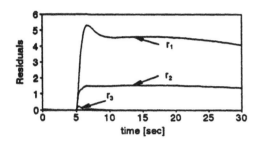

$(v_1 = 1 , v_2 = 1 , v_3 = 0 , s_1 = 1, s_2 = 1)$

(a) faults m_1 and m_2

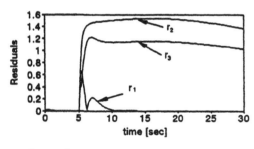

$(v_1 = 0 , v_2 = 1 , v_3 = 1 , s_2 = 1, s_3 = 1)$

(b) faults m_2 and m_3

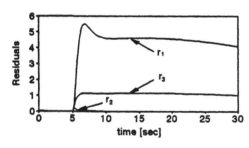

$(v_1 = 1 , v_2 = 0 , v_3 = 1 , s_3 = 1, s_1 = 1)$

(c) faults m_3 and m_1

Fig. 1. FDI results for two simultaneous faults

5. CONCLUSIONS

This paper proposed an FDIO that accentuates the effect of the preselected fault on the residual. The basic concept of this proposition is the use of the unknown disturbance modelling approach together with the algebraic decoupling approach. An important contribution of this paper is to resolve one of the major practical difficulties concerning about the existence condition of all the observer based FDIS. The FDIO may also regenerate the fault modes that can be used to accommodate the fault. The FDIS with the FDIO offers the designer additional degree of freedom such as the number of faults to be rejected by an algebraic approach and can be constructed to detect and isolate simultaneous faults more than the number of outputs. A numerical example shows the applicability of the proposed FDIO for FDIS.

REFERENCES

Frank, P.M. and J. Wünnenberg (1989). Robust fault diagnosis using unknown input observer schemes. In: *Fault Diagnosis in Dynamical Systems: Theory and Applications* (R.J. Patton, P.M. Frank and R.N. Clark, (1st Ed.)), pp. 47-98. Prentice Hall, New York.

Hou, M. and P. C. Müller (1994). Fault detection and isolation observers. *Int. J. Contr.*, 60, pp. 827-846.

Koenig, D., S. Nowakowski and A. Bourjij (1996). New design of robust observers for fault detection and isolation. *Proc. 35th IEEE Conf. Decision and Control*, Kobe, Japan, pp. 1464-1467.

Koenig, D., S. Nowakowski and T. Cecchin (1997). An original approach for actuator and component fault detection and isolation. To appear in *Proc. of the IFAC Symposium "Safeprocess97"*.

Kudva, P., N. Viswanadham and A. Ramakrishna (1980). Observers for linear systems with unknown inputs. *IEEE Trans. Automat. Contr.*, 25, pp. 113-115.

Lee, K.S., S.W. Bae and J. Vagners (1996). On the fault detection isolation systems based on GOS using functional observers. *Proc. 35th IEEE Conf. Decision and Control*, Kobe, Japan,, pp. 1181-1183.

Massoumnia, M.A., G.C. Verghese and A. S. Willsky (1989). Failure detection and identification. *IEEE Trans. Automat. Contr.*, 34, pp. 316-321.

Saif, M. and Y. Guan (1993). A new approach to robust fault detection and identification. *IEEE Trans. Aero. and Elec. Sys.*, 29, pp. 685-695.

Viswanadham, N. and R. Srichander (1987). Fault detection using unknown input observers. *Control Theory and Adavanced Technology*, 3, pp. 91-101, 1987.

Yang, H. and M. Saif (1996). Observer design and fault diagnosis for retarded dynamical systems. *Proc. 35th IEEE Conf. Decision and Control*, Kobe, Japan, pp. 1149-1154.

RESEARCH CHALLENGES AND OPPORTUNITIES IN REMOTE DIAGNOSIS AND SYSTEM PERFORMANCE ASSESSMENT

Jianjun Shi and Jun Ni

S. M Wu Manufacturing Research Center
The University of Michigan
Ann Arbor, MI 48109-2117, USA
Phone: 313-763-5321, Fax: 313-764-3451, e-mail: shihang@engin.umich.edu

Jay Lee

National Science Foundation
4201 Wilson Blvd., 585, Arlington, VA 22230, USA

Abstract: The framework for remote diagnosis and system performance consists of two major aspects: 1) The system architecture for hardware configuration; and 2) The system software and algorithms for data analysis. This paper presents the architecture, major functional components, and enabling components for remote diagnosis. Some significant research challenges and opportunities are discussed and commented on.

Keywords: process monitoring, remote diagnosis, emerging technologies, Internet.

1. INTRODUCTION

Fast advancement of Internet technologies has imposed challenges, as well as opportunities, for development of remote diagnosis and performance assessment. Research on the internet-based process monitoring and diagnosis is still in its initial stage. Although there is an awareness of the importance of remote diagnosis systems, some confusion still exists, in both academia and industry, on what the research challenges and opportunities are. This paper attempts to express the authors' views and understanding of the problem.

The urgency of developing a remote diagnosis system stems from the promise of globalize integrated manufacturing with high demands on the overall system performance of the manufacturing facilities. The development of such a system has three main motivational forces behind it: manufacturing industry, process/machine designers, and academia. 1) The current high accuracy and productivity requirements in manufacturing have brought about more complex equipment. In-process/in-tool sensing has been widely used in machine tools. How to use in-process sensing most

effectively is a challenge. Moreover, due to the lack of diagnosis expertise in many manufacturing facilities, information collected via in-process sensing is often wasted in practice. A remote diagnosis system, which centralizes in-process sensing information on a server, performs the diagnosis tasks, and broadcasts the results to the appropriate users, is surely an appealing system to the manufacturing industry. 2) Meanwhile, the process/machine designers are required to design machine tools and processes with the highest reliability and performance. In order to accomplish this, understanding the current system performance in real production environments is essential. Internet-based remote diagnosis provides the designer with such an opportunity. More importantly, the system can link many similar processes at different sites, collecting information more efficiently and economically. 3) In facing these new challenges, academia needs to redefine its research thinking to accommodate holistic research in a production environment - factory as laboratory.

This paper presents the authors' views and understanding of remote diagnosis and performance evaluation. The common architecture and key

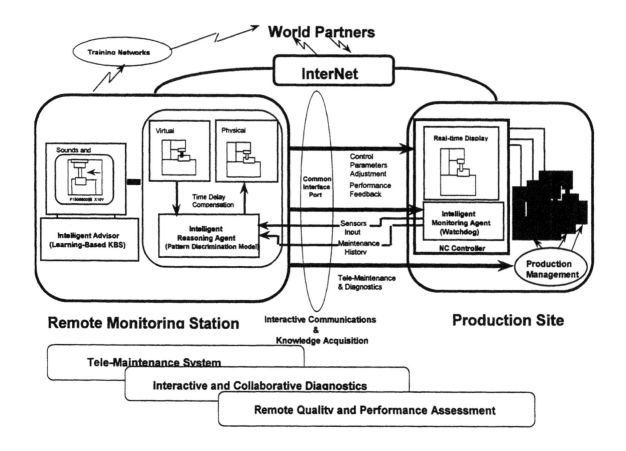

World Partners

Training Networks

InterNet

Virtual

Physical

Sounds and

F15000000S X10Y

Time Delay
Compensation

Intelligent Advisor
(Learning-Based KBS)

Intelligent
Reasoning Agent
(Pattern Discrimination Model)

Common
Interface
Port

Control
Parameters
Adjustment

Performance
Feedback

Sensors
Input

Maintenance
History

Tele-Maintenance
& Diagnostics

Real-time Display

Intelligent
Monitoring Agent
(Watchdog)

NC Controller

Production
Management

Remote Monitoring Station

Interactive Communications
&
Knowledge Acquisition

Production Site

Tele-Maintenance System

Interactive and Collaborative Diagnostics

Remote Quality and Performance Assessment

Fig. 1. An overview of the remote diagnosis and performance evaluation system

components in the system are discussed, followed by the major research challenges and opportunities.

2. A FRAMWORK FOR REMOTE DIAGNOSIS AND SYSTEM PERFORMANCE ASSESMENT

Service and maintenance are important practices to maintain the manufacturing productivity and customer's satisfaction. The recent rush to embrace highly sophisticated software-based manufacturing technology for manufacturing globalization activities has further increased the use of relatively unknown and untested technology. Difficulty in identifying the causes of system failures has been attributed to several factories, including system complexity, uncertainties, and lack of adequate troubleshooting tools. The remote diagnosis and system performance assessment, as part of the Tele-service Engineering, is an emerging field which addresses "*service*" issue for manufacturers and customers. With the growing manufacturing globalization activities, companies are looking for ways to assess the performance of their manufacturing operations and products in remote sites. Digital maintenance diagnostics and maintenance tools such as "watchdog" type information mechatronics with integrated media will improve the effectiveness of the engineering practices on maintenance activities. Research activities are underway by the authors to focus on

research activities on remote testing & performance evaluation techniques, tele-maintenance technologies, and remote diagnostics technologies.

An overview of the proposed remote diagnosis and performance evaluation system is shown in Fig. 1. In this section, the major functional and enabling components will be discussed for the remote diagnosis. Then, significant research challenges and opportunities will be discussed and commented on.

2.1 Major functional components

The system architecture is shown in Fig. 2, which indicates the major functional components in the remote diagnosis system. In the figure, the manufacturing process contains in-process and/or in-machine sensing system. A data acquisition system is used to collect the sensing information. After obtaining the data, data pre-processing will be conducted to extract features from the data, and perform some simple on-line process monitoring. The extracted features, as well as some raw process data, will be transferred to the centralized location through Internet and/or Intranet. The data and features of the signal will be saved and managed in a database. In the central computing server, various complex data analysis tasks will be performed for system diagnosis, performance assessment, maintenance scheduling, and broadcasting of the

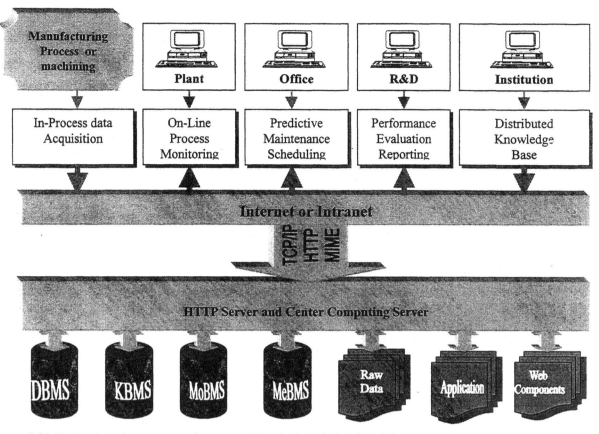

DBMS: Database Management System; KBMS: Knowledge Base Management System;
MoBMS: Statistical Model Base Management System; MeBMS: Methodology Base Management System;

Fig.2 System Architecture and Major Components in Remote Diagnosis

analysis results. The central data analysis and process performance assessment will be conducted by statistical analysis modules. Depending on the complexity of the system, a knowledge based system should be developed to take into consideration of historical information and for experience sharing. The knowledge base can be distributed to various sites, and a central decision making strategy should be developed to coordinate these activities.

One of the key components in remote diagnosis is the statistical analysis module as shown in Fig. 3. The module consists of four levels of data analysis tools: original data information, modeling techniques, estimation and diagnosis methodologies, and system evaluation algorithm and indices.

The original data include four different data sets: 1) in-process/in-machine sensing data, which reflects the process operating conditions and performance; 2) historical data containing information on past process/machine faults, patterns, and maintenance records; 3) accelerated testing data, obtained from the lab analysis and testing, which represents reliability information; and 4) process/product design information, which comes from the designer. It should be pointed out that all the information is available on the Internet in the server.

The modeling techniques include statistical modeling, on-line process modeling (time series model, transfer functions, state-space model, etc.), reliability modeling and physical engineering modeling.

The performance evaluation module contains various tasks, including process degradation monitoring, reliability analysis, fault diagnosis and root cause determination, and predictive/proactive maintenance. There are many statistical analysis tools, e.g. Principal Component Analysis (PCA), Factor Analysis, Wavelet analysis, and Hypothesis testing, that can be used in this module. Many of these have been developed and implemented in real applications.

2.2 Enabling components

There are several enabling components in a typical remote diagnosis and performance assessment system as shown in Fig. 4.

1) Information Mechatronics (Watchdog Neural Chip): The assessment of machine's performance information to operators in remote sites requires an integration of many different sensory devices. A "watchdog" agent, a neural computer, has been developed by the author

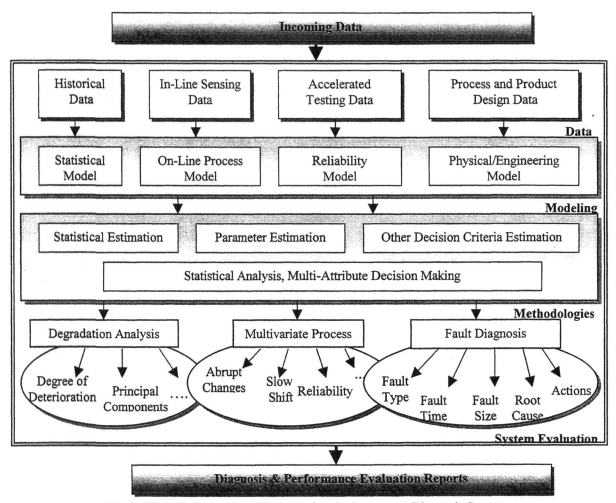

Fig. 3 The Statistical Data Analysis Module in the Remote Diagnosis System

to provide on-line composition and reasoning. In addition, this mechatronics chip could be connected with a telephone jacket so that performance information could be accessed and evaluated from a remote site.

2) Knowledge Learning and System Failure Recovery Strategy: Knowledge-intensive intelligent tools are used for acquisition and organization of data in machine and manufacturing processes to track the behavior of the machine. A "watchdog" chip will serve as a "blackbox" of machine and is able to keep the signatures of major components. In case of failure, operators can access the "blackbox" and obtain the last several minutes of information about the behavior of machine. As a result, system can be recovered rapidly and return to the normal operational mode. These knowledge also can be shared with other user sites.

3) Tele-Maintenance and Collaborative Diagnostics: Multimedia-based Tools are required to support remote users for maintenance assistance. Interactive and collaborative tools will enable the technical personnel to perform diagnostics from a remote site.

2.3 General research challenges and opportunities

There are various schools of thought, in both academia and industry, on the specific research challenges and opportunities in remote diagnosis. In this section, we propose our viewpoint on and understanding of these issues, with focus on the need of "remote" diagnosis and performance evaluation. The major technical issues have been roughly divided into ten categories, outlined as follows.

1) Sensing system standardization: In-process and in-machine sensing is the foundation for remote diagnosis and performance evaluation. Due to the complexity of a process/machine, different types of the sensors may be used, requiring different data acquisition protocols and systems. As remote diagnosis practice expands, more machines/processes will be linked within the remote diagnosis system. Consequently, sensing system standardization will have a large impact on implementation of the technology. The standards should include the selection of sensor type for typical signals, sensor signal output ranges, protocols, etc. There are several industrial standards currently available. Further efforts, which are likely to be conducted by some consortium, are needed.

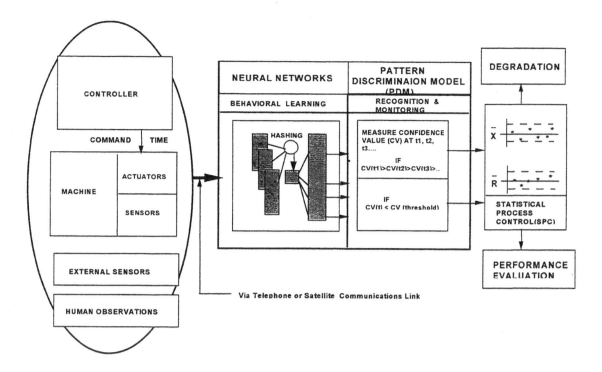

Fig. 4. Enabling components in a typical remote diagnosis and performance assessment system

2) Data compression and feature extraction: The sensing data will be transferred to a central server through the Internet. If transferring the raw data directly, long time delays due to heavy traffic may be experienced. In addition, more storage space may be required in the central server. One area that needs to be addressed is data pre-processing for data compression and feature extraction. It should be noted that the data compression task will be different from the conventional approach in image analysis and signal processing, even though there are many similarities and some techniques can be borrowed. The research emphasis here is on how to combine the engineering knowledge and diagnosis requirements in the data compression and pre-processing stages. "Engineering feature based data compression" should be developed, which considers the important features in diagnostic analysis and system performance assessment. Examples in this category include (1) identifying engineering model parameters from the data and transferring only the model coefficients to the server; (2) extracting features from the original sensing signature and adjusting the threshold based on the interested signal information; (3) transferring data continuously to the server from low to high wavelet coefficients and developing stopping criteria based on the decision making strategy. Once again, this research cannot be treated as a pure signal processing task, and needs to combine the knowledge and algorithms in diagnosis and performance measure. Engineering knowledge plays a critical role in this research. In addition, the signal uncertainties and risk assessment should be considered in the task also.

3) Adaptive sensor fusion and affordability: One physical machine faults (e.g. an unbalanced shaft) may generate different symptoms (e.g. vibration, temperature changes, motor load variation, etc.) and can be measured by different sensors (e.g. an accelerometer, thermal couples, motor current, etc.). Similarly, one sensor may sense different types of machine faults occurring simultaneously, and its sensitivity may vary as the operating condition changes. Thus, adaptive sensor fusion should be emphasized, which will improve the reliability of the diagnosis strategy. However, affordability should also be emphasized in the research. The issues of affordability will lead to cheaper sensing techniques, less demands on sensing data accuracy and high requirements on the noise rejection capability of the algorithms. In addition, optimal, objective-oriented sensor placement strategy needs to be investigated. Remote diagnosis provides a better opportunity to develop the aforementioned techniques by implementing knowledge and experience sharing and self-learning.

4) Task allocation: Task allocation is another research topic closely related with remote diagnosis and performance evaluation. According to the requirements on the response time to faults, the monitoring and diagnosis tasks can be classified as immediate response (e.g. tool breakage, collision, etc.), intermediate response (tool wear, temperature compensation, etc.) and slow response (machine wear and degradation, environment changes, etc.). According to the information required and the complexity involved in decision making, the tasks can be classified as single variable process change detection, multivariate analysis, and integrated decision making. A study needs to be conducted to classify all tasks into various categories. As a result, the concept of "watch dog" should be used to develop "intelligent" sensors which perform simple, on-line and real-time process change detection for the tasks requiring immediate response. Meanwhile, the remote diagnosis efforts should be directed towards more complex diagnosis tasks and degradation monitoring requiring multivariate data and information integration. A thorough study of task allocation is critical to the success of the remote diagnosis efforts and will have great impact on the overall system performance, cost, complexity, etc.

5) System reliability and dependability: In the system development, a backup strategy should be considered for overall system reliability improvement and for determining what to do if the Internet system malfunctions. Under various conditions, the system should be able to operate and perform basic functions, albeit with deteriorated performance. For example, if the Internet is not available for data transfer and remote diagnosis, local data processing should be executed to conduct essential tasks that are nominally carried at the remote site. How to design a redundant system with minimum cost and high performance is a challenge. This is also related with task allocation and analysis, simplified diagnosis algorithms, decision making under incomplete information, etc.

6) Tele-maintenance and collaborative diagnostics: A major advantage of implementing remote diagnosis is tele-maintenance, collaborative diagnostics, and "information and experience" sharing. Those can be achieved from two main aspects: (1) fault condition data collection: the remote diagnosis system provides an opportunity to accumulated more machine/process fault conditions. Thus, a better diagnosis algorithm can be developed from the fault conditions in various remote sites; (2) fault diagnosis: the information is available in the server and can be accessed by experts at various locations. Thus, knowledge distributed over various sites can be integrated to perform more complex collaborative diagnosis. However, significant research efforts on how to manage the information and distributed decision making components are required. Topics such as distributed AI, competitive decision making, risk management, etc. should be studied.

7) Self-learning and supervised learning/information assessment: Even though the topics of supervised learning and self-learning have been studied by various researchers, they become more critical in a remote diagnosis environment. Due to the nature of remote diagnosis, process data from many different locations can be accessed. Updated knowledge will be acquired much faster than with traditional diagnosis techniques. Thus, supervised learning (or self-learning) is more critical. Furthermore, it is anticipated that fewer well-trained supervisors may perform the task due to the availability of new information. Another challenge is how to assess new information before using it for the purpose of learning. The information assessment stage can be an integrated as part of the supervised/self- learning research.

8) Integrated performance assessment: The remote diagnosis system will provide several categories of information together, such as on-line process/machine sensing data, historical fault/degradation data, machine design information, etc. All information should be integrated together when conducting a performance evaluation. The concept of "Machine Physiology" can be developed to monitor machine degradation without having historical machine fault condition information. New indices of machine performance should be developed. In addition, knowledge of machine performance can be learned and modeled, eventually being used in *machine performance compensation.* Thus, the performance of a typical machine may not necessarily degrade over time, but instead be improved by using learning-modeling-compensation techniques. The remote diagnosis system provides a feasible scenario to conduct such research.

9) Information filtering and accumulation / Data mining: The remote system can accumulate huge amounts of information. Process faults and other interesting features in the data may not known priori analysis. Thus, data techniques will provide more opportunities for feature extraction, especially for new features not identified in the past. How to accumulate this information and filter out unnecessary information will be a challenge also.

10) Reconfigurablity and transferrability: The remote system is a complex system involving intensive hardware and software development. The systems developed for different applications should share common modules. Further more, the computer/internet technology and the diagnosis methodologies are advanced rapidly, the developed system should be able to incorporated those advancement easily without major modifications. Thus, reconfigurablity and transferrability are important in all aspects of the system development.

3. SUMMARY AND CONCLUSION

The remote diagnosis and performance evaluation is an emerging field and has a broad applications. The paper presents our understanding and views of remote diagnosis and performance evaluation. The research challenges and opportunities are summarized with focus on the topic of "remote" diagnosis.

DFKN: A MODEL TO ASSIST IN THE CAPTURE AND UTILISATION OF DESIGN KNOWLEDGE

Xiao Hui Wang[*#], **Dominique Deneux**[*], **Rene Soenen**[*], **Fu Tong**[#]

*: *LAMIH-URA CNRS N°1755- Group de Recherche en Génie Industriel et Logiciel*
University of Valenciennes- BP 311 - 59304 Valenciennes Cedex - France
Phone:(33)03 27 14 13 47 - Fax: (33) 03 27 14 12 88
E-mail: xiaohui@univ-valenciennes.fr
[#]:*Dept. of Computer Science, University of Shanghai, Jiading*

Abstract: The process of acquiring and representing the engineering design knowledge is very important for Re-design. To address this problem, a hybrid knowledge model - dynamic fuzzy knowledge networks (DFKN) - is proposed. The DFKN is a structured networks composed of knowledge blocks. In a knowledge block, fuzzy concepts can be described based on the fuzzy theory; The experiential relationships can be represented in a meaningful way at the attribute level and achieved by way of connectionist computing at the value level. Through adjusting the weight of linkage in value level with artificial neural networks training algorithm, the experiential knowledge can be efficiently captured and reused.

Keywords: knowledge representation, fuzzy hybrid systems, rule_based systems, connectionism.

1. SOME IMPORTANT CHARACTERISTICS IN DESIGN

Engineering design is a complex activity and involves a great amount of knowledge. These knowledge are characterised by the following major properties (X.H.Wang et al, 1996):
- High uncertainty, imprecision, fuzziness;
- Experiential relationships existing between criteria, constraints and design alternatives;
- Negotiation between multiple criteria;
- Knowledge adjustment based on post-design evaluations.

These important characteristics require that an efficient knowledge model must provide the ability to represent inexact knowledge (fuzzy concept, experimental relationships), to carry out approximate reasoning, and to adjust the knowledge structure based on examples.

2 DFKN: A HYBRID KNOWLEDGE MODEL FOR ENGINEERING DESIGN

To address the problem of design knowledge capture, a hybrid knowledge model-dynamic fuzzy knowledge networks (DFKN) is proposed taking into account the above mentioned requirements.

2.1 Definition of the fundamental structure of DFKN

DFKN is a network. Its fundamental structure is defined as follows.

Definition 1. A *knowledge network* (KN) is a pair KN =(N, E); where, N is the set of nodes n_i, E is the set of edges e_{ij}.

Definition 2. A *node* n_i of KN represents the element of an object and is defined by a triple: $n_i = (I_i, f_i, \mu_i)$, where, I_i is the information of n_i, f_i is the set of operations on n_i. $\mu_i \in [0,1]$ represents the truth value or degree of membership of n_i.

Definition 3 An *edge* e_{ij} of KN represents the relation and is defined by a pair $e_{ij} = (n, \mu_{ij})$, where: n is the name of the relation between n_i and n_j, $\mu_{ij} \in [-1, 1]$ is the strength of the relation between n_i, n_j. denoted $\mu(n_i, n_j)$.

Based on the fundamental structure, a Knowledge system can be described in terms of nodes and edges.

2.2 Definition of fuzzy objects

The elements of an object are called atoms. An atom is either an identifier of the object or an attribute, or an attribute value: number, linguistic value, function, or a physical structure which needs not be further divided for deep consideration. So,

Definition 4 An atom is defined as an element of an object when representing and processing knowledge.

According to the definition, in the field of engineering design, the BNF definition of atom should be:

  <Atom>::= <Number>| <Ling_value>| <Symbol>
       | <Structure>
  <Number>::=Integer| Real| Vector| Interval|
       Predicate| Expression
  <Ling_value>::= "light"| "moderate"| "easy"| ···
  <Symbol>::=Name| Identifier| ···
  <Structure>::= Function| Procedure| Structure_net|
       Table| Set

In DFKN, an atom with its degree of membership is represented by one node of fundamental structure. Objects, attributes and the domain of value can be described based on atoms.

First we give the definition of a fuzzy set.

Definition 6 a *fuzzy set* is defined as the set of atoms with their degree of membership in the given domain.

A fuzzy net V can be denoted as:

$$V= (x/\mu(x) \{, x/\mu(x)\})$$

here x is an atom, $\mu(x)$ is the truth value of atom x.

The domain of value is a fuzzy Set. In DFKN, the fuzzy sets are represented by a set of nodes as shown in figure 1. Each node is called value node.

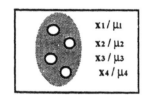

Fig. 1. Representation of the domain of value

Objects are described by attributes. An attribute can be defined as below:

Definition 7 An *Attribute* is a pair: *Attr= (A, V)*. A is an attribute identifier, *V* is a fuzzy set representing the domain of value.

An *Attribute* can be denoted as

$$Attr(A(t), V)$$

Here, *t* is the attribute type (defined later).

In DFKN, an attribute is represented by an attribute node (identifier), a set of value nodes (fuzzy set), and an edge from attribute node to the set of value nodes.. It is shown as figure 2.

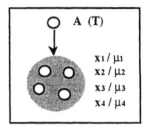

Fig 2. Representation of an attribute

According to the characteristic of attribute value, any attribute can be considered as one of the four types: numerical attribute (*N_Attr*), linguistic attribute (*L_Attr*), structural attribute (*Stru_Attr*), solution attribute (*Solu_Attr*).

Definition **8** A *Numerical Attribute* (*N_Attr*) is an attribute which values are numeric.

Definition 9 A *Linguistic Attribute(L_Attr)* is an attribute which values are linguistic.

Definition 10 A *Structural Attribute(Stru_Attr)* is an attribute which represents an object and its values represent its components.

Definition 11 A *Solution Attribute (Solu_Attr)* is an attribute which represents a problem and its values represent the alternative solutions.

Based on the attribute, an object can be described as below:

Definition 12 A object is a pair: *Obj= (O, S)*, O is an object idntifior, *S* is a set of attributes: $A_1, A_2, ...A_n$.

An object can be denoted as:

$$Obj(O, (A_1, A_2, ...A_n)$$

In DFKN, it is represented by an object node (identifier), a set attribute nodes, and the edges from objects to attribute nodes. It was shown in figure 3.

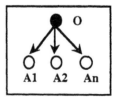

Fig 3. Representation of an object in DFKN

It is obvious that an object can be described based on the fundamental structure of DFKN.

2.3 Representation of fuzzy relations

The relationship between two attributes can be a dynamic experiential relationship or a definite functional relationship according to the property of the linkage, and namely "dynamic rule" and "functional link".

Dynamical rule and functional link

Assume that A and B are two attributes belonging to one or two objects, then:

Definition 13: A *Dynamical_Rule* (DR) from A to B is the experiential causal relationship between two attributes A and B. (1) The strength of relationship stands for the relative importance and varies with the environment; (2) The influence of A upon B is calculated through its value relationship; (3) The value_value relationship varies also with the environment.

A and B are called *Condition Attribute* and *Consequent Attribute* respectively. A dynamical rule can be denoted as:

$$DR(A, B, \mu)$$

Other relations at attribute level are definite functional relationships.

Definition 14: A Functional_Link (FL) from A to B is a kind of definite functional linkage between two attributes A, B. (1) its does not vary with the environment; (2) The information transition is achieved by the linkage at the value level; (3) The linkage at value level is fixed and does not vary with the environment.

According to this definition, the relationships between the information and the approaches by which the information are calculated, retrieved, or represented is a definite_link. For example, the relationship between the price and the method of calculation is a kind of definite_link. The relationships between the linguistic values and numerical value, the relationship between a design object and its direct graph representation are also definite_links.

A functional link can be denoted as:

$$FL(A, \otimes, B)$$

$\otimes$ is the semantics of functional_link.

A and B are called *Function attribute* and *Consequent attribute* respectively.

In DFKN, an attribute relationship can be represented by a edge connecting two attribute nodes.

Fuzzy net, logic link, and transfer link

At the value level, corresponding to the attributes relationships, there exist two kinds of relationships: experiential linkage corresponding to the dynamical

rule, and information transfer corresponding to the functional link. Namely fuzzy_net, and transfer_link respectively.

Assume that, A, B are two attributes, Val(A), Val(B) are the sets of all values of A and B, respectively. X and Y are subset of Val(A) and Val(B), $x \in$ Val(A), $y \in$ Val(B). Based on assumptions, the value relations are defined.

The fuzzy net achieves the experiential mapping from conditions attributes values to consequent attribute values.

Definition 15: A *Fuzzy_net* (Fn) is a triple: Fn=(X, F, Y). F is the fuzzy relation in the domain $X \times Y$ (Cartesian product of X and Y), $|X|$=m, $|Y|$=n, μ_{ij} (i=1,2,...,m, j=1,2,...,n), $-1 \leq \mu_{ij} \leq 1$. (1) it is established under the direction of dynamic rule and works in a connected way; (2) The strength of relationship stands for the subjective preference and is often adjusted by post evaluation.

A fuzzy net is denoted as:

$$Fn(X, F, Y)$$

If the relationship between attributes are non-linear, we can add an hidden attribute for the non-linear transition.

It is obvious that the fuzzy set is the connectionist representation of the production rule. Compared with rule, the advantage of fuzzy net is the ability to adjust the relative importance so as to obtain appropriate input/output mapping.

To realize the functional link, at the value level, the information transition between an information object and the approaches to obtain it are necessary. This relationship is called transfer_link.

Definition 17: A *Transfer_link* (Tl) is defined as the relationship between two values x and y iff it transfers information from x to y. It can be denoted as:

$$Tl(x, "\rightarrow", y),$$

Based on the above definition, greater knowledge unit, "knowledge block", can be described.

In DFKN, a value relationship can be represented by an edge connecting two value nodes.

2.4 Knowledge system

Selecting the best applicable design alternatives and updating the value of constraints can be achieved by considering all the conditions based on the experiential relations, referring to definite method through functional links. So as to describe these objects and relationship, the knowledge blocks are presented.

Definition 18: A *Knowledge Block* (KB) is defined as a 5-tuple: k_block = (*I, C, R, r, K*), *I* is the considered attribute on which we must make decision, called Core Attribute; C is the set of condition attributes and function attributes of I; R is the set of attribute relationships arriving to I; r is the set of value relationships between C and I; K is the set of identifiers of the objects which hold I and C.

It is notable that the KB is the elementary and independent reasoning unit in DFKN.

Figure 4 ullstrates two KBs: A, B:

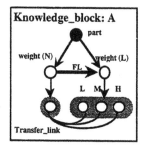

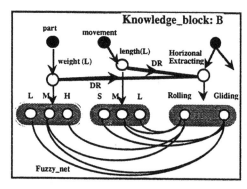

Fig 4. Two knowledge blocks

Knowledge block A llustrates the functional link between the numerical attribute, *weight*(N), and the linguistic attribute *weight*(L) which holds linguistic value: "L" ,"M" and " H". Through this link, the numeric value can be used to calculate the truth value of the linguistic values. Knowledge block B illustrates the experiential relationships between two linguistic attributes: *length*(L) and *weight*(L), and core attribute, *horizontal_extracting*. Through this relationships, the selection of the alternatives of *horizontal_extracting* can de decided based on the two conditions: *length and weight.*

A knowledge base consists of KBs.

Definition 19 A *Knowledge base* is defined as the set of all the knowledge blocks in the given domain.

$$\text{K\_base} = (KB \{, KB\})$$

By now, it has been shown that a knowledge base consisting by objects and relations can be described based on the fundamental structure of DFKN.

2.5 Representation of engineering design knowledge

In the field of engineering, the objects (functional requirements, criteria/constraints, design parameters, both experiential relationships and algorithm can be represented by DFKN.

The criteria and constraints can be described by N_Attr and L_Attr, for example:

Ex1: Part: ((Weight(N), '3') (Weight(L), (heavy/0.5, medium/0.6, light/0.0)).
Ex2: Part: ((Weight(N), ">3") (Weight(L), (heavy/1.0, medium/0.7, light/0.0)).
Ex3: Part Assembly: (adjustable(L), (easy/0.2, medium/0.8, difficult/0.0))

The functional, physical, behavioural knowledge can be described by structure attributes with a hierachy.

Ex4: Carriage:(function(stru), "Index", "Guide", "Positioner", "jointing"))

The design process knowledge are described by the knowledge blocks.

3 CONTROL AND REASONING MECHANISM

All the reasoning activities are supervised in the blackboard control architecture with a hierarchical structure, where the knowledge blocks are used as the knowledge sources. Independent knowledge blocks enables the designer freely to use the appropriate technology to arrive at the best solution and simplify the management of knowledge base. The reasoning in blackboard can be devided into the reasoning of attribute level and the computing of value level.

3.1 Reasoning at attribute level

Suppose that B is the set of considered attributes, I is the stack containing the attributes to be considered. $C(A)$ is the set of condition or function attributes of A. $T(A)$ means the type of A; Val(A): the set of val_nodes of A. R is set of attribute relationships arriving at I.

The reasoning algorithm is described below:

```
I=φ; B=φ;
Based on the user's inputs, some initial attribute
nodes (function and constraint) are fired and pushed
into I.
repeat {
3: choose a node A from I , move it to B;
   if T(A)=="Struc_Attr"
   then {push Val(A) into I. go to 3;}
   else{
       Search for a KB, G. whose core attribute is A.
       if failure
       then
         if A ∈ B goto 3;
         else Assigned_by_User();
       else {
           For every X ∈ C(A),
```

```
            if X ∉ B,
                push X into I; goto 3;
            else {
                for every I∈ R
                { if I is Dynamical_rule
                      Calcul_by_Dr();
                  if I is Functonal_link
                      Calcul_by_Fl();
                }
              }
            }
          }
        }
      until I==φ;
```

The function "*Calcul_by_Dr()*" and "Calcul_by_Fl()" are achieved by the functions at value level.

3.2 Computing in value level

One of the most important operation in this level is "*calcul_by_Fn()*" which calculates the degree of membership for every values of the core attribute. For every dynamical rule, suppose that the fuzzy net is $Fn(X, Y, F)$, we can use the following formula to calculate the degree of membership $\mu(y_j)$ $\varphi=1,2,..,n$.

$$\mu(y_j) = f(\sum_{i=1}^{m} x_i \mu_{ij}) \qquad (1)$$

$$f(x) = \frac{1}{1+e^{k(x-\phi)}} \qquad (2)$$

k, and φ are set up based on application.

Another function in value level is "*Get_by_tl()*" which gets the value from other value node.

4. LEARNING OF DFKN

On early stages of design, it is difficult to acquire the explicit knowledge rules, instead, easy to find some examples. Based on examples, the initial knowledge structure can be adjusted by following ways:

- Adjust the dynamical_rules;
- Adjust the fuzzy_nets;
- Adjust the membership function of fuzzy concept according to the different views of users.

4.1 Adjustment of dynamical rule

The initial DFKN is created by an expert using common and experiential knowledge. Based on design decisions and decision rational, some constraints and design alternatives are added into knowledge blocks, or some dynamical rules are reinforced or weakened. When a new dynamical rule is added, the corresponding fuzzy net is built based on design rational with the subjective initial value .

4.2 Learning of fuzzy net

Based on the rationale and on the final decision, the fuzzy nets can be adjusted using the incremental training algorithm of artificial neural networks, e.g. BP.

The following tables show the result of training fuzzy net in the knowledge block A in figure 2.4.. Table 1 shows four kinds of constraints conditions and the corresponding decision desired by experts and adapted from successful examples. Table 2 shows the initial value of the fuzzy net between constraints and alternatives. Table 3 and table 4 show the values of the trained fuzzy net with the different limitation: $1 \geq \mu \geq 0$ and $1 \geq \mu \geq -1$.

Table 1: Four conditions and decision examples

Criteria/Constraints		Design alternatives	
Weight(kg)	Length(mm)	Rolling	Gliding
5	100	0.50	0.50
3	450	0.70	0.30
5.5	90	0.45	0.55
4.5	110	0.60	0.40

Table 2: Initial fuzzy net between constraints and design alternatives

Criteria/Constraints		Design alternatives	
Attribute	Value(L)	Rolling	Gliding
Weight	L	0.000	0.300
	M	0.150	0.150
	H	0.300	0.000
Length	S	0.000	0.700
	M	0.350	0.350
	L	0.700	0.000

Table 3 The training results of fuzzy net between constraints and alternatives (BP, $1 \geq \mu \geq 0$)

Criteria/Constraints		Design alternatives	
Attribute	Value(L)	Rolling	Gliding
Weight	L	0.000	0.646
	M	0.222	0.001
	H	0.394	0.000
Length	S	0.000	0.419
	M	0.167	0.533
	L	1.00	0.000

Table 4 The training results of fuzzy net between constraints and alternatives (BP, $1 \geq \mu \geq -1$)

Criteria/Constraints		Design alternatives	
Attribute	Value(L)	Rolling	Gliding
Weight	L	-0.512	0.630
	M	0.162	-0.025
	H	0.895	-0.770
Length	S	-0.208	0.611
	M	0.330	0.524
	L	0.922	-0.067

Fuzzy_net with an hidden transition attribute can be viewed as a network with one level of hidden nodes and can also be trained by the artificial neural network algorithm (e.g. BP).

The result shows that the experiential relationships represented by dynamical rule and fuzzy net can be adjusted based on successful examples.

4.3 Adjustment to membership of fuzzy concept

Sometimes, the membership of a fuzzy variable requires to be adjusted. The algorithm for finding a curve or a line through given points can be applied for smoothing the membership functions of a fuzzy concept.

5. APPLICATION

Inspired from a French project SODA (Cocquebert, 1994) dedicated to analyse the experiential knowledge required to design assembly tooling system for the aerospace industry, figure 5 shows the different steps from a typical functional requirement, "*Horizontal extracting*" to the selection of a typical solution to this problem, "Carriage". In detail, in knowledge block A, the linguistic value of *weight* can be obtained by its numerical value through functional link; In knowledge block B, the technological solution, "*Rolling*" is chosen based on two constraints: weight and length. In knowledge block C, to satisfy the "*Rolling*" the technical solution "*carriage*" is chosen based on three constraints: *type*, *location*, and *quantity* of the positioning. In knowledge block D, the structure of "*carriage*" is obtained without any constraints.

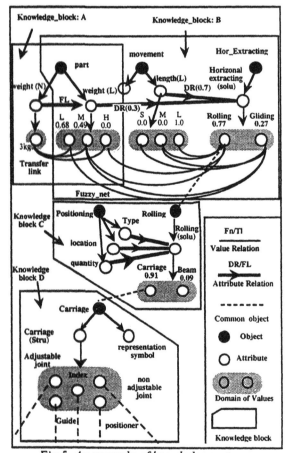

Fig 5. An example of knowledge system consisted of four KBs

The idea which is illustrated in the above figure is that the transition from one step of design to the following one can depend on an on-line estimation of the fuzzy value of design alternatives of an design object. This value can result from the global evaluation of a variety of other fuzzy concepts, representing the respective points of views of several actors. So, the decision of advancing to the next step is a concurrent decision, taking into account every actor's own requirement.

These actions globally tend to provide solutions to functional problems, based on formally (catalogue/standards) or informally (experience) known solutions. Based on the fact that some dynamical experiential relations or regularities exist between the constraints and design alternatives, we can construct the knowledge blocks, which includes two parts. The first one represents the constraints, while the second one represents the design objects and its alternatives. Between them exists a linkage network that connects constraints with alternatives. Depending on the customer's requirements and the actor's (one of the various designer's) experience, some constraint nodes can be fired. The best choice (the best compromise between different points of views) can be obtained through the linkage.

6. CONCLUSION

The DFKN is composed of knowledge blocks. In knowledge block, the uncertain, imprecise, fuzzy information of objects can be described effectively with the help of fuzzy theory, Both experiential and algorithmic relationships between the constraints and the design alternatives can be represented in a meaningful way at the attribute level and achieved by way of connectionist computing at the value level. Through adjusting the linkage of the value level with ANN training algorithm, the experiential knowledge can be effectively and efficiently captured and reused. This paper illustrates the use of a hybrid intelligence model to capture and utilise the engineering design knowledge. The experience gained applying the new method on the SODA project proved its ability to provide a significant support in representing and managing on innovative design knowledge. The main benefits of this approach are to formally relate the requirements with the suggested solution, to permit the evaluation and comparison of several candidate solutions, and to adjust the knowledge based on the user's choice.

REFERENCES

Wang, X.H., D.Deneux, R.Soenen, F Tong (1996), A Fuzzy Knowledge Model for Supporting Concurrent Design, pp.896-899, *CESA'96 IEEE-SMC. Computational Engineering in Systems Applications*, lille -France.

Cocquebert, E., C.Vat (1994): "Design of Assembly Systems and Tooling". *Final Report of the SODA Project* (MESR N° 92P0382), Dassault Aviation / University of Valenciennes.

ASSESSMENT TOOL FOR RECYCLING ORIENTED PRODUCT DESIGN AND DISASSEMBLY PROCESS PLANNING

Young-Kyu Kim, Sung-Woo Kweon*
Jürgen Hesselbach, Martin Kühn**

LG Electronics
Manufacturing Technology Center (MTC)
391-2, Ga Eum Jeong-Dong, Changwon City, Gyeong Nam, Korea

***Institute of Production Automation and Handling Technology*
Technical University Braunschweig
PO Box 3329, D - 38023 Braunschweig, Germany

Abstract: High disassembly costs limit an effective recycling of electrical and electronic appliances. Since these costs are fixed at the design stage and arise at the disassembly stage both fields need support in terms of a tool. This paper provides a close lock at a new developed method. The method measures the design performance and computes data for the planning stage by forming disassembly segments, predicting disassembly times and costs, determining the disassembly sequence and evaluating the product attributes such as product structure, joining techniques, materials, further utilization, segments and disassembly.

Keywords: Criterion functions - Design - Environmental coefficients - Evaluation - Sequences - Software tools.

1. INTRODUCTION

Due to several reasons the recycling of electrical and electronic appliances is necessary. The diminishing of natural resources, the shrinking of dumping grounds, the existence of toxic materials or the growing waste require to utilize parts, to refurbish materials and to dispose toxic materials of the products again. But the scale of recycling depends on the economic efficiency. Considering the product life cycle two stages are mainly responsible for the economic efficiency: The design stage and disassembly stage.

Although the costs for recycling and disassembly of a product are mainly determined at the design stage the products have not been designed to be easily and efficiently disassembled. Design engineers often are no specialists in disassembly and recycling requirements. Nevertheless their design must consider the ease of disassembly and recycling. For this purpose design engineers need a tool supporting and directing them to an environmentally oriented design.

The practical disassembly is faced with a lot of problems. A huge variety of different appliances are delivered to the disassembly companies. The appliances are dismantled in the order they are delivered without considering their differences associated with the disassembly operations. The differences of the appliances are related to the product kind, the manufacturer, the age, the model or size. For this reason the practical disassembly of one appliance differs from the disassembly of the other. Another problem tackles the lack of information about materials, joints and quality of the current parts. Due to these problems actually there is no real disassembly planning. For improvements in the planning phase data of the disassembly such as required times or optimal sequence are necessary (Subramani and Dewhurst, 1991).

A reduction of costs and times during the total life cycle of electrical and electronic devices requires the simultaneous working and intensive information exchange between design and planning stage. On the one hand designers need knowledge about the rationalization of the processes. Therefore information feedback from processes in an early design stage is the key to improve design quality. On the other hand to reduce the flow times and costs it is suitable to generate the disassembly instructions during the design stage and to communicate product information to the planning stage.

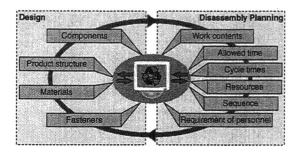

Fig. 1. Design and planning supported by ATROiD

To support both the design of electronic products and the disassembly process planning and to solve the arising problems LG Electronics - MTC and the Institute of Production Automation and Handling Technology have developed a software tool to assess the design of products and the disassembly process of electrical and electronic products (Fig. 1).

2. ASSESSMENT TOOL

The assessment tool embodies five different output modules. Based on a determination of the disassembly level the tool provides modules for predicting the time, costs and sequence of disassembly and another module to evaluate the product attributes related to disassembly and recycling. The results are obtained automatically from product information stored in a database. The method have been incorporated into a software program named ATROiD.

2.1 Disassembly Level

Disassembly differs from assembly in the number of operations. As several components or subassemblies can often be recycled in the same way there is no need to separate them. Therefore the objects of disassembly are not equal to objects of assembly. The starting point of all calculations is the determination of components or subassemblies that have to be disassembled. After the identification of the components suitable for disassembly and

recycling major units which are called segments are formed. Segments consist of one or several components which are joined with each other and which can be recycled in the same way (Hesselbach, et al., 1996). To determine the segments several criteria are considered:

- Compatibility of plastics: Functional reasons often require to use not only one homogeneous material. Because certain kinds of different materials are not suitable for a common recycling process the compatibility of these materials has to be reviewed. Components which cannot be recycled together due to the incompatibility of their materials have to be allocated to different segments.

- Valuable components: Components that can be reused in their virgin function form own segments.

- Economic advantage: Each disassembly operation is associated with additional costs. On the other hand with further disassembly degree the material fractions becoming purer and the obtained profits higher. So, costs and profits work in opposite direction and have to be analyzed for each component to find out the best disassembly level.

- Hazardous components: Components containing hazardous substances such as capacitors, picture tubes or batteries form segments and have to be dismantled to avoid a contamination of the environment.

- Priority: Often the formed segments cannot be dismantled because of obstacles and other components that have to be dismantled prior. So all determined segments have to be checked concerning components that block the access.

2.2 Disassembly Sequence

For disassembly tasks the best disassembly sequence of the product is found out. Based on a created model which includes geometrical and technological data of the product a mathematical algorithm based on two steps is applied. In the first step all feasible sequences are generated and represented in a sequence graph.

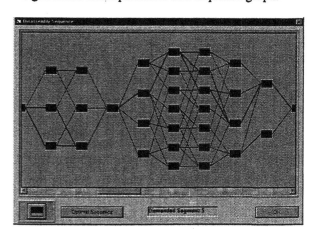

Fig. 2. Disassembly sequence

This graph states which operations can be done after other operations have been finished. Based on the theoretical possible sequences the second step involves a successively reduction according to optimization criteria which results in a best sequence. Fig. 2 depicts the sequence graph of a TV set. The lines of the graph represents disassembly operations, the boxes symbolized disassembly states.

2.3 Disassembly Time

The necessary time to perform the disassembly tasks is determined. For that purpose different disassembly methods are analyzed and the method which guarantees a fast separation of the joints is selected.
The method considers different phases of the disassembly task. The total disassembly process is subdivided into the following time phases:

- Handling Time includes the time of all motions required to grasp, orient and arrange the tool in the correct position for detaching the joining elements and components.
- Standard Time includes the time of motions required to separate the joints.
- Transition Time includes the time of motions to take off the previous joining element and to place the tool for the removing of the next joining element if more than one joining element has to be detached with the same tool.
- Taking Off Time includes the time of all motions to place the joining elements and components at a defined place.

For each phase of time the investigated procedure calculates the time on the base of the MTM method (Boks, *et al.*, 1996). The time needed depends on parameters which are specific to the disassembly task. Also, product attributes which significantly influence the time are considered. Essential parameters that impact the time are the kind of fasteners to join the parts and the tool which is chosen for disassembly. The predicted times are displayed alternatively for each component or segment in a bar graph.

2.4 Disassembly Costs

Direct costs are especially incurred by the disassembly process and the disposal of normal and toxic waste. Because of the huge variety of products caused by different sizes, models, variants and producers the disassembly operations are difficult to be automated and have to remain mostly manual. Therefore, labor costs are the largest costs driver

and are directly affected by disassembly time and the hourly labor costs.

Another kind of important costs are the fees to landfill segments of the product. These fees are calculated in dependence on the kind and amount of material.

2.5 Disassembly Evaluation

Besides the assessment of the process, the software tool evaluates the properties of the product as well. The most important aspects which significantly affect the ease of disassembly are taken into consideration. All in all about 29 different criteria evaluate the product. The criteria are divided into six groups of the same contents (Fig. 3):

- the materials of all components of the product,
- product structure,
- joining techniques,
- segments,
- disassembly operations and
- further utilization.

Scores are calculated for each criteria based on multiple-stage algorithms.

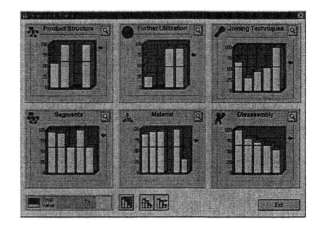

Fig. 3. Criteria values

2.6 Data Model

The integrated data model consists of two parts, a general and a product specific database. Both databases contain disassembly related information.

- Product database: Because at an early design stage of a new design no detailed information about the components are available the tool has to accept conceptual product information as evaluation inputs. The product model incorporate an abstraction of the part's characteristics. Only rough information about subassemblies, parts, their materials and fasteners are necessary. In the case of design changings it is possible to use existing product data for modyfing the changes.

- General database: Information of product attributes is necessary but not sufficient. The general database contains additional information such as economical data and technical requirements of further recycling steps.

3. PRODUCT DESIGN

The design tasks generally concerns the determination of product structure, the choice of materials as well as the selection of fasteners. All these decisions affect the disassembly and their environmental effects. To produce products with greater ease of disassembly and recycling the design stage can be supported by the assessment outputs of the developed tool.

The method is used to analyze disassembly behaviour during the design phase and helps designers assess the impact of their proposed design on disassembly difficulty. The tool provides a numerical rating of disassembly ease by calculating criteria values. These scores enable the design engineer to identify weak points. The tool is also valuable in helping to compare design alternatives. Different design alternatives can be compared quantitatively with each other and the best solution can be chosen.

Because the objective is to reduce disassembly times and costs the early selection of materials and fasteners, a product design should be based on relative economics. So, besides the criteria values the predicted disassembly times and costs are measures of difficulty. The higher the disassembly time, the more difficult the disassembly operation will be. Each fastener is evaluated for disassembly ease and assigned to a rating of difficulty corresponding to predicted disassembly times.

Once identified, the weak points of the product design can be improved using information tables and design guidelines. In this way the tool facilitates also the redesign (Hesselbach, *et al.*, 1997).

4. DISASSEMBLY PROCESS PLANNING

The disassembly planning comprises the preparation of the disassembly processes and facilities on basis of the product design. For the planning task relevant data are necessary. The ATROiD-tool enables to achieve the required information for the following planning tasks:

- The work contents for each model is established by calculating the necessary disassembly operations.

- Segments which are the objects to disassembly are formed.
- The entire work contents is splitted into working cycles to perform the disassembly task.
- The sequence of the working cycles is automatically generated.
- The allowed time for each working cycle is determined by the disassembly time module.
- The required resources, especially tools, are generated and assigned to the working cycles.
- Job contents are allocated to the disassembly stations.
- The compensation of cycle times for different disassembly stations and a capacitance balance is supported.
- The time determination allows the planning of the required personnel.

The generated informations that are important to perform the disassembly task can be summarized in a disassembly working plan (Fig. 4).

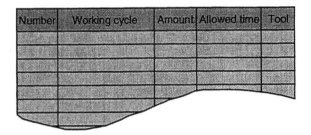

Fig. 4. Disassembly working plan

The disassembly working plan contains disassembly instructions such as a short description of each working cycle, the order, amount, allowed time and tools.

5. CONCLUSIONS

A useful and powerful tool in the development of electrical and electronic devices considering environmental aspects and the planning of the disassembly facilities has been developed. The tool monitors all parameters relevant to the design and planning task such as the level, time, costs, sequence of disassembly and the disassembly evaluation. In this way the tool provides an integrated framework to support both the design and the disassembly planning stage.

Applied at the design stage the evaluation method benefits the improvement of the design. The assessment of product attributes highlights potential problems with the product design. The design can be optimized in less time.

REFERENCES

Boks, C.B., W.C.J. Brouwers, E. Kroll and A.L.N. Stevels (1996). Disassembly Modeling: Two Applications to a Philips 21'' Television Set, *Proc. of the 1996 IEEE Int. Symposium on Electronis and the Environment*, pp. 224 - 229.

Hesselbach, J., M. Kühn, Y.-J. Kim and Y.-K. Kim (1996). Assessment of the Product and Process to Improve Disassembly and Recyclability for Design for Environment (DFE), *Proc. CARE INNOVATION '96*, Frankfurt.

Hesselbach, J., C. Herrmann and M. Kühn (1997). Eco-Potential as a Tool for Design for Environment, *Proc. 4th Int. Seminar on Life Cycle Engineering*, Berlin 1997.

Subramani, A.K. and P. Dewhurst (1991). Automatic Generation of Product Disassembly Sequences, *Annals of the CIRP*, **Vol. 40/1/1991**, pp. 115 - 118.

QUALITATIVE ANALYSIS OF QCD AND PPO PROPERTIES FOR REALIZATION OF CONCURRENT ENGINEERING

Hyowon Suh and Hyoungryul Sohn

Industrial Engineering
KAIST, KOREA

Abstract: Concurrent engineering is a systematic approach to manage the product development space for a QCD-effective product specification. The realization of concurrent engineering is not so simple because the product development space has a coupled heterogeneous information such as product, process, organizations, resources and others. To reach a QCD-effective product specification, how the properties of product(P), process(P), and organization (O) affects QCD must be analyzed. In this paper, the qualitative analysis of QCD and PPO is discussed, and the approaches to improve QCD in concurrent engineering are proposed. QCD is classified into the QCD of the development process and QCD of the manufacturing process. In PPO analysis, the relationship between product, process and organization is defined according to a QCD viewpoint. This analysis will be the basis for   quantitative or mathematical approaches of concurrent engineering.

Keywords: concurrent engineering, quality, cost, delivery, product-process-organization.

1. INTRODUCTION

Concurrent engineering is a systematic approach to manage the product development space for a QCD(quality, cost, delivery)-effective product specification (IDA 88). The realization of concurrent engineering is not so simple because the product development space has a coupled heterogeneous information such as product, process, organizations, resources and others. To reach a QCD-effective product specification, how the properties of product(P), process(P), and organization (O) affects the QCD must be analyzed. Products have various types according to the specification's complexity, processes mainly depend on a product complexity, and organization is involved by a process. Previous works focused on localized issues of concurrent engineering so that it is of no practical use (Andrew 1993, Cutkosky 1993, Gregory, etl, 1995, Hisayuki and Yosihisa 1995, Lawson 1994 , Kusiak 1990, McGuire, and Kuoldka, 1993, Suh, 1991).

To represent the product development space this complex information, the PPO model was proposed (DICE 1988). The proposed model discussed the framework of information management aspects of PPO model (figure 1). Here, we need to discuss the QCD of PPO because CE pursues the improvement of QCD values. To discuss the QCD of PPO, the role of PPO in manufacturing needs to be investigated. The definition of the manufacturing system is shown in figure 2 (DeGarmo, et al., 1988). By the system definition, manufacturing system is composed of machines, peoples and computers for performing a certain function or process to produce a product as an output. Here, we can define a PPO: product as an output, process as a function, organization as resources such as machines, computers and peoples. Thus, PPO is basic components for describing the manufacturing system including input/output. In addition, the function or process of a manufacturing system can be projected according to the time horizon such as the product development, manufacturing planing, manufacturing, sales and distribute, and field operation (figure 3).

2. PRODUCT, PROCESS, ORGANIZATION

Product: The information of products is the specification of a product from development, intermediate specification from planning and physical parts. Among these, the specification is more focused in concurrent engineering because it greatly influences the downstream processes. The specifications are generally represented by the geometry and technical attributes describing the manufacturing process aspects.

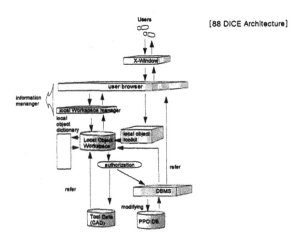

Figure 1. DICE Architecture

Figure 3. PPO in Manufacturing System

Process: Among the manufacturing processes, the product development is generally divided into 9 steps, which are strategic planning, functional requirement's analysis, basic design, engineering analysis, detail design, prototyping, manufacturing design, pilot test, and final production design. These steps have weighting according to the acquisition of product specification and product classification. The acquisition is classified into the copy of others and the original development, and the classification is the basic development, series development, model change, minor change, etc.. Thus, the process of development is typically classified as one of them.

Organization: The organization represents the operational group of resources according to their physical characteristics. The development organization is traditionally bounded to the development or design sectors in a company, but recently the organizations for CE include all the upstream and downstream sectors, related to all the product life-cycle issues.

Integration: The integration of product, process and organization for CE is necessary by the concept: "Entities of a *product* is generated by the *processes*, performed by resources at *organizational* sectors." In CE, the *product* specification at the development

process, influence the downstream manufacturing *processes* and *organizations* in general. Thus, as the information of product development increase, the relationship between products, processes and organizations becomes more complex. Thus, PPO information must be closely integrated each other. These PPO model and integration is the key technology of the product data management (PDM) system implementation (figure 4). Otherwise the information can not be managed so that the QCD will be out of control.

3. QUALITY, COST, DELIVERY

The QCD is the major goal of CE as well as manufacturing. Thus, the management of QCD is fundamental in operating the company and product development. The QCD of product development is not so simple because there are two aspects of QCD to be managed. One is QCD of the product development process, and the other is that of product specification to influence the QCD of downstream processes.

The quality is generally classified into a customer requirement, functional requirement, and fault-free requirements regardless of product, process and organization. The cost in manufacturing is divided into material cost, labor cost and overhead. The labor and overhead cost are from the process and the resources. In the product development process, the cost incurred is mainly the process cost. However, the material cost is sometimes incurred by making the prototype, which is generally expensive compared to the general manufacturing cost for a single part. In addition, the cost of

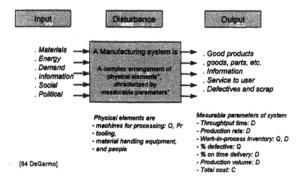

Figure 2. Definition of Manufacturing System

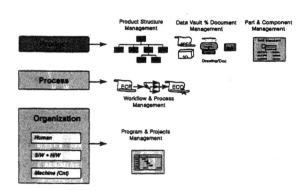

Figure 4. PPO and PDM Functions

manufacturing is greatly affected by the product specification generated in the product development. The delivery is the summation of time period of chained unit process for making products. The cycle time of unit process is composed of acquisition process, pure processing, supplementary processing and idle of resource. Among these, except the pure processing, all others need to be eliminated if possible. In addition, the pure processes are sequentially chained in the previous engineering. However, it can be concurrently chained with the aid of information sharing or process restructuring. By doing this, the overall delivery can be minimized.

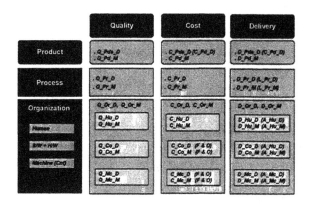

Figure 6. Parameters of QCD of PPO

4. QCD AND PPO

As shown in the figure 5, the product, process and organization have their own contribution to quality, cost and delivery respectively. In addition, the QCD of process and organization decides the QCD of product, the final outcome of manufacturing. The QCD of PPO is also represented as parameters for the analysis as shown in figure 6, and their relationships are also characterized as in figure 7.

Quality of Product (Q_Pd): Among the qualities of product, process and organization, the quality of product is the final objective to be obtained. In product development, the quality of product is classified into the customer requirements, specification for requirements, and the product specification for downstream processes. In CE, to improve the quality of product, deployment of the customer requirements over the development process (QFD), and evaluations of the effectiveness of the product specification over the downstream processes (DFX) are necessary.

Quality of Process (Q_Pr): The quality of process is the origin of the quality of product (Q_Pd). This Q_Pr is the results of the quality of organizations such as engineers, application software, system hardware, machines, etc.. The other effects on the quality of processes are the structuring or interfacing of processes and their resources for the dedicated functions. Sometimes, there are many differences

between the output of the predecessors and the input of the successors, which causes the serious quality problems. As the ISO9000 series, the documentation of each process is a measure of Q_Pr. Another measure is a rate of error of communication between tandem processes. To improve this Q_Pr, firstly, the quality of organization themselves should be retained. Secondly, the function-oriented aspects of the organization needs to be well exploited for the dedicated process, and finally, the interface between the processes and resources respectively must be well structured.

Quality of Organization (Q_Or): The quality of organization is the capabilities and the failure rate of resources such as engineers, software, hardware, machines, etc.. The quality of resources is similar to the quality of product, but this product is for producing another products. The quality of engineers has somewhat different aspects. Their quality is aptitude to given functions, level of education, level of experience, level of training and level of comprehension of functions. The Q_Or brings delay of process such that breakdown of computer causes the delay of engineering processes. The practical quality is rather difference from the planned level, which is unexpected results and hard to manage. For the proper management of Q_Or, the capabilities of development engineers are consistently evaluated and they are re-trained for a required job.

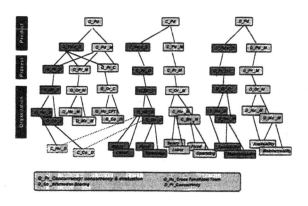

Figure 7. Relationships of QCD of PP

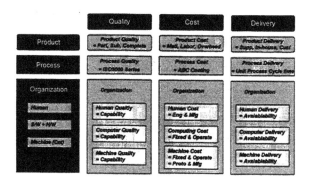

Figure 5. QCD vs PPO

The improvement approach for Q_Or is the preventive maintenance. The preventive maintenance of engineers is education and training, and that of software, hardware, and machines is regular check to prevent breakdown or level down of the capability level.

Cost of Product (C_Pd): The cost of product is the total cost of product-related activities and materials. These can be classified into the cost of sales product, cost of completed products, and product development cost. These costs also include all the cost incurred by quality problems, delay problems and other problems, and is one of the key value of management in enterprise level. The product development cost is major one to be considered in CE, but the manufacturing cost is more important in CE because the major manufacturing cost is decided during product development. Among the manufacturing costs, the direct material cost is major cost of C_Pd because other costs such as process or overhead cost can be considered as organization or resource cost. To improve the C_Pd, the geometry design, the material selection, manufacturing process selection and other decisions needs to performed integratedly and concurrently. This will minimize the re-work or iteration of engineering cycle.

Cost of Process (C_Pr): The cost of process can be referred to the accumulated cost of unit process including the material, operating and organization costs. However, the pure process cost is the operating cost except the resource-related costs. It is generally the in-direct or overhead cost in manufacturing. These costs are hard to trace so that the activity-based costing (ABC) is strongly recommended (Horngren 1997). In managing the process cost, the proper allocation of the resources to the required functions is important because relationship between the job and resource capabilities is critical. The cost of engineers are different according to the position, experience, or technologies they have. To improve these costs, to assign a right engineer to a right function is necessary. Moreover, it needs to make the high-cost engineers carry out the creative functions resulting in high value, and avoid idle time or non-value-added functions.

Cost of Organization (C_Or): The cost of organization is that of resources belonging to the organizations, and is largely fixed cost and operating cost. These costs for manufacturing are relatively high, but, for the engineering process, is not so high because most of them are for computing environments and engineers. However, this cost is not negligible because the computing hardware and application software are not so cheap. The fixed cost can be deployed according to the specified period with depletion. These costs are tends to be calculated by the purchasing prices, which is hard to be allocated to the each process or products. Thus, ABC is also recommended for this. To improve this cost, low investment and operating cost of resources is recommended. But the practical recommendation is to assign the right resources to the right function at the right time, so that the maximum effectiveness of the resources can be obtained.

Delivery of Product (D_Pd): The delivery of product is the period of the generation of physical entities related to the products. The major deliveries are those of out-making parts, out-buying parts, factory-making parts, packaging-products, etc. The delivery of product to a customer is the most important. In CE, the delivery of product specification is considered as the delivery of a product. Another focus on delivery of CE is that of product in manufacturing (D_Pd_M). Another product-related delivery is the delivery of planning. The delivery of product specification and planning is synchronized with the delivery of process. To improve the deliveries, we should concentrate on the two aspects: one is to manage the process directly related to products and the other is to eliminate the obstacles to hold the product or product-related information without any progress in processes.

Delivery of Process (D_Pr): The delivery of process is the period of each unit process. There are mainly two processes. One is the necessarily chained process for a product, which is main process, and the other is supplementary process for a development or manufacturing. All the unit processes are chained sequentially or concurrently for a complete process. The main process has synchronizing nodes with the delivery of product (D_Pd), either product-related information or physical products. Which means the delivery of process (D_Pr) directly influences the delivery of product (D_Pd), except the D_Pd influenced by the outside effects. In CE, the delivery of process has synchronizing nodes with the delivery of product-related information. The measure of D_Pr is somewhat difficult without tangible materials or specified information. To improve this delivery, several approaches are proposed such as segmentation, compression, paralleling, deletion, merging, etc.. The deliveries of processes can be continuously managed while that of products is sometimes not-controllable.

Delivery of Organization (D_Or): The delivery of organization is rather considered as the availability of resource, which mainly decides the delivery of process (D_Pr). There are two categories of this, the delivery of newly equipped resources and that of existing ones. The availability of human resources is simply the portion of available or not. However, the availability of existing systems is the portion of available, not available but being repaired, and not available as being breakdown. Thus, according to the availability, the approach for improving QCD should be different, and it relies upon the cost of organization (C_Or). For CE, the deliveries of computing environments and engineers are critical. In addition, the deliveries of the prototyping resources are also important. To improve this, the preventive maintenance is fundamental, so that the breakdown of resources can be preventive and the supporting materials can be reserved.

5. CONCLUSIONS

In this paper, the qualitative analysis of QCD and of PPO is discussed, and the QCD is represented as parameters. Firstly, the necessity of PPO definition for manufacturing system is discussed, and secondly, the characteristics of quality(Q), cost(C) and delivery(D) of product(P), process(P), and organization(O) respectively are also analyzed. Finally, the relationships between them are characterized for CE. Thus, this study supplies the basis for managing the product development and its performance, QCD, as well as navigating the relationship of PPO as shown in figure 8 and figure 9. This QCD and PPO analysis is also the key information for implementation of product data management (PDM) system. The QCD of products is mainly the result of that of processes, and the QCD of processes is also that of organizations. Thus, the information chain of QCD, either static or dynamic, needs to be stored as well as QCD itself. The qualitative analysis of QCD will be the basis of the quantitative analysis. As the further research, the functional relationship of QCD needs to be defined along with the interactions between PPO.

REFERENCES

Andrew. (1993). Design Data Storage and Extraction Using Objects, *Concurrent Engineering: Research and Application*, Vol. 1, No 1, pp. 32-38.

Cutkosky, M. R. and Engelmore, R. S. (1993). "PACT : An Experiment in Integrating Concurrent Engineering Systems", COMPUTER.

DeGarmo E. Paul, Black J Temple and Kohser Ronald A. 1988, *Materials and Processes in Manufacturing*, Macmillian Publishing Company.

DICE Sigarch. (1988). *Red Book of Functional Specifications for the DICE Architecture*, DICE Working Draft.

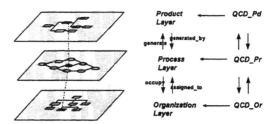

Figure 8. PPO Relationships

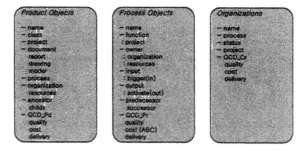

$QCD\_Pd = f (QCD\_Pr, QCD\_Or), QCD\_Pr + g (QCD\_Or)$

Figure 9. PPO Objects

Gregory R. Olsen, *etl*. (1995). Collaborative Engineering Based on Knowledge Sharing Agreements, *Concurrent Engineering: Research and Applications*, Vol. 3, No. 2, pp. 145-160.

Hisayuki Masui and Yosihisa Udagawa. (1995). A Study of Work Flow Model and Its Application to Concurrent Engineering, *Concurrent Engineering: Research and Applications*, Vol. 3, No. 2, pp. 81-92.

Horngren, Charles T. H, Foster George, and Datar, Srikant, M., (1997), *Cost Accounting*, Prentice Hall International, Inc.

IDA Report (1988). The Role of Concurrent Engineering in Weapon System Acquisition, *Institute for Defense Analyses*, Alexandria, Virginia, Report R-338.

Lawson, M. (1994). A Survey of Concurrent Engineering, *Concurrent Engineering : Research and Application*, Vol. 2, No. 1, pp. 1-6.

Kusiak, (1990), Concurrent engineering: Decomposition and Scheduling of Design Activities, *Int. J. Prod. Res.* Vol. 28, No. 10, pp. 1883-1900.

Mc Guire, J. G. and Kuoldka, D. R. (1993). "SHADE : Technology for Knowledge-based Collaborative Engineering", *Concurrent Engineering : Research and Application*, Vol. 1, No. 3, pp. 137-146

Suh, hyowon. (1991) Feature generation for Concurrent Engineering Environment, PhD.Thesis, West Virginia University, Morgantown, USA,

Multi-disciplinary Team Cooperative Work Based on Contract Net in Concurrent Engineering Environment[*]

Qian Zhong   Yingping Zheng

The Institute of Automation, Chinese Academy of Sciences
Beijing, 100080,P.R.China

Abstract: How to facilitate cooperation of multi-disciplinary teams is a key problem in concurrent engineering(CE) environment. In this paper, we discuss this problem from non-technology enabling sub-system aspects. A kind of task coordination architecture based on contract net is proposed to facilitate the cooperation of multi-disciplinary teams. The major elements of this paper include decomposition and allocation of tasks, organizational structure, task coordination relationship, mechanism of organization and management and communication etc..

Keywords: Concurrent engineering (CE), Cooperation, Coordination, Distributed artificial intelligence (DAI)

1. Introduction

The complexity and creativity associated with engineering design requires diverse problem solving techniques and the integrated contribution of many individuals distributed by geographical and functional perspectives. Thus cross-functional, multi-disciplinary product development team may be the most appropriate and feasible organizational structure for manufacturing companies to implement concurrent engineering (CE), which has been proved by some industrial practices.

Effective coordination and cooperation of team members is critical to the success of product development process since the distributed activities of the members are typically highly interdependent e.g. due to shared resources, input-output relationships and so on. There have been some works of coordination technology currently available for cooperative design support, such as process management, conflict management and memory management. Klein has proposed an integrated approach to cooperative design coordination---iDCSS. Full details have been provided in the

reference (Klein 1994). All these works are focused on the technology enabling sub-system, but ignore the non-technology enabling sub-system e.g. organizational structure, management strategies and cooperative mechanism etc.. In 《 21st Century Manufacturing Enterprise Strategy 》 (1992), it has been proposed that non-technology enabling sub-system is of same importance as technology enabling one.

In this paper, we discuss the cooperative work support from non-technology enabling sub-system aspects. As organizations shift to a team-based structure, some new organization design methods have been proposed in reference, such as matrix method(Adachi 1994), case-based method (Carley & Lin 1994) and so on. These method can suit the high requirements of CE, but they lack sufficient support to facilitate cooperation. There are many interesting challenges we face to facilitate the cooperation of team members (W.-L.Le. et.al.1994). The four problems mentioned below are most important. The first one is to choose and model the team-based structure problem involves the distribution of the global organizational objectives to teams, it is an

---

[*] Supported by 863 High-tech Project on CIMS and National Natural Science Foundation in China

important point for the modeling of cooperation. The third problem is associated with the decision behaviors in team-based organizations, because what's best for the team may not be what's best for the organization. The fourth problem is about communication. Cooperation would require communication. A cooperative system would presumably have to pay a high price in communication, and it would presumably be inefficient compared to a non-cooperation system for the same kinds of problems (in computer system communication usually requires more time than computation does). So, how to keep down the time and cost of communication is a key problem in cooperation.

The main purpose of this paper is to propose a kind of task coordination architecture concerned such problems mentioned above to facilitate cooperation of teams in CE environment. This architecture inspired by distributed artificial intelligence (DAI) techniques is established based on contract net and some mechanism.

The paper is organized in 6 sections. In section 2, following the introduction, we shortly review contract net system. Section 3 presents how to establish this architecture, major elements of it include decomposition and allocation of tasks, organizational structure and task coordination relationships. In section 4, some mechanism for organization and management are proposed. Section 5 discusses the communication problem. Finally, section 6 is the conclusion.

2. Contract net system

The main reason for current interests in DAI techniques is the increasing complexity and creativity associated with the problem which requires diverse problem solving techniques and cooperative endeavor carried out by multi-expertise. Decker(1987) provides a very usefully review of DAI techniques, the taxonomy has four dimensions:

1) The level of decomposition(granularity);
2) The distribution of expertise;
3) The methods for achieving distributed control;
4) The process of communication.

Smith and Davis (1983) discuss cooperation as a method to use when each agent has different knowledge (artificial intelligence systems are often built from so-called cooperating experts, where each expert has some knowledge for its particular domain).They advocate the use of cooperative frameworks that minimize communication, allow load balancing, and distribute control, while also maintaining coherent behavior. As a solution, they propose a *contract net*, where agents negotiate about which should do what. When the problem is too large, the agent partitions it, and requests bids from other agents in the system to perform the parts. When the

agent lacks expertise, the whole problem is offered to others. Other agents who have the expertise needed and do not currently engage in processing their own problems may make bids to solve the problems offered. The original agent then selects one or more from the bids offered, contracts with them and assumes control for solution the task.

This kind of architecture seems to be useful both as an approximate description of many human distributed systems and as a normative model for such systems. And the communication costs of this structure may be lower than in many others. So, we can explore it to facilitate the cooperation and coordination of cross-functional multi-disciplinary teams or members.

3. Task coordination architecture

In multi-disciplinary team, the distributed activities of each members are highly inter-dependent by very complex relationships among them, it is easy to make conflicts and disorder. Effective management of these activities is the key point to facilitate cooperative work of teams. Bailetti et.al.(1994) proposed a coordination structure approach to the management of projects. This approach needs to be extended and fined in order to solve the problem mentioned above more efficiently.

A feasible solution to facilitate cooperation is to establish a kind of task coordination architecture that can represent the task relationships among teams or members clearly. This architecture should have such functions or abilities as follows:

1) describing the inter-dependent relationships among teams clearly, and providing efficient management to these relationships.

2) reconfiguring itself as problems arise and change to deal with the uncertain tasks.

3) providing convenient communication protocol to keep down the communication costs and save the communication time.

As a solution, this task coordination architecture should be task driven and human centered, the process to establish this architecture should be based on negotiation. Contract net is not only a problem-solving protocol but also communication protocol, it is most suitable to be used to establish the task coordination architecture. There are four steps to establish this architecture:

Step 1:   Decompose and allocate a task;
Step 2:   Select an organizational structure;
Step 3:   Determinate task coordination architecture;
Step 4:   Manage and schedule of tasks

We give details in following sections.

3.1 Decomposition and allocation of tasks

For the complexity of product development problem and the limitation of recognition and computation abilities of human being, the first step to establish the task coordination architecture is the decomposition and allocation of tasks to teams. We use contract net protocol to partition the global organizational tasks.

In this architecture, the product development problem is called a global organizational task, the agent responsible for this task is called "a manager". He partitions the global task into several tasks and makes a *task announcement* (usually by broadcasting to the entire system) that contains eligibility requirements and task descriptions. Agents that have ability and resources to deal with this task may return *bids* for the contract, which include the reason and his solution. The manager then evaluates the bids and makes an *award* based on them. Then he selects one or more from the bids offered to be "a contractor". The manager and contractors are linked by the contract, thus the task relationships between a manager and a contractor is determined too. If a contractor can not complete the task by himself, he will subdivide the problem to other sub-contractors, then he is also a sub-manager. The decomposition and allocation process will not end until all sub-tasks can be solved, so it is an iterative process and can be illustrated by a task tree. The root node indicates the global organizational task, the leaf nodes indicate tasks, and the lower level nodes of one leaf node indicate the sub-tasks, and so on. It is illustrated in figure 1.

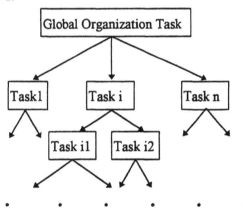

Figure 1   Task relationships tree

3.2   *Organizational Structure*

Task coordination architecture needs some form of organization to solve the task and guide the communication of goals and results. The organizational structure in this architecture is a team-based structure. There are three kinds of team in the organization to implement CE, all are task driven. The first is called "a general team", which is formed by manager, engineers and perspectives, and the manager is the leader. The main objective of this team is to manage the global organizational task, resolve design conflicts, evaluate design alternatives

and so on. Thus a perfect and successful design plan could be achieved. The second   is called "a global team", which is formed by sub-managers, who are also contractors of upper lever. These teams are mainly responsible for task coordination. Each member in global team not only has his own task, but also has a common goal determined by contract previously.   The third kind of team is "local team", all members of this team have a common goal which is also the goal of the team.

The organizational structure of this system is a kind of hierarchical structure. General team seems to be the manager of the global organizational tasks, global teams seem to be task coordinators, and local teams seem to be task processors. Thus it can be reorganized when new problems arise or global organizational tasks change. Decker(1987) calls this a dynamically opportunistic organization. This seems to be the form that most human organizations actually take, and has considerable possibilities for self-organization and taking advantage of new opportunities. The organizational and management mechanisms to form these teams are given in next section.

3.3 Task coordination relationships

While the process of decomposition and allocation of tasks is finished, the task coordination architecture is established automatically, and the task coordination relationships among teams are determined automatically by the contracts. Thus the task relationships among general team, global team and local team can be represented clearly according to the relationships between task and sub-task, their relationships are determined by the process of offering-bidding-contracting.

Contract net is a task-decomposition protocol based on negotiation, so the task relationship among teams can be decided by negotiation process. At the same time, the relationship of sharing limited resources is determined too. If a kind of consumed resource is available by one agent, then he can not contract with two managers who need this kind of resource, which can avoid resource conflicts.

Therefore, the task coordination architecture based on contract net gives a clearly description of inter-dependent relationships among teams and solves the problem of decomposition and scheduling of design tasks. And this kind of architecture can not only dynamically decompose problems because they are designed to support task allocation, but also provide dynamically opportunistic control, which makes it possible for the system to reconfigure itself as problems arise or situation changes.

The activity being modeled by contract net is *task sharing*, where agents help each other by sharing the computation involved in the sub-tasks for a problem. Another feature of this architecture is that it can be task driven while the net is lightly loaded and driven by the availability of nodes when it is heavily loaded. This allows the system to adapt to conditions surrounding it by smoothly flowing between being task bound and availability bound.

Therefore, there are four characteristics of this kind of task coordination architecture:

1) task driven and human centered

2) emphasizing logic dependent relationships but not time order relationships

3) resolving resource conflicts concurrently while assigning tasks to contractors

4) providing dynamically task distribution and dynamically opportunistic control based on negotiation.

3.4 Management and scheduling of tasks

The global organizational task can be partitioned into so many tasks and sub-tasks (in design process, they are called design activities), task implementation needs some time and resources. When task coordination relationships among teams are determined, it is necessary to manage and schedule of these task to minimize the total task processing time, which can be realized by planning the tasks, assigning the limited resources and optimizing the flow of information in a global view.

For the new characteristics of tasks in CE, traditional project management techniques have limitations to be used in CE, we propose a feasible and systematic method to manage and schedule of tasks. The main issues include extended A-O-N method to establish the network of tasks, more feasible assessment of task time distributions, formulation of scheduling problem and some heuristic scheduling rules. The full details are provided in other papers (Zhong & Zheng 1997 a, 1997b).

4 Mechanism for organization and management

Cooperative work system needs some effective mechanism for organization and management to stimulate the cooperativeness, activeness and creativeness of team members. We can explore incentive control (Zheng et.al. 1984) method to make such mechanism.

4.1 Mechanism for organization

When the organization is set up and a team is formed, an individual who wants to be a member of the organization should obey following organization mechanism:

1) The benefits of an individual depend on the team's benefit, everyone should share interests, risks and responsibility.

2) Decision is made due to the preference of global goals, but not individual's preference.

3) Individual should sacrifice personal goal to achieve an optimal solution of the global goal.

4) Each member can not take the place of other member.

5) The relationship among members is cooperative but not competitive and antagonistic.

4.2 Mechanism for management

There are some difficulties for cooperation of multi-disciplinary team for their different professional knowledge and terminology. To overcome these difficulties and facilitate cooperative work among them, the organization needs some management mechanism, which is called "control structure" in DAI techniques. Thus the mechanism for management should consist of :

1) the method to determine the organizational structure and individual's position

2) the rule of task decomposition, task announcement and bids.

3) the approach of evaluation of design results and work progress

4) the rule of reward and punishment

5) the communication paradigm and protocol to remove the common communication barriers and reduce communication cost

6) the structure for dynamically opportunistic control and self-organization.

For an organization cannot foresee all the problems that it will encounter, nor all the resources that it will command, it must rely upon self-organization to solve its task. A critical problem that the manager should consider is how to provide dynamically opportunistic control to organization. Thus point 6 can not be ignored.

5. Communication

Communication is the cement of the organization, and the greater the need for coordination and cooperation, the greater the necessity for communication. However, communication requires more resources, more time and higher costs, it is desirable to keep communication at a minimum.

The discussions of communication problem in this paper can be delineated into three areas : paradigm, content and protocol.

5.1 Paradigm

The paradigm by which communication takes place in cooperation system is either shared global memory, message passing, or some combination of the two. In task coordination architecture, both shared global memory and message passing communication are used in a team. And message passing communication is used between general team, global teams and local teams. The communication through shared global memory is realized by a hierarchical network of blackboards (as same as hierarchical organization), and the communication through message passing is realized by e-mail. Thus all members in the organization can communicate each other synchronously and asynchronously.

5.2 Content

A difficult aspect of communication is the ability to deal with uncertain and incomplete information. In this architecture, we use "the functionally accurate cooperation approach" to tolerate inconsistency and to let network function. To do so, the system should make it possible to exchange partial solutions, so that one or more agents have enough information to resolve the inconsistencies.

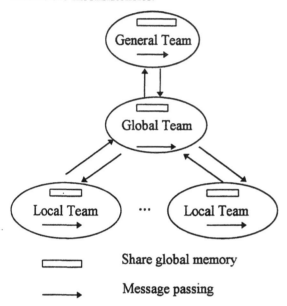

Share global memory

Message passing

Figure 2. illustrates the communication relationships among multi-disciplinary teams.

5.3 Protocol

The protocol of communication in this architecture is contract net, which is not only a communication protocol but also a problem-solving protocol. Such protocol can reduce the communication costs.

The communication among teams is illustrated in figure 2.

6. Conclusion

The task coordination architecture based on contract net can facilitate the cooperative work of multi-disciplinary teams from following aspects:

1) Decomposition and allocation of tasks to teams effectively
2) Description the task and resource relationships among teams clearly
3) Reconfiguring and self-organization
4) Coordination and cooperation
5) Responsibility sharing, risk sharing and interests sharing
6) Efficient communication and low cost.

A prototype of such cooperation system have been implemented by using Lotus Notes and Visual C++ on Windows NT LAN network.

References

Adachi, T., Shih, L. & Enkawa, T.,(1994), " Strategy for supporting organization and structuring of development teams in concurrent engineering", The international Journal of Human Factors in Manufacturing, 4(2), 101-120

Bailetti, a., Callahan, J. & Dipietro, P.,(1994), "A coordination structure approach to the management of projects", IEEE Trans. on Engineering Management, 41(4), 393-403

Carley, K. & Lin, Z., (1995), "Organizational Design Suited to High Performance Under Stress", IEEE Trans. on System, Man and Cybernetics, 25(2), 221-230

Davis, R. & Smith, R.G., (1983), "Negotiation as a metaphor for Distributed Problem Solving", Artificial Intelligence, 20, 63-109.

Decker,K.S., (1987), " Distributed problem-solving Techniques: a Survey", IEEE Trans. on Systems, Man, and Cybernetics, SMC-17, 729-740.

Iacocca Institute, 1992, "21 Century Manufacturing Enterprise Strategy", Legigh University.

Klein, M. (1994), "iDCSS: Integrated Workflow, Conflict and Rational-based Concurrent Engineering Coordination Technologies", International Conference on Concurrent Engineering: Research and Applications

Pan,Y., Zhong,Q. & Zheng,Y.P., (1996), "Multi-disciplinary Team Decision-making Mode and Group Design Support in Concurrent Engineering", accepted by System Engineering Theory and Application (Chinese edition)

W.-L.Le. et.al., (1994), "Assignment of Objectives and Incentive Systems in Team-based Organizations", Proceedings of 2nd International Conference on Concurrent Engineering: Research and Applications.

Zheng,Y.P., Basar,T. & Cruz, Jr., J.B., (1984), " Stackelberg Strategies and Incentives in Multiperson Deterministic Decision Problem", IEEE Trans System, Man & Cybernetics, 14(1): 10-24.

Zhong, Q. & Zheng, Y.P., (1997a), " Management and scheduling of design activities in concurrent engineering environment", accepted by International conference on Computer Integrated Manufacturing , Singapore, Oct.1997.

Zhong, Q. & Zheng, Y.P., (1997b), "A Heuristic Policy for Scheduling of Concurrent Design Activities", Studies in Informatics and Control Journal, Vol. 6, No.3, Sept. 1997.

SCHEDULING METHODS FOR LOT PRODUCTION IN MULTIVOLUME JIT PRODUCTION SYSTEMS

Jae Kyu YOO * Itsuo HATONO * Shinji TOMIYAMA *
Hiroyuki TAMURA *

*Department of Systems and Human Science,
Graduate School of Engineering Science, Osaka University,
Toyonaka, Osaka 560, JAPAN
Internet: yoo@tamlab.sys.es.osaka-u.ac.jp*

Abstract. This paper deals with scheduling methods in JIT Production that includes lot processes. In general, the delays often occurs in multivolume JIT production systems that include lot processes, because set-up time is increased in the multivolume JIT production. To cope with the difficulty, we propose three scheduling methods to decrease delays and the work-in-process-inventory in lot processes. Furthermore, we evaluated the methods by computer simulations.

Keywords. Multivolume JIT production, Scheduling, Lot production, Signal Kanban

1. INTRODUCTION

In many cases, there exist the processes such as press processes in the production systems whose setup times for changing job types are very long. In this paper, we call such processes *lot processes* whose setup times are very long and in which many jobs are processed once until the setup are changed for the other job types. However, the number of setups of the lot processes increase, if the lot processes are included in multivolume JIT(Just-in-time) production systems(Monden 1983). This is because the volume and the operation start time of each jobs are determined by *kanbans* from the subsequent process which is not a lot process. Therefore, utilizations of the lot processes become small and the jobs are often delayed. In conventional JIT production systems, managers try to improve the utilizations by *kaizen*(Kotani 1983). However it is not easy to improve, because they cannot always reduce the setup times dramatically.

In this paper, first, to cope with the difficulties, we pro-
pose three schedule methods to generate the processing order of jobs in a lot process. Finally, by using the three scheduling methods, we try to improve the utilizations and to decrease the delay of jobs in the multivolume JIT production system including lot processes.

2. JIT PRODUCTION SYSTEM INCLUDING LOT PROCESSES

In this paper, consider a JIT production system including a press process which is a kind of lot processes. **Fig. 1** shows the flow of jobs, parts, and information such as production orders in the JIT production system. In the production system shown in Fig. 1, first, the jobs are processed using the parts stored in each process. If the number of the stored parts of the first process in the body line decreases to the appropriate value, the process requires the appropriate number of parts from the buffer of the press process using *signal kanbans*. After

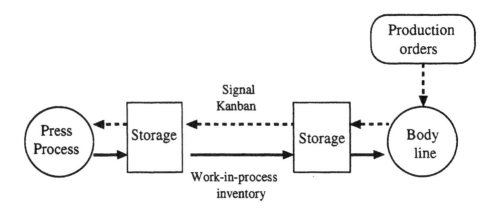

Fig. 1. Flow of production order and work-in-process inventory in the JIT production system.

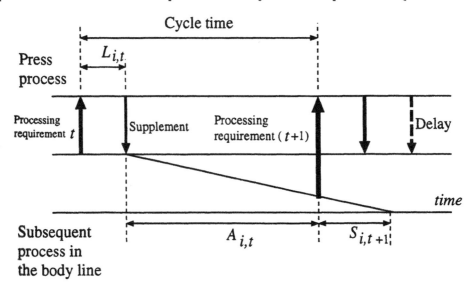

Fig. 2. Graphical explanation of control parameters of JIT production systems.

the required number of jobs are conveyed to the process, the press process produces the parts whose number is as same as that of jobs that has conveyed to the first process in the body line.

In this paper, we model the JIT production system shown in **Fig. 1** to describe the scheduling algorithms in lot processes. In modeling the JIT production system, we consider the press process and the subsequent process of the press process, which is a first process of the body line, because the other processes in the body line are not related with the press process directly in JIT production systems. To model the JIT production system shown in Fig. 1, we define several control parameters. **Fig. 2** shows the graphical explanation of each parameter(Hatono *et al.* 1996).

In Fig. 2, job i for processing requirement t are executed in the press process. We define the processing time of job i for processing requirement t is $L_{i,t}$. Where i and t mean

a job number and a sequential number assigned in order of the arrival time of each processing requirement from the subsequent process, respectively. The first process in the body line processes job i using the parts conveyed from the press process. We assume the next processing requirement $t + 1$ is sent in $A_{i,t}$. Furthermore, we assume the first process in the body line spent all parts for job i in $S_{i,t+1}$ since the process sent the processing requirement $t + 1$ to the press process.

The arrival time of each processing requirement from the subsequent process is Poisson-distributed, if $A_{i,t}$ for each i and t is different each other and the variety of the jobs are sufficiently large(Kotani 1983). Therefore, we assume the arrival time of each processing requirement Poisson-distributed in the JIT production system shown in Fig. 2.

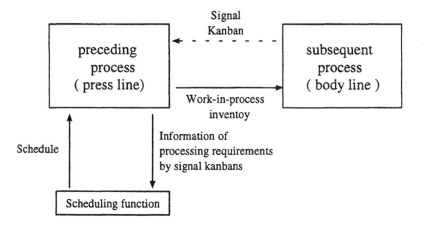

Fig. 3. System configuration of proposed scheduling methods.

3. SCHEDULING METHODS FOR LOT PROCESSES

In conventional JIT production systems, jobs are processed according to the processing requirement using kanbans. However, the number of setups in the processes becomes too large in multivolume JIT production systems including the lot processes. To cope with the difficulty, it seems that it is necessary to introduce a scheduling mechanism to the lot processes in addition to the processing requirements using kanbans (**Fig. 3**). In this paper, we propose three scheduling methods for lot processes. In the following, we describe the outline of the scheduling methods.

3.1 Method 1: Scheduling method in order to avoid delays

If $L_{i,t}$ for the production requirement i is greater than $S_{i,t}$, job i delays. Therefore, to prevent the delay of the jobs, it is necessary to generate the processing orders in order to keep the condition $L_{i,t} \leq S_{i,t}$. However it is difficult to obtain the processing orders to keep the condition $L_{i,t} \leq S_{i,t}$, because the processing requirements arrive at random in general. In this paper, we propose the heuristic scheduling algorithm in order to keep the condition if possible.

[Scheduling algorithm of Method 1]

Step 1. If the number of the processing requirements that arrived while job i is processing is 1, the processing requirement is processed next to job i.

Step 2. If the number of the processing requirements that arrived while job i is processing is greater than 1, the processing requirement whose $S_{i,t}$ and $L_{i,t}$ are smallest is processed next to job i. Goto Step 1.

3.2 Method 2: Scheduling method in order to increase throughput

In method 1, since only the jobs whose processing requirements are arrived at the press process, it seems that it is difficult to increase throughput of jobs. To cope with the difficulties, the jobs without the production requirements are processed during the time between the operation end time of the last job and the arrival time of the next production requirement job in method 2. The jobs to be processed are selected using the algorithms as follows:

[Scheduling algorithm of method 2]

Step 1. When the press machine is idle, generate a *processing jobs list*, in which the estimated processing requirements that will arrive in the near future are registered in order each estimated arrival time of the processing requirement.

Step 2. Process the first processing requirement in the processing jobs list. If processing requirements has arrived when the processing is finished, process each job corresponding to the processing requirement in order of each arrival time and goto Step 1.

Step 3. Delete the requirement from the processing jobs list. If the processing jobs list is empty, goto Step 1. Otherwise, goto Step 2.

In Step 1 of this method, we generate a processing jobs list using the algorithm in order to avoid delays of jobs as follows:

[Algorithm for generating a processing jobs list]

Step 1. If the press machine is idle, calculate estimated arrival time of processing requirement t for each job i $Y_{i,t}$ by using the equation as follows:

$$Y_{i,t} = U_{i,t-1} + A_i \qquad (1)$$

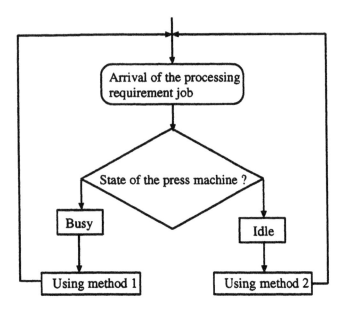

Fig. 4. Flow chart of hybrid method of method 1 and method 2.

Where $U_{i,t-1}$ denotes the conveying time of the job i for processing requirement $t - 1$. Otherwise, stop.

Step 2. Add the estimated processing requirements to the processing jobs list in order of each estimated arrival time of processing requirement $Y_{i,t}$.

3.3 Hybrid method of method 1 and 2

As described above, we can decrease the delays of the jobs, but it is difficult to increase the throughput of jobs by using method 1 if the idle times of the press process is relatively long. By using method 2, we can increase the throughput, but it is difficult to prevent the delays if the idle times of the press process is too short because we cannot produce the appropriate amount of the jobs whose processing requirement has not arrived. To cope the difficulties, we propose a hybrid method of method 1 and 2. **Fig. 4** shows the outline of the algorithm of the hybrid method. In the hybrid method, if there is no job to be processed in the lot process when a processing requirement reaches, method 1 is applied, otherwise, method 2 is applied.

3.4 Revised hybrid method

When we use the proposed methods such as method 2 and the hybrid method, it is possible to increase the work-in-process inventory of the press process, because the jobs without the processing requirements are processed in the press process. To cope with this inventory problem, we revise the hybrid method to decreases the

work-in-process inventory in the press process. The algorithm of the revised hybrid method is as follows:

[Algorithm of revised hybrid method]

Step 1. When the press machine is idle, generate a *processing jobs list*, in which the estimated processing requirements that will arrive in the near future are registered in order each estimated arrival time of the processing requirement. Let time T be the current time.

Step 2. Process the first processing requirement in the processing jobs list, if the following condition is satisfied.

$$Y_{i,t} - T > G \qquad (2)$$

Step 3. If processing requirements has arrived when the processing is finished, process each job corresponding to the processing requirement in order of each arrival time and goto Step 1. Otherwise, let T be $T + K$ and goto Step 2.

Step 4. Delete the requirement from the processing jobs list. If the processing jobs list is empty, goto Step 1. Otherwise, goto Step 2.

Where G and K denote the parameters to control the level of work-in-process inventory.

4. NUMERICAL EXAMPLES

To evaluate the efficiency of the methods proposed in this paper, we develop a simulation system of the production system shown in Fig. 3. In the simulation system, we assume that the volume of each job is between 300 and 450, and the number of parts that are needed in the subsequent lines in a day is uniformly distributed from 150 to 300. Parameter G and K is 480 minutes and $G \cdot 0.1$, respectively.

Fig. 5 shows the total sum of delay of each job when the number of job types are varied. Where "no control" means that the processing order in the press process is determined by the processing requirements from the subsequent processes. Fig. 5 shows that method 2 and the hybrid method are superior to the other methods. This is because method 2 and the hybrid of method produce the jobs without the processing requirement in the idle time of the press machine, when the number of processing requirements is small. However, in the case that the number of job types becomes greater than 9, the total sum of delay of each job increase rapidly, because the idle time of the press machine becomes too short.

Fig. 6 shows operation rate when the number of job types is varied. In Fig. 6, the hybrid method is superior to the other methods because the the idle time of the

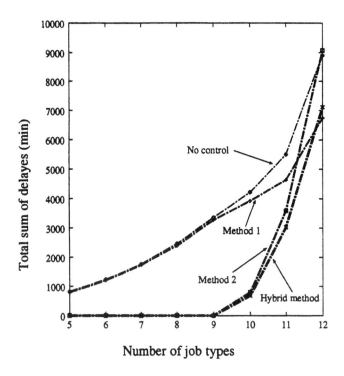

Fig. 5. Total sum of delays when the number of job types is varied

press machine is used to process jobs whose processing requirements is not arrived in the hybrid method. In the proposed methods, the operation rate of the press process does not vary with the number of job types. Therefore, we can keep the operation rate of the press process a constant value by using the hybrid method.

Fig. 7 shows average makespan of jobs when the number of job types is varied. In Fig. 7, each value of makespan in the hybrid method and the revised hybrid method is less than that in "non-control," if the number of job types is less than 11. This is because method 2 and the hybrid of method produce the jobs without the processing requirement in the idle time of the press machine, when the number of processing requirement jobs is small. Furthermore, each value of makespan in the revised hybrid method is less than that in the hybrid method. This is because the amount of the work-in-inventory decreases by using the revised hybrid method.

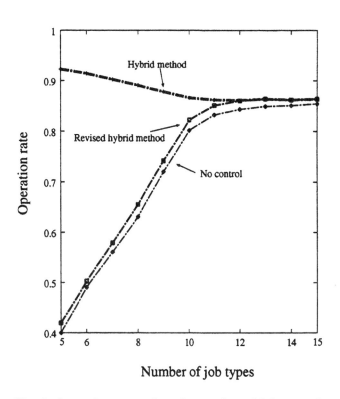

Fig. 6. Operation rate when the number of job types is varied

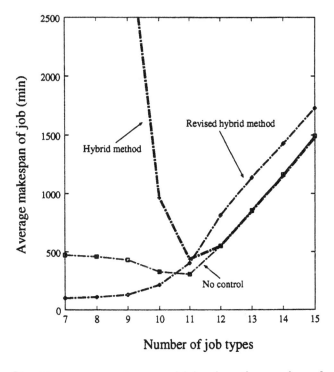

Fig. 7. Average makespan of job when the number of job types is varied

5. CONCLUSION

In this paper, we proposed three scheduling methods to schedule the order of jobs in a lot process to improve the utilizations and to decrease the delay of jobs in the JIT production system including lot processes. Furthermore, we evaluated the methods using computer simulations.

Further research might be focused on applying these methods to the real production systems and revise the algorithm.

REFERENCES

Hatono, I., J. K. Yoo, S. Tomiyama and H. Tamura (1996). Scheduling methods for lot processes in jit production system. In: *Proceedings of the 40th Annual Conference of Systems, Control and Information Engineers*. pp. 521–522. (in Japanese).

Kotani, S. (1983). Toyota production system: An integrated approach to just-in-time. *Toyota Technique*. (in Japanese).

Monden, Y. (1983). *Toyota Production System: An Integrated Approach to Just-In-Time*. Industrial Engineering and Management Press. Atlanta.

GENETIC ALGORITHMS FOR SCHEDULING OPTIMIZATION ON A MULTI-PRODUCT BATCH PROCESSING MACHINE

L He and N Mort

Department of Automatic Control and System Engineering, University of Sheffield, UK

Abstract : This paper describes loading polices for the scheduling problem on a multi - product batching processing machine(BPM) which can process a batch˙ of jobs simultaneously with a known and fixed number of jobs and their ready time. The different jobs are dispatched and sequenced in order to minimize the makespan and maximize the utilization of the servers. As an extension to the basic model of previous work (Fanti et al, 1997)) for BPM scheduling, we build up the model to schedule n jobs on m identical servers in which the optimization procedure results in a complex NP-hard combinatorial problem. Genetic Algorithms(GAs) are applied to solve this scheduling problem where we apply the features of *elitist strategy* GAs to develop a group of MATLAB functions for solving the BPM scheduling problem. This result is the optimal solution. This experiment demonstrates that GAs can provide a robust search procedure in the optimization of scheduling problem which has high dimensionality, multi-modality, discontinuity and noise (DeJong, 1975).

Keywords: Optimization, Scheduling Algorithms, Batch-Processing Machine, Genetic Algorithms

1. INTRODUCTION

The problem of scheduling optimization using GAs is an active area which is currently drawing much research attention. The Genetic Algorithm is a stochastic optimization search method taken from genetics science and the process of natural selection and evolution where individuals with higher fitness values will generate exponentially increasing copies in the offsprings (Goldberg, 1989).

Since GAs search a large amount of the population in parallel, and they have the ability to jump from schedule to schedule (Jain and Elmaraghy, 1997) under the guidance of the objective function and the corresponding fitness value, their use as a solution is proposed to avoid getting stuck at local optima. GAs are a global optimum search method, and can provide a robust search procedure (Lindfield and Penny, 1995) in the optimization of scheduling problem where traditional methods cannot provide a satisfactory solution.

As far as we know, there have been no contributions based on Genetic Algorithms to scheduling for this special Batch Processing Machine (BPM) (Fanti et al, 1997). In the next section, we apply the features of *elitist strategy* Genetic Algorithms to develop a group of MATLAB™ functions solving the BPM scheduling problem.

2. MODEL OF SCHEDULING ON A BPM

A BPM has m identical servers that can process n jobs simultaneously. The mix of n jobs is described as the vector (Fanti et al, 1997 , He and Mort, 1997):

$$y = [y_1\ y_2\ ...\ y_n] \qquad (1)$$

where y_i (for i = 1, ..., n) represents the number of i-type jobs and satisfies the constraint:

$$y_i < p \qquad (2)$$

where p is the length of the admissable trajectory.

Our scheduling objective is to minimize the $T^*(\gamma)$.

$$T^*(\gamma) = T_1(\gamma) + T^*_2(\gamma)$$

$= \tau_1 \times (p \times m - \text{sum}(y)) + \tau_2 \times (n + \Delta N') \quad (3)$

$T_1(\gamma)$ is increasing with p linearly. $T_2^*(\gamma)$ is determined by $\Delta N'$, and is non-increasing with p (non-linearly). The task of the GA is to find a schedule which contains the optimal sets of *combinations* to minimize the number of Additional Tool Change Times $\Delta N'$ with respect to the length p (He and Mort, 1997). In other words, we should apply GAs to search the *combinations* of all the jobs including as many jobs as possible to reduce the $\Delta N'$. The optimization of this scheduling problem leads to a typical NP-hard combinatorial problem .

3. GENETIC ALGORITHM FOR BPM SCHEDULING

The major difference in Genetic Algorithms from traditional optimization methods lies in that Genetic Algorithms use probabilistic transition rules together with an encoding of the parameter set instead of the parameters set themselves (Goldberg, 1989). A typical GA may be represented by the following procedure(Edwin et al, 1994):

a) An *initial population* which contains some encoding of the parameter set is randomly generated.

b) The *fitness value* of each individual is evaluated according to the *objective function* (*performance measure*)

c) The individual with a higher fitness value will have a higher probability of producing one or more offspring in the next generation.

d) Genetic operators which include *crossover* , *mutation and reinsertion* are applied to the search population .

e) Repeat step b, c and d until algorithm converges.

3.1 Representation and initiaization of population

The major difficulty in applying the GA in a scheduling problem is finding an appropriate representation for the population. The most commonly used representation in the GA is that of the binary string although other representations such as integer , or real-valued can be used (Chipperfield et al, 1994). For this BPM scheduling problem, the binary string representation of the individual is selected according to the following conditions: (Edwin et al, 1994):

a) *Completeness*: the string representation should contain all the individuals in the search population.

b) *Uniqueness*: the string maps with the individual on a one to one basis.

The second step is to generate an initial population randomly. For example, a population is composed of N_{chrom} individuals which is *bitlength* bits long. The initial population is produced by creating a ($N_{chrom} \times$ *bitlength*) matrix with the component randomly selected from the set {0, 1}.

Now we will consider the practical scheduling problem.

A. let n jobs be described as the search space *range* : *range* = y .

B. m servers are encoded in a binary string which is r (r is an integer) bits long, r must satisfy the following constraint:

$$2^r \geq m > 2^{r-1} \quad (4)$$

and the number of all binary strings having bitlength r equals 2^r. Let $R = 2^r$, then the string representation of the servers is formalized as a vector b:

$$b = [\, b_1 \quad b_2 \dots b_R \,] \quad (5)$$

if $2^r = m$, $\quad b = [b_1 \quad b_2 \dots b_m]$ $\quad$ (6)

if $2^r > m$, $\quad b = [\, b_1 \quad b_2 \dots b_m \, b_{m+1} \dots b_R \,]$ (7)

To meet the requirement of completeness and uniqueness, we give the following corresponding relation:

server j	1	2	...	(2m+1-R)	...	m-1	m
string	b_1	b_2	...	b_{2m+1-R}	...	b_{m-1}	b_m
				b_R	...	b_{m+2}	b_{m+1}

$$(8)$$

C. The bitlength of the chromosome is :

$$bitlength = r \times n \quad (9)$$

3.2 Convert binary value to real value

When the binary value of the chromosome is converted into a real value, the chromosome is examined at the intervals of r bits in length to obtain the n binary string whose position i (for i=1,... n) corresponds with the i-entries in the search space *range* and its binary value indicates the server j processing that job according to the condition (8). The element whose value is greater than zero in the loading matrix represents the i-type job. The j-th column of loading matrix is composed of the transpose of vector M_j (for j =1,..., m) indicating the i-type job processed by the server j. The real value of the individual is presented by calculating the sum of all jobs processed by every server.

$$rvalsum = [\text{sum}(M_1) \quad \text{sum}(M_2) \dots \text{sum}(M_m)] \quad (10)$$

3.3 Fitness value of individual

If a *Combination* which satisfies the following conditions is searched:

$$\text{sum}(M_j) \leq p \qquad (11)$$
$$p - \text{sum}(M_j) \leq x \quad \text{for } j = 1, \ldots m \qquad (12)$$
$$\text{and} \qquad x = p * m - \text{sum}(\textit{range}) \qquad (13)$$

the fitness value is increased by "1". Finally, the total unused machine time X is checked; if it satisfies:

$$X \leq x \qquad (14)$$

a feasible solution is obtained. Otherwise, an infeasible *Combination* is searched , "1" is subtracted from its fitness value , resulting in part of all *combinations* being rewarded.

Fitness scaling (Goldberg, 1989) is an important parameter in GAs that regulates the number of offspring that a individual will produce in the next generation. The appropriate selection of fitness scaling can improve the genetic search dramatically (Booker, 1987).

In our scheduling problem, the proportional fitness assignment (Chipperfield et al , 1994) in which the individual fitness F(f) of each individual is computed as the individual's raw performance Fs. This performs very well compared with Power Scaling which is a kind of **Linear Scaling** in the formulation $F(f) = a\,f + b$ with choosing the parameter $a = 1$, $b = 0$. The idea behind this selection is that since the chromosome with lower fitness value might contain some most successful schema, we should give them some copies to produce their offspring in order to prevent evolution from early-maturation.

3.4 Genetic operators

a) Reproduction and selection

Reproduction is a process to select the fittest individuals for breeding according to their fitness value. Stochastic Universal Sampling (Baker, 1987) is a well-known selection method to yield more robust solutions. However, in the combinatorial optimization problem whose search space is very large with much irregularity, high discontinuity and non-linearity, randomly sampling is unlikely to sample enough of the space. *Roulette wheel selection* (Goldberg, 1989) *with bias optimum* is used as the reproduction operator. Each individual is chosen in proportion to the integral part of its real expected trial at first.

Expected numbers of individual in integral phase

$$= \text{integer } \left(\frac{\textit{Fitness value of Individual}}{\sum \textit{Fitness value of Individuals}} \times \right.$$

$$\left. \text{Population size} \right) \qquad (15)$$

Then, the remaining individuals are determininistically selected from the fittest individuals. It is referred to as *bias optimum*. Without increasing the computation effort, a higher performance solution can be found by biasing the optimal individual in the searching populations. This special Remainder Sampling (see Chipperfield et al, 1994) method will be discussed in our further work.

b) Crossover

Crossover is the main operator for producing new individuals in GAs. If a new individual inherits both parents' high class merits, it will have more probability to be selected for mating. The basic forms of crossover are single-point crossover , double-point crossover and multi-point crossover (Spears and DeJong, 1991).

Studying the properties of the BPM scheduling problem, we realize that the fitness value of offspring is very sensitive to the alteration of the individual. For example, if a job is changed to be processed from one server to another one, the sum value of all jobs processed by both servers will be modified leading to the alteration of the fitness value of the new individual.

We provide the following experimental result concerning the application of the three basic crossover forms (Table 1). It shows that using both single-point crossover and double-point crossover in series speeds up the convergence procedure, and an excellent result is achieved. It is critical for success in applying GAs scheduling to the BPM problem. Actually, the combination of single-point crossover and double-point crossover is an effective method of making the same effect as the reduced surrogate crossover.

Also, this method does not have the same effect as the multi-point crossover method with crossover point equal to 3 because the mating rate (matenum) is smaller than one. The idea behind this method is that the crossover *with different choice of crossover points* can explore much more of the search space rather than rewarding the premature local optima, thus making the search procedure more robust. So, different crossover methods that make the evolutionary procedure converge well may be needed for our future research work in this area.

Table 1 Results of different crossover points

crossover point	convergence		
	state	speed	result
Single-point	stable	slow	average
Double-point	stable	slow	average
Multi-point	unstable	slow	poor
S + D	stable	fast	excellent
D + M	unstable	slow	poor
S + M	unstable	slow	poor

c) Mutation

Mutation is used as a secondary operator which is randomly applied with small probability. It is a safety policy to be used sparingly with reproduction and crossover against premature convergence to a local optimum (Edwin et al, 1994). In this BPM scheduling problem, to make the evolutionary procedure stable, mutation is applied by permitting only *One* job randomly changing to be processed from one server to another server, thus expecting that the schedule with higher fitness value would be favoured. The mutation probability represents the frequency of applying the mutation operator whose value (*mu*) is determined by the following condition

$$mu \cong \frac{1}{bitlength} \qquad (16)$$

The following experiment result gives the evidence

Table 2. Convergence property resulting from different mutation rates

Mutation Rate	Result
mu<= 0.01	No convergence to global optimum
mu= 0.02	Convergence to global optimum
mu>= 0.03	Unstable convergence procedure

When the value of mutation probability is chosen as 0.02, the mutating number of the chromosome bit equals 1.26 (mu × bitlength = 0.02×63 =1.26), very close to 1. So mutation is one of the main methods of moving from a local optimum to a global optimum.

3.5 Choosing genetic algorithm parameters

Schaffer *et al* (1989) conclude that the optimal parameter settings vary with different practical problems. The successful application of GAs may require for a large number of experiments to get satisfactory parameter settings.

In this problem, the main parameters are population size (N_{chrom}), maximum generation (Maxgen), fitness scaling , proposition of population for mating (matenum) (Lindfield and Penny, 1995) and mutation probability (*mu*).

We have determined the selection method of fitness scaling and mutation probability from our previous discussion. The population size is determined by the principle that at least one *combination* should be searched at the initial population, otherwise the search stops and initializes a new population. In a population with 100 individuals, about 7-8% individuals have the fitness value equal to 1. Then the population size N_{chrom} is decided as 40 to follow the principle of both providing enough search space and speeding up the evolutionary computation.

To keep the search space state stable, we mate only a proportion of the population. In this case , matenum is chosen as 0.6 or 60%.

To terminate the Genetic Algorithm, the best individual is expected to be found before the end of the search generation. In our BPM scheduling problem, after the proper selection of the genetic operator and other parameters, the best result can be achieved within the maximum generation which has a recommended value of 300, thus Maxgen = 300. Also, we suppose that GAs should converge well within a reasonable number of generations, otherwise the genetic operators and their parameters should be adjusted .

3.6 Complete GAs in MATLAB function

Ba. [Initialization of parameter]

Bb. [Initialization of population]Randomly generate binary population size N_{chrom} with bitlength *bitlength*, using MATLAB function chromosome = *genbin* (bitlength, Nchrom).

Bc. [Generational loop] Repeat step **Bd, Be, Bf, Bg, Bh, Bi** until Maxgen =300.

Bd. [Convert the binary value to real value] MATLAB function rval (i , :) = *binv* (chromosome , range , i) convert binary chromosome to real value in the search space *Range*. The loading matrix is recorded by matrix M_j for chromosome *i* , the real value is calculated by the sum of M_j.

Be. [Assign the fitness value to the entire population] MATLAB function fitV = *fitness* (chromosome, range, p, x, rval) computes the fitness value of each chromosome in the population.
Bf. [Record the best chromosome in the population]

Bg. [Reproduction] MATLAB function matepool = *selectg* (chromosome , range , fitV) select individuals for breeding according to the fitness value using RWS with bias optimum method.

Bh. [Crossover] MATLAB function chrom = *matesome* (matepool, matenum) perform single-point crossover and function chrom = *mateD* (chrom, matenum) perform double-point crossover with the proportion of population *matenum* for mating.

Bi. [Mutation] MATLAB function chromosome = *mutate1*(chrom, mu) perform the mutation with the mutation probability *mu*.

Bj. [Display the loading matrix for best individual] Perform MATLAB function Bestind = binvreal (chromosome, range , best) to display the loading matrix.

Bk. [Display the evolutionary procedure and record the optimal solution] MATLAB function figure; plot (Best+1),'*'; axis ([0,300,0,7]); xlabel ('generation'); ylabel ('Number of Non ATCT')

(* Note : Number of Non ATCT = fitV+1; ΔN' = m - (fitV + 1), if the condition (14) is satisfied.)

3.7 Computation result

The best solutions are recorded in table 3. All the loading matrices show that the optimal solution for this BPM scheduling problem has ΔN' = m-(fitV+1)=7-(5+1)=1. Also, we notice that the job [714] cannot be *Combined* - it means that the best fitness value (fitV) is 5 -- only 5 *combinations* can be searched. Referring to our previous study (He and Mort, 1997), the same results for the minimization of the makespan (**298.248**) and the maximization of the machine utilization (**93.139%**) are obtained by applying GAs in this BPM scheduling problem. Figure 1 displays a typical evolutionary procedure.

Figure 1 A typical evolutionary procedure

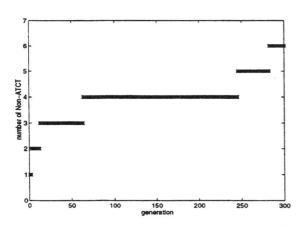

Table 3 loading matrix *M*, fitness value fitV and ΔN'

M(1)	231	497	176	144	0	380	688
	114	310	216	6	0	12	714
	12	0	153	660	0	282	170
	454	0	266	0	0	128	50
rvalsum	[811	807	811	810	~~0~~	802	~~1622~~]
M (2)	231	176	380	144	216	688	497
	170	12	153	660	0	6	310
	282	714	12	0	0	114	0
	128	50	266	0	0	0	0
	0	454	0	0	0	0	0
rvalsum	[811	~~1406~~	811	804	~~216~~	808	807]
M (3)	0	380	231	144	497	176	216
	0	50	170	6	310	153	688
	0	114	282	660	0	12	714
	0	266	128	0	0	12	0
	0	0	0	0	0	454	0
rvalsum	[~~0~~	810	811	810	807	807	~~1618~~]
M(4)	497	12	714	454	216	144	380
	310	231	0	176	688	660	153
	0	170	0	50	0	6	12
	0	114	0	128	0	0	266
	0	282	0	0	0	0	0
rvalsum	[807	809	~~714~~	808	~~904~~	810	811]
fitV	5						
ΔN'	1						

Note: For *M*(1), the job [688], [714], [170] and [50] share to be processed in M_5 and M_7. For *M*(2), the job [176], [12], [714],[50], [454] and [216] share to be processed in M_2 and M_5 , and so on.

This example represents a multi-modal combinatorial optimization problem, simple GAs can provide a number of unimodal solutions to this particular type of problem. Hence the choice of the final solution is decided by the users according to the requirement of particular problem .

4. CONCLUSION AND FUTURE WORK

In this paper , we have investigated the application of Genetic Algorithms to solve the scheduling problem on a batch-processing machine. Genetic Algorithms demonstrate a powerful capability to provide the robust search procedure for the complex NP-hard combinatorial problem. Optimal solution is achieved after the careful selection of the genetic operator and the parameter settings.

Our future research work lies in:

1) How to choose the appropriate selection methods and genetic operators or create their variants to obtain the optimization solution in respect to different manufacturing scheduling problems.

2) How to apply GAs to solve uniquely these combinatorial multimodal optimization problems.

5. REFERENCES

Baker.J.E (1987). Reducing bias and inefficiency in the selection algorithm. *Proceedings of the Second International Conference on Genetic Algorithms*, pp14-19

Booker, L. (1987). Improving Searches in Genetic Algorithms. *Genetic Algorithms and Simulated Annealing*. LAWRENCE DAVIS (ed.) (Cambridge, MA:BBN Laboratories) pp 61-73.

Chipperfield. A., Fleming, P.J., Pohlheim, H.,and Fonseca. C (1994). *Genetic Algorithm Toolbox for use with MATLAB*. University of Sheffield, ACSE Department.

DeJong, K.A (1975). An analysis of the behaviour of a class of genetic adaptive systems. *Doctoral Thesis, University of Michigan.*

Edwin, S.H. Hou, Ansari. H, Ren H (1994). A Genetic Algorithm for Multiprocessor Scheduling, *IEEE Transaction on Parallel and Distributed Systems*, **5, 2**. pp223-229

Fanti, M.P, Maione, B, Piscitelli, G and Turchiano, B (1997). Heuristic scheduling of jobs on a multi-product . batch processing machine, *International Journal of Production Research*, **34**, pp2163-2186.

Goldberg, D.E. (1989). *Genetic Algorithms in Search, Optimization, and Machine Learning*, Reading , MA: Addison Wesley.

He, Liwen and Mort, N (1997). Approximate Optimization Algorithms for Scheduling on a Multi-Product Batch Processing Machine, submitted to *International Journal of Production Research.*

Jain, A.K., and Elmaraghy, H.A (1997). Production scheduling/rescheduling in flexible manufacturing. *International Journal of Production Research.* **35**, pp281-309.

Lindfield, G. and Penny, J (1995). *Numerical Methods Using MATLAB*, Ellis Horwood.

Schaffer, J.D., Caruana, R.A., Eshelman, L.J., and Das. (1989). A study of control parameters affecting on-line performance of genetic algorithms for function optimization (Ed). *Proceedings of the Third International Conference on Genetic Algorithms*, Schaffer J.D.Morgan Kaufmann Publishers pp51-66.

Spears, W.M and DeJong, K.A (1991). An Analysis of Multi-Point Crossover, *Foundations of Genetic Algorithms*, J.E.RAWLINS (Ed.), pp301-315.

AN INTEGRATED, HIERARCHICAL FRAMEWORK FOR PLANNING, SCHEDULING AND CONTROLLING JOB SHOPS

Vandaele, Nico J.

*UFSIA, University of Antwerp
Prinsstraat 13, 2000 Antwerpen, Belgium
phone: 32-3-220 41 59   fax:32-3-220 47 99
e-mail: nico.vandaele@ufsia.ac.be*

Abstract: In this paper an explanation is proposed for the fact that job shop scheduling in real-life environments is so complicated. A typical job shop has a large variety of machines mostly organised as a functional layout. The demand for a large variety of products is irregular, both in timing and in quantities. Each product type is manufactured in batches which flow through the shop, each one following its own routing. At each machine multiple operations for various products can take place with a setup in between batches requiring two different types of operations. The objective is to satisfy as good as possible customer demands: meeting due-dates while trying to control lead-times and work-in-process and securing smooth operations on the aggregate level as well as on the day-to-day execution level. We show how a new integrated, hierarchical framework for planning, scheduling and control tries to cope with the job shop realities. Guaranteeing the practicality of the approach is anything but simple.

Keywords: integrated planning methodology, hierarchical manufacturing models, lead-time estimation, detailed scheduling.

1.INTRODUCTION

The approach taken in the literature is often a fragmented approach, concentrating on only a few dimensions in order to allow proper modelling and solution procedures. For example, some research focuses on the stochastic behaviour of the job shop while others study only a deterministic version of the problem. If we deal with realistic job shops, the ultimate approach will be a combination of several interacting approaches linked within an integrated framework. In terms of the dimension and the type of decision, different approaches, models and techniques must be used to match the specific needs of the respective decision to be taken. In this sense, we deal with an intelligent manufacturing system for planning, scheduling and controlling. We developed such a new integrated model which we implemented in a metal working company.

2. THE REALISTIC JOB SHOP

The basic conflict of the job shop is that a large number of parameters (demand, production and shop characteristics) must be taken into account, so that solution procedures are faced with combinatorial explosion. On the other hand, the realistic job shop is very dynamic and volatile in nature. Almost all parameters should be considered as stochastic variables. And unfortunately, the environment can change extremely fast due to all kinds of disruptions, changes and external events. As one can expect, a deterministic model can do some good in scheduling but often lacks the flexibility towards stochastic behaviour and unforeseen events. On the other hand, stochastic models incorporate variability in the job shop parameters, but these models cannot add much scheduling detail and suffer under transient circumstances. Even if we were able to model the job

121

shop in all its facets, we will probably lack a solution procedure which is able to solve such a huge monolithic model. The time to come up with a good (on-line, real-time) schedule mostly cannot exceed a couple of minutes. Therefore we spend some time in explaining the fundamental conflicts behind the job shop scheduling problem.

3. THE FUNDAMENTAL CONFLICTS

3.1 *The time horizon.*

The purpose of a schedule is to direct the activities of the job shop in the near-by future. For direct execution (picking, job assignment, ...) usually a couple of shifts or production days is enough. But in terms of material requirements (ordering, fabrication, ...) probably a couple of weeks is more appropriate and for general capacity measures (in the area of staffing, additional shifts, capacity extension, outsourcing, ...) probably months are the right unit of measure. Besides the fact that solving the job shop problem over the entire horizon is an impossible task, there is little virtue in having a detailed plan for a particular shift in a week somewhere six months from today, not to mention the quality of the information (forecasts, uncertain orders,...) whereon that schedule is built. On the other hand, at the current moment there is definitely a need for information from that point in time in order to be able to anticipate particular trends or events. In other words, the scheduling procedure must be able to look into the future and to grasp important information to induce actions in the periods to come. Shifts in the demands for the products, in the product mix and in ordering patterns can cause serious distortions from the schedules if anticipatory action is omitted. Also upcoming capacity issues (holidays, workforce extension or contraction, changing shift patterns, machine additions or removals, quality improvements, engineering changes, major overhauls, ...) can have a drastic impact on the schedule's ability to meet the goals of the job shop and to meet customer demands.

Therefore it is wise to distinguish between a scheduling horizon and a planning horizon. Although both horizons serve the same job shop environment and try to accomplish a common objective (e.g. due-date performance), the scheduling horizon deals with imminent issues covering the short term. The planning horizon handles significant upcoming issues from the mid-long term and analyses general trends, while the scheduling horizon will be managed by what is traditionally known as a scheduling procedure. The planning horizon will use a type of aggregate planning technique. In between the scheduling and planning horizon there is an area with

different shades of grey: what about a planned introduction of two new products next week? Therefore an additional shift is planned for the second and the third week from now. This fact is probably not urgent enough to be taken into account for the current (one week) schedule. However it is definitely stringent enough not to let it only be on behalf of the aggregate model used to analyse the planning horizon.

The development of a monolithic model, incorporating all the aspects of both horizons, together with their interrelationships is merely wishful thinking. A hierarchical approach seems to be much more suitable but then links between the various levels are necessary. Most hierarchical models look at the production facility from a deterministic point of view and they stress the material planning with less attention for capacity planning. Other traditional approaches also tried to avoid an overwhelming monolithic model: Material Requirements Planning (emphasising material), Just-In-Time (emphasising capacity), Theory Of Constraints (emphasising bottlenecks) and Finite Scheduling (emphasising deterministic schedule generation). The ultimate objective of job shop scheduling is to integrate both material and capacity at each decision level. In this context, a popular link between decision levels is to do a lot of what-if analyses before descending to a lower level of decision. Intelligent judgement of the outcomes of the different scenario's can help to define a good compromise, which will serve as a framework for lower level decision making. Another idea is deployed in Load Oriented Planning: information stemming from a long term view (higher decision level) is devaluated before this information is included in the short term decision process: the more remote the information (e.g. batch processing times) the less influence this information has on the short term decision level.

3.2 *Data requirements*

The amount of data needed to model the job shop problem is enormous. There can be hundreds of products, thousands of orders, tens of machines, hundreds of operations and numerous other things which need to be taken into account. Therefore aggregation is needed in order to be able to build a model which allows us to analyse the data and draw general conclusions about the performance of a particular instance of the job shop. Finally, numerous (instance) evaluations will be executed during a solution procedure, so that computing considerations (in terms of memory requirements and execution time) cannot be neglected. For the scheduling horizon, the total execution time to establish a schedule can mostly not exceed some minutes.

Intelligent and creative programming helps a lot but in the first place the solution procedure for detailed scheduling must be well designed and developed in terms of the current computing capabilities. However, a couple of hours is probably justified to solve an aggregate model serving the entire planning horizon. Even after aggregation of the data, this can only be realised with a cautious developed model and solution procedure.

The design principle is to emphasise on a (different) subset of parameters in each planning horizon. This makes us think of a hierarchy of decisions each based on a particular type of aggregated or disaggregated data. The longer the horizon, the more aggregate the relevant information will be. For the near-by scheduling effort, there is a need for very detailed data.

3.3 Centralised or decentralised decision making

In the set of decisions to be taken in order to solve the job shop problem, some decisions require an overall, central view of the job shop problem while others must necessarily be taken at a more decentralised level, with much less consideration about the overall problem setting. However, it is again merely a matter of judgement: it is hard to find one type of decision, taken at a decentralised element of the problem, which has absolutely no influence on other elements of the decision problem.

The lot sizing decision is a good example. If the lot sizing decision is included in the detailed scheduling problem, then the combinatorial nature of the job shop scheduling problem explodes towards another order of magnitude. On the other hand, if the lot size is already determined centrally, the setup time can easily be included in the batch processing time. By lifting an issue (like the lot sizing decision) to a central decision level, one limits the solution space for the more decentralised (hierarchically lower) decisions. In this sense, a degree of global optimality is sacrificed. The benefit however is a much easier problem at lower levels. Also, a centrally determined lot size (e.g. on the aggregate, planning horizon level) preserves some kind of stability through time. If that decision was taken at the level of the scheduling horizon, the lot size would probably be too sensitive and subject to continuous change favouring local or temporarily optimality criteria. The truth lies again somewhere in between: the analysis in the planning horizon can give some long term indication for a good lot size, while in the scheduling horizon it is allowed, within certain limits, to deviate from this long term lot size in order to respond to specific short term needs.

Another example can be found in the procedures for detailed scheduling. A general (global optimisation) procedure will, in terms of optimality, outperform a collection of local decision rules (e.g. dispatching rules). However, the price for global optimality is mostly expressed in terms of huge memory requirements and extremely long execution times. Here also, a compromise is often the only practical outcome. Nice suboptimal procedures can be found along the lines of concentrating on (multiple) bottleneck detection (the centralised view) which offer a good solution with moderate computing effort. The non-bottlenecks play their role in the decentralised decision area. Given the limitations imposed by the centralised decisions concerning the bottleneck, the decentralised decisions can operate on their own behalf.

3.4 Level of optimality and/or modelling exactness

As far as modelling exactness is concerned, it could be the ambition to model each aspect of the real world job shop in the smallest possible detail. This is impossible. Such a huge and complex problem as the job shop scheduling problem needs some simplification for the sake of modelling itself. Simplification can take many forms: aggregation, approximation, concentration on a particular aspect (e.g. bottleneck operations), etc. In this way some information is lost along the way. In the end, the model should resemble reality up to a fairly high degree and should serve the objective for which it is developed.

When optimality is the issue, it can be argued that the model (which is now a simplification of reality) must be solvable. The main principle should be that optimal procedures should be used as long as their computing efforts (memory, execution times) are acceptable for practical application. If this turns out to be impossible, then a good suboptimal procedure is the only way out. In this sense, our hierarchical approach, which will be discussed in the sequel, complies to this statement. On the aggregate level (planning horizon) an approximate queueing network is used. However the optimisation routine of the queueing network finds the optimal lot sizes. At another point in the decision process, customer orders must be grouped into manufacturing orders. A dynamic program solves this approximate problem optimally. At the detailed scheduling phase, a suboptimal extended shifting bottleneck procedure is used, but the recursive single machine problems are solved optimally. An additional factor in this discussion is the degree of numerical exactness of the solution procedure. It should be clear that scientific standards mostly overshoot practical relevance. There is a trade-off between numerical exactness and practical relevance. There is no need for a lot of decimals in most real life settings. However, the latter hinges on the practical situation on hand and it is difficult to establish general guidelines.

Once the problem is modelled in such a way that it is also solvable (sub)optimally, great care has to be taken whether the model is still relevant for the problem on hand. For each phase in the hierarchical decision process, a suitable and solvable model must be developed in such a way that it can handle the decision process in a practical setting. It is a matter of using the right model with the right solution procedure for the right purpose.

3.5 Degree of stochasticity

Stochasticity is the final, but probably the most difficult issue of reality to cope with. The knowledge that each piece of data is subject to change, both in terms of the planning horizon as well as the scheduling horizon, confronts us with a tremendous problem: despite all the effort and judgement, the obtained result (e.g. a schedule) can become much less useful if some unexpected changes or events occur. A hectic way of adjusting to each minor or major change in the data is not only undesirable, but also impossible. The search for such a procedure that serves this purpose, could turn out to be very disappointing. A possible deployment is the following. Stochastic behaviour of certain parameters (demands, processing times, ...) can be captured in probability distributions. Due to the fact that a lot of data is needed to get an idea of the stochastic behaviour of the parameters, such an analysis is only valuable when dealing with the planning horizon. In such a way, an 'average' picture in an aggregate way, can be made of the job shop. Some long term relationships and trends can be analysed.

The instantaneous events can be considered as realisations of the stochastic processes involved. As long as we stay on the aggregate level, this makes sense. However, the ultimate objective is to cope with every-day operations, being a permanent sequence of instances. Our detailed schedules should therefore be adjusted instantly, which is utopian. Therefore, a hierarchical decision framework should be designed in such a way that the inherent changes in the parameters do not hurt the validity of the aggregate model and that the damage to the detailed schedule is controllable. There are two important observations at this point. One observation is that a schedule is not useless if only certain start and finish times are not valid anymore. The sequences embedded in the schedule will continue to be of some practical value at least for a limited period ahead. Once this period is past, a recalculation of the schedule is mandatory. A second observation deals with the nature of probability distributions. Suppose

that on the aggregate level it is possible to obtain (approximated) lead time probability distributions. By defining a lead time as a lead time percentile W_P, one expects that the realisations of the lead time will fall within this lead time percentile P% of the time. Compared with the average lead time, the lead time percentile contains a safety margin (safety time) to cover up with the natural fluctuation of the lead time. If this lead time percentile is used for scheduling purposes, it can be expected that the schedule experiences some stability due to the incorporated safety time.

Unfortunately, stochastic processes themselves are subject to change and show sometimes transient behaviour. Great care must be taken in order to judge the validity of average results under transient circumstances. A rule should be: adapt the aggregate models as soon as possible when serious trends in particular parameters are detected. The difference between normal stochastic behaviour and transient behaviour is not easy to detect on a small set of parameter instances. As a consequence, the adaptation of aggregate models towards shifts in the parameter characteristics is usually rather slow. Also the inclusion of unique events is not straightforward. This is again the grey area. As a start, some what-if analyses could be conducted in order to try to grasp the consequences of transient behaviour and unique events as if these should apply to the entire planning horizon. Reality will be somewhere in the middle of these extreme results. Profound insight and careful judgement should help the planner to draw the right conclusions.

In the hope to remedy the shortcomings of the traditional approaches, we developed a procedure named ACLIPS: A Capacity and Lead Time Integrated Procedure for Scheduling.

4. THE ACLIPS APPROACH

Based on these arguments, our implementation consists of four major phases: a lot sizing and lead-time estimation phase, a tuning phase, a scheduling phase and an execution phase. Each phase addresses a typical set of decisions and focuses on the relevant time horizon. A choice is made on the correct degree of data aggregation while using a proper model and incorporating a practical solution procedure. In addition, at each level both capacity and material decisions must be integrated.

4.1 The lot sizing and lead-time estimation phase

The major reasons to look first at the lot sizing decision are the stability of the detailed schedule and the ease of detailed scheduling by considering the lot sizing decision as a centralised decision. The lot size will be determined on the level of the planning horizon, where a huge amount of data, both demand and job shop characteristics, have to be aggregated. An approximate queueing model is used, in which all operations and arrival streams are stochastically represented. The individual customer orders for a particular product are grouped into manufacturing orders. The ideal manufacturing lot sizes, minimising expected lead time, are the outcome of the queueing network. At this point we build on the observation that there is a convex relationship between the lot size and the lead time (see Karmarkar, 1987, Lambrecht, Chen and Vandaele, 1996 and Lambrecht and Vandaele, 1996). In order to be able to do this, all parameters must be written as a function of the lot size. For a detailed discussion, we refer to Vandaele, 1996. The approximate queueing model based on aggregate data is valid for the planning horizon; it takes care of congestion phenomena (the corrupting impact utilisation combined with stochasticity) which quantifies the queueing delays. In this way the lot size is considered as a centralised, long range decision variable. In order to let the detailed scheduling routines take advantage of particular short term events, the customer orders will be grouped into manufacturing orders trying to approach these target lot sizes as close as possible. Given the time varying nature of the booked customer demands, the manufacturing orders will actually differ from manufacturing order to manufacturing order, but on the average we aim for lot sizes minimising the expected lead time (and work-in-process).

4.2 The tuning phase

This phase handles all major capacity problems which require management intervention. Therefore we provided the opportunity to conduct a large number of 'what-if' analyses, based on the queueing network. It must be stressed that all serious capacity problems must be solved before diving into the detailed scheduling. The tuning phase must cope the capacity/inventory(lead time) trade-off and manages the link between the aggregate (stochastic) view and the detailed (deterministic) view of the job shop.

4.3 The scheduling phase

This phase consists of three important decisions.

The grouping of the customer orders is done by use of a dynamic program with an objective minimising the number of inventory days, given a fixed number of setups. In this way we comply with the target lot sizes obtained above.

The second decision has to do with lead time off-setting and release date determination. For each manufacturing order, the release date is set equal to the due-date minus the lead time estimate of the manufacturing order (a grouping of booked customer orders). The estimate of the lead time is equal to the expected lead time plus a safety lead time. The safety lead time depends on the customer service. The lead time estimate is such that we expect to satisfy customer orders on time, P% of the time. This of course requires knowledge of the probability distribution of the lead time. Therefore, an estimate of the variance of the lead time is mandatory (also an outcome of the queueing model). In our approach, the detailed real-time scheduling required to manage incoming customer orders, is integrated with planning through the target lot sizes and the lead time off-setting. There is an amount of safety time which can absorb some of the unpredictable events. Also the determination of the release date implies that orders will be released with respect to the load and status of the job shop. In this sense, higher congestion results in longer lead times and early release. On the other hand, lead times and work-in-process are under control because unnecessary early releases are avoided. Another point is that as long as the order is not physically released, the planning and scheduling decisions, including the manufacturing lot sizes, can be altered without much difficulties.

The third major decision concerns the sequencing policy. Within the time windows (expected lead time plus safety time; one for every manufacturing order) all operations have to be sequenced in detail. We opted for the extended shifting bottleneck procedure (see Ivens and Lambrecht, 1996). The basic shifting bottleneck procedure has to be adapted so that it can be used to sequence the operations for our general job shop environment including assembly operations, release dates, due-dates, overlapping operations, multiple resources (machines and labour force), setup times, calendars and many other real life features. This sequencing application can clearly be interpreted as a deterministic real-time scheduler. Due to the safety time, incorporated in the time windows, we expect the sequence to be valuable, even if we proceed through time. Of course, real-time scheduling being a very dynamic process, will need frequent rescheduling. It is hoped for that the estimates of the time windows are accurate enough so that most of the due-dates are finally met. At this point we see the usefulness of the safety time: it allows the detailed scheduler to solve resource conflicts and it absorbs the natural fluctuation of the job shop parameters (stochasticity). In addition to this, the safety time helps to overcome unforeseen events. It is important to stress that, although the safety time can be used for this purpose, it was not the explicitly modelled (unexpected events). Therefore if safety time is consumed by this type of events, one should be aware that corrective measures must be taken to restore the used safety time if one wants to guarantee the same customer service as before. If operations take less time than planned or if advantageous events occur, it is obvious that the safety time builds-up again.

4.4 *The execution phase*

This phase takes care of implementing the schedule and, as time goes by, produces the necessary feedback, useful for updating the status of the data.

5. AN IMPLEMENTATION

ACLIPS is embedded in user-friendly software is implemented in a metal working company where it has already showed lots of opportunities. In order to give an idea of the real-life problem dimensions: 80 machines, 500 products and 10,000 customer orders on a yearly basis (for more details, see Lambrecht, Ivens and Vandaele, 1996).

6. CONCLUSION

As a matter of conclusion, we like to put our ACLIPS approach next to the dimensions outlined in section 3:

The time horizon is present in the methodology: the lot sizing and lead time estimation phase focuses on the planning horizon while the scheduling phase concentrates on the scheduling horizon. In order to overcome the difficulties with transient behaviour and unique events, judicious use of the tuning phase is a way out.

The date requirements are indeed different from phase to phase. The lot sizing phase uses aggregate data. In this way general trends and conclusions can be drawn. In the detailed scheduling phase the use of disaggregated data is mandatory. In the tuning phase both data types are used to find out and correct the stringent capacity problems.

It is clear that the lot sizing decision is considered to be a central decision variable. However, at the operational level, it is allowed to deviate wisely from this target lot size. The detailed schedule is generated centrally but in terms of execution, slight deviations are allowed. However, it is advised that the detailed schedule is updated as soon as possible towards the deviating decisions to capitalise the consequences of the decentralised decisions.

The queueing network is an approximation, but its solution procedure is optimal. We can afford an optimal procedure here because we can allow a couple of hours for a complete reoptimisation of the entire shop. However, an evaluation can be executed in less than a second. The latter is very beneficial for conducting what-if analyses. The customer order grouping procedure is exact and optimal, given the assumption of the fixed number of setups. It takes only a few seconds due to a very efficient dynamic program. The ESBP is a heuristic but it is based on an optimal procedure to solve the subproblems (one-machine problems). A schedule is constructed from scratch in a couple of minutes.

The degree of stochasticity is included at the aggregate level where we model the probability distributions of the shop parameters. At the detailed scheduling level, we use the deterministic and combinatorial ESBP. However, due to the safety time, this deterministic scheduling procedure is not immediately vulnerable to the dynamics of the shop floor. The schedule can stand unexpected events and natural variability in the shop parameters.

We believe that in order to solve the job shop scheduling problem, a combination of techniques and modelling tools copes best with the different dimensions of this complex problem. We gave some general arguments which, we believe, can be useful for other complex problems too.

REFERENCES

Ivens P. and Lambrecht M. (1996). Extending the shifting bottleneck procedure to real-life applications. *European Journal of Operational Research*, **90**, 252-268.

Karmarkar U.S. (1987). Lot sizes, lead times and in-process inventories. *Management Science*, **33**, 409-423.

Lambrecht M.R., Chen Shaoxiang. and Vandaele N.J. (1996). A lot sizing model with queueing delays: the issue of safety time. *European Journal of Operational Research*, **89**, 269-276.

Lambrecht M.R., Ivens P.L. and Vandaele N.J. (1996). ACLIPS: A capacity and lead-time integrated procedure for scheduling. *Management Science*, under final review.

Lambrecht M.R. and Vandaele N.J. (1996). A general approximation for the single product lot sizing model with queueing delays. *European Journal of Operational Research*, **95**, 73-88.

Vandaele N.J. (1996). The impact of lot sizing on queueing delays: multi product, multi machine models. *PhD Thesis*, KULeuven, 243 pp.

HOIST SCHEDULING PROBLEM: STATE-OF-THE-ART

Christelle Bloch[1], Astrid Bachelu[1], Christophe Varnier[1], Pierre Baptiste[2]

1: LAB / ENSMM / UFC-UMR CNRS 6596, 25 rue A. Savary, 25 000 Besançon
(France)
2: KAIST, Dept Comp. Sc., 373-1 Koosung-Dong, Yusung-Gu, Taejon 305-701 (Korea)

Abstract: In automated plating processes, hoists ensure the transfer of products between workstations. Scheduling their movements is known as the Hoist Scheduling Problem (HSP). It belongs to the class of strongly NP-hard problems (Lei and Wang, 1989b). Many studies, dealing with different complexity levels of the problem have been published, and the following classification should give the reader quite a large overview of them.

Keywords: scheduling algorithms, industrial production systems, robotic manipulators control, optimisation problem, classification

1. INTRODUCTION

Electroplating facilities are made up of tanks, between which hoists transfer products according to a given route. The Hoist Scheduling Problem (HSP) consists in scheduling the hoist movements to maximise the throughput, while respecting the following specific constraints: processing times are strictly bounded, no wait and no storage between tanks is allowed, hoists share a single track and must not collide, transport time cannot be neglected. The following classification should give the reader a survey of the various studied cases found in literature. First, the problem will be briefly described. Then, four main approaches will be presented: a predictive one determines the optimal cyclic scheduling of a line producing a single type of pieces; the second and the third ones solve the problem either in advance or in real-time when parts are not identical; finally, layout is considered.

2. PROBLEM DESCRIPTION

The complexity of the problem depends on kind of line, processing specifications and production mode.

2.1 Different types of facilities

Parts, held in "carriers", undergo given sequences of treatment through workstations, from a loading buffer to an unloading one. Hoists transport carriers, one at a time, from tank to tank.

In the simplest kind of line each tank processes one job at a time and a single hoist is used. But, more generally, some tanks may be duplicated and several hoists are used, while preventing them from colliding. At least, loading and unloading may be executed in the same station (called single I/O station), which further increases complexity .

2.2 Processing requirements

Parts are processed in tanks according to strict specifications: the order in which stations must be visited and the bounded soak time in each of them. In the simplest case, tanks are sequentially visited. But, practically, the given route makes some tanks, called multi-function tanks, be visited several times. At least, in order to simplify the problem, some authors assume that processing times are fixed.

2.3 Existing production modes

Single-product mode: In mass production , the line is dedicated to a single type of pieces. Then, the problem is called: "Cyclic Hoist Scheduling Problem" (CHSP) and consists in finding the sequence of moves that hoists must periodically repeat. The time to execute such a sequence is called the period. The optimal schedule can be computed off-line by minimising the cycle length (i.e., the average time to perform a job). The cycle is said to be n-periodic if n jobs are introduced into, and consequently n jobs are removed from, the line in each cycle. Thus the cycle length is obtained by dividing the period by the cyclic degree n.

Sometimes, the line is used to process large batches of identical parts. This kind of production seems like several successive CHSP and transitional periods must be studied; otherwise operators must empty the line before treating a new batch.

Multi-product mode: When jobs are different, two cases may arise :

If production forecasts are known (i.e. predictive approach), the schedule can be found beforehand. Usually, the entering sequence that minimises total completion time is determined, possibly by defining identical batches of non-identical parts that periodically enter the line. Thus, a cyclic schedule can be found off-line.

In the opposite case (i.e. reactive approach) the problem is called " Dynamic Hoist Scheduling Problem " (DHSP) as products randomly enter the line. Consequently, authors mainly use either expert scheduling systems or heuristic methods.

2. 4 Conclusion

Most of the published works deal with the single product case, fewer focus on DHSP and layout. So, the paragraphs below are organised as follows. Section 3 provides a review on CHSP. Section 4 makes a survey of approaches based on the multi-products assumption. At least, section 5 presents works studying either the layout of the stations or the optimal number of hoists to be used.

3. CYCLIC HOIST SCHEDULING PROBLEM

CHSP studies can be classified in two main categories, according to the chosen model :
- in the "carrier model", the products being simultaneously on the line are considered,
- in the "cycle model",the order of hoist movements in the resulting period is focused on.
  Generally, authors solve the basic problem and later bring extensions to solve more complex ones. The basic problem is the search of the optimal 1-periodic schedule in a line composed of one hoist, no duplicated and no multi-function tank. Consequently, tanks are numbered according to the given route. In a general way, authors define an integer linear model and use a solving procedure based on branch and bound principle. The two following parts will first present variables and associated constraints. Then authors' solving proposals and, finally, extensions, will be examined.

3.1 Carrier model

The variables of the problem are the transfer starting dates and the period. (ÊThey are written in bold characters in the following formulation.)

nt   the number of treatments in the sequence

o_i the real time the i-th soak operation of the resulting sequence takes ($1 \leq i \leq$ nt)

m_i, M_i the soak time boundaries of the i-th soak operation

R_i the i-th transfer operation of the sequence of treatment ($1 \leq i \leq$ nt-1)

r_i the time required to execute R_i

$d_{i,j}$ the constant time a free hoist needs to move from tank i to tank j

T   the period

$t_{i,k}$ the date at which the i-th transfer operation (R_i) concerning the product k starts

k   the number of the considered products ($1 \leq k \leq K_{max}$, K_{max} is a constant which represents the maximum number of products on the line at the same time)

NB : K_{max} is fixed by authors in the basic case but can be a variable to be determined in other cases.

The cyclic structure imposes that a product k enters exactly k-1 periods later than the first one and that all the transfers are performed during the cycle:

$$\forall\ i,\ 1 \leq i \leq nt\text{-}1,\ k > 1,$$
$$t_{i,k} = t_{i,1} + (k\text{-}1)*T \qquad (1)$$

So, solving the problem consists in determining T and the $t_{i,1}$ variables (simply called t_i). Such a model is composed of nt variables. The first strong constraint represents the bounded soak times :

$$\forall\ i,\ 1 \leq i \leq nt\text{-}1,$$
$$m_i \leq t_{i+1,k} - (t_{i,k} + r_i) \leq M_i \qquad (2)$$

This constraint is independent of k. As all the products are identical, it only has to be set for the first one. Besides, limited capacity of tanks, associated with the cyclic assumption, implies that during a cycle, each tank has one product dropped in and another one removed from it (but not necessarily in this order). So, the period must be longer than all the soak operations :

$$\forall\ i,\ 1 \leq i \leq nt\text{-}1,$$
$$T \geq t_{i+1,k} - (t_{i,k} + r_i) \qquad (3)$$

The hoist holds one job at once and must have enough time to move from the tank it has just filled to the one it must empty. So, disjunctive precedence constraints are set between hoist operations :

$$\forall\ i,j,\ 1 \leq j < i \leq nt,\ 1 < k \leq K_{max},$$
$$t_{j,k} - t_i \leq r_i + d_{i+1,j}$$
$$\text{or}\quad t_i - t_{j,k} \leq r_j + d_{j+1,i} \qquad (4)$$

Many solving methods rest on that model:
Shapiro and Nuttle, (1988) solve a benchmark presented by Phillips and Unger, (1976) but consider separated loading and unloading tanks. The starting

dates t_i are deduced from the vector $S = (s_1, ..., s_{nt})$ of the resulting soak times. At each node of the search tree, a product j is added and relative orders are defined between this product and in process jobs (from 1 to j-1). An order is represented by a vector $V_h = (v(1), ..., v(nt-2))$ with $1 \leq h \leq j-1$; $v(i) = k$ means that the i-th transfer of the product j follows the k-th transfer of the product j-h and precedes the k+1-th transfer of the product j-h. At this step a linear program checks the feasibility of relative order $[V_1, ... V_{j-1}]$ regarding the current constraints and S. When $Vp(1)$ is equal to nt this means that a solution composed of p-1 products has been found.

Lei and Wang, (1994) define a schedule function $\phi(i,j)$ which assigns an order number to the i-th transfer operation of the product j. This number fixes the place of a transfer operation in the schedule. So, at each step of the search, they determine which move can take place at the next free row in the sequence. To select moves, they also use temporal windows which express the earliest and the latest starting date to perform a move.

Baptiste, et al., (1994) have developed an original approach by using Constraint Logic Programming (CLP). They use a depth first search that consists in fixing a disjunction (i.e. to choose one of the mutually exclusive inequations (4) and checking the consistency of the current problem using a complete constraint solver). The solving method goes on until all disjunctive constraints are fixed. If the consistence test fails, the system backtracks.

3.2 Cycle model

This model considers the starting date of the transfer operations executed during a cycle and T. Besides, boolean variables are used to specify precedence relations between transfer operations.

τ_i the starting date of the i-th operation transfer (R_i) in the cycle

$x_{i,j}$ the boolean variable expressed as:

$$x_{i,j} = 1 \text{ if } \tau_i > \tau_j, \text{ else } x_{i,j} = 0$$
$$\text{with } x_{i,j} + x_{j,i} = 1 \qquad (5)$$

As a single cycle is considered, it is no use setting constraints (1). Besides, equation (2) and (3) become:

$$\forall i, 1 \leq i \leq nt-1 ,$$
$$m_i \leq \tau_{i+1} - (\tau_i + r_i + x_{i,i+1}*T) \leq M_i \quad (6)$$

$$T \geq \tau_{i+1} - (\tau_{i,k} + r_i + x_{i,i+1}*T) \qquad (7)$$

Using $x_{i,j}$ simplifies constraint (4) expression:

$$\forall i,j, 1 \leq i \leq nt, 1 \leq j \leq nt, i \neq j,$$
$$\tau_j - \tau_i \geq d_{i+1,j} + r_i - x_{i,j}*T \qquad (8)$$

Phillips and Unger, (1976) were the first to be interested in the CHSP. They have used this model to represent a complete example, called the Phillips benchmark. The latter deals with a basic line supplied with a single I/O station and is considered as the reference problem. Phillips and Unger have developed a Mixed Integer Programming method based on a branch and bound procedure to solve it.

Armstrong, et al., (1992) define a schedule (M,T) where $M = (m[1], ..., m[nt])$ is a sequence of transfer operations, m[i] being the i-th move in a cycle, where $T = (t[0], ..., t[nt])$ is the relative starting time of the moves. The authors progressively build the schedule and use a parameter called the Minimum Time Span (MTS) that gives a tight lower bound on the time required to complete any partial sequence of moves. This bound takes into account both the hoist moving time and the job minimal processing time.

It seems that Hanen and Munier, (1994a) mix the "cycle model" and the "carrier one". They have defined a relation between variables τ_i and t_i:

$$\tau_i = t_i + k_i * T \ (1 \leq i \leq nt) \qquad (9)$$

k_i is called occurence of the transfer operation R_i. Indeed, a transitional period is needed to make all the products enter the line and so k_i "pseudo periods" elapse until the hoist executes R_i for the first time. Hanen and Munier wanted to develop a CHSP specific algorithm instead of using classical linear programming. Thus, they have studied the structure of the problem solutions and have presented a solving method based on the graph theory. They have also used time windows to build the cycle (called ñthe motifî) progressively.

Chen, et al., (1995) solve the problem with two branch and bound procedures. The first one finds an initial state of the line and the second one determines the corresponding sequence of movements. In the first step, they have defined a useful bound K_{max}. The program lists the feasible distribution and represents one distribution in a vector $Cn = (c(1), ..., c(n))$ where c(i) is equal to 1 when a product is in tank i at the initialisation stage and equal to 0 otherwise. The second program manages conflicts between two operations. A bi-valued graph is used to check the feasibility and estimate the period.

3.3 Extensions of the basic problem

Phillips and Unger, (1976) extend their model by considering multi-function tanks. Shapiro and Nuttle, (1988) have considered duplicated tanks. They have indicated that it might be difficult to include the n-periodic assumption. Although their solving method is efficient for simple cases, Chen, et al., (1995) have proposed no extensions.

Other extensions developed in many papers are the n-periodic and the multi-hoist case. Armstrong, et al., (1992) have proposed to extend their model to the multi-hoist problem (Lei, et al., 1993) while precising this requires additional constraints to prevent the hoists from colliding. Both Manier, et *al.*, (1994) and Lei and Wang, (1989a) solve the n-periodic problem, but they restrict their search to low degrees (n<4) to limit the number of variables. They consider complex lines with duplicated tanks, multi-function tanks and several hoists. In this case, moves must be assigned to hoists. Manier, *et al.*, (1994) and Hanen and Munier, (1994b) solve this problem in a less restrictive way than Lei does: they allow hoists to move on the same part of the line (overlapping zone) whereas Lei and Wang, (1991) make consecutive hoists share only one common tank (non-overlapping zone).

Some papers consider fixed processing times, and develop polynomial algorithms to solve the problem (Hanen and Munier, 1994, Son, *et al.*, 1993, Levner, *et al.*, 1995). This can also be used to get a first solution for the complete CHSP.

At least, an other extension consists in determining a transitional period between two cyclic productions (Varnier and Baptiste, 1995a). Knowing two fixed cycles, an algorithm, based on the work of Manier, schedules the hoist moves to perform progressively moves of the second cycle instead of those of the first one. So simultaneously, the first type of product leaves the line when the second one comes in, while ensuring that soak times boundaries are respected.

4. MULTI-PRODUCT CASE

The following sections present a survey of reported works solving multi-product HSP either in advance or in a reactive way. Most of these papers rest on heuristic algorithms and consider a basic line.

4.1 Predictive approach

When products to be processed during the next several hours are supposed to be known, the problem can be solved off-line.

Fleury, *et al.*, (1996) assume that they know the next 24 hours production. Kangaroo algorithm and hill climbing are used to choose the best entering sequence. A Multi-Agents simulation evaluates the makespan and schedules all movements. A specific constraint is added: the number of carriers is fixed, which means that empty carriers can not leave the facility and must be considered in the schedule. Soak time constraints may be violated.

Ptuskin (1995) considers fuzzy processing times. An entering sequence of n non-identical parts, the order of which is supposed to be known, is periodically repeated. Each part entering date Vi in the sequence, a period R and exact soak times must be determined. All the tanks are sequentially visited by all the parts, according to various processing times. The algorithm uses n sub-CHSP solvings to find a set of common periods and a rule called "Fuzzy Prohibited Intervals Rule" representing the hoist disjunctive constraints, to determine variables Vi.

4.2 Reactive approach:

When products randomly arrive, the line must be dynamically controlled. Most of the methods proposed to solve this problem are either based on expert scheduling systems or on heuristics:

Expert scheduling systems: Thesen and Lei (1986, 1990) present a rule based expert system. Some decision rules permit to choose the heuristic to be applied according to the line current state. The used heuristics rest on priority levels associated with each in-process carrier (according to its current treatment and soaking time) and on hoist assignment rules. This kind of control system can quite easily drive several hoists.

Sun, *et al.*, (1994) develop a simulation system, based on Thesen and Lei's dispatching rules, in order to help schedulers by letting them "watch" the system's operations before they are implemented. When loading a job, they check fewer future assignments than Thesen and Lei do. Hence, they can not ensure that no defective job will be produced. Besides,they consider multifunction tanks, which is not examined by Thesen and Lei, but they assume that processing times are fixed.

The main drawback of such approaches is that they do not necessarily satisfy processing time constraints. That is why some authors have chosen to use heuristic algorithms to solve DHSP.

Heuristic approaches: All of them determine a new hoist movement schedule before letting the new entering product n be processed, which guarantees that no product will become defective. They are all based on a generalised "carrier model" that differs in the way constraint (3) is expressed. Indeed, the period T can not be considered anymore and r_i must be replaced by $r_{i,k}$, where k is the considered product. Besides, S[i,k] is the tank in which the i-th soak operation on job k is performed. So, constraint (3) becomes:

$$\forall\ i,j,k,l,\ 1 \leq i < nt_k,\ 1 \leq j \leq nt_l,\ k \neq l,$$
$$\text{such as } S[i,k]=S[j,l]$$
$$t_{j,l} - t_{i-1,k} \leq r_{i-1,k}$$
$$\text{or } t_{i,k} - t_{j-1,l} \leq r_{j-1,l}$$
$$(9)$$

The published works either keep the partial schedule related to the already in-process products or not.

Yin and Yih (1992) make the moves related to job n fit in the existing schedule. First, the timing of the product n transfers is computed from an initial entry time and minimum soak times, without considering the in-process carriers. Then, tanks and hoist constraints satisfaction is checked by looking for overlaps in the Gantt Chart. When a treatment of product n is responsible for a constraint violation, either its processing time is extended, or the product entry time is updated.

Yih (1994) improved this algorithm by using all the processing time tolerances: the order of the previous schedule is kept, but move starting dates may be changed.

Cheng and Smith (1995) applied a constraint satisfaction problem solving model for deadline scheduling to the problem Yih had set out. They used a heuristic procedure, generically called "precedence constraint posting", that relies on a temporal constraint graph representation.

Ge and Yih (1995) do not hold the previous schedule. Their heuristic method is based on a depth first branch and bound search that terminates on discovery of a feasible schedule . At each node, the algorithm chooses the next move to be performed, by trying first to make the new product enter. If it is not possible, the search order is defined from the time each job can still remain in its current tank before becoming defective. The feasibility of each sub-branch is checked by solving a linear programming problem based on the "carrier model".

Lamothe (1996) uses a depth first branch and bound search that gives priority to tank constraints. The search order is also defined according to increasing operation ready dates. Each sub-branch is evaluated by finding the longest path between two nodes on a PERT graph. If a positive circuit is detected, this proves the constraint system's inconsistency. Moreover, a dynamic backtracking procedure is used. It stores information from previous DHSP solvings, named Nogoods to explain why the backtrack is necessary. Thus, when a new job n arrives, Nogoods can be used to modify the previous search-tree (dealing with n-1 products). Up to now, Lamothe (1996) is the only author who solves DHSP in complex lines (with duplicated tanks, multi-function tanks and n hoists).without producing defective jobs.

Bloch, *et al* , (1996) compare the solutions given by hill climbing, simulated annealing, taboo search, kangaroo method and genetic algorithm. The previous schedule is partially called into question, since two adjustable points are defined to specify which portions of the sequence can undergo changes. The criterion evaluation rests on the same longest path assessment as the one presented in Lamothe (1996) apart from the fact that no information is stored when an inconsistency is detected.

5. LAYOUT

The layout of the tanks and the number of hoists are generally considered as fixed data. Yet, few authors have studied the relationship between layout of the tanks and productivity. Other ones have determined the optimal number of transporters. All of them consider a single type of product.

5.1 Layout of the tanks

Grunder, *et al.*, (1997) underline the relationship between the physical layout of the stations and the productivity of a treatment line. They focus on saturated single-hoist production lines. (In such lines, there are n-1 products simultaneously processed in a line of n stations.) They first consider a particular class of layout that is commonly used by industrial users and show that this class does minimize the hoist moving time during a cycle time. Then, they demonstrate that this particular class of layout is, however, not dominant in all cases for maximizing the productivity of the line. Finally, they propose a branch and bound algorithm to determine the optimal layout for maximizing the productivity in such lines.

5.2 Optimal number of hoists

Armstrong, *et al.*, (1995) determine the minimum number of hoists so that all the transport operations can be executed while satisfying all the constraints and avoiding traffic collision. The set of operations is partitioned in groups, each of them being served by a single hoist. The optimal solution, that corresponds to maximum group sizes, is obtained by solving linear programming subproblems, the duals of which are structured as shortest path problems.

Kats and Levner (1995) find the optimal number of hoists needed to meet a given schedule for all possible periods. They determine the minimal number of hoists as a function of T. They consider the basic case with fixed processing times.

6. CONCLUSION

This survey shows that a lot of different cases have been solved. Yet, a wide research field has still not been explored, particularly concerning multi-products cases, layout, and the simultaneous assignment and scheduling problem. Moreover, papers dealing with similar problems could have been quoted, such as Su and Chen (1996) or Rochat (1995). To permit the reader to place these works among the other described approaches, they have been included in the synoptic classification graph that is provided in appendix.

REFERENCES

Armstrong, R., L. Lei, and S. Gu (1992). A bounding scheme for deriving the minimal cycle time of a single-transporter n-stage process with time-window constraints. In: *GSM Working paper 92-07*. Rutgers University.

Armstrong, R., S. Gu and L. Lei (1995). A greedy algorithm to determine the number of transporters in a cyclic electroplating process, In: *Proc. of the symp. on Emerging Technologies and Factory Automation (ETFA 95)*, **vol 1**, pp 460-474. Paris.

Baptiste, P., B. Legeard, M-A. Manier and C. Varnier (1994), A scheduling problem optimisation solved with constraint logic programming, In: *Proc. of the 2-th int. Conf. on the practical application of prolog*, pp 47-66. London.

Bloch, C., C. Varnier and P. Baptiste (1996). Applying stochastic methods to the real time hoist scheduling problem. In: *Proc. of the computational engineering in systems application multiconference (CESA'96)*, **vol 1** pp 479-484 . Lille.

Chen, H., C. Chu and J-M. Proth (1995), Cyclic hoist scheduling based on graph theory", In: *Proc. of the symp. on emerging technologies and factory automation (ETFA 95)*, **vol 1,** pp 451-459. Paris.

Cheng, C-C. and S.F. Smith (1995), A constraint-posting framework for scheduling under complex constraints", In: *Proc. of the symp. on emerging technologies and factory automation (ETFA 95)*, **vol 1,** pp 269-280. Paris.

Fleury, G., J-Y. Goujon, M. Gourdand and P. Lacomme (1996), A hoist scheduling problem, containing a fixed number of carriers, solved with an opportunistic approach, In: *Proc. of the computational engineering in systems applications multiconference (CESA'96)*, **vol** 1, 473 - 478. Lille.

Ge, Y. and Y. Yih (1995), Crane scheduling with time windows in circuit board production lines, In: *INT. J. PROD. RES.*, **vol 33**, n° 5, pp 1187-1199.

Grunder, O., P. Baptiste and D. Chappe (1997), The relationship between the physical layout of the workstations and the productivity of a satured single hoist production line, In: *Int. J. of Prod. Res.*, to appear.

Hanen, C. (1994a), Study of a NP-hard cyclic scheduling problem, In: *European J. of Operations Research*, **n° 72**, pp.82-101.

Hanen, C. and A. Munier (1994b), Pilotage periodique d'une ligne de galvanoplastie ^ plusieurs robots, *Research report LITP*.

Kats, V. and E. Levner, The constrained cyclic robotic flowshop problem : a solvable case, In: *Intelligent scheduling of robots and flexible manufactruring systems* (Levner, (ed.)), pp 115-128. Holon.

Lamothe, J., C. Correge and J. Delmas (1995), A dynamic heuristic for the real time hoist scheduling problem, In: *Proc. of the symp. on emerging technologies and factory automation (ETFA 95)*, **vol 2**, pp 161 - 168. Paris.

Lamothe, J., C. Thierry and J. Delmas (1996), A multi-hoist model for the real time hoist scheduling problem, In: *Proc. of the IMACS int. Multiconf. on computational engineering in systems applications (CESA'96)* , pp 461 - 466. - Lille.

Lei, L. and T.-J. Wang (1989a), On the optimal cyclic schedules of single hoist electroplating processes", In: *Working paper #89-0006*, publisher, Rutgers university.

Lei, L. and T.-J. Wang (1989b), A proof : the cyclic hoist scheduling problem is NP-complete, In: *Working paper #89-0016*, Rutgers university.

Lei, L. and T.-J. Wang (1989b), The minimum common cycle algorithm for cycle scheduling of two material handling hoists with time window constraint, In: *Management Science*, **vol 37**, n°12, pp 1629-1639.

Lei, L., R. Armstrong and S. Gu (1993), Minimizing the fleet size with dependent time-window and single track constraints", In: *Operations Research Letters*, **vol 14**, pp 91-98.

Lei, L. and T.-J. Wang (1994), Determining optimal cyclic hoist schedules in a single hoist electroplating line, In: *IIE Transactions*, **vol 26**, n° 2, pp 25-33.

Levner, E., V. Kats and V-E. Levit (1995), An improved algorithm for a cyclic robotic scheduling problem, In: *Intelligent scheduling of robots and flexible manufactruring systems* (Levner, (ed.)), pp 129-141. Holon.

Manier, M-A., C. Varnier and P. Baptiste (1994), A multi-hoist scheduling problem approach, In: *proc. of the 4-th workshop on project management and scheduling*, pp 110-115. Belgium.

Phillips, L.W. and P.S. Unger (1976), Mathematical programming solution of a hoist scheduling program, In: *AIIE Transactions*, **vol 8**, n° 2, pp 219-225.

Ptuskin, A-S. (1995), No-wait periodic scheduling of non-identical parts in flexible manufacturing lines with fuzzy processing timesÓ, In: *Intelligent scheduling of robots and flexible manufactruring systems* (Levner, (ed.)), pp 210-222. Holon.

Rochat, Y. (1995), A genetic approach for solving a scheduling problem in a robotized analytical system", In: *Intelligent scheduling of robots and flexible manufactruring systems* (Levner, (ed.)), pp 191-209. Holon.

Shapiro, G. and WH. Nuttle (1988), Hoist scheduling for a PCB electroplating facility, In: *IIE Transactions*, **vol 20**, n° 2, pp 157-167.

Song, W., R-L. Storch and Z-B. Zabinsky (1993), An algorithm for scheduling a chemical processing tank line, *Production planning and control*, **vol 4**, n° 4, pp 323-332.

Song, W., R-L. Storch and Z-B. Zabinsky (1995), An example for scheduling a chemical processing tank line", In: *Proc. of the symp. on emerging technologies and factory automation (ETFA 95)*, **vol 1**, pp 475-482, Paris.

Su, Q. and F.F. Chen (1996), Optimal sequencing of double-gripper gantry robot moves in tightly-coupled serial production systems, In: *IEEE transactions on robotics and automation*, **vol 12**, n° 1, pp 22-30.

Sun, T_C., K-K. Lai, K.Lam and K-P. So (1994), A study of heuristics for bidirectional multi-hoist production scheduling systems", In: *Int. J. of production economics*, **vol 33**, pp 207-214.

Thesen, A. and L. Lei (1986), An expert system for scheduling robots in a flexible electroplating system with dynamically changing workloads, In: *Proc. of the second ORSA/TIMS conf. on FMS/ operations researsch models and applications*, pp 555-566.

Thesen, A. and L. Lei (1990), An expert scheduling system for material handling hoists, In: *J. of manufactoring system*, **vol 9**, n° 3, pp 248-252.

Varnier, C. and P. Baptiste (1995a), A CLP approach for finding a transition schedule between two cyclic mono-product productions in electroplating facilities, In: *Proc. of the int. Conf. on industrial engineering and production management (IEPM '95)* , **vol 1**, pp 194 - 203. - Marakech.

Varnier, C., O. Grunder and P. Baptiste (1995b), Improving the productivity of electroplating lines by changing the layout of the tanks, In: *Proc. of the symp. on Emerging Technologies and Factory Automation (ETFA 95)*, **vol 1**, pp 441-450. Paris.

Yih, Y. (1994), An algorithm for hoist scheduling problems", In: *Int. J. Prod. Res.*, **vol 32**, n° 3, pp 501-516.

Yin, N-C. and Y. Yih (1992), Crane scheduling in a flexible electroplating line: a tolerance based approach, *J. of electronics manufacturing*, **vol 2**, pp 137-144.

APPENDIX

The synoptic classification graph. To simplify, only the first author and date are mentioned.

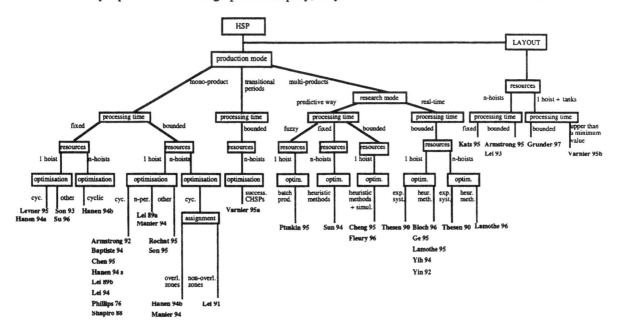

Classification of different publications on HSP

Abbreviations:

batch prod.: batch production
cyc.: cyclic
exp. sys.: expert system
heur.meth.: heuristic methods
n-per.: n-periodic

non-overl. zones: non-overlapping zones
overl. zones: overlapping zones
simul.: simulation
success. CHSPs: successive CHSPs

A DESIGN APPROACH THAT INTEGRATES
THE SAFETY AND DEPENDABILITY CONCEPT

Laurent Grudzien, Laurent Jacquet and René Soënen

LAMIH - URA CNRS N°1775 - Groupe de Recherche en Génie Industriel et Logiciel
University of Valenciennes - BP 311 - 59304 Valenciennes Cedex - France
Phone: (33) 03 27 14 13 79 - Fax: (33) 03 27 14 12 88
E-mail: lgrudzien@univ-valenciennes.fr

Abstract: Design a system that is dependable is today a constraint appearing in any customer requirements. It is a stake that can't be ignored and is a part, among others, of the performances that a product must reach during its running phase. For that, it must be taken into account and integrated at the design phase of it. Thus, after a presentation of the safety and dependability concept and an analysis of the problematic, the paper suggests a design approach that makes a distinction between the product model and the design process. This approach also integrates, in a concurrent engineering organisation, the safety and dependability concept.

Keywords: specification, design, product model, design process, safety, dependability, maintenance.

1. INTRODUCTION

Nowadays, the industrial competitivity is supported by the quality and the technological level of products (systems). In the same way, some parameters as product innovation degree, product functionalities, design delay respect... must be considered. The previous characteristics can be reached through the automation and the technical optimisation of these products (systems). This objective requires to use sophisticated and complex technologies. Thereby, the slightest system failure can have, in these conditions, critical indeed even catastrophic effects.

In this context, the safety and dependability aspects become a design stakes because they allow to decrease economical, human and environmental risks. The dependability characteristics that a system have to respect during the running phase must be defined in the design phase. This paper is relative to this problem. It explains a design procedure that integrates the safety and dependability aspects. The presentation of this design procedure will be done through three parts.

In the first part of the paper, we present the safety and dependability domain. It groups together characteristics as reliability, maintainability, availability and safety. These characteristics have to be defined at the design phase and maintain during the running phase of the product.

In the second part, we briefly describe the existing design approaches found in the literature. However they don't take the dependability and safety aspects into account, significant criteria of product quality.

With the objective to solve this problem, we present, in the third part of the paper, the design approach defined in the laboratory. This design approach integrates, in a simultaneous engineering organisation, the dependability and safety concepts. This approach makes a distinction between the product model and the design process. We will explain essentially the product model which is described through five modelling levels and more precisely the dependability concepts that are taken in consideration at each modelling level with the aim of defining a safe product.

2. THE SAFETY AND DEPENDABILITY DOMAIN

To face competition, each industry must be equipped with very high performance products (systems). For

135

this, they are automated, integrate different technologies, that lead to an increase of their complexity. It is therefore necessary to master this complexity (Moreau, 1995) and to see to it that the systems will be dependable, the delays and the costs reduced. To reach these objectives, it is necessary to:
— minimise non-expected stopping (failures) by a reliability improvement,
— minimise intervention times by taking into account maintainability criteria,
— increase availability, concept that covers the reliability and maintainability ones,
— put the system in fail-safe state when a failure occurs,
— make the system run even in a degraded state.

It is consequently necessary to obtain a system that is dependable i.e. (Leroy and Signoret 1992) a system that realises the mission for which it has been designed, without incident putting its profitability in question and without accident bringing into play safety. It is so important to define, from the requirements definition, the safety and dependability characteristics that the system will have to meet.

2.1. The safety and dependability concept

The safety and dependability concept relies on the notion of risk which can be considered as a two dimensions entity: occurrence probability of a dread event and gravity of its consequences. This general notion allows to characterise catastrophic events that question safety (low occurrence probability, high consequence) as events that question the installations production (high occurrence probability, low consequences).

Risk can be expressed as:

$$R = f(Pr , G) \quad (1)$$

where Pr represents the probability (or frequency) of appearance of system failure and G the importance (gravity) of the consequences of the considered misfunction. Thus, in accordance with the objectives pursued during the study (reliability, maintainability, availability or safety, the four dependability characteristics), this one can take a different way. But, the other characteristics must not be neglected because, by improving one aspect, one risks to damage the others. It is so essential to take into account these different aspects simultaneously in order to find a compromise, particularly between safety/reliability on the one hand, and availability/productivity on the other hand. It is so advisable to have at one's disposal methods and tools that allow to contribute to these objectives achievement.

2.2. Determination of the dependability characteristics

Reliability (Arinc, 1995). It is defined as the probability that a system or product will perform in a satisfactory manner for a given period of time [0,t] when used under specified operating conditions:

$$R(t) = Prob\ (E\ not\ failed\ during\ [0,t])$$

The reliability evaluation is generally measured by:
— the Mean operating Time To First Failure (MTTF) which is the mathematical expectation of the running time before the first failure

$$MTTF = E(T) = m = \int_0^\infty (1 - F(u))du = \int_0^\infty R(u)du \quad (2)$$

— the Mean operating Time Between Failures (MTBF) that corresponds to the mathematical expectation of the running time between failures

$$MTBF = \int_{-\infty}^{+\infty} t.f(t)dt = \int_0^{+\infty} R(t)dt \quad (3)$$

Maintainability (Blanchard, et al., 1995). It is defined, for an entity under specified operating conditions, as the probability that a given maintenance operation should be realised during a given period of time (0, t), when the maintenance is ensured in given conditions and with the use of prescribed procedures and means, as:

$$M(t) = Prob\ (maintenance\ of\ E\ is\ achieved\ at\ time\ t)$$

The Mean Time To Restoration (MTTR), that represents the mathematical expectation of the time before restoration, can be associated to the maintainability. This time is mathematically defined by:

$$MTTR = \int_0^{+\infty} (1 - M(t))dt \quad (4)$$

Availability. It can be defined as the probability that an entity is in an availability state under specified conditions, at a given time, by supposing that the supplying of necessary external means is ensured:

$$A(t) = Prob\ (E\ not\ failed\ at\ time\ t)$$

To improve an entity availability, it is necessary to reduce the number of its stopping and especially the ones that have for origin a failure (reliability) but also reduce the time spent to correct them (maintainability).

In the case where the entity is unrepairable, the availability A(t) is equal to its reliability R(t).

The Mean Up Time (MUT), mathematical expectation of availability time and the Mean Time Between Failures (MTBF), mathematical expectation of the time between failures can be associated to the availability. For a great number of systems, the difference between MTTF and MTBF is very small. Thus, we consider:

$$MTTF \approx MTBF = \int_0^{+\infty} R(t)dt \quad (5)$$

Safety. It is an entity ability to avoid, under specified conditions, to make critical and catastrophic events

appear (Villemeur, 1988). Safety and risk are two notions closely linked. In systems safety, the risk relative to a dread event occurring during a dangerous activity is defined by two parameters:
— the occurrence possibility of a dread event (causes probability),
— the consequences gravity which, finally, correspond to deaths, serious injuries, destruction, a mission loss, ...

Synthesis. The results obtained by this dependability analysis are varied: probabilistic parameters linked to the safety or to the economy of the installation (availability rate, MTBF, MTTR, ...), give prominence to preponderant parameters beside the envisaged risk (critical path), evaluation of different possible designs, ... All these results are finally decision aided elements that will allow to establish specifications for sensible components, tests programs for new components, maintenance strategies to put in place or the risk level of the system and/or of its entities.

3. THE DESIGN DOMAIN

3.1. The design approaches

Two design approaches can be distinguished in the literature. The axiomatic one (Suh, 1990) is based on the definition of several domains (customer, functional, physical, ...) and axioms allowing to have a good design.

The second approach, the algorithmic one (Pahl and Beitz 1984) is composed of a phases set (specification, detailed design, ...) and steps that aim to define progressively the system design.

3.2. Synthesis

Even if the axiomatic approach has the advantage to fix the framework of the design, it nevertheless not precises how to act inside this framework. For the algorithmic approach, its sequential organisation as its lack of precision on the different steps that compose each phase, make it go against the concepts defined by the concurrent engineering. In addition, these two approaches don't take into account the design for exploitation concepts and especially the concepts related to the maintenance.

To make up for these problems, a design approach has been developed at the laboratory. This one is "locally algorithmic", non monotonous, and is made of an axiomatic component. This approach makes a distinction between two aspects: product model and design process. This approach is described in the next part.

4. PRODUCT MODEL AND DESIGN PROCESS

These two notions have been defined (Krause, *et al.*, 1993) as a logical accumulation of relevant information concerning a given product during its life cycle, for the product model. The design process, which is also commonly referred to as product development work flows or product modelling processes, represents the product modelling processes consisting of a set of technical and management functions required to transfer initial idea to final product. We call a product any system that is of use to somebody.

The approach proposed by the laboratory is relatively similar. For the first aspect, this one takes into account the product modelisation during the design through the product model. This product model is described by five representation levels. Each level makes reference to a set of entities or concepts that allows to model different aspects of the product. The second aspect explains, as for it, the dynamics that exists between each representation level of the product through the design process.

The five representation levels that compose the design approach, will now be described.

4.1. Presentation of the different representation levels

The five representation levels of the product model (Fig.1) represent a part of the information that are essential for the product model definition. These information are generated during the design process. Thus, whatever the adopted design process, the concepts associated to each representation level will have to give information at a moment or another. The representation levels that have been identified are presented in the next paragraphs.

Needs representation level. This level allows to model, in a formal way, the customer needs. This point of view represents the customer needs transcription, expressed in a literal form, in a set of requirements (functions) specifying the services that the product must satisfy, and the set of constraints (functions) that it must respect. This point of view models in fact the different functions or missions that a product must carry out (perform) and the constraints it must respect especially in safety and dependability terms (product reliability, safety of human beings and goods, ...).

At this level are defined the functions the product will have to meet by respecting a certain number of constraints. For this reason, the concepts "Service functions" and "Global constraints functions" have been proposed (Jacquet, *et al.* ,1996). The first one models the services the product must fulfil, the second models the constraints it has to respect. Thus, a product can be expressed as a doublet:

$$P = < FS , Cg > \qquad (6)$$

with   P: product to design
           FS: set of service functions
           Cg: set of global constraints

Cg can be defined as a set of characteristics to which a value is attributed:

$$Cg = (\{ Charact.1, Value1\}; \{Charact.2, Value2\} ; \ldots ; \{Charact.N, ValueN\}$$

Each of these characteristics can be seen, by the different designers, as constraints they will have to respect at each representation levels of the product. They can also be seen as objectives to reach at the product performances level.

In relation to this formulation, the global constraint Cg is defined as a restriction in the set of acceptable values. We note so:

$$T_{|Cg} \, (fsk) = T \, (fsk) \tag{7}$$

the restriction of T to Cg.

It is also at this level that are defined the dependability requirements of the system. They group together the reliability, maintainability, availability and safety characteristics. They can be expressed in a qualitative or a quantitative form.

The global constraint, taken into account in the product modelisation context, is the dependability one. This concept is defined by a quadruple:

$$D = \{R,A,M,S\}$$

with   R: product reliability characteristic
         A: product availability characteristic
         M: product maintainability characteristic
         S: product safety characteristic

The meeting of the dependability concept (or objective) will be possible by the use of reliable components (low failure rate), use of redundancy, by the setting up of preventive maintenance strategies, but also by the definition and the implementation of supervision and diagnosis means in order to detect abnormal situations.

Functional requirements representation level. The objective of this level is to model the set of solutions which can be able to support each service function. Even, these solutions must respect the constraints specified by the client. This level defines the behaviour the system must adopt to accomplish its mission and the principles necessary to support the behaviour. In this aim, three concepts are associated to this level (Jacquet, 1996): "operational function", "operational principle" and "solution principles".

The operational function concept model the operations which are necessary to define the behaviour of the system. The second one precises the principle (magnetic, mechanical, electrical ...) able to support each operation. The last one models the conceptual solutions which can be associated to each service function.

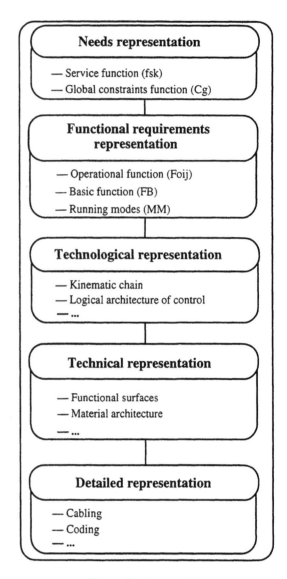

Fig. 1. Product model

The mapping between service functions and operational functions is realised by the transformation T :

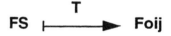

with: FS: set of service functions fsk
         Fo_{ij}: set of operational functions of level "0" (i=j=o)
         T: transformation operator

Even, an operator of transformation "tr" allows the mapping between operational functions of level "0" to operational functions of low level until the definition of elementary operational functions.

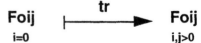

The set of global constraint functions which are not satisfied can also be defined as a restriction of the set of admissible values. The restriction of "tr" to "Co" is noted as :

$$tr_{|Co}(foij) = tr(foij) \tag{8}$$

with: Co: set of global constraints no satisfied

$$Co \subseteq Cg$$

The degraded running modes is defined from the sequence of operational functions which represent the normal running mode of the system. Then, it is necessary to determine means which will permit to detect and diagnose these degraded running modes.

"Functional failure mode" concept. The exploitation designer starts the design study at this representation level when all the operational functions are defined. First of all, he contributes toward the choice of the operational principles associated to the operations. Then, he defines the system running modes. In this objective he will determine unusual running modes of the system.

Unusual running modes definition. In this objective, we determine all the failure or degraded running modes that can occur in the system. However, at this level of representation, only the information about the operations that the system must satisfy are available. Therefore, only the functional failure modes of the system can be defined.

First of all, the operational functions supposed to incorporate a risk for the realisation of the system mission are identified. Then, we define the failure modes of the elementary operational functions. At last, the effects generated when a failure mode occurs, are determined.

Four failure modes consequences on each elementary operational function can be listed:
— *no function*: the function doesn't run. In this case the function is unavailable,
— *function loss*: the function is active and it is stopped by a failure. In this case the function is unreliable,
— *degraded function*: the function is active but its performances are not nominal. The risk of this failure mode is that it can generate operational error with high consequences (human and material safety),
— *untimely function*: the function is active by itself at any time or the function effect are not in accordance with the required effect.

The representation model used to present this concept is the Failure Modes and Effect Analysis (FMEA) array . The algebraic form of the information contained in the Failure Modes and Effect Analysis array, is as following:

$Mof = \{m_1, m_2, m_3, m_4\}$ =set of functional failure modes

$$Card(Mof) = 4$$

with: m_1 = function loss
m_2 = no function

m_3 = degraded function
m_4 = untimely function
Caf = set of causes
Eff = set of effects

A functional failure relation has been defined as :

$$Rd(fo_{ij}) \subseteq (\phi(Mo_f) \times \phi(Ca_f) \times \phi(Ef_f)) \tag{9}$$

with Rd (fo$_{ij}$): functional failure relation of fo$_{ij}$
Ø: power set operator
x: Cartesian product
fo$_{ij}$: operational function of level i and row j

The definition of these failure modes will add new states, on the operational function sequence, as safety modes, that represent the running modes of the system.

"Running mode" concept. The running mode, which will be necessary to take into account in the next design phases, are highlighted from the definition of the system states generated by a failure. In the following sections, the running modes that can be found when one used the system, are presented.

Running modes of a system. Any system is made in order to bring services to the user. In this aim it is often necessary to use procedures. These procedures characterise the different running modes of the system. The running modes (states that a system can reach) are grouped into in three families (Adepa, 1981) :
— normal running modes,
— stop modes,
— failure modes.

The first family is composed by normal, verification, test, preparation and closure running modes. The second one is constituted by all the modes which lead the system in a none running mode (stop in initial state, after the cycle end...). In the last family, are found all the modes which lead to service loss in consequence of a failure. It is composed by the following modes: emergency stop, diagnosis, failure treatment, degraded running modes.

Representation model associated to this concept. The representation model associated to the running mode concept is the states graph. The states associated to the dependability objectives to reach, contribute to highlight the necessity to determine supervisor or redundancy means. The choice between one of these two solutions is linked to the failure mode, so to the event generated by the failure mode (unavailability, no safety, no reliability). From this graph, can be defined the redundancy functions (active or passive) and the functions that need instrumentation (control and/or supervision instrumentation).

Algebraic form of "Running mode" concept. For the set of operational functions, a set $\emptyset_{Eo}$ of the states φ_{eo} of fo$_{ij}$ input flow with i, j > o and a set $\emptyset_{So}$ of the states φ_{so} of fo$_{ij}$ output flow have been

defined. Two sets can be added:
— a set $\emptyset_{deg}$ of degraded states φ_{deg} of fo$_{ij}$ output flow. $\emptyset_{deg}$ characterises the decreasing performances of fo$_{ij}$ function;
— a set $\emptyset_{fail}$ of failure states φ_{fail} of fo$_{ij}$ exit flow. $\emptyset_{fail}$ characterises a loss of fo$_{ij}$ function.

The running mode concept as an operator that associates an fo$_{ij}$ exit flow to an fo$_{ij}$ entry flow by using procedures. These procedures allow to avoid an unusual running mode of the fo$_{ij}$ function. This operator is defined as :

$$MM : \{\emptyset_{Efoij}\} \longrightarrow \{\emptyset_{Sfoij}\}$$

$$\varphi_{efoij} \xrightarrow{\ mm\ } \varphi_{sfoij}$$

with: $\{\emptyset_{Efoij}\}$/ set of states φ_{efoij} of fo$_{ij}$ input flow
$\{\emptyset_{Sfoij}\} = \{\emptyset_{deg}, \emptyset_{fail}, \emptyset_{test} ....\}$= set of unusual states φ_{foij} of fo$_{ij}$ output flow
MM: set of running modes "mm"as:

$$\forall mm \in MM$$

$$mm(\varphi_{efoij}) = \varphi_{sfoij}$$

For an operational function, the mapping between the entry and the exit flow is made by an operation "o" associated to the operational function. When this operation is not carry out in "normal conditions", the exit flow is so not in accordance with the required exit flow, and unusual running modes occur.

Technological representation level This level presents the technological structure of the system. It allows the representation of the system from the point of view of the different designers (Automation, Mechanical and Maintenance teams...). From the principle solution and the material means (resources) associated to them, detection and diagnosis means which must be applied to the resources are defined.

Technical representation level This level models the technical structure of the system. The proportioning and the realisation techniques are considered here. All the material components that compose the system are chosen. As information about failure rate are available, reliability, safety and maintainability mathematical models can be elaborated. Therefore, the control that the dependability requirements are respected, can be made. Constraints about the use of standard components rather than prototype components, maintenance procedures, ..., can be defined.

Detailed representation level This level presents the detailed specifications of the system (morphological aspect of components, coding, cabling...).The final structure of the product is determined in the end of this stage. The dependability characteristics of the system are verified (in term of safety an d reliability). Means which allow to carry out a better maintainability have been defined (detection means, failure diagnostic means...).

5. CONCLUSION

Dependability is a real stake nowadays because it is very important to master the risk (economical, human, environment). Therefore, the dependability aspect must be integrated in the design and the exploitation phases of modern systems.

In this paper, a design approach that integrates, in a concurrent engineering organisation, the safety and dependability concept has been presented. The dependability concepts that the designer must take into account in order to obtain a dependable product have been described. They have been defined for two of the five representation levels of the product model.

This proposition has been validated through the study of an assembly system (Dspt8, 1997). It will be illustrated during our presentation through the study of a double-entrance security door example. This product allows the entrance of only one person and forbids entrance of persons that have metal objet. One objective is to show the interactions between designers and the calling into question that can occur when the dependability requirement are not satisfied.

REFERENCES

Adepa (1981). *Le GEMMA, guide d'étude des modes de marches et d'arrêts*. ADEPA, Collection Génie Productique.

Arinc Reseach Corporation (1995). *Product reliability, maintainability and supportability handbook*. Edited by M. Pecht. CRC Press.

Blanchard B.S., D. Verma and E.L. Peterson (1995). *Maintainability, a key effective serviceability and maintenance management*. Wiley & sons.

Dspt8 (1997). *Scénario d'ingénierie communicante pour les systèmes intégrés de production*. Rapport final du projet DSPT8 en productique.

Jacquet L., G. Doriry and R. Soënen (1996). « Towards a specification method for an automated system : the third phase of the method ». In: Proceedings of the World Automation Congress (WAC'96), pp. 253-258. Montpellier, France.

Krause F.L., F. Kimura, T. Kjellberg and S.C.Y. Lu (1993). « Product modelling », In: CIRP annals manufacturing technology, Vol. 42/2, pp.695-706.

Leroy A. and J-P. Signoret (1992). *Le risque technologique*. Col. Que Sais-Je ?, PUF.

Moreau J-C. (1995), «Automatismes et maîtrise de la complexité». In *Les nouvelles stratégies techniques — La puce à l'usine*. Collection F. R. Bull. Masson.

Pahl G. and W. Beitz (1984). *Engineering design*, (Edited by Ken Wallace). Published by The design council. London.

Suh N. P. (1990). *The principles of design*. Oxford series on advanced manufacturing. Oxford university press.

Villemeur A.(1988). *Sûreté de fonctionnement des systèmes industriels*. Col. Direction des Etudes et Recherches d'EDF, Eyrolles.

TOWARDS A HOLON-PRODUCT ORIENTED MANAGEMENT

Eddy BAJIC, Frédéric CHAXEL

Research Center for Automatic Control of Nancy
CRAN - CNRS URA 821
Faculté des Sciences, BP 239
Vandoeuvre les Nancy, FRANCE
Tel/Fax : 00.3.83.91.20.16 Email : bajic@cran.u-nancy.fr

Abstract : This paper describes research currently being carried out in the field of holonic Manufacturing Systems as a step towards holonic concept partial validation. The originality of the presented work deals with the idea that process management may be achieved not only by machines, operators, but also by part and product itself, carrying and managing its own information. Presented research work deals with STEP methodology and formal models to provide a product-holon behavior validation case for manufacturing control, based on information management. This paper demonstrates the feasibility of high level information structuration, based on product data definition, with advanced management technics on attached escort memories. Distribution of the information system onto each individual product requires tools and methods to manage and provide access to product related information for resources, automated systems and human operators. Automatic identification technology using electronic data carriers, Radio Frequency transponder, can provide such functionality and act as product information vectors to provide a generic support for holon-product implementation.

Keywords : Information technology, Process control, Data models, Holonic manufacturing, STEP, EXPRESS, Product Data Management

1. INTRODUCTION

Since the last decade, tremendous changes have occurred in the field of manufacturing systems considering production control and organization.
To manage product diversity and rapid changes in production demands, research effort of the eighties have been focused on manufacturing systems architectures satisfying productivity, flexibility, openness characteristics for integration aspect and cost reduction, widely based on automation increase (Dilts, 1991).
Originally represented by the CIM concept (Computer Integrated Manufacturing), new approaches have emerged to remedy the limits hit by such hierarchically and centrally controlled organization suffering from rigidity to fast production changes and adaptations.
Then, priority have been put on decentralization and distribution of decision-making activities and information flow with an expected downsizing of manufacturing systems. Fully distributed structures were proposed (Duffie, 1986) involving concurrent mechanism for job decision and coordination support, based on network communication procedures, contract nets. In this heterarchical architectures were promoted to provide manufacturing systems with decentralization, modularity, robustness and cooperative functionnalities.

Beyond that approach characterized by system-based functionnalities, strong demands appears for new formalizations of manufacturing systems to break off from conventional system organization to satisfy autonomy and individuality along with cooperation capabilities amongst all components of a manufacturing system. Emphasis was put on coordination, self-organization, hyper-flexibility, adaptation and part-orientation.

2. NEXT GENERATION OF MANUFACTURING SYSTEMS

Several new paradigms for manufacturing system description were formulated drawing inspiration from the real world mainly over the two aspects of social organization of human societies and natural rules and environment.

Biological metaphor have been proposed by Okino (1994) to fit autonomous distributed manufacturing system description, so called *Bionic Manufacturing System* (BMS) in CAM-I/Japan.
BMS concept lays on a system theory involving self-organization rules over distributed components with quasi-life functions copying genetic behavior, allowing freely adaptive and flexible connections in response to changes of system conditions. Ueda (1994) specified a model of BMS as a pseudo ecosystem built around works and manufacturing cells thus mimicking biological organisms carrying DNA (Deshoxyribo Nucleic Acid) gene type and BN (Brain and Neuron) type behavior and information. In a Bionic world environment, works are considered as organisms that grow up to become or generate products, and manufacturing cells are other type of organisms that process works. DNA type information is supporting inherited information and transformation as "growing up" objectives of products, while BN type information governs works activities. In this, a manufacturing system control behavior should be described by evolution strategies with mathematical model based expressions.

A social organization metaphor has been derived from the Hungarian journalist and philosopher Arthur Koestler (1989) proposition, specified early in 1971 in the prophetic book "The ghost in the machine", as the holonic concept. Based on the observation of self-regulation capabilities of social organization, a social holonic behavior involves very few characteristics supported by elementary entities named *Holon*, acting and cooperating to assume living of complex and ruled organization named *Holarchy*. Koestler adds that holonic argument can be applied to biological, social or cognitive hierarchy which manifests rule-governed behavior and structural consistency.
The word Holon is build over the Greek root "Holos" meaning "Whole" completed by termination suffix "on" for "Particle" as encountered in the word neutron, to represent behavior properties of "a whole

by itself and a part within other wholes".

The holonic metaphor have been applied to manufacturing system organization by the International joint program IMS (Intelligent Manufacturing System) with the thematic research consortium HMS (Holonic Manufacturing System) as a promising response to provide production systems with robustness and adaptability to condition changes and disturbances, modularity and flexibility capacities.
Underlying principles of holonic application to manufacturing systems have been defined as follows by the HMS consortium (Van Brussel, 1995) :

Holon : An autonomous and co-operative building block of a manufacturing system for transforming, transporting, storing information and physical object.
Holarchy : An assembly of holons which act in cooperation having a specific set of objectives and common goal.
Autonomy : Ability of an holon to control its own execution plan associated with its own strategy.
Co-operation : Capabilities of systems entities to communicate, negotiate and execute actions plans in order to reach an objective.

The previously presented paradigms are actually drawing up the next generation manufacturing systems to come in the next decades. Scientist community research activities have to cope with formalization and resolution of the aiming capabilities of the real world to validate and prove efficiency within the production world.
The following paragraphs will focus on a validation approach of the holon concept supported by part product and resources cooperating in a manufacturing environment.

3. PRODUCT ORIENTED MANAGEMENT

A common characteristic is revealed in both approaches BMS and HMS, with the duality of elementary organisms performing in complex systems. An holon consists of an information processing entity allied to a physical processing entity, as also biological organisms are unified entity of information and substance. This characteristic contributes to the emergence of an autonomous behavior appearance in complex systems.

Three types of basic building blocks in an holonic manufacturing system are: resource holons, order holons also call task holons and product holons (Bongaerts, 1995), co-operating to perform production (Fig 1.)

A resource holon consists of a physical part, namely a manufacturing resource in the HMS, powered by physical capabilities. It is an abstraction for the production means such as production machines, conveyors, ...

An order holon is an entity responsible for performing the work in a suitable and successfulness way. It captures all information related to a job (Bongaerts, 1995).

Finally a product holon can be seen as an entity created by real or forecasted market demand. It holds all the necessary knowledge to assure the realisation of a product.

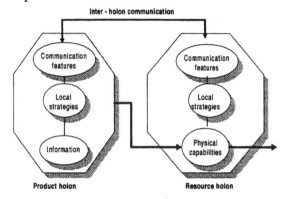

Fig 1. Resource and Product Holons structure.

Product· information and knowledge is basically defined at the design level in words of functionality, shape, quality, manufacturing operations sequence, ... This knowledge is constructed step by step and structured through a life cycle product information model.

This product knowledge must be shared by all the actors tools used for the product design. In this field some work has be done in order to define both standardised product model, like STEP for instance, and integrated information infrastructure like AIT Integration Platform or Esprit NEUTRABAS platform.

At the manufacturing stage, this knowledge is conceptually duplicated to fit to each individual product or holon. One can say that product definition data set is cloned, and then follow it's own life.

One of the major problem lies in the management of this information set associated to each holon. For instance Duffie (1986) proposed some computers, connected with each others, in order to play the role of products data manager. Some other authors (Upton, 1992) are proposing the use of electronic tags, but without any regards about integration problems.

In the following proposed approach, a product or any other holonic entity object carry its own information on a programmable tag which can be accessed in read/write mode by each user involved in its development process. Advantage of this approach is in machine capacities to focus on operations control (transport, measures, ...) and not on coherency

maintenance of product related information, this point is conceptually assumed by each individual product. A dialog can be imagined inbetween machine and product as follows : *"Part 17843, where are you ?"* say the machine , *"I am currently at Machine 4 for a 10 minutes operation and I expect to be soon served by yourself "* replies the product.
This leads to the emergence of claims contracts between customers and suppliers in a concurrent client/server formalism.

Unfortunately, electronic tags are by now not easy to integrate in an enterprise information system. It is for instance impossible to read or write structured information on these systems because they use elementary data handling protocol instead of high level and semantic information manipulation dialogue. In this field, the MONOLIN project (MObile NOdes in Logistics and Industrial Network, EP 6936) develops the basic requirements for a standard identification systems network interconnection interface allowing integration : the Escort Memory System - Application Programming Interface so called EMS-API (Monolin, 1994).
Significant validation of a such approach can be foreseen, taking into account that ANSI X3T6 subcommitte is actually working on an Radio Frequency identification devices (RFID) interoperability standard, information structuration will follow (X3T6, 1996).

An Holon-Product Information management approach is proposed in the following chapter as a logistic support to manufacturing system control. Holon-Product specification will assume both vertical integration (link with the design level) and the horizontal integration by supplying methodology and support tool allowing that each part carrying electronic tag, can act as a communication vector of process information system, in charge of the overall application management, and information coherency, consistency and reachability.

4. PRODUCT-HOLON INFORMATION MODELLING

One of the most significant approach today in product modelling is the development of ISO 10303 standards called STEP (STandard for Exchange of Product model data) which define models, database access and neutral data files format for representation and exchange of product data. The goal is to define complete models for product life-cycle in a CIM context, as well as the means for exchange of data between enterprise functions along a product life-cycle definition (from CAD to production, maintenance, ...).

EXPRESS is a formal modelling language, object oriented, that models the knowledge about information used by an information system. It provides the words, syntax and grammar needed to describe an application field. EXPRESS is designed to satisfy basic requirements as modelling of the information and processing objects; definition of the constraints rules; definition of the operations performed by objects; be readable by an human operator as also be automatically computed.

5. PRODUCT-DRIVEN MANUFACTURING

5.1. *Product/Process Interrelationship*

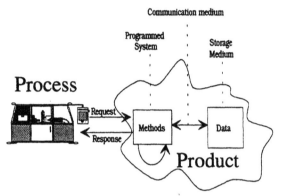

Fig 3. Process/product relationship based on the client/server model.

In the presented model, every product is acting as a real actor within its process able to :
- manage itself its characteristics like shape, design version, ...
- has knowledge of its evolution through its operation sequence
- store its history in the holonic process.

Figure 3 points out the concept adapted from the client/server model we chose in order to respond to these objectives. In such a case each product is the manager of its own information and gives them to the process after it received a service request : *"What is the next operation to be proceed on yourself"* for example.

On the basis of a client/server relationship between product and process, user requests - i.e. process requests - are sent to an interface supporting product access methods in an object-oriented interrogation form such as *Object_Selector.Message (Parameters)* or *SystemMessage (Parameters)* :

- COLOR.GETVALUE ( );
- COLOR.PUTVALUE (Red);

The data storage medium, in our case, consists of identification tags carried by products or by pallets and also an optional network database. Explanation about the choice of two possible storage medium is

made in sub-paragraph 5.3.

5.2. *Product Model Translation for Manufacturing Control Services*

The major problem, related to this model (Fig 3.) is the integration of tags in the information system which necessitates :

- reference conceptual models to put the information about the products themselves on tags;
- tools to implement these models and to access the information.

For the purpose of the presented work, it is assumed that a logical data model of product for manufacturing operations exists and is defined in the EXPRESS language (this model could be a STEP application protocol or a specific model).

Figure 4 shows the STEP-based methodology allowing translation of product model, described in the EXPRESS formalism (entities, attributes, functions, rules) :

- to a product-information access library supporting a direct requesting of target

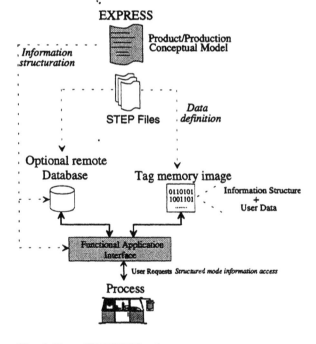

Fig 4. From EXPRESS schema to customers requests management at the shop floor.

application on the EXPRESS schema data;
- to the optional database structure definition.

The product definition within the STEP neutral file is used :
- to create the tag memory image, for the first state of a product associated to the current phase of its life cycle in the manufacturing

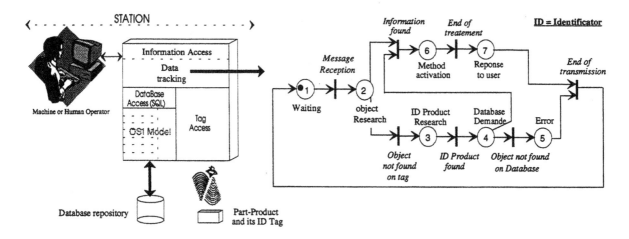

Fig 5. State graph of product data remote access from a process point.

environment;
- if needed, to populate the previously created database schema.

5.3. Remote Database Reachability

Tags are of course limited in capacity, and storing complex structures needs more memory than only storing the contents (Attributes value). Today, tags tend to have higher storage capacity (up to 128 Kbytes) but in some cases (like automotive industries) this may not be enough, so if it is relevant for the application a part of the data could be stored on other medium.

A network database in client/server mode seems to be the natural architecture to manage the overflow of data :
- we need to have a persistent computer storage in order to manage EXPRESS schema population and translation;
- it stores and manages high volume of data;
- time response is efficient for most of the applications;
- enterprises are already trained with such an architecture.

The presented approach has the same philosophy as distributed database (Morris, 1992) that is, "users do not have to know where data are stored to access to them". The system performs data locating, formatting and transferring through the structure (see Fig 5.). In fact, when we populate the schema, we specify where EXPRESS entities must be (on tags or remote database). Afterwards, the user can send a request to the system, without specifying the location of the objects he wants to access.

The search for physical storage of objects is made after each user request reception, as shown in the state graph model exposed in the Fig 5.

5.4. Application Tool

A prototype Tag-STEP (see Fig 6.), running under MS-Windows, uses a standard computer to generate a MobIle DAtabase Nodes structure. It is connected to a relational network database management system (ORACLE 7) through SQL_NET-TCP-IP-Ethernet protocols.

The EXPRESS schema is converted in a relational schema which has to be put on the ORACLE database. In this case each EXPRESS entity is

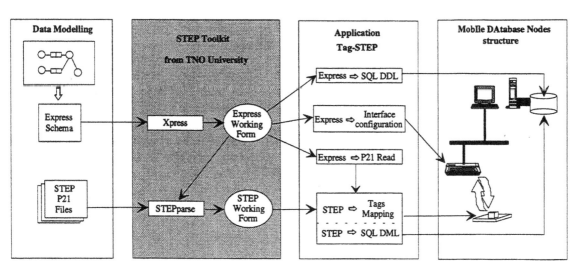

Fig 6. STEP based methodology for holon product definition and management

translated to a relational entity - i.e. a table - and to a view in order to manipulate sub-typing. The schema is then converted as a C program to generate the Functional Application Interface as explained before. This program is able to read or write data independently on tags or on the database. To populate the schema we use a STEP neutral file provided by the product definition phases.

6. CONCLUSION

The presented work is a step toward holonic concept validation. This new approach for information management and decision making necessitates methodological framework for Information System implementation as well as interfacing tools based on international standard definitions. Dealing with STEP formal models, the paper has demonstrated the feasibility of high level information structuration, based on product data definition, with advanced management technics on attached escort memories.

Emerging technologies and standards for intelligent automatic identification systems are coming up with local processing capabilities, and also full object language characteristics of Express with methods and algorithms, should allow to upgrade this methodological approach for a full holonic characteristic implementation.

Expected benefits in manufacturing systems organisation and management are:
- better data distribution in the overall manufacturing system allowing real time tracking of products. Each holon-product is an element of a distributed manufacturing database allowing immediate consultation by other process actors;
- simplification in the production machine structure and software capabilities. A generic interface is defined to link different kinds of users to the structure in the product coupled memory. There will be a gain in the production system modularity and modifiability, ...

One of the most significant extension of this concept concerns it's application to cover the full product life cycle data management from production to after sale services and recycling. This is actually in progress in the field of a research contract between the CRAN and a French cars manufacturer, to provide a holonic approach car management all along the vehicle life-cycle (Lonc, 1996).

Basically, STEP could be the general framework used for vehicle data management and as recommended by AIT research program. Linking between different phases - i.e. different data models - will be realised through models described in EXPRESS-X, for the creation of views on a model and mapping data from one model to another.

REFERENCES

Bongaerts L., Valckenaers P., et al. (1995) Schedule Execution for a Holonic Shop Floor Control System, *In Proc. ASI'95*, Lisbon, Portugal

Dilts D.M., Boyd N.P., (1991). Whorms H.H. The evolution of control Architecture for Automated Manufacturing systems. *In Manufacturing Systems*, Vol 10, N° 1, pp 79-93.

Duffie N. A. , Piper R. S. (1986). Non-hierarchical Control of Manufacturing Systems. *In Manufacturing Systems*, Vol 5, N° 2, pp 137-139.

Goh A. , Hui S.C. , Song B. , Wang F.Y. (1994) A study of SDAI implementation on object-oriented databases. *In Computer Standards & Interfaces*, Vol 16, 1994, pp 33-43

Koestler A. (1989) The ghost in the machine, *Arkana books*, London.

Lonc B., Bajic E. (1996) "Design and exploitation of communicating escort memories for automotive applications". In Proc. Int. conf. on Advanced microsystems for automotive applications, VDI-VDE-IT, pp. 142-150, Dec. 96, Berlin, Germany.

Monolin Esprit 6936 (1994) Deliverable 06 - MONOLIN Implementation Guide : Implementation Context and General Overview

Morris K.C. , Mitchell M. (1992) Database Management Systems in Enginneering, *NISTIR 4987*, NIST, Gaithersburg, Maryland.

Okino, N (1994). Bionic manufacturing system. *In Manufacturing Systems*, Vol. 23, pp. 175-187

Ueda, K (1994). Biological-oriented paragidgm for artifactual systems. *In Proc. Japan USA symposium on flexible automation*, Kobbe, pp 1263-1266

Upton D.M. (1992) A Flexible Structure for Computer-Controlled Manufacturing Systems. *Manufacturing Review*, Vol 5, N°1, pp 58-74

Van Brussel H., Valckenaers P., Bongaerts L., Wyns J. (1995). Architectural and system design issues in holonic manufacturing systems. *In Proc. IMS 95*, Bucharest, Romania, pp 142-145.

Weston H., Clements P., Murgatroyd I. (1994) Information Modelling Methods and Tools for Manufacturing Systems. *In Proc. ISATA 27th* pp 227-234, Aachen, Germany

X3T6 Standards, ASC X3T6 96-100, RFID Systems - Long Range Active RF tags, Vol. 1 Communications standards, Draft version 1 - release 5, November 1996

TECHNICAL FAULT AND QUALITY MANAGEMENT - TERMINOLOGY OF FUNCTIONS, PURPOSES AND MEANS

Dirk van Schrick

Safety Control Engineering Group
Safety Engineering Department
University of Wuppertal, Germany
Email: schrick@wrcs1.urz.uni-wuppertal.de

Abstract: The paper provides a description of a new and comprehensive model of concepts termed *Technical Fault and Quality Management* and focusses, after outlining its basic structure, the *Technical Management Model* and the *Management System Model*. Some aspects of these partial models of concepts are explained in more detail where the relation between the means of the Technical Management sections and the Management System functions are in the center of interest. The relation between Technical Diagnostics and five Quality Management elements is discussed also.

Keywords: Terminology, Management, Fault , Quality, Functions, Purposes, Means.

1. INTRODUCTION

In many engineering and non-engineering fields concepts just as their terms and definitions constitute the basis for making oneself understood to each other when talking about things of the non-linguistic reality. During congresses or workshops as well as with the study of technical literature, it is noticeable all the time that contributions on terminology in the field of topics of interest are very rarely. For example, in the field of automated manufacturing system this fact is reflected in form of talking and writing about different subjects such as modeling, scheduling, planning or diagnosing of the production process and system. Contributions in which no methods and no applications but concepts termed or defined and their integration into a general context are dealt with can be found only here and there. Currently, the IFAC Technical Committee SAFEPROCESS discusses the terminology in the field of supervision, fault detection and diagnosis. The following terminological approach is a basic contribution to this discussion.

Within the fields *process monitoring, fault diagnosis and control* or *quality process control and ma-* nagement, for instance, a great number of concepts can be found which built up on the basic concepts *failure* and *fault*. These concepts are often one part of more complex concepts expressed by two-constituent linguistic items, such as composite words. Within the context of this paper, these concepts and their thematically related concepts form the basic set of concepts for developing a complex adaptive model of concepts. The *Technical Fault and Quality Management Model (TFQM-Model)* does not comprise terms and definitions of concepts but, within a limited context, both the concepts of interest and their mutual relations that depend on different given aspects. A complex modelling process led to the TFQM-Model intended as a first step for discussing terminological problems within the IMS community and the diagnostic and quality community world wide. Thus, a terminology in sense of a fixed system of terms for the communication upon this field of interest can be developed to reduce communicational barriers.

The TFQM-Model consists of three parts termed *Elements for Consideration Model (EC-Model)*, *Technical Management Model (TM-Model)* and *Management System Model (MS-Model)*. A *Fault*

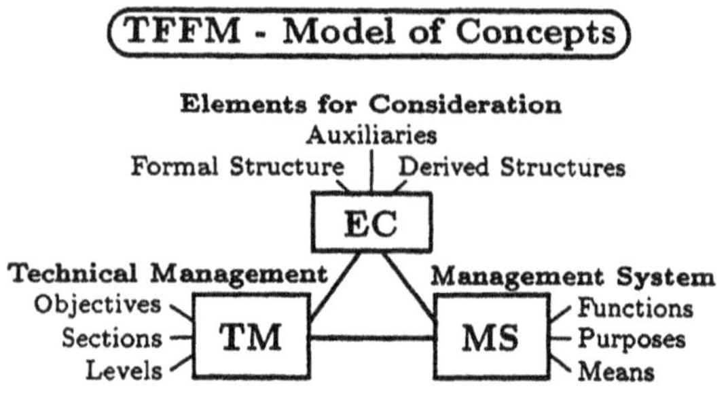

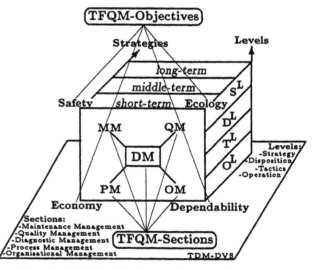

TFFM - Model of Concepts

Elements for Consideration
Auxiliaries
Formal Structure | Derived Structures

EC

Technical Management / \ Management System
Objectives   Functions
Sections — TM —— MS — Purposes
Levels   Means

Fig. 1. TFQM-Model of Concepts; Aspect: Structure

Cause and Effect Schema (FCES) is the basis for the EC-Model and for the link between the EC-Model and both the TM-Model and the MS-Model. Therefor, the concept *failure* relating to a state-oriented view represents the link to the TM objectives *safety, economy, ecology* and *dependability*, and the concept *fault* relating to a function-oriented view represents the link to the MS purposes *fault consideration, fault avoidance, fault surmounting* and *fault control*. The link between the TM-Model and the MS-Model is that a management system is indispensable to implement the management for approaching its objectives. The engineering-oriented part of the MS-Model is termed *Diagnostic System Model (DS)* and is of interest in this contribution. Consequently, the three partial models of concepts EC, TM and MS are standing in a mutual relation resulting in the TFQM-Model structure shown in Fig. 1.

The contribution to the IMS'97 workshop is structured as follows: Section 2 provides an outline of the TM-Model, Section 3 provides an outline of the MS-Model and Section 4 contains a detailed description of the relation between the means of the Technical Management sections and the Management System functions emphazising the relation between Technical Diagnostics and five Quality Management elements.

In van Schrick (1996), a first version of a German-language model of concepts has been introduced where the German-language EC-Model has been described in detail in van Schrick (1997a). English-language model descriptions can be found in van Schrick (1997b) on condition monitoring concentrating on the FCES and the TM-Model, in van Schrick (1997c) with respect to the diagnostic system concentrating on diagnostic purposes and functions, in van Schrick (1997d) focussing systems diagnosis concentrating on system's insufficiencies and diagnostic means and in van Schrick (1997e) reference is made to 23 definitions of concepts relevent to the SAFEPROCESS community (Patton, 1997) where the set of concepts has been both considered as a sub-set of the TFQM-Model and structured with respect to various aspects.

2. TECHNICAL MANAGEMENT

The comprehensive concept *management* can only be reflected correctly with the term "management", if it is evident that *to manage* means the handling comprehensively in all sections and on all levels of an organization and not the activities restricted to top managers. In this case, not only the failure- and fault-related and therefore the diagnosis-related matters are concerned, but also all matters with respect to maintenance, quality, process and organization. The concept *technical* refers to technical insufficiencies such as faults and failures and to tasks for the fulfilment of purposes solved with technical means. The tasks are distributed to the organization and concern technical and organizational fields. Moreover, organization-related not desired phenomena affect an item during the pre-operational and the operational phases and become noticeable as performance insufficiencies of that item that have ideally to be detected as early as possible to be able to react as fast and safe as possible.

All activities of the overall management function of an organization that determine the policy, objectives and responsibilities, and implement them by appropriate technical means within the management system is understood as *Technical Management (TM)*. To provide a basis for discussion and modelling, a partial conceptual model termed *Technical Management Model (TM-Model)* is introduced. Figure 2 illustrates this partial model of concepts with the help of the main aspects *ob-*

Fig. 2. TFQM - Buildung; Aspect: Organization

jectives, sections, levels and *strategies* in form of a multi-level TFQM building. The essential concepts will be discussed in the following.

148

2.1. TM Objectives

The TFQM building reflects the objectives *safety, ecology, economy* and *dependability* with regard to the principle of structure (sections and levels) and the principle of temporality (strategies) according to a general characterization of an organization. In the context of an economical and ecological promising operation, the avoidance of local and global accidents is not only an economic(al) and ecological objective, it concerns the operation under the aspects *safety* and *dependability* too.

The concept *safety* is associated with the use of an item in health- and life-critical situations. Safety is a principle of the society and is one aspect of its requirements for an item. The concept is based on the concepts *risk, limit risk* and *danger* where the risk quantity R is the parameter defining the state of an item. Depending on R being less or greater than a limit risk R_L, the concept *safety* or *danger* is applicable. In addition, the concept *hazard* refers to the source, the origin, of danger. Consequently, the concept *safety* refers to a state in which the risk of harm to person or damage is limited to an acceptable level (DIN 1995).

In contrast, the concept *dependability* is associated with the reliance into the provided service of an item. Dependability is a principle of the user and is one aspect of his requirements for an item. The concept consists of both the concept *availability* and the concepts *reliability, maintainability* and *maintenance-support*. The latter mark the availability influencing factors. Consequently, the concept *dependability* can be seen as a collective concept used to describe the availability performance and its influencing factors (DIN 1995).

Besides these differences, the concepts *safety* and *dependability* share some common properties. Both concepts refer to the element for consideration *failure* expressing the link to the EC Model. With respect to a specification, the concept *failure* is associated to an indicator for the undesired event *transition* of an item from the state *conformity* to the state *nonconformity*. Both concepts refer to requirements for technical items to be fulfilled by appropriate means to approach to the TFQM objectives. Considering a quantitative description, appropriate mathematical tools are required to describe safety and dependability where the latter, often regarded synonymously to the concept *reliability*, is a time-dependent function, for example. According to DIN (1995), both concepts subsume time-related aspects of the concept *quality* associated with the totality of characteristics of an item with respect to its ability to satisfy stated and implied needs. The content of the concept *item* is that what can be individually described and considered, essentially processes and products. The totality of *requirements*

for quality has to be fulfilled by means of *quality engineering*, e.g. the sectional means stated in the following sub-section.

2.2. TM Sections

Regarding Fig. 2, the central TFQM section is represented by the concept *Diagnostic Management (DM)*. The DM subjects are technical insufficiencies, such as component faults or production disturbances, that have to detected, diagnosed and controlled by suitable technical means. There are three engineering-oriented and one organization-oriented TFQM sections associated with the DM.

The concepts *Maintenance Management (MM), Quality Management (QM), Process Management (PM)* and *Organisational Management (OM)* refer to these sections where the concepts *Entity Maintenance (EM), Quality Control (QC), Process Control (PC)* and *Enterprise Controlling (EC)* refer to the sectional means. In addition to QC, the concept *QM means* includes the concepts *Quality Planning (QP), Quality Assurance (QA)* and *Quality Improvement (QI)*, cf. DIN (1995). With respect to the functions, sectional means, strong relations exist between section DM and MM, QM, PM and OM. This will be shown later.

2.3. TM Levels

All operational techniques, activities and actions have to be applied on all TFQM levels showing the different views onto an organization. A general accepted schema for structuring is a model of levels comprising the levels *operation, tactics, disposition* and *strategy* (Erdmann, Schnieder and Schielke, 1994). With respects to the aspect *function*, these levels can be assigned to the levels of an organization, e.g. in manufacturing, production engineering or administration. According to the TFQM philosophy, all levels are affected to reach the TFQM objectives, but both the operational and the tactical level are of main interest due to the aspect *technical*.

With respect to the aspect *temporality*, short-, median- and long-term TFQM strategies can be stated which contribute to approaching to the objectives, such as planned or unplanned maintenance, to minimize the causes and effects of manufacturing insufficiencies, for instance.

3. MANAGEMENT SYSTEM (MS)

In general, the concept *system* refers to elements and relations between these elements forming a structure and showing a behaviour. The elements are sub-systems interacting with each other according to the definition of the relations. Depending on the level of consideration, each element, i.e. each sub-system, can be considered itself as a system being in contact with its environment. Similar to Erdmann, Schnieder and Schielke (1994), the concept *system* can be defined with respect to the concepts *structure*, *causality*, *temporality* and *decomposition* where the latter subsumes the concepts *hierarchy*, *differentiation* and *aggregation*. Systems can be natural, i.e. physical or biological, or artificial, i.e. interactive or other, and, seen from a different point of view, the concept *system* refers to a totality of technical, organizational or other means for an independent fulfilment of tasks.

In the TFQM context, the totality of organizational structure, procedures, processes and resources needed to implement the TFQM is the subject of the concept *Management System (MS)*. The MS should be as comprehensive as needed to meet the objectives of the organization. With regard to the concept *system* given above, the MS is an artificial, interactive system consisting of two essential parts, the *Technical Management System (TMS)* and the *Organizational Management System (OMS)*. The TMS itself consists of the sub-systems *Maintenance Management System (MMS)*, *Quality Management System (QMS)*, *Diagnostic Management System (DMS)* and *Process Management System (PMS)* necessary to implement the TM sections to which the sectional TM means are associated. The OMS necessary to implement the OM section by the sectional means EC, of course, consists also of appropriate sub-systems such as the *Logistical Management System (LMS)* or the *Administrational Management System (AMS)*. But in the context given, no details on the OMS are of interest.

Intended as a support for the organization approaching its objectives, the TMS has to provide appropriate functions realized with suitable means to accomplish the TFQM purposes. The MS-related basic concepts *functions*, *purposes* and *means* will be explained in the following.

3.1. Functions

The concept *function* refers to a task conditioned by the purpose of use. To derive appropriate reactions such as maintenance activities or safety-related actions, the functions to be provided by the TMS have to enable the operator or the PMS to gain and to process information on faults being or emerging in a technical item or on fault causes acting onto it and possibly influence the manufacturing process. The functions can be understood as parts of a control loop illustrated in Fig. 3. Within the TFQM context, there exists a multi-level schema for structuring the function-related concepts where the meta-function concepts *monitoring*, *diagnosis*, *prognosis* and *reaction* are

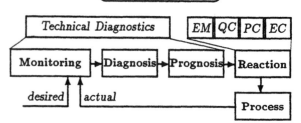

Fig. 3. TFQM - Control Loop; Aspects: Functions, Sectional Means

in superordinate position with respect to different standard-function concepts, e.g. *evaluating*, *optimizing* or *protecting*. These concepts refer to basic-function concepts such as *actuating*, *displaying*, *intervening* or *switching*. Within the meta-function *monitoring*, for instance, a basic-function subset is necessary to form the control error to be processed with the help of diagnostic and prognostic meta-function both comprizing also suitable basic-functions for maximizing fault information. The meta-functions *monitoring*, *diagnosis* and *prognosis*, termed *diagnostic functions* and to be provided by the *diagnostic system*, are realized by the TM means *Technical Diagnostics (TD)* provided by the TM section *Diagnostic Management (DM)*. Considered as the controller in the control loop, the meta-function *reaction* is realized by the sectional TM means EM, QC, TD, PC and EC.

With respect to quality control, the feedback is realized not only with process information but also with product information. In general, the implementation of the meta-functions requires both technical and organizational activities and actions within the TMS and OMS. The latter, for instance, has to provide organizational means for the accomplishment of purposes such as long-term *fault prevention*, medium-term *fault-effect minimization* and short-term *fault-correction*.

3.2. Purposes

The TFQM functions accomplish purposes helpful with approaching to the TFQM objectives. With regard to the causes of insufficiencies of an item the concept *purpose* very often refers to the element for consideration *fault* expressing the link to the EC Model. The large number of concepts found in literature can be structured by means of the relations *superordination - subordination* resulting in the taxonomy shown in Fig. 4 neglecting the symbol *F* for the term *Fault*. The superordinated concepts are *fault consideration*, *fault avoidance*, *fault surmounting* and *fault control*. Both concepts *fault surmounting* and *fault control* subordinate the concept *fault recognition* that subsumes the fault-related concepts *detection*, *dia-*

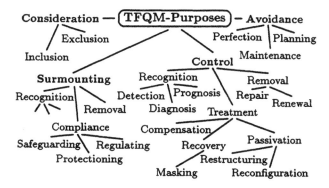

Fig. 4. TFQM - Taxonomy; Aspect: Purposes

gnosis, prognosis directly associated to the three meta-function-related TD concepts. Excluding the concepts *fault consideration* and *fault recognition*, the totality of purpose-related concepts is associated to the fourth meta-function-related concept *reaction*. With respect to the aspect *structure*, the concept *fault intolerance* subsumes the concepts *fault avoidance* and *fault surmounting*. The latter refers, for example, to classical sensor- and signal-based emergency actions. With respect to the concept *fault control*, the concept *fault tolerance* subsumes the concepts *fault recognition*, *fault treatment* and *redundancy* where all three concepts are aspects of the concept *fault-tolerant control*. The concept *fault avoidance* is associated with short-time pre-operational activities for fault prevention, for example by perfection. The medium-term and long-term operation-related activities and actions are aspects of the concepts *fault removal* and *fault treatment*. The latter refers to the concepts *fault causes* and *fault effects* where only the concept *fault passivation* is associated to the concept *fault cause*.

3.3. Means

For the realization of effective functions purposive means have to be provided. The concept *means* refers to the concepts *operation, description, realization, techniques* and *tools* and is one aspect of the concept *resource* subsumed under the concept *management system*. There is a conceptual relation to all phases an item passes through such as *planning, design, test, operation, maintenance* or *disassembly*. Actually, more and more attention is paid to compatible means for description used in all phases and resulting in a large number of promising quantitative and qualitative model-based techniques and relevant information-based tools.

4. TM-SECTIONS AND MS-FUNCTIONS

In the previous sections the TM-Model and the MS-Model have been outlined. In the following, two different relations will be described: firstly,

with regard to the sectional means, the relation between both models where the TM section concepts are related to the MS meta-function concepts *monitoring, diagnosis, prognosis* and *reaction* symbolized by M, D, P, R; secondly, with regard to technical means, the conceptual relation between the sectional means TD and QC described by five QM elements.

4.1. Sectional Means

The four symbols stated are used to label the levels of a function-related model of levels. With respect to the concepts of the TM sections and sectional means, functional concepts can associated to them on each level. Table 1 illustrates the applied model of levels where the index refers to the TM sections, - symbolizes that a suitable concept is unknown and a symbol in round brackets means that the concept is used seldom, but is often subsumed by the concept *monitoring*. The table shows that in all management sections the function *monitoring* is essential whereas the function *reaction* can be found in all but the DM section.

Regarding the MM section that refers to the condition of an item, the concept *monitoring* is replaced non-synonymously by the concept *inspection*, often with reference to *condition-based maintenance (CBM)*. Sometimes it is used as a superordinating concept for the concepts *diagnosis* and *prognosis*. The concept *reaction* consists of the concepts *prevention* and *correction*, the latter subsuming the concepts *repair, renewal* and *rework*.

Regarding the QM section where its means *quality control* refers to processes during the production, maintenance or installation phase, the concept *monitoring* is replaced non-synonymously by the concept *surveillance* including, with respect to an item, the concepts *monitoring* and *verification*, both associated to process parameters and product characteristics, and the concept *analysis of records*. The concept *diagnosis* subsumed by QC refers to the determination of fault causes. The concept *reaction* consists of number of concepts re-

Sections		MM	QM	DM	PM	OM
Means		EM	QC	TD	PC	EC
Functions	M	M_M	M_Q	M_D	M_P	M_O
	D	(D_M)	D_Q	D_D	-	(D_O)
	P	(P_M)	-	P_D	-	(P_O)
	R	R_M	R_Q	(R_D)	R_P	R_O

Table 1. TFQM - Sections; Aspect: Functions

fering to activities and actions for parameter and characteristics control, e.g. the concepts *corrective actions* and *process control*.

Regarding the DM section refering to an item's condition, state and function, the concept *diagnosic functions* is the basis for the underlying model of levels and need not to be discussed further. The concept *reaction* is seldom used in this context, but some concepts subsumed by it are included in the concepts *process supervision* or *plant safeguarding* in sense of monitoring and emergency reactions.

Regarding the PM section refering to processes during the operational phase, the concept *monitoring* possesses its fundamental content consisting of the concepts *measuring*, *observation* both associated to the item's state and function expressed by process and plant parameters. The concept *reaction* consists of a number of concepts describing standard and basic functions mentioned earlier.

Regarding the OM section refering to organizational processes, the concept *monitoring* is used in its fundamental sense too. The concept *reaction* consists of a number concepts, such as *replanning*, describing functions of no interest in this context.

Summarizing, it is obvious that the concept *diagnostic system (DS)* refers to each of the three diagnostic function-related levels in each TM section focussing *monitoring*. But intuitively, this concept refers to the TD means implemented.

4.2. Technical Means

With respect to technical means, a different kind of relations can be stated considering requirements for 20 QM elements, specified by a QA model according to DIN (1994), with regard to their relevance to Technical Diagnostics. Essentially, there are 5 elements, labeled with their number in the standard, with a relation to TD. Concepts in round brackets refer to other sectional means. The concepts are *process control (9)*, subsumed by *QL* and refering to *monitoring (and control)* of suitable process parameters and product characteristics, *inspection and testing (10)*, refering to *monitoring* product characteristics, *control of inspection, measuring and test equipment (11)*, refering to *monitoring* (and *calibration and maintenance*) of equipment and devices, *corrective and preventive action (14)*, subsumed by *QL* and refering to *diagnosis* (and *elimination*) of causes of actual and potential faults, and *statistical techniques (20)*, refering to *QL* and to *establishing, monitoring, verifying* process capability and product characteristics.

Summarizing, it is obvious that TD is essential to QC. With respect to processes, TD can provide an important contribution to establish *quality (control) loops* resulting in qualitatively highly valuable products based on processes controlled.

5. SUMMARY AND FUTURE WORK

With respect to a conceptual description, the TM-Model and the MS-Model have been considered and described as a independent parts of the complex model of concepts TFQM. Two kinds of relations between these partial models have been explained. Current work is directed to a bilingual description of the TFQM emphasizing the relation to German-language and English-language standards within the European Community.

REFERENCES

DIN (ed.) (1994). DIN EN ISO 9001. *Qualitätsmanagementsysteme; Modell zur Qualitätssicherung/QM-Darlegung in Design, Entwicklung, Produktion, Montage und Wartung*, Beuth Verlag GmbH, Berlin.

DIN (ed.) (1995). DIN EN ISO 8402. *Qualitätsmanagement; Begriffe*, Beuth Verlag GmbH, Berlin.

Erdmann, L., Schnieder, E. and Schielke, A. (1994). Referenzmodell zur Strukturierung von Leitsystemen, *Automatisierungstechnik*, at 42, No. 5, 1994, pp. 187-197.

Patton, R.J. (ed.) (1997). *Proc. SAFEPROCESS'97*, August 26-28, Hull, England.

van Schrick, D. (1996). Ein Begriffsmodell für ausfall- und fehlerrelevante deutschsprachige Termini der Automatisierungstechnik, in: VDI/VDE-GMA (ed), *VDI-Berichte GMA-Kongress'96*, No. 1282, VDI Verlag GmbH Düsseldorf, pp. 743-752.

van Schrick, D. (1997a). Begriffsmodell zu den Betrachtungsgrundlagen Fehler, Ausfall, Störung und Versagen, in: VDI/VDE-GMA (ed), *VDI-Berichte GMA-Aussprachetag Sicherheitstechnik und Automatisierung*, No. 1336, VDI Verlag GmbH Düsseldorf, pp. 3-12.

van Schrick, D. (1997b). Technical Diagnostic Management: Terminology of Objectives, Functions, Purposes, Means. Accepted for Presentation and Publication in: *Proc. COMADEM'97*, June 9-11, 1997, Helsinki, Finland.

van Schrick, D. (1997c). The Diagnostic System - A Termininological Approach. Will be published in: *Proc. SYSID'97*, July 8-11, 1997, Kitakyushu, Japan.

van Schrick, D. (1997d). System Diagnosis - Terminology of Purposes, Functions, Means and Methods. Will be published in: *Proc. ASCC'97*, July 22-25, 1997, Seoul, Korea.

van Schrick, D. (1997e). Remarks on Terminology in the Field of Supervision, Fault Detection and Diagnosis. Will be published in: *Proc. SAFEPROCESS'97*, August 26-28, 1997, Hull, England.

WORKPIECE SHAPE AND SETUP IN MILLING

P. Risacher   **J.Y. Hascoët**   **F. Bennis**

Institut de Recherche en Cybernétique de Nantes
IRCyN (ex. LAN) UMR CNRS 6597 - École Centrale Nantes
1, Rue de la Noë, BP 92101, 44 321 Nantes cedex 3 - France
E-mail : Jean-Yves.Hascoet@lan.ec-nantes.fr

Abstract : This paper deals with the positioning of a workpiece on a milling machine. The problem is to orient the workpiece, so as to minimize the number of machining sequences, for reducing costs. This work is also to study the choice of machine configuration. A new method is shown for constructing the set of solution - machine choice and workpiece setup.

Keywords : CAD/CAM, CNC Machining, Manufacturing systems, Mechanical engineering, obstacle detection and avoidance.

1. INTRODUCTION

In order to use NC machining for the manufacturing of a workpiece, a great deal of preparation work has to be done. Since this is a very time-consuming job, many attempts have been done in order to automate the process. There are many works on the toolpath generation, on the choice of milling tool and other parameters. There is always a common hypothesis : the choice of the milling machine and the workpiece setup are already done. But if some areas of the workpiece are not accessible to the tool, a toolpath may not be feasible. This depends on both the milling machine type and the workpiece setup. In actual CAM (Computer Aided Manufacturing), this problem is only taken in account by the operator know-how. A complex part may need several attempts before a solution is found. A less satisfactory solution is to mill the part in several sequences. The workpiece is setup several times until all surfaces have been milled. But the milling machine is not productive during the setup time, and the numerous clampings lead to high machining uncertainties.

This paper is concerned with the search of milling machine type - 3, 4 or 5-axes - and the setup of the workpiece on the machine. There are some works on related problems. Spyridi and Requicha (1990) use visibility as a tool for the setup of a workpiece on a coordinate measuring machine. Woo and von Turkovich (1990) introduce the problem for the milling.

2. A TOOL : THE VISIBILITY

The problem is to find the setup of a workpiece on a milling machine. In terms of kinematics, this problem can be divided into three subproblems. The workpiece is machined by the milling-cutter and the cutter moves regard to the machine-tool table; visibility is a tool for describing those two interactions. The orientation of the workpiece on the machine table is directly related to our problem of piece setup.

2.1 Visibility of a workpiece

There are a number of surfaces S_i $(i=1 \rightarrow n)$ to mill on a workpiece. Some tool directions cannot be used for the milling of a surface S_i. The visibility is defined as the set (noted V_i) of tool directions allowing the milling of the whole surface S_i. Notice that in this paper the word *surface* means simply an area to be milled on the workpiece ; it has no relation with the CAD definition of the workpiece.

For a freeform surface, the computing of V_i is reduced to a visibility calculation. Actually, when the milling-cutter collides a portion of the workpiece, this obstruction blocks the visibility from the surface. More elements on this calculation can be found in (Hascoët, *et al.*, 1997; Risacher, *et al.*, 1997). Additionally, for machining a feature like a slot, some technological constraints can further restrict the set of allowed tool directions. For example, for drilling a hole, the tool direction must be the same as the hole axis. Consequently, the *visibility* of a surface S_i is the set V_i of tool directions allowing the milling, considering the risks of collision between the tool and the finished part, as well as technological constraints linked to a particular feature. With this formulation, visibility is a powerful tool for describing possible machining of a piece.

Figure 1. Visibilities of a pocket and a drilled hole, and their representations on the sphere.

A visibility is a cone bounded by the extreme tool directions. Without loss of generality, visibility cones are intersected with a sphere of unit ray, centered at the apex. The unit sphere is named Gauss sphere and noted S^2. Then, a visibility is represented as a surface on S^2; it is under this form that they will be manipulated from now on. Since the work is done on a spherical surface, it is useful to introduce the notion of angular distance. O is the center of the gauss sphere ; the *angular distance* between two points P and Q of S^2 is the angle between the vectors OP and OQ ; its value is computed by d(P,Q)=Arccos(OP.OQ).

2.2 Visibility of a milling machine

A milling machine morphology does not permit the use of any tool directions. The *visibility* of a milling machine is defined as the set of tool directions allowed by the machine kinematics. This approach was proposed by Woo (1990; 1994) for milling machine. On a 3-axes milling machine, the tool direction is fixed; the visibility is a single point on the Gauss sphere.

A 4-axes NC machine is a 3-axes machine with a rotating table (figure 2). This degree of freedom in rotation allows the tool to rotate all around the workpiece during the machining. The spindle can be tilted manually; after this setting, the angle between the spindle and the plate axis has a fixed value α. Of course, this manual setting must be done before the machining sequence. The visibility of this machine is a small circle on the Gauss sphere S^2.

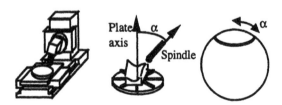

Figure 2. A 4-axes milling machine visibility.

2.3 Resolution of the problem

For milling all surfaces in a single sequence, the visibility V_F of the machine tool must be in contact with all visibilities V_i of the surfaces S_i to be machined. This is expressed as : $\forall i \in \{1...n\}$, $V_F \cap V_i \neq \emptyset$. Then, there is always a tool direction allowed by the machine kinematics that can be used for the milling of a surface S_i; different tool directions can be used for the machining of different surfaces. Finding a workpiece orientation so that all surfaces S_i could be machined in a single sequence, is a problem that can now be reformulated as follows : *Find an orientation of the machine visibility V_F on the sphere S^2, so that V_F is in contact with all surfaces visibilities V_i (i=1→n)* (see figure 3).

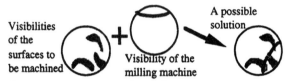

Figure 3 : How to search a solution.

However, under this formulation, solving this problem is fairly complex. Woo and von Turkovich (1990) shows a solution in 3-axes milling; he also solves the problem of the 4-axes milling, with some restricting hypothesis : the spindle is perpendicular to the plate axis (α=90°) and the workpiece visibilities are simplified. See also (Hascoët, *et al.*, 1997).

3. A NEW METHOD

3.1 Principle

In the following, a new method for solving the problem is shown. Instead of searching directly the set of solutions allowing the milling of the whole piece, the sets O_i of solutions for the machining of a single surface will be computed separately for each surface S_i (i=1→n). This is done with a simple transformation of the visibility V_i of a surface ; the transformation used depends on the machine visibility V_F. Then the solutions for the whole piece are computed by intersecting the obtained sets : $O=\cap O_i$ (i=1→n). If the intersection is empty, then the milling of the whole part in a single sequence is not possible on the studied type of milling machine. In this paper, this method is presented for 4-axes milling, for any value of the angle α.

3.2 Calculation in 4-axes milling if the tilting spindle is fixed

At first, assume that the spindle axis setting has been done; the choice of this setting will be studied later. The value of α is fixed, and the machine visibility V_F is a small circle of angular ray α. An orientation of this small circle must be found on the Gauss sphere. The pole of a small circle is the top of the segment of the sphere limited by this circle. Due to the circle symmetry property, the orientation of the small circle is completely defined by giving its pole. In what follows, the poles of the solutions circles are searched rather than the circles themselves.

The first problem is to orient the piece on the machine, so that a single surface S_i can be machined. The small circle must intersect the visibility V_i. Then there must be a point M belonging to V_i, so that the angular distance between the circle pole and M is exactly equal to α. In mathematical terms this is expressed as :

$$O_i = \{P \in S^2 \text{ so that } \exists M \in V_i, \ d(P,M) = \alpha\}.$$

In the usual case, the visibility V_i is a single surface on the sphere S^2. The set O_i is obtained by offsetting the limit of the visibility by a value α (figure 4a). If the visibility shape is indented, some holes can appear during the process ; however, they do not affect the reasoning nor the following treatment. If the angular distance between any two points M,N belonging to V_i is always less than 2α, then the visibility is said to be *small* with respect to the angle α. In this case, a hole must be removed in the offsetted set ; this hole is the intersection of the sphere segments of ray α, with their poles belonging to V_i. Notice that if V_i is small, there is no hole due to an indented shape.

Now let us examine some special cases. A visibility which is reduced to a line or a point on the Gauss sphere is only a particular case of the previous study. For a visibility composed of several separate subsurfaces, a subset is computed for each separate surface ; O_i is the Boolean union of all of them (that is true since the machine visibility needs only to be in contact with at least *one* point of V_i).

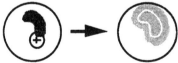

Figure 4a. Searching the circles in contact with a visibility by offsetting the visibility.

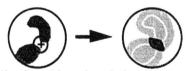

Figure 4b. Searching the circles in contact with several visibilities by intersecting enlarged visibilities.

Now, a setup of the workpiece on the machine is to be found, in order to machine all surfaces S_i in a single sequence. The small circle must intersect *all*

visibilities V_i : $\forall i \in \{1...n\}$, $V_F \cap V_i \neq \emptyset$. It is easily shown that the poles verifying this condition are in the intersection of all sets O_i of solution for a single surface S_i (figure 4b).

In a few words, searching the possible settings of the piece on the machine is done by enlarging all visibilities V_i, then by computing the intersection of those sets. Each point of the resulting set is the pole of a small circle of angular ray α in contact with all visibilities; it corresponds to a solution of orientation of the piece on the machine-tool.

3.3 How to set the spindle ?

The way for finding all possible orientations of the workpiece on a 4-axes milling machine has been shown when the spindle setting has already been fixed. However, the mechanical engineer tries to answer simultaneously to both questions : How to set the spindle ? How to set the workpiece ? Those two questions cannot be dissociated. The basic solution is to try one after another all possible adjustment of the spindle. This is valid if the setting is discrete, e.g. from 5° to 5°. It is not applicable if the setting is continuous.

In 4-axes milling, each potential solution is described by three parameters as follows :
• workpiece orientation (angles ψ, θ giving the position of the pole of the small circle)
• angle α between spindle axis and plate axis (angular ray α of the small circle).
In terms of visibilities, a solution corresponds to a small circle. The position of its pole on S^2 defines the workpiece orientation, and the circle angular ray is linked to the spindle axis setting. For a fixed angle, the solutions are obtained as a surface on the unit sphere. However, the choice of a unit ray is arbitrary; the sphere ray $R\alpha$ may depend on the value of α. Then there is a system of concentric spheres. The set of solutions is a volume composed of successive "layers", each layer corresponding to a setting α of the spindle. Now, the solutions are shown as a volume; each point of this volume corresponds to a solution - spindle axis setting and workpiece orientation (figure 5).

The same methodology as before is applied. A volume O_i of solutions is constructed separately for each surface S_i. This volume is computed directly from the visibility V_i. The Boolean intersection $O = \cap O_i$ ($i=1\rightarrow n$) of all those volumes is the set of solution for milling the whole piece in a single sequence. Additional indications on the computing of O_i can be found in (Risacher, *et al.*, 1997). In some very constrained cases, the resulting set O is empty; it means that the piece is not machinable in a single phase with 4-axes milling.

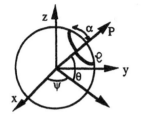

The machine visibility :
A small circle $\mathcal{C}$ of pole P
and angular ray α

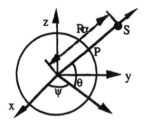

The solution point
for the same circle :
The point S on lineOP
at a distance $R\alpha$ from O

Figure 5. Relation between a small circle and a solution point.

3.4 An extension to 5-axes milling

This method can be extended to 5-axes milling. This machine has a rotating plate that can turn 360° around the z-axis. The workpiece is mounted on the plate. The main difference with 4-axes milling is that instead of a manual setting, the spindle tilting angle α is controlled by the numerical command; therefore the angle between the spindle axis and table axis can vary during the machining sequence. The useful run of the spindle is limited by two stops α_1 and α_2, so that the condition $\alpha_1 < \alpha < \alpha_2$ is respected. The machine visibility is usually a spherical band on the Gauss sphere S^2, denoting the added degree of freedom (figure 7). The machine described here is not the only possible morphology. For example, in some case, it is the table instead of the spindle which can be tilted ; while this can affect the definition of the angle α, the method remains basically the same.

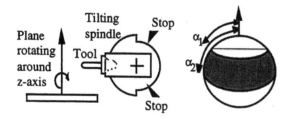

Figure 7. A sketch of a 5-axes milling machine and its visibility.

A new method has been shown for solving the problem of workpiece setup in milling. This method can also be used for choice of the machine morphology and spindle setting. The method has been applied it in the case of 4-axes milling. A similar solution exists for 5-axes milling.

4. SOME PRACTICAL EXAMPLES

In this section, the use of the setup method is illustrated by a few examples. A number of different situations will be studied, involving 3-, 4- and 5-axes milling. Some aspects of setup in milling will be discussed.

4.1 Setup in 3-axes milling

The first workpiece illustrate how to find a setup for 3-axes milling. On such a milling machine, the tool direction is fixed for the whole milling sequence. A single tool direction must be found, so that all the surfaces are machinable. This tool direction must belong to all visibilities V_i. Therefore, the possible tool direction belongs to the set given by : $V = \cap V_i$ $(i = 1 \rightarrow n)$. V is the set of tool directions allowing the milling of the whole part in a single setup; if $V = \varnothing$ then 3-axes milling does not provide a solution for this workpiece. Solving the setup problem needs to rotate the part, so that the spindle axis belongs to the intersection V.

This method is applied to two workpieces. The first one has a freeform shape similar to a wave (figure 8a). The area under the crest (the wave trough) is a recess very difficult to access by the tool; its visibility is very reduced. The corresponding visibility and a 3-axes milling solution are shown on figure 8b. The workpiece of figure 9 is a cube with two shallow pockets. Independent machining of one pocket is easy; however since they are located on two perpendicular faces of the cube, a special setup should be found if the machining is to be done in only one sequence.

8a The workpiece with the wave.

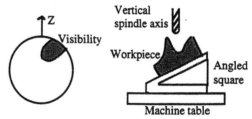

8b The wave trough visibility and a 3-axes solution.

Figure 8. The wave example.

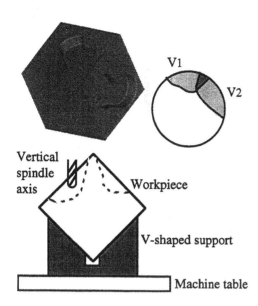

Vertical spindle axis

Workpiece

V-shaped support

Machine table

Figure 9. A cube with pockets

Notice that for many workpieces, the supporting face for clamping are often fixed by the workpiece shape and specifications. Supporting faces are usually chosen with technological criterions. A large plane will give a good support as well as easy positioning and clamping. A reference face for many toleranced features is also a good choice as a supporting face, as this will reduce machining uncertainties. However, this does not mean that the setup problem is solved. In order to improve accessibility to the milling tool, the workpiece should sometime be tilted with respects to the milling machine principal axes. Some special mounting (as the angled squares and V-shaped wedge shown on figure 8 and 9) can be used for this. An indexing table is also another more expensive solution.

4.2 Four-axes milling

The workpiece chosen for illustrating 4-axes milling is a distorted version of the previous wave freeform surface, with four hollow shapes in the corners, numbered from 1 to 4 on figure 10a. The four recesses are the only surfaces to be milled. This workpiece is not machinable in 3-axes milling, since the intersection of the visibilities is empty (figure 10b). This part will need at least a four axes milling machining. This is easily shown by visibilities analysis. The choice of a (small) circle in contact with all four visibilities V_1 to V_4 is extremely restricted. The figure 10 displays one example of a solution. It is easy to see that all others solution are approximately similar. The construction method shown in paragraph 3 will give all the possible circles.

Notice that in this example, there is no great circle that allows the milling. This means that the usual configuration of a 4-axes milling machine (with the spindle axis perpendicular to the plate axis) cannot be used. However, the workpiece could be milled on

a 4-axes milling machine with an indexing spindle. If such a machine-tool is not available, using a 5-axes milling machine could be a solution, even if this machine is more expensive.

This work focus on the accessibility problem for the milling tool. When a multiaxes machine is needed, the rotational fourth and fifth axes are used for the tool axis positioning. However, the simultaneous control of the four or five axes is not always needed, except if toolpaths continuity is required. See (Lee and Suh, 1995, 1996) for an interesting work on additionally-five-axes milling machine (*i.e.*, a 3-axes milling machine with an indexing table instead of the rotating axes).

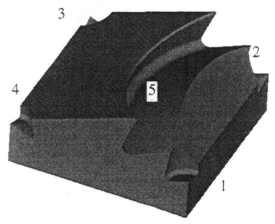

10a. Workpiece with wave and hollow corners.

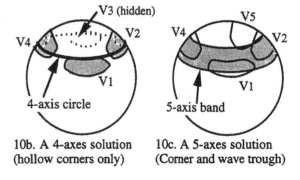

10b. A 4-axes solution (hollow corners only)

10c. A 5-axes solution (Corner and wave trough)

Figure 10. The wave and the corners.

4.3 Five-axes milling

Now, assume that the wave trough (numbered 5 in figure 10a) is added to the workpiece area to be milled. This new feature should be machined in the same sequence as the four recesses. None of the previous small circles is in contact with the visibility V_5. Therefore, a 5-axes milling machine is needed. An example of solution is given in figure 10c.

In the last example, only the recesses numbered from 1 to 4 are to be milled. Two vertical walls have been added to the workpiece (figure 11); those surfaces are not to be machined. However, they act as obstacles for the milling tool. While the four recesses are left untouched, the visibilities V_2 and V_4 are much reduced. There is no circle that are in contact with all the four visibilities. Some authors (Woo and von Turkovich, 1990; Woo, 1994) use simplified visibilities ; their visibility computation rely on a local analysis around the point to be milled. However, this example shows that surfaces not to be milled have a great influence on setup and must therefore be taken in account.

walls.

11a. Workpiece with walls.

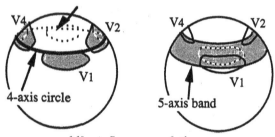

4-axis circle

5-axis band

11b. A five-axes solution.

Figure 11. A workpiece with walls.

5. CONCLUSION

The visibility allows the characterization of constraints linked to the workpiece and to the milling machine. A new method has been shown for determining if a piece is machinable in only one sequence. It was applied in the case of 4-axes NC milling. This is of great interest when trying to choose on which type of machine a piece could be milled. The possible workpiece setups and spindle setting are obtained as a result of the algorithm. This method is displayed with a number of examples. Current work focus on a interface for displaying the sets of solution, which interpretation is not always easy.

REFERENCES

Hascoët, J.Y., F. Bennis and Ph. Risacher (1997), "Choice of machine tool configuration", *Kluwer Acad. Publisher* (1997).

Lee, J.J. and S. H. Suh (1995). Five-axis machining with three-axis CNC machine. *Journal of Korean Institute of industrial engineers*, **21**, pp. 217-237.

Lee, J.J. and S. H. Suh (1996). Flank milling of ruled surface with additionnally-five-axis CNC machine. *Proc. Int. Conf. on Integrated Design and Manufacturing in Mechanical Engineering*, Nantes (France) pp. 375-384.

Risacher, Ph., J.Y. Hascoët, F. Bennis (1997). An optimized subdivision of workpiece surfaces for visibility computing. *2nd International ICSC Symposium on Intelligent Industrial Automation*, Nîmes (France).

Spyridi, A.J. and A.A.G. Requicha (1990). Accessibility Analysis for the Automatic Inspection of Mechanical Parts by Coordinate Measuring Machines. *Proc. 1990 IEEE Int. Conf. on Robotics & Automation*, Cincinnatti (Ohio USA), pp. 1284-1289.

Woo, T.C. and B.F. von Turkovich (1990). Visibility Map and its Application to Numerical Control. *Annals of the CIRP*, **39/1** pp 451-454.

Woo, T.C. (1994). Visibility Map and spherical algorithms. *Computer Aided Design*, **26/1** pp. 6-16.

FEATURE-ORIENTED PROGRAMMING INTERFACE
OF AN AUTONOMOUS PRODUCTION CELL

Nils Brouër, Manfred Weck

Aachen University of Technology
Laboratory for Machine Tools and Production Engineering (WZL)
D-52056 Aachen, Germany
N.Brouer@wzl.rwth-aachen.de

Abstract: Following the recent trend towards small lot production and large numbers of product variations, an "autonomous production cell" (APC) is being developed at Aachen University of Technology. The article describes the basic concepts of the APC with special focus on it's programming interface. To enable a high-level data interface between programming department and numeric control, a feature-based approach is used, which is compatible with the ISO STEP standard.

Keywords: Autonomous control, Numeric control, CAD/CAM, Factory automation, Programming languages.

1 INTRODUCTION

In recent years, there has been a strong tendency towards small lot production featuring a large number of product variations. These circumstances result in a demand for a new generation of flexible manufacturing units which are able to satisfy today's need for "agile" manufacturing. At the Aachen University of Technology, an "autonomous production cell" (APC) is being developed, which aims at solving these problems.

In this context, autonomy is defined as the independence of the manufacturing unit from other units within the company. By equipping the machine with intelligent functions like chatter avoidance or process control, a safe production process can be realised. Autonomy is reached by combining this with a man-machine interface which enables the machine operator to make use of his expertise to maximise the productivity of the system.

Autonomy is also reached by improving the interfaces between the cell and surrounding units within the company thus enabling a better information transfer to the cell. One of these interfaces is the programming interface. In this paper, the work which has been done concerning the programming interface is being described.

2 BASIC CONCEPTS OF THE AUTONOMOUS PRODUCTION CELL (APC)

2.1 Autonomy of the APC

At Aachen University of Technology, a group of about 20 scientists from 10 different research institutes is currently developing the new concept for manufacturing systems called the autonomous production cell, see (Weck, et al., 1996).

The idea behind this project reflects the experience that the highly automated systems of the 80's designed for unmanned operation have not delivered the promises they made. In essence, they are hard to control and maintain. The complexity of these systems results in long down-times and high costs.

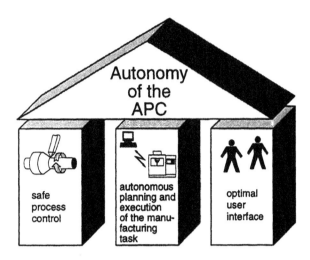

Fig. 1: Autonomy of the APC

Rather than trying to eliminate the worker at the shopfloor, a different approach is taken in this concept. Here, autonomy does not necessarily mean automation. While those tasks which can be handled better by the machine are automated, others are explicitly left to the worker.

In this project, a "manufacturing cell" consists of a single machine including necessary handling systems and a worker. The goal is to develop the necessary technical means to enable such manufacturing cells to undertake complex machining tasks largely independent of other units within the company. The three main aspects of autonomy are shown in Fig. 1. To reach autonomy, first a safe process has to be guaranteed.

The second aspect of autonomy as it is defined in this context is the integration of the tasks of planning, machining and quality control at the shopfloor. Finally, additional measures are undertaken to provide the user with specific means to interact with the cutting process. This includes viewing of the process as well as intervening with it. Whereas many complex systems do not allow any user interaction, this concept makes it possible for the worker to put his knowledge and his abilities into use for the company.

2.2 Process monitoring and control at the APC

The architecture of the autonomous production cell and its control reflect the basic concept outlined above. To assure the failure-safety of the system, several process monitoring and process control modules are being incorporated into the control of the APC. The modules perform, among others, the following tasks:

· chatter avoidance,
· compensation of thermal errors,
· compensation of static errors,

· general process monitoring.

Each task is covered by a dedicated model, called a "process model". More details on these process models can be found in (Kaever et al., 1997).

2.3 Open control

To realise the integration of process models into the control, an open control system is needed. While first tests were undertaken using a commercially available control system offering limited openness, further implementations require a completely open control. A suitable open control architecture was developed in the European project "Open System Architecture for Controls within Automation Systems" (OSACA), see (OSACA 1996). WZL is one of the major partners of the OSACA consortium and has developed a fully OSACA-compatible NC. This NC will be the basis of the APC control. Fig. 2 shows the structure of the OSACA-compatible controller.

The process models form so-called "architecture objects" which use the OSACA communication mechanisms to communicate with each other and the NC-kernel. Within OSACA, a library for communication between modules on the same as well as different CPUs and operating systems was developed. This library will be used for the APC-control as well.

3 NC PROGRAMMING INTERFACE

As mentioned in section 2.1, there are several aspects of autonomy in the APC. One of them is the autonomous planning and execution of the manufacturing task. The NC programming language forms the interface between planning department and the shopfloor. By improving this interface to enable a better transfer of data between the cell and surrounding units (like the planning department), the autonomy of the APC is aided. This chapter contains a detailed description of the proposed data interface.

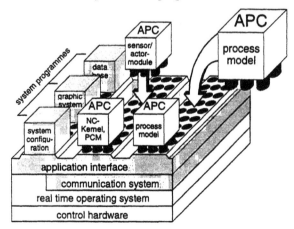

Fig. 2: OSACA-compatible APC-control

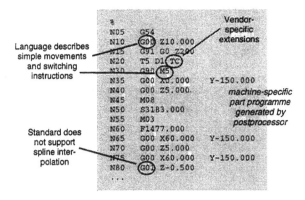

Fig.3: Present state of data exchange using ISO 6983

3.1 Present situation of NC programming

Currently, the exchange of data between planning and shopfloor is mostly done using the ISO standard 6983, see (ISO6983, 1984). This standard is very well established world-wide, it nevertheless has a number of shortcomings (fig. 3).

One of the main disadvantages of ISO 6983 is the low level of data it describes. A part programme describing a manufacturing operation consists mainly of two types of information: (a) commands describing simple movements of the tool, which are given using G0, G1, G2, ... statements, and (b) switching instructions for the logic control of the machine, e.g. M4 (spindle on), M7 (coolant on), M30 (end of programme). Movements describe the toolpath, traversed either in rapid feed (G0) or cutting feed (G1, G2, G3). No higher-level information about the manufacturing process can be described. It is for example not possible to include information whether a roughing or a finishing operation is in process, or whether a hole or a pocket is being cut.

Another shortcoming is the fact that the ISO standard does not support higher-level interpolation like splines. This is a major obstacle which inhibits efficient milling of freeform surfaces. While these surfaces can be described very efficiently using splines, they are being approximated using very small linear segments in the control. This leads to very large part programmes, which can easily reach several Megabytes of data, and bad surface quality, especially in high speed milling (MATRAS, 1996).

Because of these and other limitations of the standard, control and machine tool makers have introduced their own extensions to ISO 6983. This does not only include additions to the standard like M-commands for additional functions (e.g. pallet changers) or cycles, which are used like makros for repetitive operations, but also commands which replace statements set in ISO 6983 (e.g. use of "TC" instead of M6 for tool change). Sometimes, even controls of the same control manufacturer have an incompatible set of commands. These vendor-specific extensions prohibit the interchange of part programmes between different controls or machines. This leads to the situation that the manufacture of a part is rather delayed than transferred from one machine to another if the original machine breaks down.

3.2 Data model for NC programming

A new data model for NC programming should satisfy a number of demands. It should support a higher level of data and should e.g. enable the description of manufacturing operations like "mill pocket 3" or "cut thread of hole 7". For improved surface quality and reduced data volume, the use of splines for geometric data should be possible. It should make use of existing standards, especially the emerging STEP standard for product model data (ISO10303, 1994). Ideally, it should support the use of CAD data without conversion. This would ensure that no geometric data is lost in the CAD-CAM-CNC chain. To take into account the fact that usually part programmes have to be adapted at the shopfloor to adjust cutting speeds, the interface should support such modifications and especially enable the transfer of data from shopfloor back to the planning department. Finally, the data model should form a well-structured, hierarchical interface.

The above-mentioned constraints have been taken into account during the design of the data interface used with the autonomous production cell. In the next section, the new interface is described in detail.

3.3 Structure of the programming interface

The new interface as shown in figure 4 consists of three parts: the main programme file A, the technology description B and the geometry description C. Usually, the three parts would be represented physically by three separate files. This is not necessary, however: the data may be represented in just one file or as many files as needed. References between separate files have to be given, however.

To enable a direct use of CAD data, the new data interface uses STEP data for geometry representation in part C. STEP is the acronym for ISO 10303, the "standard for exchange of product model data". It consists of several parts covering a scope ranging from the description language EXPRESS to application protocols (AP) which define data for specific applications like shipbuilding or automotive industry. To be independent from specific applications, parts 42 and 43, which describe basic geometric elements, are used in the programming

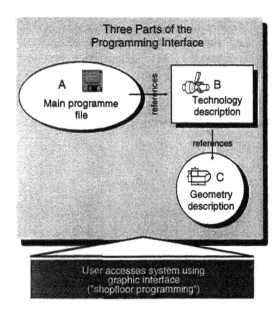

Fig. 4: Structure of the Programming Interface

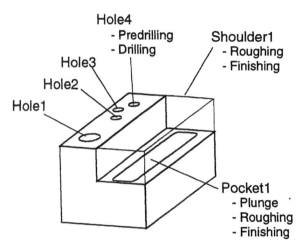

Fig. 5: Prismatic Workpiece

interface. Physical coding is done according to the rules defined in part 21. For 2.5D milling, manufacturing features are used which are similar to the form features described in part 224.

Part A is the main programme file. This file contains the sequence of all operations needed to manufacture the part. The operations are called "workingsteps" and will be described in detail in the next section of this paper. In part A, a list of all workingsteps in the order of execution is given. Intelligent controls are able to insert connecting movements between these workingsteps as well as approach and lift movements to the surface to be cut. Thus, by simply deleting a workingstep from the list or exchanging the order of workingsteps in the list, the manufacturing process can easily be adapted to new demands.

All information necessary for the part which is neither geometry given in part C nor contained in part A will be specified in part B. This includes specifically the detailed description of the workingsteps introduced in part A. All technological information like cutting speeds, tools used, etc. is given in this part. Part B also contains the connection between the workingsteps and the geometric elements which are described in part C. For milling, workingsteps are defined for manufacturing features like holes or pockets or regions of a freeform surface. A detailed view of this is given in the next section of this paper.

The three parts which comprise the complete data interface will usually all be coded in ASCII. In cases where it is desired that the data cannot be altered once it is generated, binary coding may be used instead. Coding in ASCII does not mean, however, that a programme is supposed to be written using an ASCII text editor. Because of the complexity of the

interface, this is a very error-prone way to generate a part programme and should thus be avoided. Instead, part programmes will be generated by CAM or shopfloor programming systems. While ISO 6983 is also a programming language used to manually code part programmes, the presented approach is rather a data interface for communication between two computers, i.e. the CAM system and the CNC.

3.4 Sample applications: 2.5D- and 3D-milling

The data interface described in the previous sections presents a novel approach which may be used for various machine types and technologies. At WZL, detailed definitions of the technological data set needed have been done for 2.5D- and 3D-milling. For these applications, an interpreter has been developed, which, as part of a CNC, reads a part programme according to the new syntax and generates the necessary tool movements.

Fig. 5 gives an example of a prismatic workpiece to be cut using 2.5D-milling. The volume to be cut is described using several manufacturing features. In this case, there are four holes, a shoulder, and a pocket to be machined. For each feature, at least one manufacturing operation is required. Holes may be predrilled, then drilled, the shoulder needs a roughing and a finishing operation, and the pocket needs a plunge, a roughing, and a finishing step. As already mentioned in section 3.3, these manufacturing operations are called "workingsteps" in the data interface described here.

The interface is based on an object-oriented description of all technological and geometrical data. Fig. 6 gives a simplified view of a subset of the data structure. The data definition was done using EXPRESS, the language defined in part 11 of STEP. In Fig. 6, fat lines represent inheritance between two elements (or "entities", as they are called in

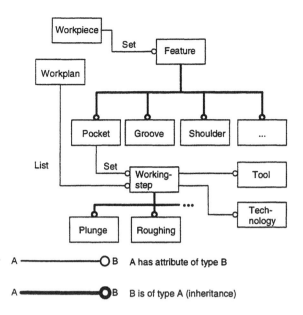

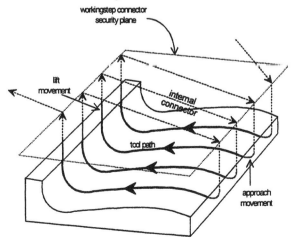

one set of technological parameters. The figure shows the tool paths of the workingstep as well as approach and lift movements and internal connectors.

Today, for five axis milling a part programme is typically generated as follows: the CAM system takes as input the surface data from the CAD system and calculates cutting paths. From these, axis movements are generated by the post processor, taking into account the kinematics of the machine and the tool selected. Changes of the tool type or size usually require a new postprocessor run, which may take several hours. The axis movements are then coded in ISO 6983 resulting in part programmes with a size of up to several Megabytes.

In the proposed data interface, not only axis movements, but cutting paths (cutter contact points) described as curves on a surface are transferred. This makes it possible to integrate the tool correction in the CNC thus enabling the change of the tool diameter at the shopfloor. An intelligent control will also automatically generate approach and lift movements as well as internal connectors thus enabling a change of order or omission of workingsteps.

3.5 Integration into the CAD/CAM chain

As already mentioned in chapter 3.4, it is important to notice that the proposed data interface is not a programming language meant for manual programming using a text editor, but rather an interface between two computers: the programming system and the CNC. Since every company has got it's own way to organise NC programming, it has to be guaranteed that this is still possible with the new interface. Fig. 8 shows possible scenarios for the use of the proposed interface.

Starting with CAD workpiece data from the design department, a company may decide to add technological data in the planning department using a CAM system. Output of this CAM system would then be done using the syntax of the new data interface. Alternatively, a company may use shopfloor

Fig. 6: Object-oriented Description of Data

EXPRESS), thin lines represent attribute relationships.

The figure shows that a workpiece has a set of features. A feature can be of different type, e.g. pocket, groove, shoulder or other type not shown here. Attached to each feature is a list of workingsteps, which -in case of a pocket workingstep- can be of type plunge, roughing, etc. Tool and technology definitions are attached to each workingstep. The workplan is defined as having a list of workingsteps.

Fig. 7 shows an application involving the milling of a freeform surface. This part of the new data interface was developed in the European project "OPTIMAL" (see Optimal, 1996). Since it is not possible to describe features for freeform surfaces, a different approach is taken here: the surface is divided into so-called "regions". One or more workingstep is defined for each region. As in 2.5D-milling, a workingstep describes the milling of a surface with one tool and

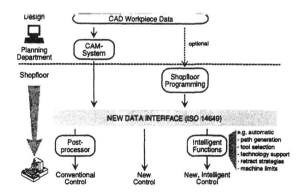

Fig. 7: Freeform Surface

Fig. 8: Possible Scenarios of the Proposed Interface

163

programming at the machine tool. In this case, the shopfloor programming system would generate part programmes according to the new syntax. Geometrical information may be input at the shopfloor also or CAD files may be referenced. During the last years, much research has been done in the field of feature-based CAD or process planning (see e.g. Champati, et. al. (1996)). The feature-based programming interface would be an ideal extension of this work.

Depending on the control, different alternatives exist for executing a part programme according to the new data interface. For conventional CNCs, which are able to understand ISO 6983 only, a post processor is needed to generate this code from the new data interface. Post processors would also enable a migration from present-day ISO to the new interface without the need to replace all existing controls. Unlike conventional controls, new controls would be able to run part programmes written in the new syntax directly. For added functionality, controls would be able to calculate approach and lift movements as well as internal connectors and advanced cycles.

Because of the added data included with part programmes of the new syntax, additional intelligent functions may be realised in the control. Besides automatic path generation already mentioned, controls might be able to automatically select tool types (e.g. automatically select 8 mm drill for an 8 mm hole) and give technological support for the programmer. The latter function is possible because both the material of the rawpiece as well as the cutting tool is given in the programme. Because of the full geometric description of the workpiece, retract strategies in case of tool breakage may be calculated by the CNC. Finally, a CNC might "know" the limits imposed by machine stability or the power rating of the main spindle and thus prevent the execution of part programmes with technological parameters which exceed this limit. These are some examples of highly innovative functions made possible by the new data interface.

3.6 Standardisation of the data interface

The data interface as described in the previous paragraphs has been developed with major participation of WZL. A prototype implementation of a controller able to execute a part programme according to the new syntax has been implemented in the OSACA-based control at WZL. WZL has taken this approach and introduced it into the ISO working group concerned with standardisation of a new data interface for device control (ISO/TC184/SC1/WG7). The draft standard developed by WG7 has largely been influenced by this approach, see (ISO14649, 1997). Currently, work is undertaken to finish this standard for milling and extend it to other technologies.

4 CONCLUSION

A new data interface for NC programming has been demonstrated. This interface is being integrated into the autonomous production cell (APC) developed at Aachen University of Technology. Geometric data from CAD represented in STEP format is used within the interface. The data structure has been developed for 2.5D- and 3D-milling applications. Whereas the 2.5D application is based on manufacturing features, so-called "regions" are defined for freeform surfaces. By improving the data interface between the cell and surrounding units, the autonomy of the APC is aided.

5 ACKNOWLEDGEMENTS

The work on autonomous production cells is funded by the Deutsche Forschungsgemeinschaft (DFG) as Sonderforschungsbereich SFB368. Funding for the project OPTIMAL was given by the European Union within the third framework programme of ESPRIT.

REFERENCES

Champati, S., Lu, W.F., Lin, A.C. (1996). Automated Operation Sequencing in Intelligent Process Planning: A Case-Based Reasoning Approach. *Int. J. Adv. Manuf. Technol.*, **12**, 21-36.

ISO6983 (1982). *Numerical control of machines - Program format and definition of address words*. ISO, Geneva.

ISO10303 (1994). *Standard for the Exchange of Product Model Data*. ISO, Geneva.

ISO14649 (1997). *Data model for CNC controllers (committee draft)*. ISO, Geneva.

Kaever, M.; Brouër, N.; Rehse, M.; Weck, M. (1997). NC Integrated Process Monitoring and Control for Intelligent, Autonomous Manufacturing Systems. In: *Proc. of 29th CIRP International Seminar on Manufacturing Systems*, May12-13, 1997, Osaka, Japan. (in print)

MATRAS (1996). *Extended CNC-Internal Velocity and Acceleration Control*. Deliverable 4.2 of ESPRIT working group MATRAS, Brussels.

OPTIMAL (1996). *Definition of the Interface and the Interface Levels*. Deliverable 1.2 of ESPRIT working group OPTIMAL, Brussels.

OSACA (1996). *Project Report OSACA phase I and II*, ESPRIT III EP 6379 & EP 9115, Brussels.

Weck, M., Brouër, N., Herbst, U., Michels, F., Prust, D. (1996). Autonome Produktionszellen - Sichere Prozeßbeherrschung und höhere Werkstückqualität. *wt - Produktion und Management*, **Vol. 86**, No. 5, pp 243-247.

A STUDY ON THE DESIGN AND OPERATION OF MODULAR CELL MANUFACTURING SYSTEMS

Woo, Kwang Bang, Kim, Sung Soo, and Hur, Kyeon

Department of Electrical Engineering, Yonsei University, Seoul 120-749, Korea

Abstract - In this paper, a new powerful genetic algorithm(GA) is proposed to be used in scheduling of the modular cell manufacturing systems. The proposed GA is to perform a uniform crossover based on the nucleotide(NU) concept, in which DNA and RNA consist of NUs with a concrete way to vary the probabilities of crossover and mutation dynamically for every generation. The efficacy of the proposed GA is demonstrated by its application to the unimodal , multimodal and nonlinear control problems, respectively. Simulation results show that in the convergence speed to the optimal value, the proposed GA was superior to existing ones, and the performance of GAs with varying probabilities of the crossover and the mutation was improved as compared to GAs with fixed probabilities of the crossover and mutation. Through a series of experiments conducted on a randomly generated MCMS scheduling problem of practical complexity, the proposed genetics-based scheduler strategy is shown to be promising and worthy of more in-depth investigation compared to the previous strategies.

Keywords: Manufacturing systems, Genetic algorithms, Scheduling algorithms

1. INTRODUCTION

The customers are changing, and so is the nature of their demand. Customer demand is becoming individualistic, leading to product variations. The multitude of product variations, the complexity and exacting nature of these products, coupled with increased international competition, the need to reduce the manufacturing cycle time, and the pressure to cut production costs, require the development of manufacturing technologies and methods that permit small-batch production to gain economic advantages similar to those of mass production while retaining the flexibility of discrete-product manufacturing(Singh, 1996). One approach used to gain these benefits has been to develop a modular cell manufacturing system designed for modular products in which reconfigurability that is an important characteristic of modern manufacturing systems is emphasized.

Modular cell manufacturing is an application of group technology in manufacturing in which all or a portion of a firm's manufacturing system has been converted into cells. A manufacturing cell is a cluster of machines or processes located in close proximity and dedicated to the manufacture of a family of parts. The parts are similar in their processing requirements, such as operations, tolerances, and machine tool capacities. The primary objectives in implementing a modular cell manufacturing system are to reduce setup times (by using part family tooling and sequencing) and flow times(by reducing setup and move times and wait time for

moves and using smaller batch sizes) and therefore, reduce inventories and market response times. In addition, cells represent sociological units conducive to teamwork. This means that motivation for process improvements often arises naturally in manufacturing cells(Heragu, 1994; Singh, 1996). Modular cell manufacturing systems permit machine flexibility. That is, an operation on a part can be performed on alternative machines. Consequently, it may take more processing time on a machine at less operating cost, compared with less processing time at higher operating cost on another machine. Production of parts with minimum processing cost and quick delivery of parts are the two important criteria from the manufacturing management point of view. The other consideration from the cell operation point of view is balancing of workloads on the machines. Because the machines are flexible, and operation can be performed on alternative machines. Therefore, a part can be manufactured along a number of processing routes.

A number of techniques have been developed for GT(Group Technology) and MCMS(Modular Cell Manufacturing Systems)design. Most of them are primarily concerned with the identification of part families and machine cells and do not take into consideration other practical design constraints which include available capacity of machines, safety and technological requirements, number of machines in a cell and number of cells, machine utilization rate, etc.. In addtion to the design constraints, there are other planning issues that need to be addressed. One of the more important ones includes scheduling of jobs in individual cells. Since there are fewer jobs in each cell of MCMS when compared to the entire system, solving scheduling problems optimally in a MCMS may not be impossible. In this paper, is mainly focused the operational level in MCM, specifically the scheduling problem of MCMS(currently assumed to be well designed by the proposed desin techniques).

2. SCHEDULING OF MCMS

MCMS's represent the state-of-the-art in the design of manufacturing systems and may be another form of FMS's(Flexible Manufacturing Systems). Such systems seek to facilitate the efficient processing of families of parts with demands in the low- to medium-volume range. Each cell in the system consists of one or more workstations. Typically, each station is capable of performing two or more manufacturing operations. A given operation can usually be performed by more than one station and these stations may be located in different cells. A job consisting of a single part or several units of a particular type of part, that are demanded as a batch, may be entirely processed using the capabilities of a single manufacturing cell. Alternatively, it may require processing at workstations located in multiple cells. Typically, all part units in a job require several operations to be performed in a particular sequence called the "operational precedence requirement" for that part/job. Each cell is usually equipped to perform a set of operations needed by a "family" of parts (i.e., parts belonging to different jobs that share common operational requirements).

At the operational level in MCMS installations, scheduling decisions must be made on routine basis and must account for several complicating issues. They are as follows: 1) demands for batches of parts of varying sizes, with each batch possibly having a specific due dates, 2) the possible grouping of parts from different batches (based on homogeneity of processing requirements) into part families to reduce set-up times, 3) the prioritizing of part families, and the members of each family, within the schedule, 4) alternative ways of routing parts within the system, given that the processing rate of different workstations and transfer times between stations on different routes vary, 5) unexpected turbulence due to sudden loading of one or more high-priority jobs for immediate processing(i.e, job influxes) and/or deletion of one or more existing jobs from a fully or partially prepared schedule(i.e, job cancellations)

Given its complexities, there are two broad ways in which the MCMS scheduling problem can be handled. One strategy is to construct a schedule dynamically(i.e., on line and in real time). This is the usual approach followed when a job list, related specifications and criteria for schedule generation are not all available in advance. As and when jobs become available (perhaps in a queue), scheduling decisions are made based on an assessment of the current and anticipated MCM states. A second approach is to divide the

scheduling exercise into two stages, comprised of a static scheduling stage followed by a dynamic rescheduling stage. Essentially, static scheduling is based on hard and soft constraints such as system topology and job specification, and objectives of the scheduling exercise, provided all of these are known a priori. The static scheduling phase is done ahead of run time. If there are no shocks and/or turbulence at run time. If there are no shocks and/or turbulence at run time, the static schedule may be implemented as is. Otherwise, it becomes necessary to adjust the static schedule in an on-line , real-time fashion to account for prevailing conditions. Such adjustments are collectively referred to as dynamic rescheduling. The static MCMS scheduling problem and the point how to rapidly reschedule in the next stage are focused on here.

3. GENETIC ALGORITHM AND ITS APPLICATION TO SCHEDULING

3.1 NU-based Genetic Algorithm

In this section, a powerful NU-based GA and its application to the scheduler are predented.

GAs can find solutions to linear and nonlinear problems by simultaneously exploring multiple regions of the state space and exponentially exploiting promising areas through mutation, crossover, and selection operations. Genetic algorithms have proven to be an effective and flexible optimization tool that can find optimal or near-optimal solutions. Unlikely many other optimization techniques, GAs do not make strong assumptions about the form of the objective function. Generally, the crossover is operated on one point or two points or multi-points (randomly) selected in the genes. And a further generalization of one-point, two-point, and multi-point crossover is the uniform crossover which is shown in Fig. 1. Each of these crossovers is particularly useful for some classes of problems and quite poor for other problems but the uniform crossover is generally known to be superior to one- or two-point crossover. (Goldberg, 1989;Michalewicz 1992)
However, the natural genetics in which GA gets ideas reports that genetic operations in human-beings occur in different way from that of genetic operations in GA. DNA contains four such bases-the purines adenine (A) and guanine (G) and the pyrimidines cytosine (C) and thymine (T). The RNA molecule, makedly similar to DNA, usually consists of a single chain. The RNA chain contains ribose sugars instead of deoxyribose In RNA, the pyrimidine uracil (U) replacles the thymine of DNA. DNA

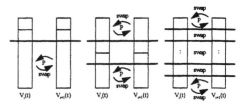

Fig. 1. Various crossover techniques.

and RNA are made up of basic units called nucleotides. In DNA, each of these is composed of a phosphate, a deoxyribose sugar, and either A, T, G, or C. RNA nucleotides consist of a phosphate, a ribose sugar, and either A, U, G, or C. Nucleotide chains in DNA wind around one another to form a complete twist, or gyre, every ten nucleotides along the molecule. The two chains are held fast by hydrogen bonds linking A to T and C to G-A always pairs with T(or with U in RNA); C always pairs with G. Sequences of the paired bases are the foundation of the genetic code. This real mechanism makes us propose a new uniform crossover operation shown in Fig 2.

```
subprocedure PROPOSED UNIFORM
             CROSSOVER
begin
    form nucleotide sequences corresponding to the
    genes/gene groups;
    i = 0, j = 0
    while i < M do
    begin
        while j < N do
        begin
            if nucleotide_{i,j} and nucleotide_{i+1,j} make a
                nucleotide pair
            then swap the information between
                    them;
            else
                preserve their own information;
            j = j + 1;
        end
        i = i + 2;
    end
end
```

where nucleotide$_{i,j}$: the jth nucleotide of the ith chromosome
> M : chromosome size
> N : nucleotide sequence size

Fig. 2. Proposed uniform crossover procedure.

The above procedure is described in terms with Fig. 3 and Fig. 4. The first step to perform the proposed uniform crossover operation is to select the nucleotide type. We consider only A,C,G,U in Fig. 3 and 4 but they can be generally expanded to n.

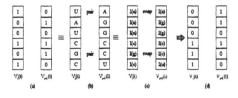

Fig. 3. A uniform crossover with uniformity 1.

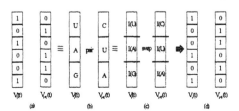

Fig. 4. A uniform crossover with uniformity 0.5.

And the second step is to make particular pairs like A-U, C-G shown in Fig. 3 and 4. If we only consider when n is equal to 2, the proposed uniform crossover looks like the uniform crossover proposed by Ackley, Syswerda, and Spears(Ackley, ;Syswerda, 1989;Spears, 1991) at the first glance, the very point that our uniform crossover exchanges units of nucleotides rather than units of bits and gives the probability of crossover 1/n urges authors to say that the proposed uniform crossover is different from the previously researched uniform crossover.

The next genetic operation, mutation, is usually performed on a bit-by-bit basis at some fixed rates. However, as the ideal mutation rates are not known(Bäck,1993)dynamic mutation procedure that means the mutation rates vary randomly every generation, which leads to increase the chance of the advent of new individuals is suggested.

As the performance assessment of the proposed NU-based GA described in detail(Kim, 1996) shows that it is much more powerful than the previously researched GAs both experimentally and theoretically, we can be sure that our GA can be a powerful tool in implementing a good scheduler of MCMS which requires a rapid rescheduling.

3.2 Application to MCMS scheduling

Consider a simple modular cell with one loading dock, one unloading dock, and two machines(workstations) W1, W2 and W3. Each machine is connected with another machine as well as the loading dock and the unloading dock by a dedicated path(e.g., by using an Automated Guided Vehicle(AGV)). It is assumed that, at the manufacturing level, the system has three operations: OP1, OP2, OP3 and OP4. The operation times in min for various operations on different machines are given in Table 1:

Table 1 A modular cell manufacturing system

Workstations	OP1	OP2	OP3	OP4
W1	-	15	10	-
W2	13	-	12	6
W3	7	11	-	14
Mean	10	13	11	10

Besides, the loading and unloading operations are labeled as LOAD and UNLOAD respectively. The operation times in min for various operations on different machines are given.

In Fig. 5 and 6 a flow chart of scheduling procere based on genetic algorithm and its recombination procedure are shown.

The most important thing to note in GA application to scheduling problems is how to represent individual genes, which means that the solution of the problem depends highly on the representation of information on schedule(genes).

Three jobs, J1, J2 and J3 need to be scheduled

such that the makespan is minimized. Each job has a strict sequence of operations to be performed. The order of operations for each job, job sequences and related schedules are given below:

Table 2 A job related data

	FIRST_OPERATION	SECOND_OPERATION
Job_1	OPERATION_2	OPERATION_3
Job_2	OPERATION_1	OPERATION_4
Job_3	OPERATION_4	OPERATION_2

SCHED1 = <JOB1,JOB2,JOB3>
SCHED2= <JOB1,JOB3,JOB2>
SCHED3 = <JOB2,JOB1,JOB3>
SCHED4 = <JOB2,JOB3,JOB1>
SCHED5 = <JOB3,JOB1,JOB2>
SCHED6 = <JOB3,JOB2,JOB1>

We represent the genes with multiple arrays to efficiently solve the problem here as follows.

$$OP\_1 = \{1, 3, 7\} \qquad (1)$$

Eq. (1) means that unit operation is performed in workstation 3 and the time to spend is 7. Therefore, the jobs can be represented as follows.

$$JOB\_1 = \{\{2,1,15\},\{3,2,12\}\} \qquad (2)$$
$$JOB\_2 = \{\{1,2,13\},\{4,3,14\}\} \qquad (3)$$
$$JOB\_3 = \{\{4,2,6\},\{2,3,11\}\} \qquad (4)$$

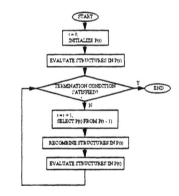

Fig. 5. Flow chart of a scheduling procedure based on genetic algorithm.

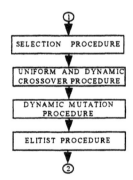

Fig. 6. A recombination procedure of Fig. 5.

Let the scheduling method based on the previous representation of genes be described now. As Fig.7 shows that the jobs can be duplicated, a new crossover mechanism called uniform subtour chunking crossover is needed and it is described in Fig.8.

The second genetic operation, mutation should be modified like crossover operation to avoid meaningless schedules and it's procedure is illustrated in Fig. 10. Through the above procedure the following Gannt chart (Fig.9) and job schedule can be obtained.

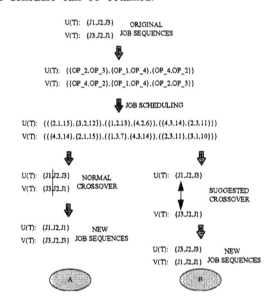

Fig. 7. An illegal crossover.

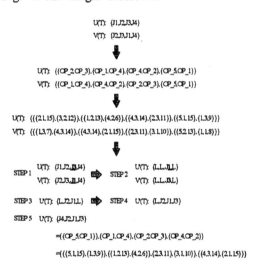

Fig. 8. A uniform subtour chunking crossover procedure.

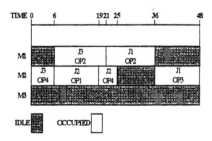

Fig. 9. A Gannt chart of new U(t).

168

ORIGINAL
U(T): {J1,J2,J3} JOB SEQUENCE

↓

U(T): {(OP_2,OP_3),(OP_1,OP_4),(OP_4,OP_2)}

↓ JOB SCHEDULING

U(T): {((2,1,15),(3,2,12)},{(1,2,13),(4,2,6)},{(4,3,14),(2,3,11)}}

↓ ONE GENE IS SELECTED
BUT NOT VARIED.

U(T): {J1,J2,J3}
U(T): {(OP_2,OP_3),(OP_1,OP_4),(OP_4,OP_2)}

↓ MUTATION

U(T): {((2,3,11),(3,2,12)},{(1,2,13),(4,2,6)},{(4,3,14),(2,3,11)}}
U(T): {((2,1,15),(3,1,10)},{(1,2,13),(4,2,6)},{(4,3,14),(2,3,11)}} POSSIBLE SCHEDULING
U(T): {((2,3,11),(3,1,10)},{(1,2,13),(4,2,6)},{(4,3,14),(2,3,11)}}

Fig. 10. A right mutation procedure.

Table 3 A new job schedule

operation_4(station_2)
operation_2(station_1)
operation_1(station_2)
operation_4(station_2)
operation_2(station_1)
operation_3(station_2)

4. Experimental Results

4.1 Simulation

Consider a more realistic manufacturing cell than the one examined above, with on loading dock, one unloading dock, and four machines: M1, M2, M3 and M4. Each machine is connected with another machine as well as the loading dock and the unloading dock by a dedicated path(e.g., by using an Automated Guided Vehicle(AGV)). It is assumee that, at the manufacturing level, the system has four operations: OP1, OP2, OP3 and OP4. Besides, the loading and unloading operations are labeled as LOAD and UNLOAD, respectively. The operation times in min for various operations on different machines are given in the table. The time to load or unload a part is assumed to be one min. The time to transfer a job between a machine and the loading dock(or unloading dock) or another machines is assumed to be two min. Five jobs, J1, J2, J3, J4, and J5 need to be scheduled such that the makespan is minimized. Each job has a strict sequence of operations to be performed. The order of operations for each job is given:

Table 4 Processing time for operating of each machine

Workstations	OP1	OP2	OP3	OP4
W1	30	23	16	–
W2	–	26	–	32
W3	26	–	18	34
W4	34	–	20	38
Average	30	24.5	18	34.7

Table 5 Job-related data

Job Number	Operation sequences	Parts to be produced
J1	OP1, OP2, OP3	A
J2	OP3, OP4, OP1	B
J3	OP2, OP3, OP1	C
J4	OP4, OP3, OP2	D
J5	OP2, OP1, OP4	E

Table 6 Order size of the part types to be produced

Parts	Customer1	Customer2
A	5	–
B	–	5
C	5	–
D	5	–
E	–	5

4.2 Performance Assessment

The above system was implemented using C programming language on a pentium PC running LINUX operating systems. The performance measures are average run time, utilization, wait time, and average sum of completion time. The performance of the proposed GA was assessed with respect to the above performance measures being compared with that of FBS(Filtered Beam Search)-driven scheduler. FBS is a refinement of the beam search method. It uses one other construct called the "filter width" to further prune the state space and generates a 'good' schedule quickly by controlling the amount of search required.(Clyde,et al., 1993).

The results depicted in Figs. 11, 12, 13, and 14 confirm that the proposed method is superior to FBS in terms of selected performance measures.

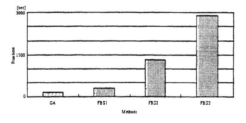

Fig. 11. Average run time of each method.

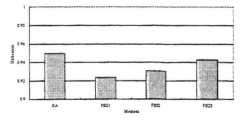

Fig. 12. Average utilization of each method.

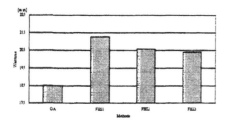

Fig. 13. Average wait time of each method.

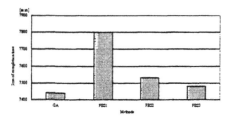

Fig. 14. Average sum of completion time for all products of each method.

Table 7 Simulation results

Performance measures	GA	FBS
Ave. run time[sec]	153.2	1500.1
Ave. utilization	0.95	0.93
Ave. wait time[min]	185.2	207.6
Ave. sum of completion time[min]	7440.0	7604.4

5. Concluding Remarks

This paper presents a new scheduling method based on a genetic algorithm to be used in modular cell manufacturing systems. A genetic algorithm with a uniform crossover using a technique to vary the crossvoer and mutation probatilities were proposed. Then a scheduling method was proposed to implement rapid rescheduling necessary for modular cell manufacturing systems. The proposed GA utilizing the concept of nucleotides varies the crossover and mutation probabilities dynamically for every generation. The performance of the proposed scheduling method, was evaluated using its application to modular cell manufacturing system composing of four machines; it was compared to the FBS(Filtered Beam Search) by run time, utilization, wait time and sum of the completion times.

The results of computer simulation are as follows:

(1) The proposed genetic algorithm is applicable and effective to the problem of requiring the real time processing due to the short run time.
(2) A new scheduling method improved utilization by reducing the wait time, and the completion times, and its scheduling speed was faster than the FBS method by avoiding the illegal schedules from the search space.
(3) The unifom subtour chunking crossover and the proposed mutation techniques were implemented to remove the illegal schedules from the search space.
(4) The dynamic variation of the crossover and mutation probabilities in the GA improved the global performance compared to the fixed GA.

Through an implementation and a series of experiments, both the efficacy and the high performance of the proposed scheduler are demonstrated and its practical utility is promising.

References

Ackley, D. *A Connectionist Machine for Genetic Hillclimbing*, Kluwer Academic Publishers.
Bäck, T(1993) Optimal Mutation Rates in Genetic Search. *Proc. of 5th Int. Conf. on GA*, pp.2-8.
Clyde W. Holsapple, Varghese S. Jacob, Ramakrishnan Pakath and Jigish S. Zaveri(1993). A Genetics-Based Hybrid Scheduler for Generating Static Schedules in Flexible Manufacturing Contexts. *IEEE Trans. on SMC* **Vol. 23, No. 4,** pp.953-972.
Goldberg, D.E.(1989) *Genetic Algorithms in Search, Optimization, and Machine Learning,* Addison-Wesley.
Heragu, S. S.(1994) Group Technology and Cellular Manufacturing. *IEEE Trans. on systems, MAN, and Cybernetics,* **Vol. 24, No. 2.**
Kim, S.S.(1996) A Scheduling of the Modular Cell Manufacturing System Using Genetic Algorithm *Ph.D. Thesis, Dept. of Electrical Engineering, Yonsei University,* Seoul.
Michalewicz Zbigniew(1992) *GENETIC ALGORITHMS + DATA STRUCTURES = EVOLUTION PROGRAMS,* Springer Verlag, Berlin.
Singh, N(1996) *Systems Approach to Computer Integrated Design and Manufacturing,* John Wiley & Sons, Inc.
Spears, W.M. and De Jong, K.A.(1991) On the Virtues of Parameterized Uniform Crossover. *Proc. of 4th Int. Conf. on GA*, pp.230-236.
Syswerda, G(1989) Uniform Crossover in Genetic Algorithm. *Proc. of 3rd Int. Conf. on GA,* pp.2-9.

DECOMPOSING A JOBSHOP INTO FLOWSHOP STRUCTURES USING GENETIC ALGORITHMS

Xianyi Zeng and Michel Happiette

GEMTEX   ENSAIT
2, place des Martyrs de la Résistance 59070 Roubaix France
Tel: (33) 20.25.64.64   Fax: (33) 20.24.84.06
E-mail: Xian.Zeng@univ-lille1.fr

Abstract: This paper presents a method for decomposing a Jobshop production structure into a number of Flowshop structures so that the optimal layout of production can be found in each Flowshop. This decomposition is considered as a clustering problem, which is solved using a genetic algorithm. It permits to optimize a specified objective function related to the overall number of machines used and the number of additional machines inserted for obtaining admissible solutions of layout of production.

Keywords: decomposition, manufacturing systems, optimization, classification, genetic algorithms.

1. INTRODUCTION

In manufacturing systems, Flowshop and Jobshop are two structures widely used to model a production workshop (Charalambous and Hindi, 1993), (Fisher, et al., 1983). In a Jobshop structure, the processing order of each item on machines is well defined and it may be different for different items. But in a Flowshop structure, this order is identical for all items.

A great number of successful algorithms have been proposed to solve scheduling problems in Flowshop structures (Blazewicz, et al., 1994). In real applications, Jobshop scheduling problems, which are considered more complex, can be solved using Group Technology (GT). GT has been recognized as the key to improve productivity, material-handing, management and control of a typical batch-manufacturing system (Harhalakis, et al., 1990). The basic idea of GT is to decompose the manufacturing system into subsystems, by classifying items to be processed into famillies, and machines into machining cells, based on the similarity of items. In our previous paper (Zeng, et al., 1996), an general

scheme was proposed for decomposing the original complex Jobshop problem into simpler Flowshop scheduling subproblems. This decomposition permits to bring out a double clustering of machines and products in order to define homogeneous production subsystems. The combination of the solutions to these subproblems leads to the general solution to the original scheduling problem.

Each decomposed Flowshop structure corresponds to a linear production family (an ordered subset of machines) where the layout of production for all items belonging to this family is realized without any backward move of flow of products, i.e., the number of machines used in this linear family is minimal (Happiette and Staroswiecki, 1990). In this paper, such a structure is considered as an optimal layout of production. It has the sequence of the range of production of each item in this family included in the sequence of the corresponding layout of production.

In order to generate such a linear production family with minimal number of machines, we built a tree structure which is conditioned at each of its nodes by applying two rules to reduce the expansion of the searching space.

The double clustering of machines and products should be designed so that both the optimal layout of production and admissible scheduling solutions can be found in each Flowshop structure. For simplicity, only the optimization of layout of production is taken into account in this paper. In this case, the objective function of our clustering problem can be defined as linear combination of the overall number of machines used in all decomposed Flow-shop structures and the number of additional machines inserted for obtaining admissible solutions of layout of production.

In this paper, the clustering problem has been solved using a genetic algorithm. Genetic algorithms are stochastic search methods based on the principle of natural genetic systems (Michalewicz, 1994). Unlike conventional search methods, genetic algorithms deal with multiple solutions simultaneously and compute the fitness function values for these solutions. Genetic algorithms have been found to provide global near-optimal solutions for various complex optimization problems, including clustering problems.

In the genetic algorithm used in this paper, a new string representation is proposed in order to reduce the searching space by eliminating repeated solutions. The simulation results show that this string representation is much more efficient than other existing ones and it permits to converge quickly to the globally optimal solution even if the calculation of the objective function is very heavy.

This paper is organized as follows. Section 2 defines the formalism for our problem. The procedure of layout of production is described in Section 3. The objective function is defined in Section 4. The principle of genetic algorithms is presented in Section 5. The simulation results and the conclusion are included in Section 6 and Section 7 respectively.

2. FORMALISM

Each of the items to be produced is defined by an ordered set of the machines required for its manufacturing (repeat of machines is possible). It can be described by a word as follows:

$$W = \alpha_1^{n\alpha_1} \alpha_2^{n\alpha_2} \ldots \ldots \alpha_m^{n\alpha_m} = (\alpha_i^{n\alpha_i}), \ i = 1, \ldots \ldots, m$$

where α_i is a letter representing a machine for the manufacturing of the item W and we have $\alpha_i \in ALPHA$, a set containing all the letters. $n\alpha_i$ is the multiplicity of the machine α_i.

The total multiplicity of a given machine α in the word W is then

$$N\alpha(W) = \sum_{\alpha_i = \alpha} n\alpha_i$$

The set $H(W) = \{(\alpha, N\alpha)\}$ is the minimal alphabet on which the word W can be constructed.

For example, let W be the word describing the following working process: $W = a^2bcac^3d$, the corresponding $H(W) = \{(a, 3), (b, 1), (c, 4), (d, 1)\}$.

We can see from this example that each item or word can be composed of several machines of the same type. In this paper, we are interested in the possibility of building a set of Flowshop production structures where each of them is served by a linear conveyor. In this way, the statement of the problem is the following:

Searching for the possibility of grouping, within the same production structure, the machines' succession (words) relative to different items. The chosen type of conveyor then imposes a condition of compatibility between the words grouped together: the order of the machines laid out along the conveyor must necessarily correspond to the order imposed by each of the machines' succession.

According to this idea, for a given class C, the items (words) are manufactured by the minimal set of the machines corresponding to C, defined by: $H(C) = \{(\alpha, N\alpha)\}$ where $\alpha \in \{\cup M_i, \ i \in C\}$ and $N\alpha = max\{N\alpha_i, \ i \in C\}$.

For example, let $C = \{W_1, W_2, W_3, W_4\}$ with $W_1 = aba^2c$, $W_2 = bdcba$, $W_3 = adabc$, $W_4 = cbac$, the corresponding minimal set of machines and its total multiplicity can be calculated by

$H(C) = \{(a,3)(b,2),(c,2),(d,1)\}$
and $N(C) = 3+2+2+1 = 8$

These four items are order consistent because there exists a layout of machines $abdacbac$ which allows the production of all items without any backward move.

3. LAYOUT OF PRODUCTION

In order to search for the layout of production for the set $C = \{W_1, W_2, \ldots, W_n\}$, we adopt the algorithm for the detection of order consistency published in our previous paper (Happiette and Staroswiecki, 1990), (Zeng, et al., 1996). This algorithm is briefly presented below:

For each word W_i, an expansion is progressively constructed by setting at the kth position either the first character of W_i which has not already been used, or a joker. Our objective is to build a generalization which is common to all the words of C. The set is order-consistent iff all the characters have been located for each $W_i \in C$ and the obtained generalization uses only characters from $H(C)$. Our algorithm uses a searching procedure based on a tree of layout permitting, at each step, the elimination of useless nodes by applying two rules R_1 and R_2. If class C is order-consistent, on the terminal useful nodes can be found all the possible solutions of layout associated with C. If C is not order-consistent, all the terminal nodes on the tree of layout are useless and there does not exist any solution of layout associated with C.

The two rules R_1 and R_2 are given as follows:

R_1: Let s_k be a node such that: $\exists(\alpha, W_i) \in ALPHA \times C$ s.t. $N\alpha(C/s_k) < N\alpha(Wi/s_k)$. The success condition cannot be satisfied by any terminal node of the tree having s_k as a predecessor; s_k will be labeled as useless and its successors will not be explored.

R_2: The node s_k is useless if there exist two words W_i, $W_j \in C$ and two letters α, $\beta \in ALPHA$ s.t. $(\alpha, 1)$, $(\beta, 1) \in H(W_i/s_k)$ and $H(W_j/s_k)$ with $\alpha < \beta$ in W_i/s_k and $\beta < \alpha$ in Wj/s_k ("<" is the order between two letters in a sequence).

According to R_2, one couple of subsequences $(\alpha\ \beta)$ and $(\beta\ \alpha)$ belonging to W_i and W_j respectively leads to a fail or a backward move.

Remark: In the rules R_1 and R_2, s_k is the common generalization of the expansions of the first sequences of the words in C. C/s_k represents the set of machines in the remaining resources of $H(C)$ and W_i/s_k represents the set of machines in the remaining needs of W_i.

For the example described in Section 2, the procedure of searching for layout of production is shown in Fig. 1. Two solutions of layout (*abdcabac* and *abdacbac*) can be found from Fig. 1. Each of them corresponds to a Flowshop structure. For the solution *abdcabac*, we get the following Flowshop: $M_1=a$, $M_2=b$, $M_3=d$, $M_4=c$, $M_5=a$, $M_6=b$, $M_7=a$, $M_8=c$. The corresponding four items to be produced can be rewritten as: $J_1=(O_{11}\ O_{12}\ O_{15}\ O_{17}\ O_{18})$, $J_2=(O_{22}\ O_{23}\ O_{24}\ O_{26}\ O_{27})$, $J_3=(O_{31}\ O_{33}\ O_{35}\ O_{36}\ O_{38})$, $J_4=(O_{44}\ O_{46}\ O_{47}\ O_{48})$.

If no solution of layout is obtained for a given set of items of production, i.e. all the terminal nodes in the exploration tree of layout fail, we add a new machine so that the procedure of layout of production can

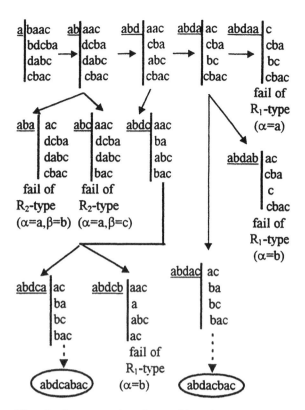

Fig. 1. An example of searching for solutions of layout of production

continue. This new machine is selected so that the number of machines of the same type occuring in fails (R_1-type or R_2-type) is maximal. In the example of Fig. 1, both a and b are of the most frequent occurrence in the fails of the tree of layout. So, we can add a machine of type a or of type b in order to make the procedure of layout of production continue.

4. OBJECTIVE FUNCTION

The objective function of our clustering problem is based on the procedure of layout of production (tree of layout), presented in Section 3. This procedure does the generalization for a class of items (words) and returns two following values: 1) the minimal number of machines used in this generalization to obtain admissible solutions of layout; and 2) the number of additional machines inserted. The objective function has to be designed to minimize the overall number of machines used for manufacturing all items and the overall number of additional machines inserted into all classes. For a given class of items, the number of additional machines can not be too great because the size of each Flowshop is limited. According to these principles, we give the definition of the objective function as follows.

Assuming that there exist K classes of items in the present partition: $P=\{C_1, C_2, \ldots, C_K\}$. The corresponding objective function is defined by

$$f(P) = \sum_{i=1}^{K} [\|H(C_i)\| + \beta \cdot NA(C_i)]$$

where $H(C_i)$ is the minimal alphabet on which the items of C_i can be constructed in the tree of layout of production. $NA(C_i)$ is the minimal number of additional machines inserted in order to obtain admissible solutions of layout.

For the class C_i, if some solutions are obtained from the terminal nodes of the tree of layout, then $NA(C_i)=0$ and no additional machine is needed and we have $NA(C_i)=0$. Otherwise, as presented in Section 3, an additional machine is inserted to $H(C_i)$ so that the procedure of layout continues. In this case, we have $H(C_i)=H(C_i)+1$ and $NA(C_i)=1$. This procedure is repeated until admissible solutions can be found on the tree of layout.

5. GENETIC ALGORITHM FOR CLUSTERING

As described in Section 4, our objective is to find the best partition of items so that the overall number of machines used for manufacturing all items is minimized and the overall number of additional machines inserted into all classes is not too great. In practice, this clustering problem is very complex because the searching space is rather large and a lot of local minimum exist in this space. So, a robust and empirical optimization technique is needed to find the globally optimal partition.

In this paper, the clustering problem is solved by a genetic algorithm where the objective function f is taken as fitness function. The clustering problem is solved by finding the minimal value of the function f, which corresponds to the globally optimal partition of items.

The main principle of a genetic algorithm is described as follows.

It begins with an initial set (also called population) of randomly generated potential solutions to an optimization problem. The value of a fitness function is evaluated for each solution, and the "best" solutions are selected for survival according to the selection probabilities. For each solution, the selection probability is defined as function of its fitness value. Then, the genetic algorithm manipulates these selected solutions in its search for better solutions. Each solution is encoded into a binary string, so that new encoded solutions can be generated through the exchange of information among surviving solutions (crossovers) as well as sporadic alterations in the bit string encodings of the solutions (mutations).

Typically, a genetic algorithm is characterized by the following components (Michalewicz, 1994):

1) a string representation for the feasible solutions to the optimization problem;
2) a population of encoded solutions;
3) a fitness function that evaluates each solution;
4) genetic operators that generate a new population from the existing population.

To solve clustering problems with a genetic algorithm, we must define string representations to encode partitions in a way that allows manipulation by genetic operators. Two typical string representations can be found in literature. The first is based on approximating boundaries of classes by piecewise linear segments in the feature Euclidean space (Bandyopadhyay, et al., 1995). This method is available only when patterns can be represented by multi-dimensional vectors and dissimilarity can be measured by the Euclidean distance. The second string representation permits to encode a partition as a string of length n where n is the number of patterns. The ith element of the string denotes the group number assigned to the ith pattern (Murthy and Chowdhury, 1996). Under this representation, one partition corresponds often to several strings, which enlarges the searching space. For example, for a set of 5 patterns ($n=5$), the number of partitions is 52 but the number of strings attains $5^5 = 3125$. Moreover, a non uniform probability distribution can be found in the space of partitions, i.e. some partitions have higher probabilities being selected than some others. To solve these two drawbacks, we defined a string representation based on relationship of pairs of patterns. Under this representation, a bit is 1 if the two patterns in the corresponding pair are in the same class and 0 if they are not in the same class. This representation permits to encode a partition in an unique string but some strings are not valid for representing partitions because bits in a significant string are constrained by transitive relations. For example, if W_1, W_2 are two patterns in the same class, then the bits corresponding to the pairs $(W_1\ W_3)$ and $(W_2\ W_3)$ have to be both 1 or both 0 (W_3 is another pattern). Therefore, a validity test has to be done when new strings are generated in a population. It is shown in practice that only 10% strings are valid and much time is consumed in the validity test.

From the above discussion, we can see that a good string representation in a genetic algorithm has to be defined so that one partition is uniquely represented by one string and any string corresponds to an unique partition. The simplest way to do so is to arrange all partitions and to convert the decimal numbers denoting the orders of partitions into binary strings. This string representation has been implemented for solving our problem of decomposing a Jobshop into Flowshop structures and a good convergence behavior can be obtained. For a set of n items, all possible partitions are denoted by P_1, P_2, ..., P_m where m is the total number of partitions calculated by

$$m=h(n,1)+h(n,2)+......+h(n,n)$$

where $h(i,j)$ denotes the number of partitions having i items and j classes

$h(i,j)$ can be recursively calculated according to the following equations.

i) $h(1,1)=1$
ii) $h(i,1)=h(i-1,1)$, $h(i,j)=h(i-1,j-1)+j \cdot h(i-1,j)$
for $j \in \{2, ..., i-1\}$ and $h(i,i)=h(i-1,i-1)$
where $i \in \{2, 3, ..., n\}$

In a genetic algorithm with this string representation, the length of each binary string is $\lceil log_2 m \rceil$ and a string can be easily converted into a decimal number, which corresponds uniquely to one partition taken from $\{P_1, P_2, ..., P_m\}$.

In the next example, we try to encode all partitions for the case $n=4$. First, we calculate the values of $h(i,j)$ as follows.

Table 1 The values of $h(i,j)$'s for $n=4$

i \ j	1	2	3	4
1	1			
2	1	1		
3	1	3	1	
4	1	7	6	1

Then, the total number of partitions m=15 and the length of each string is $\lceil log_2 m \rceil = 4$.

These *15* partitions are then arranged in order and encoded as Table 2. In this table, each partition is denoted by $(j_1 j_2 ...... j_n)$ for any $j_i \in \{1, 2, ..., n\}$ and $j_i=j_k$ means that the items W_i, W_k belong to the same class. The number of classes in this partition is given by $max\{j_1, ..., j_n\}$. For example, the partition *(1111)* means that all of the four items constitute only one class and the partition *(1234)* corresponds to four classes of which each is composed of one item. From Table 2, we can see that each binary string corresponds uniquely to a partition of items.

In a genetic algorithm, genetic operators include "selection", "crossover" and "mutation". In the selection, strings are selected from a population (P_1 P_2 ...... P_s) to create a mating pool (s is the size of all populations). In our problem, we try to find the optimum by minimizing the fitness function. So, the selection probability of each string is defined inversely proportional to the fitness value, i.e.

$$\Pr ob(P_i) = \frac{f^{-1}(P_i)}{\sum_{i=1}^{s} f^{-1}(P_i)}$$

Table 2 The encoded partitions for $n=4$

Partition order	String representation	Partition
1	(0000)	(1111)
2	(0001)	(1112)
3	(0010)	(1121)
4	(0011)	(1122)
5	(0100)	(1211)
6	(0101)	(1212)
7	(0110)	(1221)
8	(0111)	(1222)
9	(1000)	(1123)
10	(1001)	(1213)
11	(1010)	(1223)
12	(1011)	(1231)
13	(1100)	(1232)
14	(1101)	(1233)
15	(1110)	(1234)

Our genetic algorithm manipulates binary strings in a limited searching space. So, the other operations "crossover" and "mutation" can be done in a classical way such as described in (Michalewicz 1994).

6. SIMULATION RESULTS

In our simulation experiments, we select the crossover probability $pc=0.75$ and the mutation probability $pm=0.05$, which makes each population strongly agitated in order to obtain diversified solutions. The number of solutions in each population is fixed to *20*. The number of generations is also fixed to *20*. Two examples are given below to illustrate the results of decomposition of a Jobshop structure into Flowshop sub structures.

Example 1:

Let $\{bac, abacd, cabd, caabd, acbdb\}$ be a set of *5* items to be manufactured ($n=5$). The number of partitions m can be recursively calculated according to the equations in Section 5. We obtain $m=52$ and the string length is then $\lceil log_2 m \rceil = 6$. The clustering results for different values of β are given in Table 3.

Example 2:

Let $\{bac, abacd, cabd, caabd, acbdb, abaac, bdcba, adabc, cbac\}$ be a set of *9* items to be manufactured ($n=9$). We obtain $m=21147$ and the string length is then *15*. The clustering results for different values of β are given in Table 4.

Table 3 Optimal partitions for Example 1 with different values of β

β	$f(P_{opt})$	$K(P_{opt})$	P_{opt}
0	9	1	{bac, abacd, cabd, caabd, acbdb}
2	14	2	{bac, abacd} {cabd, caabd, acbdb}
4	15	3	{bac, abacd} {acbdb} {cabd, caabd}

where P_{opt}, $K(P_{opt})$ and $f(P_{opt})$ denote the optimal partition found by the genetic algorithm, the corresponding number of classes and value of the fitness function.

Table 4 Optimal partition for Example 2 with different values of β

β	$f(P_{opt})$	$K(P_{opt})$	P_{opt}
0	17	2	{bac, caabd, cbac} {abacd cabd, acbdb, abaac, bdcba, adabc}
2	21	3	{bac, abacd, abaac, bdcba, adabc} {acbdb} {cabd, caabd}
4	27	3	{bac, abacd, caabd, abaac, bdcba, adabc, cbac} {cabd} {acbdb}

From these two examples, we can see that the number of classes of the found optimal partition increases with β. Big values of β lead to more decomposed Flowshop structures with smaller size and small values of β lead to fewer decomposed Flowshop structures with bigger size. In practice, the exact value of β is selected by workshop's organizers according to the real constraints to layout of production.

7. CONCLUSION

This paper presents a method for decomposing a Jobshop structure into a series of Flowshop sub structures. This decomposition is designed to minimize the overall number of used machines under the constraint that the size of each Flowshop can not be too great. A genetic algorithm is applied to find the optimal structure of the decomposition. This genetic algorithm is modified by defining a new string representation in order to eliminate repeated partitions in each population and to reduce the

corresponding searching space. The proposed string representation can be extended to other clustering problems using genetic algorithms.

The proposed method can be improved in two following aspects. 1) The scheduling criteria for items can be introduced into the fitness function because a good decomposition should be designed not only to minimize the number of used machines but also to allow admissible scheduling solutions for the items in each decomposed Flowshop. 2) In the proposed algorithm, much time is consumed in the calculation of fitness values because a tree of layout has to be completely covered for calculating only one fitness value. So, the structure of the tree of layout has to be simplified by introducing more rules and by adding some heuristic strategy to predict failure branches.

REFERENCES

Bandyopadhyay, S., Murthy, C.A., Pal, S.K. (1995). Pattern classification with genetic algorithms. *Pattern Recognition Letters*, **Vol.16**, pp.801-808.

Blazewicz, J., Ecker, K.H., Schmidt, G. and Weglarz, J. (1994). *Scheduling in computer and manufacturing systems*, Springer Verlag, Berlin.

Charalambous, O. and Hindi, K.S. (1993). A knowledge-based Jobshop scheduling system. *Production Planning and Control*, **Vol.4**, pp.304-312.

Fisher, M.L., Lageweg, B.J., Lenstra, J.K. and Rinnoy, K. (1983). Surrogate duality relaxation for Jobshop scheduling. *Discre. Applied Math*, No.5, pp.65-75.

Happiette, M. and Staroswiecki, M. (1990). An algorithm for the detection of order-consistency on a set of syntactic patterns: application to workshop organisation. *Systems Science*, **Vol.16**, No.3.

Harhalakis, G., Nagi, R. and Proth, J.M. (1990), An efficient heuristic in manufacturing cell formation for group technology applications. *Int. J. Prod. Res.*, **Vol. 28**, No.1, pp.185-198.

Michalewicz, Z (1994). *Genetic Algorithms + Data Structures = Evolution Programs*, Springer-Verlag, Berlin.

Murthy, C.A. and Chowdhury, N. (1996). In search of optimal clusters using genetic algorithms. *Pattern Recognition Letters*, **Vol.17**, pp.825-832.

Zeng, X., Happiette, M. and Vasseur C. (1996). Decomposition of Jobshop into Flowshop structures for scheduling, *Proceedings of the International Conference IPMU'96*, Granada, Spain.

ON THE APPLICATION OF MULTISTRATEGY LEARNING AND HYBRID AI APPROACHES IN INTELLIGENT MANUFACTURING

László Monostori

*Computer and Automation Research Inst., Hungarian Academy of Sciences
Kende u. 13-17, H-1518 Budapest, POB 63, Hungary*

Abstract: The application of *pattern recognition (PR) techniques, artificial neural networks (ANNs)*, and nowadays *hybrid artificial intelligence (AI)* techniques in manufacturing can be regarded as consecutive elements of a process started two decades ago. The fundamental aim of the paper is to outline the most important steps of this process, to introduce some new results with special emphasis on hybrid AI and multistrategy machine learning approaches, and to highlight future possibilities.

Keywords: control and monitoring of manufacturing processes, intelligent manufacturing systems, learning, neural networks, fuzzy systems

1. INTRODUCTION

Over the past decades, the field of artificial intelligence has made great progress toward computerizing human reasoning. *Symbolic approaches* are based on the *hypothesis of symbolic representation* - the idea that perception and cognitive processes can be modeled as acquiring, manipulating, associating, and modifying symbolic representations. *Expert systems* represent the earliest and most established type of intelligent systems attempting to embody the "knowledge" of a human expert in a computer program. Knowledge representation in these systems precedes symbolically in the form of *production rules, frames* or *semantic networks*.

A different approach to intelligent systems involves constructing computers with architectures and processing capabilities that mimic the processing characteristics of the nervous system. The technology that attempts to achieve these results is called *neural computing* or *artificial neural networks* (ANN). These *subsymbolic methods* work with *numeric*

values and seem to be more appropriate for dealing with perception tasks and perhaps even with tasks that call for combined perception and cognition.

Several techniques for integrating expert systems and neural networks have emerged over the past few years spreading from *stand-alone* models, through *transformational, loosely* and *tightly coupled* models to *fully-integrated expert system/neural network models* (Kandel and Langholz, 1992). The integration of neural and fuzzy techniques which can be considered as a "full integration" is an approach of high importance.

Most research in *machine learning (ML)* has been concerned with methods that employ a single learning strategy, that is, with *monostrategy* methods. With the growing understanding of the capabilities and limitations of monostrategy methods, there has been an increasing interest in *multistrategy* systems that integrate multiple inference types and/or computational mechanisms in one learning system. Such systems are expected to learn from wider scope of input, and be applied to a wider range of problems than monostrategy methods (Monostori and Barschdorff, 1992).

2. AI AND ML TECHNIQUES IN PROCESS MODELLING, MONITORING & CONTROL

Investigations confirmed that - similarly to our present conception of biological structures - adaptive ANN techniques seem to be a viable solution for the lower level of intelligent, hierarchical control and monitoring systems, where abilities for *real-time functioning, uncertainty handling, sensor integration,* and *learning* are essential features. Since the higher levels of the control and monitoring hierarchy require mostly symbolic knowledge representation and processing, the *integration of symbolic and subsymbolic methods* is straightforward (Monostoir and Barschdorff, 1992).

Using the former categorization, two main approaches for integrating symbolic and subsymbolic approaches have been attempted in this field:
- hierarchically connected hybrid AI system (coupled model)
- symbiotic type of hybrid AI system by integrating neural and fuzzy approaches (fully integrated model).

The pattern recognition - artificial neural networks - hybrid AI systems evolution in applications can be observed in the following, where the overlapping fields of tool condition monitoring, process modeling and adaptive control issues will be treated. Special emphasis will be laid on learning abilities, admitting that it cannot be treated separately from the other important issues (self-calibration, signal processing, decision making, fusion ability, etc. (Byrne, *et al.,* 1995)).

2.1 Tool condition monitoring (TCM)

The application of numerical *pattern recognition (PR)* techniques for monitoring purposes started with linear decision functions trained iteratively (Sata, *et al.,* 1973; Dornfeld and Bollinger, 1977; Zhang, *et al.,* 1982).

Fuzzy pattern recognition techniques proved to be efficient tools for dealing with the uncertain nature of cutting processes (Wang, *et al.,* 1985; Ko, *et al.,* 1995).

A number of *multipurpose monitoring systems* were developed on the basis of multisensor integration and parallel processing through multiprocessor systems (Weck, *et al.,* 1984; Monostori, 1986). The majority of these approaches used pattern recognition techniques for learning and classification.

Dornfeld applies ANNs for TCM (1990a, 1990b). A series of tests were made on a lathe machining hardened material to induce faster tool wear. Acoustic emission (AE) and cutting force were measured on the tool shank and on a fixture in which the tool shank was mounted. The applicability of ANNs for multisensor integration was demonstrated. The comparison of results gained by linear classifiers and BP ANNs' outlined the better noise suppression and classification abilities of neural networks.

One of the main - but often neglected - problems in monitoring of machining processes is *how to treat the varying process parameters*. Some possibilities for incorporating process parameter information into the learning and classification phases were demonstrated in (Monostori, 1993):
- networks trained under constant process parameters,
- networks trained under varying process parameters,
- networks incorporating process parameters as inputs.

Comparing the different approaches, the best generalization ability (networks' performance for patterns generated under cutting conditions outside of the region considered during the training phase) was achieved with the latter variant, i.e. the networks were able to filter out the disturbing effects of the varying process parameters, and to generalize.

Due to the numeric, real-time nature of TCM, comparing with connectionist approaches, relatively few reports are available on the application symbolic learning techniques in this field. In (Junkar, *et al.,* 1991), the identification of the plunge grinding by decision tree generation using vibration signals generated by the grinding wheel was described.

Combined structure and parameter learning technique through a neuro-fuzzy (NF) system for cutting tool monitoring was described in (Monostori and Egresits, 1995). Given 4 wear classes of milling, the task was to generate and compare ANN and NF structures which are able to reliably classify unknown patterns characterizing different wear states.

A four step learning algorithm integrating self organized clustering, competitive learning, and supervised BP learning techniques was applied for determining the fuzzy rules and the parameters of the membership (MBF) functions. The generated 14 rules have the linguistic form such as *IF F_1 is high and F_2 is high and F_3 is high and F_4 is low and F_5 is high and F_6 is high THEN output is Wear025* (average wear of teeth is 0.25 mm).

After initialization, taking all the possible combinations determined by the number of MBFs assigned to the input and output variables, 4*3*2*3*4*4 = 1152 rule nodes were generated. During the competitive learning phase 1138 (!) of them - which proved to be inappropriate or unnecessary for the given task - were deleted, resulting in a network structure with 14 rules. This *self organization* is a very important feature of the chosen model. The mean values and variances of bell shaped MBFs were adjusted to reach a kind of optimum regarding the network error for training patterns (Monostori and Egresits, 1995). The *NF technique* with structure and parameter learning showed superior performance to the BP solution and previous investigations with a commercial NF system.

2.2 Process modeling

Rangwala and Dornfeld suggest ANNs as learning structures for the lower level of an intelligent controller (Rangwala and Dornfeld, 1989). A *learning process* enables the controller to understand how input variables (such as feed rate (f), depth of cut (d), and cutting velocity (v)) affect output variables (such as cutting force (F), power(P), temperature(T) and workpiece surface finish(R)) in the case of a turning operation.

The decision-making approach of Chryssolouris et al. incorporates several process models that correlate process state variables such as surface roughness or chip merit mark to process parameters such as feed rate, cutting speed and tool rake angle (Chryssolouris and Guillot, 1988).

In (Monostori, 1993) inverse models of the milling process, i.e. separate models for three process parameters (axial depth of cut, cutting speed, and tooth feed) were generated always using the other two process parameters and force and vibration features as networks' inputs. Fairly good agreement was found between the programmed and estimated process parameter values. The inverse models were successfully used for TCM purposes too.

Bukkapatnam et al. (1995) prove the existence of low dimensional *chaos* in turning and develop a unified approach for tool condition monitoring using a combination of wavelets and fractal dimensions.

Analytical models (e.g. for forces or wear occurring in profile grinding) are available and can be adapted to different quantities by using learning strategies. These analytical models can be combined by simulation techniques, e.g. by FEM simulation to estimate the thermoelastic behavior of the workpiece. ANNs due to their high adaptivity and non-linearity

can be advantageously used also here (Westkämper, et al., 1995). The combination of multisensor integration, propound signal processing, adaptive models, process simulation can lead to the adaptive control of quality.

2.3 Adaptive control

The above described investigations for determining suitable process models for machining operations aimed at realizing powerful adaptive control schemes.

The task to be fulfilled can be formulated as follows. There exist some limitations on input variables (e.g. machine limitations), some output variables are to be kept sufficiently close to the desired values and others can have upper limits (e.g. vibration). The algorithm suggested by Rangwala and Dornfeld is based on an augmented Lagrangian method to minimize a properly selected combined performance index, which takes into consideration the above requirements. Their attractive algorithm, which implies forward and BP, has resulted in good performance using simulated data (Rangwala and Dornfeld, 1989).

As further improvement to this "batch mode" technique (where the learning and synthesis are done separately) a so called incremental approach was introduced, where the learning and synthesis phases proceed simultaneously. However, during this incremental operation only a part of the possible input space is scanned, mostly the region near to the operating point. Tested with values further from this region, the network showed poor generalization properties (Rangwala and Dornfeld, 1989).

One of the most well-known models for intelligent machining was introduced by Chryssolouris et al. (1987). The control strategy is based on the simultaneous measurement and processing of different signals (force, temperature, acoustic emission). These signals are fed into independent process models. The essential element of this structure is the synthesis of the different sensor information.

Two hybrid AI systems for control and monitoring of manufacturing processes on different hardware and software bases were described in (Barschdorff, et al., 1995). In these hybrid systems, networks outputs are conveyed to an expert system that provides process control information (Fig.1). On the base of accumulated knowledge the hybrid systems influence the functioning of the subsymbolic levels, generate optimal process parameters and inform the user about the actual state of the process.

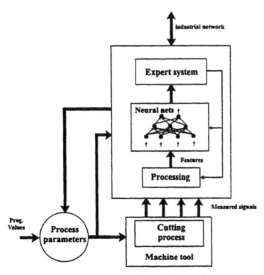

Fig. 1. Adaptive control of manufacturing processes by hierarchically structured hybrid AI systems

This *tight coupling* approach has clear advantages:
-   it fits to the monitoring-control hierarchy of manufacturing cells regarding the form and speed of information processing,
-   modular structure enabling and facilitating the use of commercial tools,
-   faster development,
-   clear interfaces,
-   relatively easy integration into existing manufacturing environment.

In the *HYBEXP* system (Monostori, 1995), an artificial neural network simulator called *NEURECA* constitutes the *lower, subsymbolic level* (A in Fig. 2). The *higher, symbolic level* is based on the commercially available *GoldWorks III* expert system shell (B).

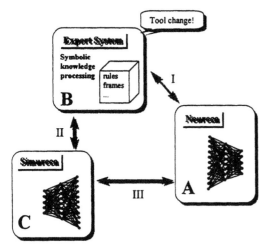

Fig. 2. Components of the *HYBEXP* PC based hierarchically structured hybrid AI system

Both the symbolic and neural subsystems are connected to the machine tool (the machine tool controller is incorporated). The symbolic part forwards (II) process parameter information (feed rate, depth of cut, cutting speed) to the machine tool (C). The generated indirect signals (e.g. force components, vibration) are measured and conveyed (III) to the subsymbolic part (A) of the hybrid system.

In the figure the machine tool is substituted by a *simulator* of the manufacturing process (called *SIMURECA*, a version of *NEURECA*), enlightening the test and demonstration of the system. On the base of accumulated knowledge and actual process parameters (cutting speed, feed rate, depth of cut), *SIMURECA* estimates the selected features of the force and vibration signals. These features are forwarded to the *NEURECA* subsystem, which fulfills the estimation or classification assignment.

Configuration of the *SIMURECA* and *NEURECA* subsystems can be initiated from the symbolic level. Type of manufacturing (e.g. turning, milling), the signals to be measured (e.g. force, vibration), the number of considered features, and the type of task (classification or estimation) to be fulfilled by *NEURECA* can be defined. All of the other tasks (e.g. activation of corresponding signal processing routines, ANNs, communication between the subsystems) proceed automatically governed by rules.

According to previous investigations (Monostori, 1993) reliable ANN models for classification of cutting tools or for tool wear estimation using indirect signals can only be constructed if they handle process parameter (e.g. cutting speed, feed rate, depth of cut) information. Therefore, the inputs of ANN models used in *HYBEXP* additionally incorporate cutting parameters to indirect signal features. Models for both classification and estimation can be used.

In both cases the results are conveyed to the hybrid part, where using additional stored knowledge (e.g. the type and number of cutting tools available, actual cutting parameters, the parts to be machined, etc.) different decisions can be made. *HYBEXP* can initiate e.g. machine stop, tool change, modification of cutting parameters (AC control) or change of parts to be machined. *HYBEXP* can work also as a *decision support system*.

Inverse dynamics of the milling process is learned by an ANN model in the adaptive control system for constant milling force under varying cutting conditions with adjustable feed-rates, proposed by Tarng and Hwang, and a fuzzy feedback mechanism is designed to perform an adaptive modification of connection weights in the ANN (1995). Automatic generation of fuzzy rule base in a fuzzy control system for constant cutting forces in turning is achieved by a genetic algorithm in (Tarng, et al., 1996).

Tönshoff and Walter described a self-tuning fuzzy controller for process control in internal grinding (1994). The authors pointed to the time consuming and subjective adjustment of a large number of design parameter in fuzzy control approaches and the black-box nature of ANN-based solutions.

Jee and Koren introduced a self-organizing fuzzy logic controller for friction compensation in feed drives (1995). The parameters of the fuzzy controller are self-tuned according to the previous performance of the controller and a friction model in the low velocity range. The proposed controller was compared with a conventional fuzzy logic controller and a PID controller on a 3-axis milling machine for contour milling. Concerning the contour errors, the new system resulted in a significant reduction.

3. CONCLUSIONS

Learning process models, cause-effect relations, automatically *recognizing different process changes* and *degradation* and *intervening* in the process in order to ensure safe processes and product qualities are sophisticated approaches with high potential. They are the subjects of intensive research and development work world-wide. The complexity of the problem and the associated uncertainties necessitate the application of *learning techniques* to get closer to *working intelligent manufacturing systems.*

In correspondence with a recently published larger survey on ML approaches to manufacturing (Monostori, et al., 1996), concerning intelligent processing of materials, the following main trends can be enumerated:

- attributable mostly to the results of ANN research, and in contrast to the situation 8-10 years ago, when they were considered as fields hardly manageable with AI and ML techniques the area of *process modeling, monitoring and control* (together with the automatic inspection, diagnostics and quality issues) came to the fore;
- in recent years *ANN based learning* is the dominant ML technique in manufacturing;
- ANNs are natural tools at *lower levels of intelligent manufacturing systems,* where abilities of sensor integration, signal processing, uncertainty handling, real-time and adaptive functioning are required, but
- ANNs find their applications on nearly every, also on *non-sensory domains* of manufacturing, as general, very effective information processing blocks with learning abilities;
- development and routine production *of programmable ANN chips* (e.g., cellular neural

networks, which are very effective tools also for visual tasks (Roska, et al., 1995) will give a new, spectacular impulse to ANN applications in manufacturing, especially with the rapid spread of *smart sensors*;
- comparing with the results of a previous survey specific to ANN applications (Monostori and Barschdorff, 1992), a significantly *larger variety of ANN models* and *learning paradigms* can be enumerated, though many applications use the back propagation approach and in its most simple setting;
- *neuro-fuzzy* approaches manifest themselves as prospective tools of numerous problems in manufacturing, *where real-time nature, learning ability, handling of uncertainty and both symbolic and numeric information* are essential requirements;
- As the considerable number of projects *with multistrategy learning* methods shows, the future belongs to these integrated approaches (e.g. Monostori and Egresits, 1997): however, in order to create such powerful, multifaceted environments the investments in growing up-to-date systems will grow higher and higher. In the future, there is not much hope to produce further novel results with prototype software systems.

ACKNOWLEDGMENTS

This work was partially supported by the *National Research Foundation, Hungary*, Grant No. 014514, and T016512 (Fundamental research for intelligent manufacturing).

REFERENCES

Barschdorff, D., L. Monostori, G. W. Wöstenkühler, Cs. Egresits and B. Kádár (1995). Approaches to coupling connectionist and expert systems in intelligent manufacturing, Proc. of the *Second Int. Workshop on Learning in IMSs*, Budapest, Hungary, April 20-21, pp. 591-608.

Bukkapatnam, S., A. Lakhtakia and S. Kumara (1995). Analysis of sensor signals shows turning on a lathe is chaotic, *Physics Rewievs E*, Vol. 52, No. 3, pp. 2375-2387.

Byrne, G., D. Dornfeld, I. Inasaki, G. Ketteler, W. König and R. Teti (1995). Tool condition monitoring (TCM) - The status of research and industrial applications, *CIRP Annals*, Vol. 44/2, pp. 541-568.

Chryssolouris, G. and M. Guillot (1988). An AI approach to the selection of process parameters in intelligent machining, Proc. of *The Winter Annual Meeting of The ASME on Sensors and Controls for Manufacturing*, Chicago, Illinois, Nov. 27 - Dec. 2,

Chryssolouris, G., M. Guillot and M. Domroese (1987). An approach to intelligent machining, Proc. of the *1987 American Control Conf.*, Minneapolis, MN, June 10-12, pp. 152-160.

Dornfeld, D.A. (1990a). Neural network sensor fusion for tool condition monitoring, *CIRP Annals*, Vol. 39/1, pp. 101-105.

Dornfeld, D.A. (1990b). Unconventional sensors and signal conditioning for automatic supervision, Proc. of the *AC'90, III. CIRP Int. Conf. on Automatic Supervision, Monitoring and Adaptive Control in Manufacturing*, Sept. 3-5, Rydzyna, Poland, pp. 197-233.

Dornfeld, W. H. and J. G. Bollinger (1977). On line frequency domain detection of production machinery malfunctions, Proc. of the *18th Int. Machine Tool Design and Research Conf.*, London, pp. 837-844.

Egresits, Cs. and L. Monostori (1997). Multistrategy learning approaches to generate and tune fuzzy control structures and their application in manufacturing, Preprints of the *Second World Congress on Intelligent Manufacturing Processes & Systems*, June 10-13, Budapest, Hungary, pp. 88-94.

Jee, S. and Y. Koren (1995). A self-organizing Fuzzy logic control for friction compensation in feed drives. Proc of the *American Control Conf.*, Seatle, Washington, June , pp. 205-209.

Junkar, M., B. Filipic and I. Bratko (1991). Identifying the grinding process by means of inductive machine learning, Proc. of the *First Workshop of the Intelligent Manufacturing Systems Seminars on Learning in IMSs*, Budapest, Hungary, March 6-8, pp. 195-204.

Kandel, A. and G. Langholz (1992). *Hybrid architectures for intelligent systems*, CRC Press, Boca Raton.

Ko, T.J., D.W. Cho and M.Y. Jung (1995). On-line monitoring of tool breakage in face milling using a self-organized neural network, *J. of Manufacturing Systems*, Vol. 14, No. 2, pp. 80-90.

Michalski, R. and G. Tecuci (1994). *Machine learning: A multistrategy approach*, Morgan Kaufmann, San Francisco, California.

Monostori, L. (1986). Learning procedures in machine tool monitoring, *Computers in Industry*, Elsevier, Vol. 7., pp. 53-64.

Monostori, L. (1993). A step towards intelligent manufacturing: Modeling and monitoring of manufacturing processes through artificial neural networks, *CIRP Annals*, 42, No. 1, pp. 485-488.

Monostori, L. (1995). Hybrid AI approaches for supervision and control of manufacturing processes, Key-note paper, Proc. of the *AC'95, IV Int. Conf. on Monitoring and Automation Supervision in Manufacturing*, Miedzeszyn, Poland, Aug. 28-29, pp. 37-47.

Monostori, L. and D. Barschdorff (1992). Artificial neural networks in intelligent manufacturing, *Robotics and Computer-Integrated Manufacturing*, Vol. 9, No. 6, Pergamon Press, pp. 421-437.

Monostori, L. and Cs. Egresits (1995). On hybrid learning and its application in intelligent manufacturing. Preprints of the *Second Int. Workshop on Learning in IMSs*, Budapest, Hungary, April 20-21, pp. 655-670.

Monostori, L., A. Márkus, H. Van Brussel and E. Westkämper (1996). Machine learning approaches to manufacturing, *CIRP Annals*, Vol. 45, No. 2, pp. 675-712.

Rangwala, S.S. and D.A. Dornfeld (1989). Learning and optimization of machining operations using computing abilities of neural networks, *IEEE Trans. on SMC*, Vol. 19, No. 2, March/April, pp. 299-314.

Roska, T., L.O. Chua and Á. Zarándy (1995). Translating neumorphic CNN visual models to the analogic visual microprocessors, Proc. of the *ISCAS 1995, IEEE International Symposium on circuits and systems*, Seattle, Vol. 2, pp. 1307-1309.

Sata, T., T. Matsushima, T. Nagakura and E. Kono (1973). Learning and recognition of the cutting states by the spectrum analysis, *CIRP Annals*, Vol. 22/1, pp. 41-42.

Tarng, Y.S. and S.T. Hwang (1995). Adaptive learning control of milling operations, *Mechatronics*, Vol. 5, No. 8, pp. 937-948.

Tarng, Y.S., C.Y. Lin and C.Y. Nian (1996). Automatic generation of a fuzzy rule base for constant turning force, *J. of Intelligent Manufacturing*, Vol. 7, pp. 77-84.

Tönshoff, H.K. and A. Walter (1994). Self-tuning fuzzy-controller for process control in internal grinding, *Fuzzy Sets and Systems*, pp. 359-373.

Wang, M., J.Y. Zhu and Y.Z. Zhang (1985). Fuzzy pattern recognition of the metal cutting states, *CIRP Annals*, Vol. 34/1, pp. 133-136.

Weck, M., L. Monostori and L. Kühne (1984). Universelles System zur Prozess- und Anlagenüberwachung, Vortrag und Berichts- band der *VDI/VDE-GMR Tagung Verfahren und Systeme zur technischen Fehlerdiagnose*, Langen, FRG, Apr. 2-3, pp. 139-154.

Westkämper, E., D. Lange and H.W. Hoffmeister (1995). Safe abrasive processes through adaptive control under highly chaotic conditions, Proc. of the *IV. Int. Conf. On Monitoring and Automatic Supervision in Manufacturing*, Miedzeszyn, Poland, Aug. 28-29, pp. 269-282.

Zhang, Y.Z., Z.F. Liu, L.X. Pau, Y.I. Liu and W.B. Yang (1982). Recognition of cutting states for the difficult-to-cut materials, Application of the pattern recognition techniques, *CIRP Annals*, Vol.31/1, pp. 97-101.

USING NEURAL NETWORKS FOR MODELING VLSI INTERCONNECTION MANUFACTURING PROCESSES

Byungwhan Kim and Kwang-Ho Kwon*

Memory R & D Laboratory, Hyundai Electronics
San 186-1, Ami-ri, Bubal-eub, Ichon-si, Kyungki-do, Korea
Tel: +82-336-30-5167, e-mail: litho7@sr.hei.co.kr
*Department Electronic Engineering, Hanseo University
360, Daegok-ri, Haemi-myun, Seosan-si, Chungnam, Korea
Tel: +82-455-60-1414, e-mail: khkwon@gaya.hanseo.ac.kr

Abstract: Achieving an optimized recipe with a good etch characteristics requires developing accurate process models. Precisely modeling plasma process is complicated due to an extremely complex nature of plasma dynamics. In this paper, back-propagation neural networks were used to build models of etch characteristics for the AlSi film etched in a $BCl_3/Cl_2/N_2$ plasma using magnetically enhanced reactive ion etching. The networks were trained on the data resulting from a 2^{6-1} fractional factorial experiment designed to investigate etch variations with the rf power, pressure, magnetic field and gas composition. A three-dimensional visualization technique has been employed to see interaction effects between parameters.

Keywords: Neural network models, backpropagation, modellings, Interconnection, Integrated circuits

1. INTRODUCTION

Aluminum and its alloy films are widely used for fabricating VLSI interconnection lines and therefore patterning of these films with high resolution is of primary importance. Reactive ion etching (RIE) technology has been the most prevalent means to achieve the level of details in defining fine line pattern. However, this pure capacitive discharge suffers from low ion density and high ion-bombarding energy, prompting equipment industry to devise efficient plasma sources. The resulting high density plasma (HDP) etcher offers several attractive features, including high degree of ionization, low plasma potentials and controllable ion energies. The HDP etchers popularly utilized in IC fabrication are electron cyclotron resonance plasma (ECR), inductively coupled plasma (ICP) and magnetized plasma (Lieberman and

Lichtenberg, 1994). Another viable alternative is a magnetically-enhanced reactive ion etcher (MERIE), which produces a dense plasma by confining energetic (ionizing) electrons using a magnetic field. The magnetically confined high flux, low energy ions provides an efficient physical sputtering advantageous in removing contamination on the wafer surface, while reducing an ion impact damage by lowering a self-dc bias.

Previously, plasma etch models have been developed using a variety of approaches, including first-principle physics (Gerodolle and Pelletier, 1991) and statistical experimental design in conjunction with response surface methodology (May et al., 1991). Modeling plasma discharge behavior from a fundamental physical standpoint has thus far had limited success. Physical models attempt to derive self-consistent solutions to fundamental equations involving continuity, momentum balance, and energy balance inside a high frequency, high intensity electric field. This is accomplished by means of computationally intensive numerical simulations which typically produce distribution profiles of plasma parameters within the reactor. Although the simulations are somewhat useful for equipment design and optimization, they are subject to many simplifying assumptions, and further the connection between microscopic plasma parameters (such as ion density or plasma potential) and macroscopic etch parameters (such as etch rate or anisotropy) has yet to be clearly distinguished. As a result, other efforts have focused on empirical approaches to plasma modeling such as response surface methodology (RSM). The RSM models reflect the behavior of a specific piece of equipment under a wide range of etch recipes, thus making them very useful for manufacturing purposes.

More recently, adaptive learning techniques which utilize neural networks combined with statistical experimental design methods have been applied to etch process modeling. Himmel and May (1993) showed that neural network-based RIE models exhibited less experimental error than their statistical counterparts, even when created from less experimental error. Kim and May (1996) introduced a new learning rule and applied to modeling silicon dioxide etch process in a chloroform and oxygen plasma. The modification consisted of implementing a new weight update scheme in conjunction with an error minimization technique conceptually similar to simulated annealing. Similarly Mocella et al. (1991) found that neural networks consistently produced plasma etch models exhibiting better fit than several variations of response surface models. Rietman and Lory (1993) modeled the gate etching of MOS transistors, successfully using data from a production machine to train neural nets to predict the amount of silicon dioxide remaining in the source and drain regions of the devices after the etch. Rietman (1996) further demonstrated the usefulness of a neural network etch model by utilizing it to predict the post-etch oxide thickness remaining after contact vias process of CMOS structures. Huang et al. (1994) presented a method to develop neural network etch models which outperform the predictions of statistical response models from limited experimental data.

The derived neural network-based etch models, in nearly every one of these studies, offered advantages in both accuracy and robustness in capturing the extremely complex nature of particle dynamics within a plasma. In this paper, neural networks are used to model the etch rate, selectivity to resist and oxide, and anisotropy. The process modeled is the removal of AlSi films by a $BCl_3/Cl_2/N_2$ plasma in a

MERIE system. The process was characterized by a 2^{6-1} *fractional factorial* experiment with three center point replications augmented by twelve face-centered points. The factors varied included pressure, RF power, magnetic field intensity, and three gas flow rates. The experimental data was used to train feed forward neural nets via an error *back-propagation* (BP) algorithm.

2. EXPERIMENTAL DETAILS

2.1 Apparatus and Technique

The 100-mm Si substrates used for this study were doped with B (0.85-1.15 Ω-cm), oriented (100), and chemically etched for 60 sec prior to chemical vapor deposition (CVD) growth using 1% $HF:H_2O$. The substrates were coated with a 600-nm-thick layer of SiO_2 grown by low pressure $CVD(SiH_4 + O_2, 420 \,°C, 240$ mTorr). Deposition of the AlSi films was performed using a Varian 3180 d. c. sputtering system. The sputtering target was a mixed alloy of Al and 1 wt.% Si. Al mixed with one percent of Si was used to eliminate a junction spiking at the source and drain regions in transistors. Approximately 1.5 μm of photoresist was spun and patterned with equal lines and spaces using g-line Nikon stepper. The developed samples were subsequently hard baked at 120 °C for 30 minutes. In a similar way, a second set of test wafers were prepared to measure simultaneously the vertical etch rates of AlSi and the resist, selectivity, and anisotropy.

Etching of the AlSi films was performed using a MERIE. Wafers were placed on a low electrode which was connected to RF power through a matching box. Electromagnetic coils were mounted around the reactor side wall. Substrate holder temperature during the AlSi etch was held at 45 °C using the circulation of cooling water.

Film thickness measurements were performed at five points per wafer using a Tencor Model α-step 200 surface profiler. The vertical etch rates were calculated by dividing the difference of pre- and post-etch thickness by the etch time. The lateral etch rates for AlSi film were measured by scanning electron microscope (SEM) images of the etched features.

Expressions for the selectivity of AlSi with respect to resist (S_{ph}) and oxide (S_{ox}) along with anisotropy (A) are given below:

$$S_{ph} = R_{AlSi} / R_{ph}$$

$$S_{ox} = R_{AlSi} / R_{ox}$$

$$A = 1 - R_{AlSi} / L_{AlSi}$$

where R_{AlSi} is the mean vertical AlSi etch rate over the five points, R_{ph} is the mean resist rate, R_{ox} is the mean oxide etch rate and L_{AlSi} is the lateral AlSi etch rate.

2.2 Experimental Design

The six factors chosen for the MERIE experiment and their respective ranges of variations are shown in Table 1. The process was characterized via a 2^{6-1} fractional factorial experiment with three center-point replications (Box and Draper, 1987). The design was further enhanced by twelve face-centered points. This *complete face-centered central composite circumscribed* design required 47 experimental runs.

Table 1 Range of input factors

Parameter	Range	Units
BCl_3	20 - 80	sccm
Cl_2	10 - 70	sccm
N_2	20 - 80	sccm
Pressure	10 - 100	mTorr
RF power	50 - 250	watts
Magnetic Field	10 - 100	gauss

3. NEURAL PROCESS MODELING

Previously, neural networks have demonstrated the capability of learning complex relationships between groups of related parameters. Such learning capabilities can be attributed to the fact that neural networks, possessing many simple parallel processing units (called neurons), crudely resemble the architecture of the human brain. These rudimentary processors are interconnected in such a way that knowledge is stored in the weight of the interconnections between them. Each neuron contains the weighted sum of its inputs filtered by a sigmoidal transfer function, endowing neural networks with the ability to generalize with an added degree of freedom not available in statistical regression techniques. The nonlinear mapping capabilities of neural networks have recently applied by several researchers in semiconductor process modeling. Each of these studies reported much success using neural network-based techniques.

In order to accurately model the MERIE process described above, the quantitative relationships which relate the six input parameters to the etch responses are encoded in the error back-propagation neural networks. The input layer of neurons receives the external information such as that represented by the six adjustable MERIE input parameters in Table 1. The output layer transmits information to the outside world and thus corresponds to the various process responses (etch rate, selectivity and anisotropy). The BP networks also incorporate "hidden" layers of neurons which do not interact with the outside world, but assist in performing nonlinear feature extraction on data provided by the input and output layers. The number of hidden layer neurons is varied to provide optimal network performance.

In the standard BP learning algorithm, termed as the *generalized delta* rule (Freeman and Skapura, 1991), the network begins with a random set of weights. An input vector is then presented to the network, and the output is calculated using the initial weight matrix. Next, the calculated output is compared to the measured output data, and the squared difference between these two vectors determines the system error. The accumulated error for all the input-output pairs is the Euclidean distance in the weight space which the network attempts to minimize. Minimization is accomplished via a gradient descent approach, in which the weights in the system are adjusted in the direction of decreasing error.

4. RESULTS AND DISCUSSION

4.1 Model Accuracy

The back-propagation neural networks were trained using the generalized delta rule as described above. The training data for these networks consisted of 32 trials from 2^{6-1} fractional factorial array plus three center point replications. The remaining twelve face-centered trials from the original etch characterization experiment were used as the test data for the models. Separating the experiment into a training set and a testing data set is necessary because the neural process models cannot be judged sorely on the basis of how well they mimic the training data. Thus, measuring model performance on the test data provides a truer indication of network prediction accuracy under process conditions other than those used to build the models. Network performance was measured by the root-mean-squared (RMS) error and the computed ones are included in Table 2.

Table 2 Network Prediction Errors

Etch Response	Error
Etch Rate	28.5(nm/min)
Resist Selectivity	4.25
Oxide Selectivity	5.58
Anisotropy	0.08

4. 2 Physical Interpretations

Three-dimensional visualization techniques have been used in order to gain insight into the underlying relationships between etch input and output parameters. This was accomplished by using the trained neural process models to generate response surfaces shown in Figures 1-6.

Etch Rate; The effects of the input parameters on AlSi etch rate are shown in Figures 1 and 2. In Figure 1, etch rate is plotted as a function of BCl_3 flow rate and Cl_2 flow rate with all other parameters set at their nominal values. Here etch rate is seen to increase with Cl_2 flow rate due to an enhanced Cl radicals. A random collision of BCl_3 additive increases an active plasma species, thus further improving upon etch rate. Figure 2 shows the variations of etch rate with magnetic field and RF power. Increasing magnetic flux density leads to a higher ion (etching species) density and thus higher etch rate. This, meanwhile, indicates that etch rate is more dominantly affected by an ion-enhanced

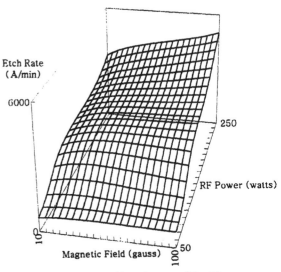

(BCl_3=50sccm, Cl_2=40 sccm, N_2=50sccm, Pressure=55mTorr)

Figure 2 AlSi etch rate versus magnetic field and RF power

chemical reaction than a physical sputtering since a self-dc bias is inversely proportional to field strength. An increased etch rate at a high power density may be explained by the removal of etch inhibitors on the surface by a more energetic ion bombardment (Bell et al., 1986).

Selectivity; As displayed in Figure 3, resist selectivity decreases drastically with increasing RF power. This mainly stems from the fact that AlSi film is etched both by radicals and by the effect of

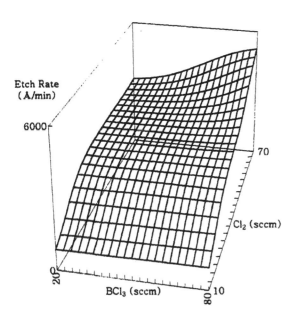

(N_2=50sccm, Pressure=55mTorr,RF power=150watts, magnetic field=55 gauss)

Figure 1 AlSi etch rate versus BCl_3 and Cl_2

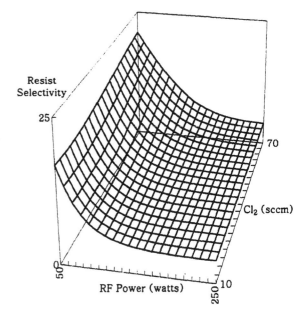

(BCl_3=50sccm, N_2=50sccm, Pressure=55mTorr, magnetic field=55 gauss)

Figure 3 Resist selectivity versus RF power and Cl_2

186

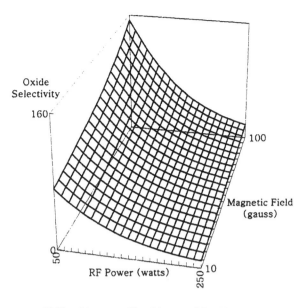

(BCl₃=50sccm, Cl₂=40sccm, N₂=50sccm, Pressure=55mTorr)

Figure 4 Oxide selectivity versus RF power and magnetic field

ion bombardment, whereas resist is etched only by the latter case (Sato and Nakamura, 1982a). At this power level, increasing Cl₂ flow rate maximizes selectivity due to a relatively higher etch rate of AlSi over resist. As shown in Figure 4, a sharp decrease of oxide selectivity with power level is a consequence of its being more influential factor than magnetic flux density. This in conjunction with Figure 2 also implies another competing effects of RF power. An increasing selectivity with magnetic field strength can be correlated by a lower ion impact energy (namely lower plasma potential).

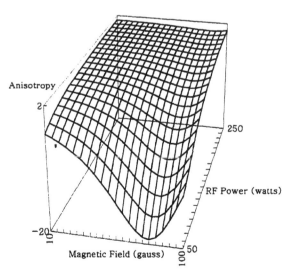

(BCl₃=50sccm, Cl₂=40sccm, N₂=50sccm, Pressure=55mTorr,)

Figure 5 Anisotropy versus magnetic field and RF power

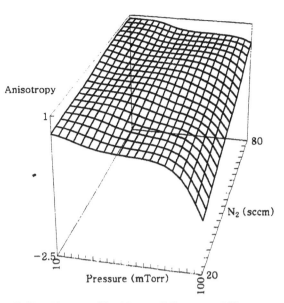

(BCl₃=50sccm, Cl₂=40sccm,RF power=150watts, magnetic field=55 gauss)

Figure 6 Anisotropy versus pressure and N₂

Anisotropy; The combined effect of RF power and magnetic field on anisotropy can be observed in Figure 5. The anisotropy increases with RF power due to an increased ion directionality in the plasma sheath. A decrease in magnetic field strength induces a more energetic ion bombardment, thereby increasing anisotropy. Its effect becomes more manifest at low power density. However, this results in a trade-off between etch rate and anisotropy compared with Figure 2. In Figure 6, anisotropy is plotted as a function of pressure and N₂ flow rate. It has been demonstrated that ion energy and current density increases in accordance with N₂ flow rate (Sato and Nakamura, 1982b). The resulting ion-assisted mechanism induces an etching only in a vertical way, thereby increasing anisotropy.

5. CONCLUSION

A magnetically-enhanced reactive ion etching of AlSi film has been characterized via an error back propagation neural networks. Data from a 2^{6-1} fractional factorial design was used to model the etch rate, selectivity and anisotropy of AlSi film in a BCl₃/Cl₂/N₂ plasma. Several tradeoffs between etch models were investigated with respect to six controllable parameters.

An increased plasma species due to the BCl₃ additive improves upon the etch rate. A high AlSi etch rate obtainable with increasing magnetic flux density is traded against a lower anisotropy. High etch directionality could be obtained by limiting the availability of active etching species and also by using high self-bias voltage.

ACKNOWLEDGEMENT

The authors would like to thank Dr. Seung S. Han for his assistance in recipe simulation and Professor Gary S. May for his helpful reviews.

REFERENCES

Bell, H., Anderson, H. and Light, R. (1986). Reactive ion etching of aluminum/silicon in BBr_3/Cl_2 and BCl_3/Cl_2 mixtures. *J. Electrochem.*, **135**, 1184-1191.

Box, G. and Draper, N. (1987). *Empirical Model Building and Response Surfaces*, Wiley, New York.

Freeman, J. and Skapura, D. (1991). *Neural Networks*. Addison Wesley, New York.

Gerodolle, A. and Pelletier, J. (1991). Two-dimensional implications of a purely reactive model for plasma etching. *IEEE Trans. Elec. Dev.*, **38**, 2025-2032.

Himmel, C. and May, G. (1993). Advantages of plasma etch modeling using neural networks over statistical techniques. *IEEE Trans. Semi. Manufact.*, **6**, 103-111.

Huang, T., Edgar, T., Himmelblau, D. and Trachtenberg, I. (1994). Constructing a reliable neural network model for a plasma etching process using limited experimental data. *IEEE Trans. Semi. Manufact.*, 7, 333-344.

Kim, B. and May, G. (1996). Reactive ion etch modeling using neural networks and simulated annealing. *IEEE Trans. Comp. Packag. Manufact. Technol.-Part C*, **19**. 3-8.

Lieberman, M. and Lichtenberg, A. (1994) *Principles of Plasma Discharges and Materials Processing.* Wiley, New York.

May, G., Huang, J. and Spanos, C. (1991). Statistical experimental design in plasma etch modeling. *IEEE Trans. Semi. Manufact.*, **4**, 83-98.

Mocella, M., Bondur, J. and Turner, T. (1991). Etch process characterization using neural network methodology: A case study," In: *SPIE Proc. Module Metrology, Control and Clustering*, Vol. 1594, pp. 232-242.

Rietman, E. (1996). A neural network model of a contact plasma etch process for VLSI production. *IEEE Trans. Semi. Manufact.*, **9**, 95-100.

Rietman, E. and Lory, E. (1993). Use of neural networks in semiconductor manufacturing processes: An example for plasma etch modeling. *IEEE Trans. Semi. Manufact.*, **6**, 343-347.

Sato, M. and Nakamura, H. (1982a). Reactive ion etching of aluminum using $SiCl_4$. *J. Vac. Sci. Technol.*, **20(2)**, 186-190.

Sato, M. and Nakamura, H. (1982b). The effects of mixing N_2 in CCl_4 on aluminum reactive ion etching. *J. Electrochem. Soc.: Solid-State Science and Technology*, **129**, 2522-2527.

EVALUATING THE CONTRIBUTIONS OF PROCESS PARAMETERS IN SLA USING ARTIFICIAL NEURAL NETWORK

*W. S. Park, *S. H. Lee, **H. S. Cho, and ***M. C. Leu

* : Graduate students, KAIST, Korea; ** : Professor, KAIST, Korea; *** : Professor, NJIT, USA

Department of Mechanical Engineering
Korea Advanced Institute of Science and Technology
Tel: +82-042-869-3253; Fax: +82-42-869-3210; E-mail: hscho@lca.kaist.ac.kr

Department of Mechanical Engineering
New Jersey Institute of Technology
Tel: +1-201-596-3335; Fax: +1-201-596-5601; E-mail: mleu@nsf.gov

Abstract: Though SLA(Stereolithography Apparatus) is being recognized as an innovative technology, it still can not be used to fully practical applications since it lacks of dimensional accuracy compared to conventional processes. In order to improve the accuracy of the SLA, this paper quantitatively evaluates how largely each process parameter of the SLA contributes to the part accuracy. For this pupose, a multi-layered perceptron is designed and trained by a set of sample patterns obtained via Taguchi's experiment planning, which estimates the dimensional errors of the test part, "letter-H" part from process parameters. Since the patterns are sparsely distributed, very careful design of the network is performed. Based upon the results obtained from the neural network estimator, a quality index of the part is evaluated, through which contribution of each process parameter is evaluated and discussed in detail.

Keywords : Rapid prototyping, Stereolithography, Process analysis, Diagnostic part

1. INTRODUCTION

1.1 Motivation

RPD(rapid product development) is an emerging concept for global marketing and manufacturing, which can be constructed through effective organization of RPD network consisting of solid modeling, CAE, CA reverse engineering, CA measurement and inspection, rapid machining, rapid prototyping, and so on. RP(Rapid Prototyping) is such technology that produces prototype parts in much shorter time than traditional machining processes. This technology includes SLA(stereolithography apparatus), LOM(laminated object manufacturing), BPM(ballistic particle manufacturing), SLS(selective laser sintering), TDP(three dimensional printing), FDM(fused deposition manufacturing), among which SLA shows the best accuracy of the shapes of parts[1,2].

In spite of its potential usage to variety of areas, SLA is being used only for a few applications since the accuracy of its products is still not high enough. Thus, many researchers have tried to improve SLA accuracy through various approaches.

1.2 Related works on SLA part accuracy

On the improvement of SLA part accuracy, there have been several researches which can be classified into three categories as follows:

1) *Accuracy improvement by resin development:* Jacobs (Jacobs, 1993) and Schulthess et al. (Schulthess,+) reported that parts build from epoxy resin show higher accuracy than those from acrylate resin.

2) *Accuracy improvement by H/W or S/W development:* Developments of higher performance servo controller and laser scanning system aided SLA to make more accurate parts (Jacobs, 1995). Ullett et al. (Ullett, 1994) proposed a new hatching

scheme to reduce warpage in SLA.

3) *Accuracy improvement by SLA parameter tuning:* Pahati and Dickens (Pahati and Dikens, 1995) found that there exits an optimal layer thickness for given hatch spacing. Chartoff et al. (Chartoff et al.) reported that shrinkage and warpage can be reduced by selecting appropriate scanning speed of laser beam.

In order to improve part accuracy of SLA, here an approach that analyzes the contributions of process parameters to part accuracy is proposed, which can be classified to be one of the third approach described in section 1.2.

In order to evaluate the contributions of SLA parameters to part accuracy, a multi-layered perceptron is utilized, which models the relationships between the dimensional errors of a standard part and the SLA parameters. To minimize the required volume of training patterns, Taguchi's experiment planning technique is used. By the experiment planning, the network can be trained by only 18 sets of patterns, which is quite notable.

By aid of the generalizing performance of the network, the contribution of each parameter is determined by using a quality index which is defined in this paper.

Through the results of the analysis, layer thickness is found to be the major parameter that affects part accuracy more significantly than any other parameters.

2. STEREOLITHOGRAPHY

2.1 Working principle of SLA

As shown in Fig. 1, SL process utilizes visible or ultraviolet(UV) laser and scanning mechanism to selectively solidify liquid photo-curable resin and form a layer whose cross-sectional shape is previously prepared from CAD data of the product to be produced. Through repeating the forming layers in a specified direction, desired 3-dimensional shape is constructed layer by layer. This process solidifies the resin to 95% of full solidification. After building, the built part is put into an UV oven to be cured up to 100%, i.e. post-curing process.

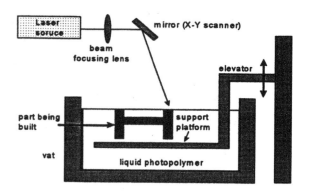

Fig. 1   A schematic drawing of SLA

2.2 Process parameters of SLA

There are three kinds of parameters in SLA: *part parameters, support parameters,* and *recoat parameters,* among which part parameters are the most important ones that affect the accuracy of built parts in SL process. Thus, through the fine selection of part parameter, SLA can build parts more accurately, which is the point of this paper. Part parameters include *layer thickness, hatch spacing, fill spacing, border overcure, hatch overcure,* and *fill cure depth. Layer thickness* is the depth of a layer, which is such region that is solidified at the same elevation. *Spacing* is the distance between a couple of adjacent strands which is the narrow region solidified by the laser scanning as shown in Fig. 2. If the strand is located at the top or bottom surface of part, spacing is called *fill spacing* otherwise *hatch spacing. Cure depth* is the depth of strands. If the strand is located at the top or bottom surface of part, cure depth is called *fill cure depth. Overcure* is the depth that a strand pierces into the lower adjacent layer. If it is located at the lateral boundaries, overcure is called *border overcure,* otherwise *hatch overcure.* In this paper, fill cure depth(DF) is selected to be 0.004 larger than fill cure depth(UF) usually.

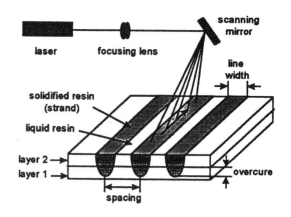

Fig. 2 Process parameters in SLA

2.3 "Letter-H" part : a standard geometry

Generally, in order to evaluate the accuracy of a 3-dimensional shape, a lot of dimensional values representing the part geometry are needed. For example, when the user-part(Gargiulo and Ed, 1991) is under test, 170 dimensions should be measured and analyzed, and six dimensions for a "letter-H" part(Pang et al., 1995). This "letter-H" part is a much more simple part than the former one as shown in Fig. 3. "Letter-H" part indicates reliably the distortion and shrinkage characteristics of SLA with its simple shape and is easy to measure, whose five dimensions are used for characterizing its dimensional accuracy. Thus, the "letter-H" part is chosen as a standardized test part. Its shape and characteristic dimensions are presented in Fig. 3. The characteristic dimensions are denoted by *H-top*, *B-top*, *Waist*, *Ankle*, and *Lateral*, among which the *Ankle* is utilized to replace the *Foot*, presented in Fig. 3, because of its poor repeatability.

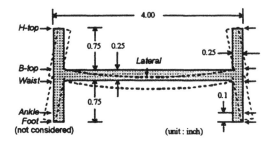

Fig. 3   CAD data(solid line) and built geometry(dotted line) of a "letter-H" part

3. PROCESS ANALYSIS USING ARTIFICIAL NEURAL NETWORK

3.1 Training the neural network using Taguchi's experiment planning

For the relationship between process parameters and part dimensions, a neural network is to be constructed and trained by experimental data, which associates six categories of dimensional errors with the process parameters. The type of neural network to be used here is a multi-layered perceptron. As shown in the Fig. 4, the network inputs, $\vec{x} = (x_1, x_2, \cdots, x_6)$, are the process parameters and the network outputs, $\vec{y} = (y_1, y_2, \cdots, y_5)$, are the dimensional errors of the "letter-H" part described in sections 2.2 and 2.3.

There are six process parameters to be tested and five part dimensions to be measured. These large numbers of process parameters and part dimensions lead us to the necessity of great numbers of experiments and measurements. For a series of full factorial experiments, $\prod_{i=1}^{6} \xi_i$ experiments and $5 \cdot \prod_{i=1}^{6} \xi_i$ measurements should be performed, where ξ_i represents the number of levels of parameter i under test. If the numbers of levels, ξ_i, are set to be three equally for all parameters, it needs to perform 3^6 (= 729) experiments and $5 \cdot 3^6$ (= 3645). Therefore, it needs to reduce the number of experiments.

Table 1   Employed orthogonal array: L18   (unit : inch)

Trials	layer thickness	border overcure	hatch overcure	fill cure depth (UF)	fill cure depth (DF)	fill spacing	hatch spacing
1	0.004	0.005	-0.004	0.003	0.007	0.003	0.002
2	0.008	0.009	-0.001	0.007	0.011	0.006	0.006
3	0.012	0.013	0.003	0.011	0.015	0.010	0.010
4	0.004	0.005	-0.001	0.007	0.011	0.010	0.010
5	0.008	0.009	0.003	0.011	0.015	0.003	0.002
6	0.012	0.013	-0.004	0.003	0.007	0.006	0.006
7	0.004	0.009	-0.004	0.011	0.015	0.006	0.010
8	0.008	0.013	-0.001	0.003	0.007	0.010	0.002
9	0.012	0.005	0.003	0.007	0.011	0.003	0.006
10	0.004	0.013	0.003	0.007	0.011	0.006	0.002
11	0.008	0.005	-0.004	0.011	0.015	0.010	0.006
12	0.012	0.009	-0.001	0.003	0.007	0.003	0.010
13	0.004	0.009	0.003	0.003	0.007	0.010	0.006
14	0.008	0.013	-0.004	0.007	0.011	0.003	0.010
15	0.012	0.005	-0.001	0.011	0.015	0.006	0.002
16	0.004	0.013	-0.001	0.011	0.015	0.003	0.006
17	0.008	0.005	0.003	0.003	0.007	0.006	0.010
18	0.012	0.009	-0.004	0.007	0.011	0.010	0.002

To set out an experiment plan reducing a number of experiments without the loss of physical characteristics of the process, Taguchi's experiment planning method(Roy, 1992) is adopted. Table 1 shows the orthogonal array and the parameter values applied to the experiments. This experiment plan reuires only 18 trials. According to the plan, each parameter is varied to three levels of values, from which the trend of second-order polynomials is extract.

The neural network estimator is trained by using 18 sets of input/output patterns. The system architecture of the network is shown in Fig.4. This network architecture known as multi-layered perceptron (MLP) is one of the most widely known networks. To train this network, the back-propagation (Haykin, 1994) learning rule with a momentum term is used in updading the weights as follows:

$$\Delta \omega_{ji}(n+1) = -\eta \frac{\partial E}{\partial \omega_{ji}}(n) + \alpha \Delta \omega_{ji}(n) \quad (1)$$

where

$$E = \sum_{i=1}^{18} \left\| \vec{e}^{\,i} \right\|^2, \quad (2)$$

$$\vec{e} = \vec{y}^{\,i} - \vec{y}^{\,i,p}, \quad (3)$$

ω_{ji} is the synaptic weight of synapse i belonging to neuron j, $\Delta \omega_{ji}(n)$ and $\Delta \omega_{ji}(n+1)$ are the incremental weight changes at step n and n+1 respectively, η is the learning rate, and α is the momentum rate. $\vec{y}^{\,i}$ is the estimated output of pattern i and $\vec{y}^{\,i,p}$ is the output pattern.

Through the tests on the proper number of hidden nodes, a single hidden layer and four hidden nodes are adopted. In learning of neural network estimator, learning and momentum rate are selected as 0.1 and 0.3 respectively. The adopted sigmoid activation function is shown below:

$$f(net) = \frac{1}{1 + \exp[-(net + \theta)]} \quad (4)$$

where $net_j = \sum_i w_{ji} x_i$, x_i is the inputs of neuron j and θ is the constant bias term.

In order to prevent both of overfitting and underfitting, the numbers of hidden layers and hidden nodes are carefully selected(Hush, 1993). Since the number of levels of each parameter is three, which is quite small, the number of hidden layers is fixed to one. And various numbers of hidden nodes are tried and checked to investigate if the energy represented as Eq. 2 converges to a specified value

that can be previously determined based upon the magnitudes of measurement error contained in training input patterns.

Fig. 5 presents the network errors evaluated by Eq. 2 while learning in case when the number of the nodes n of the hidden layer is varied from 3 to 7. Fig. 5(a) denotes the case of $\eta = 0.3$, $\alpha = 0.3$, which Fig.5(b) is that for $\eta = 0.1$, $\alpha = 0.3$.

The mean magnitude of noises contained in the training patterns are formed to be 5 μm which can be converted to be normalized error 0.015. From the figures, it is easily seen that the network slightly underfits the training patterns in the case which $n = 3$ and overfits in the cases which $n = 5, 6, 7$. Thus, the optimal number of hidden nodes is selected to be 4, i.e $n = 4$.

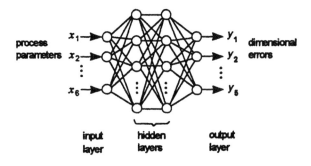

Fig. 4 Neural network estimating the dimensional errors of built parts

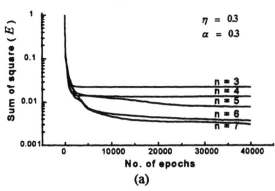

(a)

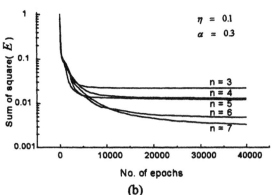

(b)

Fig. 5 Convergences of the tested neural networks

192

3.2 Evaluating contributions of process parameters to the quality

In order to evaluate the quality of a built part, it needs to define a quality index that indicates how accurately or errorneously the part was built. To this end, it is natural to define the quality index to be the sum of squares of the dimensional errors as follows:

$$J(\vec{y}) = \|\vec{y}\|^2 = \sum_{i=1}^{5} y_i^2 \qquad (5)$$

where $y = (y_1, \cdots, y_5)$ is the dimensional errors of a part. Once the neural network estimation is trained, it can be considered as a function of process parameters ;

$$\vec{y} = NNE(\vec{x}) \qquad (6)$$

where the function NNE represents the relationship between the dimensional error and process parameters. Thus, the quality index can be rewritten as a composition as follows:

$$J(\vec{y}) = J(NNE(\vec{\alpha})) \qquad (7)$$

The function representint the contribution of the ith parameter x_i can be defined over a finite range $[x_i^{min}, x_i^{max}]$ by

$$\psi_i = \sum_{j=1}^{m} \left| \begin{array}{c} J(NNE(x_1^o, \cdots, x_{i,j}, \cdots, x_6^o) \\ \\ - J(NNE(x_1^o, \cdots, x_{i,j-1}, \cdots, x_6^o)) \end{array} \right| \qquad (8)$$

In the above, the jth value of the x_i, x_{ij} is defined by

$$x_{i,j} = x_i^{min} + \frac{x_i^{max} - x_i^{min}}{m} \times j, \quad j = 1, 2, \cdots, m \qquad (9)$$

and $\vec{x} = (x_1^o, \cdots, x_6^o)$ are the nominal process parameters, which are recommended by resin suppliers and widely utilized by users, and m is the number of non-overlapping partitions $[x_{i,j-1}, x_{i,j}]$, whose union becomes the whole range of parameter i, $[x_i^{min}, x_i^{max}]$.

Though evaluating the Eq. 8, the contribution of each parameter can be determined. Fig. 6 presents the contributions of parameters which are evaluated by the above Eqs. 8 and 9. The figure shows that the most important parameter is the layer thickness and hatch overcure is the secondary. The rest of the parameters is found to be relatively less influential.

Figs. 7 present the estimated dimensional errors of

the standard H-part with respect to layer thickness and hatch overcure which are proved to be most important. As can be seen from the Figs.7 (a) & (b), the dimensional errors of the H-top and waist tend to decrease for smaller values of the two parameters. On the other hand, the ankle error shows that there exists an optimal layer thickness for given range of the hatch overcure, which gives the minimal error value.

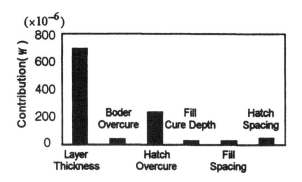

Fig. 6   Contributions of the process parameters

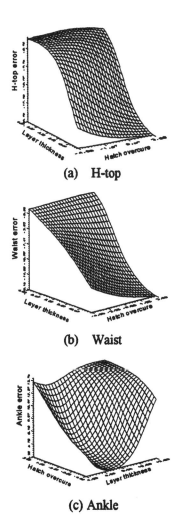

(a)   H-top

(b)   Waist

(c)   Ankle

Fig. 7 Error estimations with respect to layer thickness and hatch overcure

4. CONCLUSION

In order to evaluate the contributions of SLA parameters to part accuracy, a multi-layered perceptron is trained, which associates the dimensional errors of "letter-H" part with the SLA process parameters.

To minimize the required volume of training patterns, Taguchi's experiment planning technique is used. Since the patterns are sparcely distributed, very careful design of the network is performed, which is for avoiding overfitting and underfitting. For fitting the effects of six parameters, only 18 sets of patterns are needed, which is quite notable.

By using the trained network, a quality index is evaluated over the whole domains of parameters. And the contributions defined as a variation of the quality index over the whole domains of parameters.

Through the evaluation processes, layer thickness is proved to be the most important one and hatch overcure is the secondary. The dimensional errors of the H-top and Waist tend to decrease for smaller values of the two parameters. On the other hand, the Ankle error shows that there exists an optimal layer thickness for given range of the hatch overcure, which gives the minimal error value.

The quality index is very important, while it is not sufficiently refined in this paper. Thus, it needs to work more on this, and for the fine tuning of SLA, it also needs more training patterns planned by higher level orthogonal array of Taguchi's method.

REFERENCES

Chartoff, R. P., L. Flach, and P. Weissman(1995), Material and process prameters that affect accuracy in Stereolithography, *The sixth international conference on Rapid Prototyping*

Chen, C. C., and P. A. Sullivan(1995), Solving the mystery - The problem of Z-height inaccuracy of the Stereolithography parts, *The Sixth International Conference on Rapid Prototyping.*

Gargiulo and Ed(1991), Proc. of the 1991 North American Stereographiy User Group Meeting, Orlando, Florida, March, 1991.

Haykin, S.(1994), *Neural networks*, Macmillan.

Hush, D. R. and B. G. Horne(1993), Progress in supervised neural networks, *IEEE Signal Processing magazine*, January .

Jacobs, P. F. (1995), Stereolithography 1993: epoxy resins, improved accuracy and investment casting, *Rapid Prototyping System fast track to product.*

Jacobs, P. F.(1995)., Stereolithography and other RP&M technologies, ASME Press.

Kim, J., S. N. Hong, and I. H. Paik(1996), A study on algorithm development of offset data generation in Stereolithography, *Journal of the Korean Society of Precision Engineering*, **13**, 9, September.

Leslie, H., E. Gargreb, and M. Keefe(1995), An experimental study of the parameters affecting curl in parts created using Stereolithography, *The sixth international conference on Rapid Prototyping.*

Leu, M. C., D. H. Sebastian, and W. L. Yao(1996) Sreolithography rapid prototyping technology: characteristics, applications and R&D needs.

Nguyen, H., J. Richter, and P. F. Jacobs(1992), *Rapid prototyping and manufacturing: fundamentals of Stereolithography*, SME, Dearborn, MI.

Pahati, S., and P. M. Dickens(1995), Stereolithography process improvement, *First National Conference on Rapid Prototyping and Tooling Research.*

Pang, T. H., M. D. Guertin, and H. D. Nguyen(1995), Accuracy of Stereolithograhpy parts : mechanism and modes of distortion for a "Letter-H" diagnostic part.

Roy, R. K.(1990), *A primer on the taguchi method*, Van Nostrand Reinhold.

Schulthess, A., M. Hunziker, and M. Hofmann, New resins for Stereolithography applycations, *Rapid Prototyping Systems fast track to product.*

Ullett, J. S., R. R. Chartoff(1994.), Reducing warpage in Stereolithography through novel draw style., *Solid Freeform Fabrication System.*

Yao, W. L., H. Wong, M. C. Leu, and D. H. Sebastian(1996.), An analytic study of investment casting with webbed epoxy patterns, *ASME 1996 Winter Annual Conference and Rapid Response Manufacturing Symposium*, Atlanta, GA, Nov.

DATA FUSION OF ODOMETRY AND TRIANGULATION FOR ENVIRONMENT RECOGNITION IN MOBILE ROBOTS

S. Shoval*, A. Mishan and J. Dayan

*Faculty of mechanical Engineering, *Faculty of Industrial Engineering and Management
Technion - Israel Institute of Technology - Haifa 32000*

Abstract: Autonomous vehicles usually use more than one positioning system in order to improve their position estimate. Some positioning systems are advantageous in certain types of environments, while others are more efficient in others. This paper describes a data fusion method, where the differences between measurements are used to identify the type of terrain through which the vehicle is traveling. In this system, position estimate by *odometry* is compared to that calculated by *triangulation* and the differences are fed into neural network. This neural network, which is pre-trained by a set of different terrain's types, classifies the examined environment by matching it with the most similar environment it can "recognize".

Keywords: Mobile Robots, Neural Networks, Environment, Recognition, Data fusion.

1. INTRODUCTION

Continuous and accurate positioning of mobile robots is of a major importance for a reliable operation. Positioning systems can be divided into two categories : relative and absolute. This work concentrates on the two most commonly used positioning methods, namely *odometry* and *triangulation* that are relative and absolute methods, respectively.

In *odometry*, sensors that quantify angular position are mounted on the robot's driving wheels. The vehicle displacement along the path of travel is directly derived from a calculation which takes into consideration the robot's parameters (wheels' diameter, wheelbase etc.). This is a rather cheap and self-contained method, that can provide a continuous estimation of the robot's position. The method's most significant drawback is a time dependent boundless position error. The uncertainty of the robot's location increases with travel distance, reducing its reliability as a single positioning system. Errors in this technique are caused by the interaction between the wheels and the terrain, for example

slippage, cracks, debris of solid material, etc. [Borenstein and Feng, 1996].

The *triangulation* method is based on measuring of the planar angles between the robot and natural landmarks or artificial beacons, whose positions are known in advance. In order to guarantee an accurate localization of the robot, the beacons' configuration has to satisfy certain constraints [Shoval, 1997]. The location of the robot is calculated using one of the *triangulation* algorithms [Cohen and Koss, 1993] that mathematically relate the beacons' relative angles with the robot's actual position. The main drawback of the *triangulation* technique is the requirement for presetting the environment, which contradicts the notion of full autonomy. Another drawback is that a tilt in the robot's turret due to uneven floors or bumps along the traveled path may cause erroneous measurements resulting in faulty position estimate [Feng et al, 1994].

This paper describes a method to identify the type of terrain, based on the differences in measurements obtained by *odometry* and *triangulation* positioning systems. Knowing the environment in which the

195

robot is operating, may provide important information.

2. TERRAIN IDENTIFICATION

Figure 1 describes our method for terrain identification. Both *odometry* and *triangulation* continuously estimate the vehicle position. However, due to the different effects the terrain features have on each procedure, there are differences in the position estimate. Analyzing these differences show that each type of terrain has a typical effect. A neural network, can classify data into pre-defined clusters and the examined area can be related to the most similar terrain's cluster [Hakin, 1994].

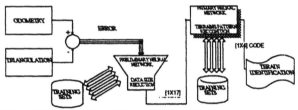

Fig. 1. Terrain identification algorithm.

The two positioning systems provide information about the robot's absolute position in the working space. The "Manhattan distance" (the Euclidean distance between the two estimates of the robot's position in the XY plane) and the differences in the robot's orientation are continuously calculated. This calculation neutralizes the effects of the vehicle's motion profile (e.g. circular motion as compared to a straight line). The signal that represents the position differences, calculated by *odometry* and *triangulation*, is defined as the *position difference signal*, given by:

$$P_E(t) = \begin{bmatrix} \Delta d \\ \Delta \theta \end{bmatrix} = \bar{P}_O(x, y, \theta, t) - \bar{P}_T(x, y, \theta, t) \qquad (1)$$

where $P_E(t)$, the momentary position difference, represents the differences between $\bar{P}_O$ and $\bar{P}_T$ position vectors calculated by *odometry* and *triangulation*, respectively.

$E_P(t)$, the square power of the position difference signal ("energy") multiplied by its first derivative, emphasizes distinctive events within the signal and is given by:

$$E_P(t) = P_E^2(t) * \frac{\partial P_E(t)}{dt} \qquad (2)$$

As $E_p(t)$ is continuously calculated and recorded at a rate of 10Hz, data-set size increases significantly with travel time.

One of the major drawback of neural network, is that the complexity increases as the input dimension increases, and therefore the solution requires large amount of computation resources. For that reason, size reduction of the input data-set to the network is required. To reduce the data-set size, a preliminary backpropagation neural network analyzes the preprocessed signal, $E_p(t)$, frame by frame. Each frame represents a small time portion (1.5 seconds) of the overall travel duration. This network is consists of four layers having [15 10 10 1] neurons, respectively. The output layer of this network, a single neuron, indicates the type of irregularity in the terrain during that time frame. If the terrain is ideal (i.e. no irregularities) the neuron's value is '0'. The network reduces the size of the data-set from a vector of [2x280] to [1x17]. Once the data-set is reduced to an acceptable size the primary neural network can identify the terrain. This network is structured in architecture of four layers of [17 10 4 4] neurons, where the final 4 neuron's layer outputs code for one of five different possible types of terrain.

3. EXAMPLES OF TYPICAL TERRAIN

In an industrial robot's working space there are several possible types of terrain such as inclined surfaces, smooth floors etc. As previously mentioned, the robot's behavior depends on the type of terrain in which it is operating. Figure 2 shows the unique effect of a *shifting terrain*. This environment is characterized by sharp orientation changes or "shifts", caused by external forces such as contact with an object. Figure 3, on the other hand, depicts the effects of a *bumpy environment*.

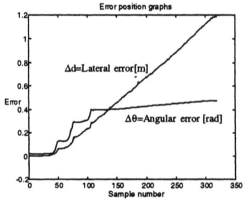

Fig. 2. Shifting environment - differences between Odometry and Triangulation.

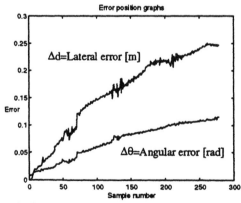

Fig. 3. Bumpy terrain - differences between Odometry and Triangulation.

The vector [0 0 2 0 0 0 0 3 0 0 0 0 0 0 0 0 0] represents the bumpy terrain shown in figure 3 as calculated by the preliminary neural network. The values '2' and '3', in this vector, indicate two bumps of mid size in the 3rd frame and one big bump in the 8th frame, respectively.

4. EXPERIMENTAL RESULTS

To train our neural networks several hundreds of experiments were conducted in our laboratory using the MRV4 (a synchro-drive mobile robot [Denning Branch, 1995]). The robot moved in two motion profiles: circular motion and movement in straight lines. Along the robot's path, three different types of disturbances were arbitrarily planted. During its travel, the robot executed the *triangulation* and *odometry* algorithms at a rate of 10Hz and stored the results. Each experimental frame duration was of 30 seconds in which approximately 300 position estimates were computed.

To test the system, the robot traveled in straight line over bumpy terrain. This environment was simulated by placing three tubes, of 0.8 cm in diameter, along the robot's path. Figure 4 depicts the *triangulation* and *odometry* position estimates in the room coordinates. The transformation of this data to the position difference plane is described in figure 5.

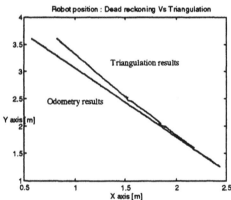

Fig. 4. Bumpy terrain - representation in room coordinates.

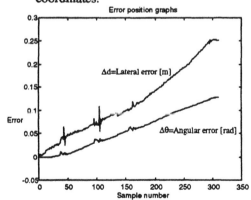

Fig. 5. Bumpy terrain - representation in the difference plan.

The first network identified three bumps of mid size in the 2nd, 6th and 10th frames yielding the following vector [0 2 0 0 0 2 0 0 0 2 0 0 0 0 0 0 0]. This [1x17] output vector was fed to the main network resulting

in the code [0 1 0 0], which indicates that the examined terrain is built of several similar bumps of mid size.

5. SUMMARY AND CONCLUSIONS

This paper introduces a mechanism which uses the differences between *odometry* and *triangulation* position estimate to identify the terrain through which the vehicle is traveling. The data excess is dealt with neural network which reduces it to an acceptable size. Another neural network maps the preprocessed signal to terrain's codes.

The environment do affects the positioning systems' results differently. This attribute was successfully used to differ between several environments and to classify them into typical clusters.

The capabilities of the presented mechanism were shown using a few representing experiments.

A mechanism that can identify the environment in a fast and accurate way, may allow the use of adaptive control for the vehicle motion as well as mapping the working space for future path planning.

In addition this mechanism may be used to bound the Odometry positioning error caused by the environment.

REFERENCES

Borenstein, J., Feng, L, (1996) "Measurement and Correction of Systematic Odometry Errors in Mobile Robots", *IEEE transactions on robotics and automation*, **Vol. 12** December.

Cohen C, Koss V. Frank, (1993) "A Comprehensive Study of Three Object Triangulation", *Proceedings of the 1993 SPIE Conference on Mobile Robots*, Boston MA, Nov.

DBI Denning Branch International Robotics, (1995) "MRV-4 Product Manual", Commercial Sales Litrature, Pittsburg.

Feng L., Borenstein J., Everett B., "Where Am I ? Sensors and Methods for Autonomous Mobile Robot Localization", The Univ. of Michigan, Ann Arbor 1994.

Hakin S., (1994) *"Neural Networks - A comprehensive Foundation"*, Macmillan College Publishing,.

Shoval S., (1997) "The Effect of Landmarks Configuration On the Absolute Positioning of Autonomous Vehicles Using Triangulation Methods", submitted *to IEEE transactions on robotics and automation*, Jan.

A TOOL CONDITION RECOGNITION SYSTEM USING IMAGE PROCESSING

Min-yang Yang[a] and Oh-dal Kwon[b]

[a] *Dept. of Mechanical Engineering, Korea Advanced Institute of Science and Technology,*
373-1 Kusung-dong Yusung-gu Taejon 305-701 Korea,
E-mail : myyang @ sorak.kaist.ac.kr
[b]*Manager, Core Technology Research Center, Samsung Electronics Co., Ltd.*
E-mail : odalkwon @sec.secns.samsung.co.kr

Abstract: In unmanned machining, when an abnormality condition like tool fracture occurs in a tooling system, a proper action to detect, interpret and determine the condition has to be made. In this paper, a method of sensing the abnormal conditions of tool on the basis of image processing is developed. To recognize the tool conditions, a pattern recognition technique using the multi-layered perceptron with the back-propagation algorithm as a neural network is applied. This paper also presents a quantitative measuring system in order to obtain the amount of flank wear and crater wear and use the data as the input of the neural network.

Key words: Neural network, Flank wear, Crater wear, Image processing

1. INTRODUCTION

Until now to improve productivity in the metal cutting process, several methods of maximum utilization of machine tools have been studied in point of CAD, CAM, CAPP, etc. and ultimately to achieve unmanned machining through automation of machine tools. Growing level of complexity and diversity in products, shortened life cycle, JIT delivery all are the factors that speed up the change of manufacturing systems to have flexibility. Flexible manufacturing systems(FMS) have been developed to respond to the demands. Also tooling management has been one of the essential aspects of an effective FMS. (Eversheim, 1991).

A successful tool wear/fracture sensor has been a long standing goal of the manufacturing community. Numerous techniques of tool wear measurement have been proposed (Li and Matthew, 1990;

Shiraishi, 1988). There are two kinds of methods: direct and indirect method. Among these methods, the computer vision technique using image analysis, classified into direct method, has many advantages for tool wear measurement that it is easy to get an accurate amount of wear, convenient for use and comparatively fast. Decision-making systems which are able to interpret incoming sensor information and decide on the appropriate control action have also been studied in accordance with the methods using the algorithms of regression, fuzzy, neural network, etc. (Matsushima, *et al.*, 1979; Rangwala and Dornfeld, 1990; Park and Ulsoy, 1993; Toshio, *et al.*, 1993).

In this paper, a method of sensing the cutting tool conditions on the basis of image processing is developed. The tool conditions including abnormalities are classified into normal, flank wear, crater wear, chipping, and fracture. To recognize the tool conditions, a pattern recognition technique using a multi-layered perceptron with the back-propagation algorithm as a neural network is applied. Since the cutting tool conditions are generally known from the worn shape of rake face and flank face of tool, some characteristic features of tool shape are extracted from them for preprocessing and used as input of the neural network. This paper also presents quantitative measuring methods by which the amount of flank wear and crater wear is measured in order to obtain the characteristic features.

2. TOOL MORPHOLOGIES AND FEATURE SELECTION

Tool wear or fracture morphologies are so various that it is hard to classify them into some types. A tool insert has two faces directly participating in cutting such as rake face and flank face and it is first step toward solving a problem to investigate these faces. To classify the state of tool, some measuring values obtaining from these faces can be used as the input of the neural network. For convenience's sake, the feature extraction from the morphologies is a useful step to determine the category to which the object belongs in the pattern classification technique.

2.1 Feature selection

Here are selected features as follows
1) Maximum flank wear : $V_{max} / 0.8$
This value represents the maximum wear of each flank face that is minor flank, corner, major flank and is normalized by 0.8, the ISO criterion of tool failure.
2) The ratio of average wear to maximum wear : V_{avg} / V_{max}
This value represents the uniformity of the wear land. When it is closer to 1, the wear land becomes flatter and to 0, the wear land sharper.
3) The ratio of the total wear width to the maximum wear distance : D / D_{max}
When this value is closer to 1, groove wear or chipping occurs and to 0, nose wear occurs
4) Average wear in the major flank face : $V_B/0.3$
This value represents the average wear land of the major flank face and is normalized by 0.3, the ISO criterion of tool wear.
5) Loop area on the rake face : Area
This value can be used to classify chipping, crater wear and fracture.
6) The ratio of the moment of area about the x to y axis : M_x / M_y
This value shows the position of the loop so that chipping, crater wear and fracture are classified by it.

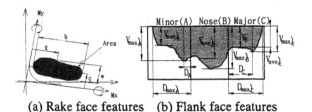

(a) Rake face features   (b) Flank face features

Fig. 1. Characteristic features of tool morphology

7) The ratio of the width to the breadth of the loop : w/b
When this value is close to 1, the shape of loop has circle and otherwise, the shape has irregular.
Among these features, the first four terms are belong to the information from flank face and the others from rake face. Fig.1. describes the features.

2.2 Background on the neural network with back-propagation

Recently artificial neural networks have been used in the areas of pattern recognition of vision, speech, linguistic symbols etc., since by using amplifiers so called "weights", they can be learned and encoded knowledge in the system like a biological brain. The neural networks can be considered as a mapping from a N-dimensional space to a M-dimensional one. As the properties of the input, output, and mapping, they can perform kinds of associative memory, filter, transformation, classification, perception, optimization etc. Recognition of cutting tool conditions is to determine the tool abnormalities from used tool patterns. In this paper, for the purpose of the study object, a supervised multi-layer perceptron with error back-propagation learning model is adapted. (Pao, 1989)

Fig.2. shows the structure of this type of a network. Let x, x' and y be the output value of input layer, hidden layer and output layer and N, N_1 and M the number of nodes respectively.

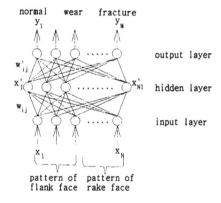

Fig.2. The structure of multi-layer perceptron

$$x'_i = S\left(\sum_{j=0}^{N-1} w_{ij} x_j - \theta_i\right)$$

where, $\quad 1 < i < N_1$

$$\qquad\qquad\qquad\qquad (1)$$

$$y_i = S\left(\sum_{j=0}^{N_1-1} w'_{ij} x_j - \theta'_i\right) x$$

where, $\quad 1 < i < M$

Here, w_{ij}, w'_{ij} are weights representing the strength of the connection between the individual processors in each layer and the parameter θ_i, θ'_i serves as a threshold or bias. For a sigmoidal activation function, $S(\cdot)$ is used as follows.

$$S(net)_j = \frac{1}{1+e^{-net_j}} \qquad (2)$$

In learning w_{ij}, the change in weights can be calculated with the expression as follows

$$\Delta w_{ji}(n+1) = \eta(\delta_j x_i) + \alpha \Delta w_{ji}(n) \qquad (3)$$

where the quantity (n+1) is used to indicate the (n+1) step, η corresponds to learning rate and α is used to momentum rate of change.

The errors of δ'_i and δ_i are given by following two expressions for the output-layer and hidden-layer units respectively.

$$\delta'_i = S'(\hat{y}_i)(d_i - y_i) \qquad (4)$$

$$\delta_i = S'(\hat{x}_i)\sum_j \delta'_j w'_{ji} \qquad (5)$$

Where d_i is the desired output, y_i is the actual output of the output-layer and $\hat{y}_i$ is the output of the output-layer before computing the sigmoidal function. S' is the derivative of sigmoidal function; that is

$$S'(\hat{x}_i) = x_i(1 - x_i) \qquad (6)$$

3. TOOL WEAR MEASUREMENT

3.1 Flank wear measurement

Flank wear can be measured using the information of the tool edge detected from a rake face image. Since even a used tool also has some parts of unworn cutting edge because of not taking part in cutting, an original cutting edge can be created by a least square fit through them.

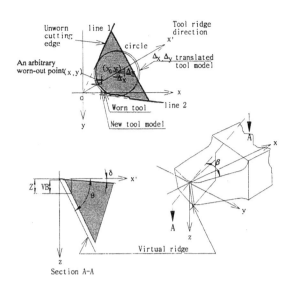

Fig.3. The mathematical model of cutting tool

Throw away type tools usually have three parts of cutting edges that is minor cutting edge, corner, major cutting edge which can be expressed with two straight line and a circle contacted between the two as shown in the Fig.3. At an arbitrary worn point(x,y), the distance between the new tool model and the worn point is the same as the displacement of tool model translated Δx, Δy and the worn point is considered with as the point on the translated tool model. The tool model translated Δx, Δy on the XY plane is as follows.
The circular equation is

$$(x - (x_o + \Delta x))^2 + (y - (y_o + \Delta y))^2 = r_o^2 \quad (7)$$

The linear equation 1,2 are

$$(y - \Delta y) = m_{1,2}(x - \Delta x) \qquad (8)$$
$$\text{where} \quad m_1 = -\cot C_e, \quad m_2 = \tan C_s$$

Suppose that the tool wear increases toward the tool ridge direction, the equations can be obtained by relation of geometry of tool.

$$\Delta x = Z\cos\theta\cos\beta$$
$$\Delta y = Z\cos\theta\sin\beta \qquad (9)$$

Where θ is the tool edge angle on the tool ridge line and β is the angle between x-axis and tool ridge line. The equations are respectively

$$\tan\theta = ((\tan\xi_e + \tan\xi_s \cos\sigma)^2 + (\tan\xi_s \sin\sigma)^2)^{-\frac{1}{2}} (10)$$

$$\beta = \tan^{-1}(\frac{\tan\xi_s \sin\sigma}{\tan\xi_e + \tan\xi_s \cos\sigma}) - C_e \qquad (11)$$

By substituting these equations to (7) and (8) in the

expression for Z, four values of Z for an arbitrary worn point (x,y) are obtained. Two of them are from the circular equation and the others from the linear ones. However, there is only one Z value satisfying the constraints of the regions in which they can be belong to. Also effective tool rake angle along tool ridge direction δ is represented by

$$\tan\delta = \frac{\tan\alpha_b + \tan\alpha_s}{\sqrt{(1+\cos\sigma)^2 + \sin^2\sigma}} \qquad (12)$$

Therefore the flank wear value V_B is as follows.

$$V_B = Z - d\tan\delta, \qquad (13)$$
$$\text{where} \quad d = \left(\Delta x^2 + \Delta y^2\right)^{1/2}$$

This equation can calculate any worn point along the whole cutting edge, therefore the wear map can be obtained from minor edge to major edge. It also gives 3-D information from 2-D image.

3.2 Crater wear measurement

Contour or edge detection is a useful technique to serve a simple analysis of image by reducing the amount of data without loss of structural information. In the case of crater wear, the crater area and its centroid and crater width can be obtained from the contour.

A problem encountered in contour detection of crater wear is the high frequency noise from the texture of rake face image. To attenuate the high frequency noise, low-pass filtering has been proved useful. However this operation usually results in blurring the image. To accomplish this task without blurring, image consolidation is applied (Ballard and Brown, 1982).

Image consolidation is to reduce the original image by subsampling it. For this purpose, the gaussian shape low-pass filter is used and then subsampling operation is applied. The subsampled image $I_r(x,y)$ is given by

$$I_r(x,y) = \begin{cases} I_1(Lx, Ly), & 0 \le x \le N/L, \ 0 \le y \le M/L \\ 0, & \text{otherwise} \end{cases}$$
$$(14)$$

where $I_1(x,y) = I_0(x,y) \otimes H(x,y)$, $I_0(x,y)$: the intensity of the original image and the separable gaussian filter $H(x,y) = H_x(x)H_y(y)$,
In order to extract the worn region from the tool image it is important to select an adequate threshold of gray level. The optimal threshold is determined with utilizing the zeroth- and first-order cumulative moments of the gray-level histogram(Otsu,1979). After thresholding, most of noises are cleaned and the contour is detected by the Laplacian zero-crossing method. Then, edge linking techniques have been employed to connect between broken pixels since the contour boundary may often be broken due to the noise in the image. To recover the accuracy, it is necessary to return to its original size. Since a pixel of reduced contour equals to LxL pixels of the original contour, each pixel in the reduced contour is transformed to the center pixel of LxL region discretely connected with linear lines making a closed loop. This closed loop has an uneven shape and smaller than the original size. Eight-neighborhood chain codes used for describing boundaries enable us to develop a new dilation algorithm. Since chain codes have information of input or output direction vector, the dilation can advance toward the normal direction between two vectors. However, since pixels are discrete points the normal direction is not expressed as one point so that some points inflate one or more points and others disappear. Since the dilation bound is constrained in the original contour of crater wear, the reduced contour dilates to the form finally.

After detecting the contour of crater wear, crater maximum depth can be measured by automatic focusing (Yang, 1996)

4. EXPERIMENTS

The proposed experimental equipment is shown in Fig. 4. The system is composed of a microscope, an illumination unit, a CCD camera, a digital image processing unit, and a personal computer. The microscope has objective lenses and a halogen lamp. The CCD camera has 512x480 pixel elements with a pixel size 17mm(H)x13mm(V).

The tool insert is located on the X-Y table of the microscope driven up and down toward Z-axis. The camera axis is perpendicular to the X-Y table. The collimated light is projected onto the rake face of a tool. An image of the rake face of the tool is captured through the CCD camera by frame grabber with 256 digitized gray levels. The image is analyzed with the personal computer.

Neural networks with two hidden layers and 25 nodes in each hidden layer were used. The number of nodes of input layer was 12 and output layer 4 so that the network had 12-25-25-4 structure. In this study, tool conditions basically divided by four: flank wear,

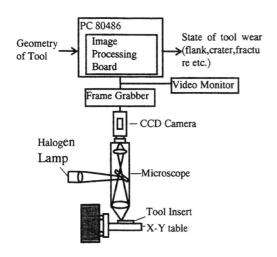

Fig. 4. Experimental equipment

crater wear, chipping, fracture. Each one also classified more detail. Table 1 shows the samples of learning patterns. Fig.5. and 6. shows the relationship between these features.

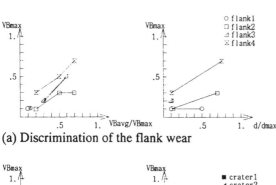

(a) Discrimination of the flank wear

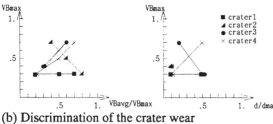

(b) Discrimination of the crater wear

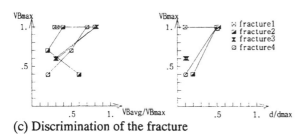

(c) Discrimination of the fracture

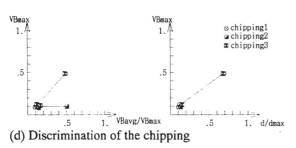

(d) Discrimination of the chipping

Fig.5. The relation of features in the flank face

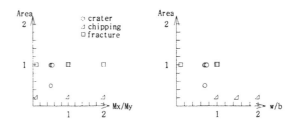

Fig.6. The relation of features in the rake face

Table 1 Learning patterns for training

	Input Pattern											Output Pattern				
	Vmax			Vavg/Vmax			d/dmax		VB	Area	Mx/My	w/b					
	min	cor	maj	min	cor	maj	dmin	dmaj									
flk1	0.1	0.1	0.1	0.1	0.1	0.1	0.1	0.5	0.1	0	0		0	0	0	0	0
flk2	0.1	0.3	0.3	0.2	0.5	0.7	0.1	0.7	0.5	0	0		0	0	0	0	1
flk3	0.2	0.5	0.2	0.3	0.6	0.3	0.1	0.1	0.1	0	0		0	0	0	1	0
flk4	0.3	0.5	0.7	0.2	0.5	0.7	0.1	0.7	0.8	0	0		0	0	0	1	1
cra1	0.3	0.3	0.3	0.2	0.5	0.7	0.1	0.5	0.3	0.5	0.5	0.7	0	1	0	0
cra2	0.7	0.5	0.3	0.4	0.6	0.8	0.2	0.5	0.3	1	0.5	0.7	0	1	0	1
cra3	0.4	0.7	0.4	0.3	0.6	0.3	0.1	0.1	0.3	1	0.5	0.7	0	1	1	0
cra4	0.3	0.5	0.7	0.2	0.5	0.7	0.1	0.5	0.7	1	0.5	0.7	0	1	1	1
chi1	0.5	0.1	0.1	0.5	0.1	0.1	0.1	0.7	0.1	0.2	2	2	1	0	0	0
chi2	0.1	0.5	0.1	0.1	0.5	0.1	0.1	0.1	0.1	0.2	1	1.5	1	0	0	1
chi3	0.1	0.1	0.5	0.1	0.1	0.5	0.1	0.7	0.5	0.2	0.1	1	1	0	1	0
fra1	1	1	1	0.3	0.8	0.3	0.1	0.5	1	1	1	1	1	1	0	0
fra2	1	0.7	0.4	0.4	0.2	0.6	0.5	0.2	0.6	1	2	2	1	1	0	1
fra3	0.6	1	0.6	0.3	0.8	0.3	0.1	0.1	0.3	1	1	1	1	1	1	0
fra4	0.4	0.7	1	0.2	0.5	0.7	0.1	0.5	1	1	0.1	0.1	1	1	1	1

5. RESULTS AND DISCUSSIONS

In this study, investigated tool materials are P20 cemented carbide inserts without chip breaker.

A series of results for tool identification is shown in the Fig.7. The first column of it shows original image and the second shows results of flank face measurement and the third shows results of rake face measurement. In the case of flank face measurement, worn tool cutting edge and created new tool edge are shown and also the wear map from minor cutting edge to major cutting edge is described with respect to maximum wear and to insert thickness. In the case of rake face measurement, worn cutting edge and contour of crater wear, chipping or fracture are depicted. The measuring values of contour that is

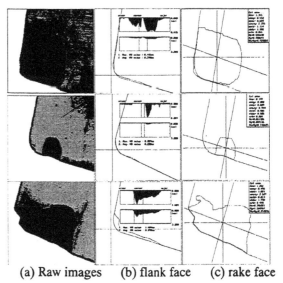

(a) Raw images　(b) flank face　(c) rake face

Fig.7. Measurement results

203

Table 2 Classification of the worn tools

Tool Conditio	Vmax			Vavg/Vmax			d/dmax		VB	Ar ea	Mx/ My	w/ b	Result of Classifying
	min	cor	maj	min	cor	maj	dmin	dmaj					
flank	0.16	0.25	0.25	0.31	0.62	0.75	0.01	0.04	0.5	0	0	0	flk2
crater	0.34	0.48	0.52	0.55	0.57	0.89	0.18	0.11	1	1.32	0.73	0.86	cra4
chipping	0.04	0.05	0.36	0.02	0.03	0.87	0.1	0.48	0.83	0.22	0.13	0.81	chi3
fracture	0.65	1	1	0.11	0.71	0.93	0.06	0.09	1	1.27	0.26	0.47	fra4

area, center of gravity, width, breadth, the moment of area are calculated and displayed in the upper right corner. As shown in the figure, it can be known that the measurement results are well accordance with the real shape. Table 1 shows the feature values calculated from the results and the classifying output. They are well classified according to the learning pattern.

6. CONCLUSION

We presented a tool condition recognition system using image processing with a neural network and measured flank wear and crater wear. Some conclusions are as follows.

This system was made up with simple implementation that it was possible to measure both of the flank wear and crater wear by one image frame obtained from rake face of tool. Flank wear was measured using mathematical tool model and geometry. Contour of crater wear was detected by image consolidation and dilation in noisy images. A series of tests measuring some insert tip was proved to be accurate and was well accordance with the real shape. For the convenience's sake of the pattern recognition, some features were selected from rake face and flank face. With these measurements, series of learning pattern sets were determined. Then, the state of tool condition was classified into 15 kinds of abnormalities using the neural network with multi-layer perceptron. Therefore the proposed system was able to recognize the tool conditions qualitatively and quantitatively.

REFERENCE

Dana H. Ballard, Christopher M. Brown (1982). *Computer vision*, pp. 102~113 Prentice Hall Inc.

Jong-Jin Park, A. Galip Ulsoy, (1993). Transactions of the ASME Journal of Engineering for Industry. *On-line flank wear estimation using an adaptive observer and computer vision, part 1 : theory*, **Vol.115**, Feb., pp. 30~36

K. Matsushima, T. Kawabata and T. Sata, (1979). Annals of the CIRP. *Recognition and control of the morphology of tool failures*, **Vol.28/1**, pp. 43~47

Li Dan and J. Matthew (1990). Int. J. Mach. Tools Manufact. *Tool wear and failure monitoring techniques for turning - A review*, **30(4)**, pp. 579-598.

M. Shiraishi, (1988). Precision Engineering, *Scope of in-process measurement, monitoring and control techniques in machining processes part1:in-process techniques for tools*, **Vol.10**, No.4, pp. 179~189

M.Y. Yang and O.D. Kwon, (1996). *Material processing technology. Crater wear measurement using computer vision and automatic focusing*, **vol. 162**, pp. 1200~1207

N. Otsu, (1979). IEEE trans, System Man Cybernet. *A threshold selection method from gray level histograms*, **9**, pp. 62~66.

S. Rangwala, D. Dornfeld, (1990). Transactions of the ASME Journal of Engineering for Industry. *Sensor integration using neural networks for intelligent tool condition monitoring*, **Vol.112**, AUGUST, pp. 219~228

Toshio Teshima, Toshiroh Shibasaka, Masanori Takuma, Akio Yamamoto, (1993). Annals of the CIRP. *Estimation of cutting tool life by processing tool image data with neural network*, **Vol.42/1**, pp. 59~62

W. Eversheim, (1991). Annals of CIRP. *Tool management : the present and the future*, **Vol.40/2**, pp. 631~639

Yoh-Han Pao (1989). *Adaptive pattern recognition and neural networks*. Addison-Wesley Pub. Co.

INTELLIGENT SELECTION OF CUTTING PARAMETERS IN MILLING OPERATION

Kyungcheul Ko and Reza Langari

Department of Mechanical Engineering
Texas A&M university
College station, TX 77843, USA

Abstract: For mass production of uniform quality demand in modern industry, computer numeric control (CNC) machining is widely used. One of evident phenomena in machining is that tool wear is not avoidable. Hence, the best management of cutting tool in mass product is the minimization of tool wear during sequential machining processes. To provide cutting information in wide cutting range, a systematic selection of cutting parameters is essential for strong foundation in advanced machining industry. When this feature is added in conventional CAD/CAM systems, it is possible that the seamless machining with safe and efficient cutting parameters state in various cutting range.

Keywords: Milling, Parameters, Decision Making, Intelligent, CAD/CAM

1. INTRODUCTION

In modern manufacturing industry, metal cutting process by CNC machining is the proper mechanism for mass production of uniform quality according to Lynch (1992). To produce new parts, the following process is needed: (1) design with engineering analysis, (2) machining process planning and (3) actual machining. Normally, the machining jobs including the process plan, tool selection, cutting parameter selection, tool trajectory plan based on the required surface tolerance is planned for CNC machining according to Chang (1989). In the machining field, the machining program expert with experience can setup the machining process plan by the analysis of drawing data and select the proper cutting parameters that satisfy the required accuracy of the final product. But those methods can not provide the systematic ways of the selection of cutting parameters for a specified quality or tolerance of surface. One of most common way of selecting cutting parameters is safety fixed setting of the feed rate and spindle speed to insure the cutting state satisfying the allowable force on tool. Empirical methods can not be applied to wide range because those methods are obtained in specific cutting range.

In metal cutting shown in Elanayar (1996), tool wear is not avoidable. Hence, the qualities of mass product made by a uniform machining program may not be uniform from the effect of tool wear by Sutherland (1986) after unit machining process. Furthermore in each stage of the machining process, unskilled operation can not guarantee the desired output quality of product. The best use of cutting tool in mass product is the minimization of tool wear amount and its rate. The proper cutting parameters with allowable reaction force between the tool and workpiece guarantees continuous machining with the minimal tool wear. Systematic or analytic ways of selecting cutting parameters have not been developed yet because of nonlinear and complex dynamic characteristics between tool and workpiece. From a dynamic standpoint, the reaction force shown in Altintas (1991) between tool and workpiece is functions of several variables: tool diameter, tool helix angle, number of flutes, tool material, axial depth of cut, radial depth of cut, feed rate, spindle speed, and workpiece material. For wide range of applicable cutting situations, the systematic mechanistic cutting dynamic model is strongly needed. The proper cutting parameters with allowable reaction force between the tool and workpiece guarantees sequential machining with minimal tool wear. In rough cutting conditions, the cutting parameters are chosen to satisfy the time requirement for job and reasonable rate of tool wear. In this paper, the intelligent selection of cutting parameters in milling operation using a mechanistic cutting model and experimental results.

2. KINEMATICS AND DYNAMICS OF MILLING OPERATION

When the tool diameter is much larger than feed rate per tooth, The circular tooth path approximation for small feed rate makes the cutting model more simply. Using this approximation, the chip thickness equation is given by Martellotti (1941).

$$t_c = f_t \sin\theta \tag{1}$$

where, f_t is feed rate per tooth (in/tooth), θ is the angle of the cutter, and t_c is the chip thickness

This approximated chip thickness equation is reasonably accurate and applied to general research work. The chip formation, axial depth of cut, radial depth of cut is illustrated in Fig. 1.

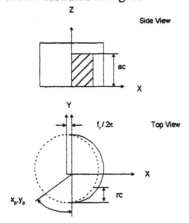

Fig. 1 The chip formation, definition of axial depth of cut, radial depth of cut, entrance angle and exit angle

As shown in Fig. 2, the cutting area enclosed by an axial depth of cut and radial depth of cut is divided by distinct point given point a, b, c, and d. At point a, the cutting flute enters the cutting zone. This angular position is called entrance angle. At point b, flute reaches axial depth of cut height and all point on flute is involved in cutting. In a case of no helix mill, a point a and b will be coincident. At point c, a point on the bottom of the cutting flute exits the cutting region. At a point d, a flute exits the cutting region entirely. Tool engagement in regions between point a and b, and c and d is partial. In a case of the face mill or zero helix angle end mill, classification of machining done by the multi-tooth in the cutting region is simple. Compared with the face mill or

zero helix angle end mill, machining by the helix angle end mill is based on the following procedure.

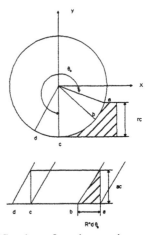

Fig. 2 Classification of cutting region

The angular distance of the previous flute at point a in Fig. 3 is given $\theta_{k+1} = 2\pi/N_f$. Also, at the same time, real angular position of the k th flute is given

$$\theta_k' = \omega t_a + \frac{2\pi}{N_f} \tag{2}$$

where N_f is number of flute

The exit angle, θ_d at point d is given as:

$$\theta_d = 2\pi + \frac{ac}{R}\tan\alpha \tag{3}$$

where ac is axial depth of cut , α is helix angle of tool and R is tool radius

As shown in Fig. 3 below, machining by two or more flutes is determined by considering the relationships in Eqn. 4 and Eqn. 5.

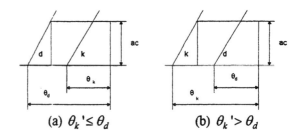

(a) $\theta_k' \le \theta_d$ (b) $\theta_k' > \theta_d$

Fig. 3 The classification of the overlap effect in cutting region

$$\theta_k' \le \theta_d \qquad \text{for overlap cut} \tag{4}$$

$$\theta_k' > \theta_d \qquad \text{for no overlap cut} \tag{5}$$

Within the cutting zone, the force components act on the finite cutting area as shown in Kline (1982). The finite cutting area is composed of height and chip thickness at a given instant .

The finite height of finite cutting area is given as:

$$\Delta h_i = R\Delta\varphi / \tan\alpha$$
$$\text{for } i = 0,...,N_d \text{ where } \Delta\varphi = d\theta_b / N_d \qquad (6)$$

Finite tangential force and radial force acting on the above elementary area is

$$\Delta F_{t_i} = cft \cdot t_i \Delta h_i \qquad (7)$$
$$\Delta F_{r_i} = cfr \cdot t_i \Delta h_i$$

where, cft is coefficient of tangential force and cfr is coefficient of radial force

Fig. 4 shows the tool geometry and finite force components on a given cutting area. These finite forces normal to the cutting flute act on each finite area within cutting zone from θ_a to θ_d. The center of finite force acting on a flute composed of finite element is obtained:

$$\theta_c = \frac{\sum_{i=1}^{N_d} \Delta F_{t_i} \cdot \theta_i}{\sum_{i=1}^{N_d} \Delta F_{t_i}} \qquad (8)$$

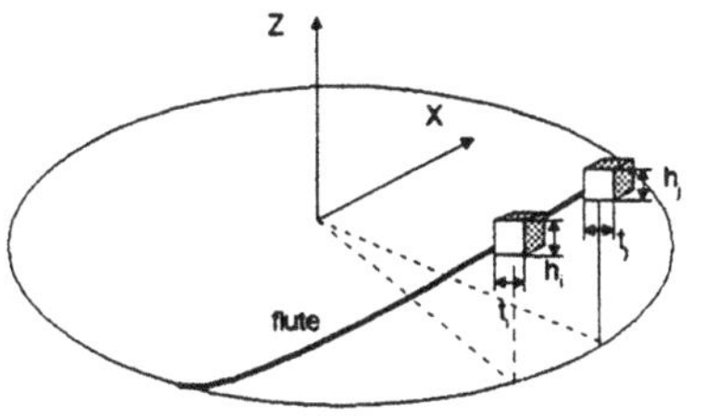

Fig. 4 The tool geometry and finite force components on cutting region

The summed tangential force through the flute is given as:

$$F_t = cft \sum_{i=1}^{N_d} \Delta F_{t_i} \cos(\theta_c - \theta_i) + cfr \sum_{i=1}^{N_d} \Delta F_{r_i} \sin(\theta_c - \theta_i) \qquad (9)$$

The summed radial force through the flute is also obtained:

$$F_r = cfr \sum_{i=1}^{N_d} \Delta F_{r_i} \cdot \left|\sin(\theta_c - \theta_i)\right| + cft \sum_{i=1}^{N_d} \Delta F_{t_i} \cos(\theta_c - \theta_i) \qquad (10)$$

The tangential and radial force components at a given angular position can be translated into x and y directional force components for comparison with the experimental data.

$$F_x = F_t \sin\theta_c' - F_r \cos\theta_c'$$
$$F_y = F_t \cos\theta_c' + F_r \sin\theta_c' \qquad (11)$$

where, θ_c' is $\theta_c - 3\pi/2$

Fig. 5 illustrates the relationship between x and y axis forces and tangential and radial directional forces.

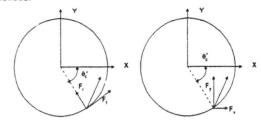

Fig. 5 The relationship between tangential and radial axis forces and x and y axis forces

3. EXPERIMENT VERIFICATION

Diverse cutting experiments by helix end mill are carried out to verify the validation of a mechanistic cutting dynamic model. For the validation of cutting dynamic model in a wide range, various cutting situations with different axial depth of cut, radial depth of cut, spindle speed and feed rate per minute are selected. The experiments are performed on a Bridgeport milling machine Torque Cut 22. The cutting force components are measured by a Kistler 3 components dynamometer model 9255B and amplified in multichannel charge amplifier model 5017A. The data are also translated to PC and accumulated in data acquisition board. A National Instrument data acquisition system is used to gather experimental data.

For data acquisition for cutting dynamic model of helix end mill, 1.905 cm diameter solid carbide end mill is used. The tool used in these experiments has 3 flutes and 30 degree helix angle, 5 degree radial rake angle and effective flute length is 3.80 cm. The test material was medium carbon steels with 210 BHN. For a given cutting condition with no overlap cutting condition and $\theta_b < 3\pi/2$, the effect of various feed rate per minute and spindle speed is investigated. Table 1 shows the different combination of feed rate per minute and spindle speed. Fig. 6 shows the comparison of y axis force profile between each experiments.

Table 1 The Effect of feed rate per minute and spindle speed at ac= 0.508 cm and rc= 0.508 cm

No.	rpm	f_m (cm)	f_m/rpm
1	1000	15.24	0.0152
2	1000	20.32	0.0203
3	800	20.32	0.0254
4	600	20.32	0.0339

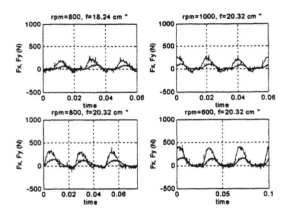

Fig. 6 The comparison of y axis force of various feed rate per minute and equal spindle speed

Based on a mechanistic cutting dynamic model, the actual cutting force profiles can be classified as three categories: (1) no overlap cutting with $\theta_b < 3\pi/2$, (2) no overlap cutting with $\theta_b \geq 3\pi/2$, (3) overlap cutting with $\theta_b \geq 3\pi/2$.

The first type of force profile can be found in the cutting situation with no overlap cutting and $\theta_b < 3\pi/2$. The maximum chip load on the tool occurs when a bottom of the flute locates at point b. At that location, tangential force reaches maximum because the chip thickness is maximum. After that point, the tangential force reduces gradually to zero. Fig. 7 shows the x and y axis cutting force profiles calculated by a mechanistic cutting dynamic model in Fig. 7 (a) and the measured experiment results is shown in Fig. 7 (b).

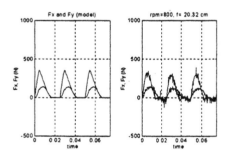

Fig. 7 The x and y axis force from mechanistic cutting model (a) and measured force (b) at ac= 0.508 cm and rc= 0.508 cm

The second type of the force profile can be found in the cutting situation under no overlap cutting with $\theta_b > 3\pi/2$. At point b, unlikely first type of force profile, the maximum chip load does not occur at point b. The maximum chip load happens after a bottom of the tool passes point b because the highest feed rate per tooth occurs when a bottom of the tool locates at $3\pi/2$. The cutting pattern is divided into up milling and down milling. Fig. 8 shows the x and y axis cutting force profiles calculated by a mechanistic cutting dynamic model of helix end mill

in Fig. 8 (a) and the measured experiment result is shown in Fig. 8 (b).

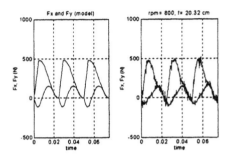

Fig. 8 The x and y axis force from mechanistic cutting model (a) and measured force (b) at ac= 0.508 cm and rc= 1.016 cm

The third type of force profile can be found in the cutting situation under overlap cutting with $\theta_b \geq 3\pi/2$. In a case of overlap cutting by two or more flutes, overall cutting force in a given instant can be achieved by the summation of involved flutes of the individual cutting force profile obtained by a single flute cutting. The cutting situation is selected as an axial depth of cut 0.508 cm and a radial depth of cut 1.27 cm. Fig. 9 shows the x and y axis cutting force profiles calculated by a mechanistic cutting dynamic model of helix end mill in Fig. 9 (a) and the measured experiment result is shown in Fig. 9 (b).

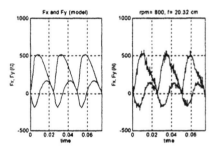

Fig. 9 The x and y axis force from mechanistic cutting model (a) and measured force (b) at ac= 0.508 cm and rc= 1.27 cm

In each cutting situation, the cutting parameters are selected to maintain reasonable magnitude and mean cutting force of x and y axis by consideration of value, feed rate per minute divided by spindle speed. In light cutting condition, allowable ranges for feed rate per minute and spindle speed are large. But in heavy cutting condition, allowable margin of feed rate per minute and spindle speed within limitation is very small. Fig. 10 shows the mean value of y axis force component for $f_m/rpm = 0.0254$ and Fig. 11 depicts the maximum value of y axis force component for $f_m/rpm = 0.0254$. Fig. 12 illustrates the mean value of x axis force components for $f_m/rpm = 0.0254$ and Fig. 13 shows the mean value of absolute value of oscillating x axis force components for $f_m/rpm = 0.0254$.

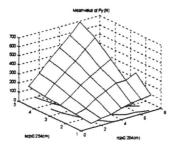

Fig. 10 The mean value of y axis force in different depth of cutting condition for f_m/rpm = 0.0254

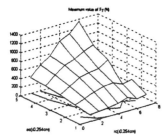

Fig. 11 The maximum value of y axis force component in different depth of cutting condition for f_m/rpm = 0.0254

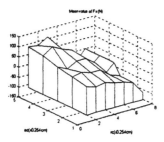

Fig. 12 The mean value of x axis force component in different depth of cutting condition for f_m/rpm = 0.0254

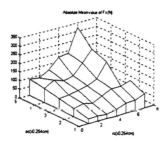

Fig. 13 The mean value of absolute x axis force component in different depth of cutting condition for f_m/rpm = 0.0254

4. CUTTING PARAMETERS SELECTION SCHEME FROM A MECHANISTIC MODEL

Based on the relationship among horse power,

machinability factor of material and metal removal rate, the feed rate per minute and spindle speed value will be suggested with respect to a given axial depth of cut and radial depth of cut. Actual feed rate per minute should be decided within the range of maximum and minimum feed rate per minute. Minimum value is given from information of the tool diameter and surface feet per minute. Range of spindle speed should be determined by the allowable force components of x and y axis force for a given axial depth of cut and radial depth of cut. With respect to y axis force, the maximum force of y axis force should be determined as limitation. With respect to x axis force, both of the maximum and vibrational effect will be x axis force's limitation. This maximum magnitude is a function of feed rate per minute and spindle speed at a given cutting situation.

For a given axial and radial depth of cut, the range of feed rate per minute and spindle speed should satisfy following constraints.

$$\max(f_m) \leq \frac{K \cdot HP_c}{ac \cdot rc} = \frac{MRR}{ac \cdot rc} \qquad (12)$$

where, HP_c is horsepower available at the cutter, K is machinability factor of material, rc is radial depth of cut and MRR is metal removal rate per minute (cm^3/min)

The limitations of feed rate per minute and spindle speed is given from machining table by Kennametal Inc. (1986).

$$f_{m\min} \leq f_m \leq f_{m\max} \qquad (13)$$

$$rpm_{\min} \leq rpm \leq rpm_{\max} \qquad (14)$$

The ratio of feed rate per minute and spindle speed satisfy the constraints of each force components.

$$\max(F_y) \leq F_{y\max} \qquad (15)$$

$$\overline{F_y} \leq \overline{F_y}_{\max} \qquad (16)$$

$$\overline{F_x} \geq \overline{F_x}_{\max} \qquad (17)$$

$$\left|\overline{F_x}\right| \leq \left|\overline{F_x}\right|_{\max} \qquad (18)$$

Each allowable ratio of feed rate per minute to spindle speed are obtained by following relationships.

$$\left(\frac{f_m}{rpm}\right)_1 = \frac{F_{y\max}}{\max(F_y)} \qquad (19)$$

$$\left(\frac{f_m}{rpm}\right)_2 = \frac{\overline{F_y}_{max}}{max(\overline{F_y})} \qquad (20)$$

$$\left(\frac{f_m}{rpm}\right)_3 = \frac{\overline{F_x}_{max}}{\overline{F_x}} \qquad (21)$$

$$\left(\frac{f_m}{rpm}\right)_4 = \frac{\left|\overline{F_x}\right|_{max}}{\left|\overline{F_x}\right|} \qquad (22)$$

Among each ratio, minimum allowable ratio will be selected.

$$\left(\frac{f_m}{rpm}\right)_{op} = min\left\{\left(\frac{f_m}{rpm}\right)_1, \left(\frac{f_m}{rpm}\right)_2, \left(\frac{f_m}{rpm}\right)_3, \left(\frac{f_m}{rpm}\right)_4\right\} \qquad (23)$$

Based on ratio of f_m/rpm, the feed rate per minute and spindle speed value will be decided. Within constraint Eqn. 12 and 13, maximum value of feed rate per minute will be chosen first and spindle speed should be selected late and satisfy constraint Eqn. 14.

$$rpm = \frac{f_m}{\left(\frac{f_m}{rpm}\right)_{op}} \qquad (24)$$

where, f_m is feed rate per minute (in/min),

When the ratio of f_m/rpm is adjusted to satisfy the fixed axial and radial depth of cut, force profile of x and y axis force in overall cutting range is multiplied by f_m/rpm. As the radial depth of cut is increased , the multiplied force profile for the increased radial depth of cut by f_m/rpm factor indicates over the limitation of specified force components. Especially, in the heavy cutting situation in big cutting area, the gradient with respect to the radial depth of cut is much greater than that of medium or light cutting situation in maximum and mean value of y axis force components.

At a cutting condition under axial depth of cut 0.762 cm and radial depth of cut 0.762 cm, f_{mmin} is given 5.08 cm/min, f_{mmax} is given 25.4 cm/min, rpm_{min} is 500 rpm and rpm_{max} is 1200 rpm. The specified values are given $\overline{F_y}_{max}$ = 325 N, $F_{y_{max}}$ = 700 N, $\overline{F_x}_{max}$ = 120 N, $\left|\overline{F_x}\right|_{max}$ = 150 N for limitation.

$max(\overline{F_y})$ is observed 249.0 N, $max(F_y)$ 400.0 N, $max(\overline{F_x})$ 60.5 N, and $max(\left|\overline{F_x}\right|)$ 79.2 N.

The allowable rate of f_m/rpm is obtained 1.30, 1.74, 1.98, and 1.89. Hence $(f_m/rpm)_{op}$ is determined as 1.50. When f_m is given as maximum value 25.4 cm/min and allowable spindle speed is given 770 rpm.

Fig. 14 shows the effect of multiplication factor in increased radial depth of cut condition.

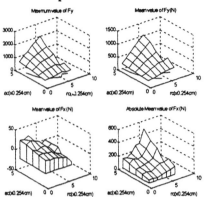

Fig. 14 The effect of multiplication factor f_m/rpm in increased radial depth of cut

5. CONCLUSION

In this paper, a systematic selection of cutting parameters in wide range of cutting situation was developed. The cutting conditions was classified by the distinct cutting profiles and verified by experiments. The dynamic cutting model in various cutting condition provides the sophisticated dynamic behavior of end mill cutting. With respect to diverse cutting condition, the best and safe parameters including feed rate per minute and spindle speed is possible by consideration of dynamic characteristics of maximum and mean value of x and y axis force components.

REFERENCES

Altintas, Y. and Spence, A. (1991). End milling force algorithm of CAD systems. *Annals of the CIRP*, **40/1**, 31-34.

Chang, C.H. and Melkanoff, M.A. (1989). *NC Machine Programming and Software Design*. Prentice Hall Press, New York.

Elanayar, S. and Shin, Y.C. (1996). Modeling of tool forces for worn tools: Flank wear effects. *ASME J. of manuf. sci. and eng.*, **118**, 359-366.

Kennametal Inc. (1986). *Kennametal milling catalog 5040*. Latrobe, PA.

Kline, W.A., DeVor, R.E., and Shareef, I.A. (1982). The prediction of surface accuracy in end milling. *ASME J. of eng. for industry*, **104**, 272-278.

Lynch, M. (1992). *Computer numerical control for machining*. McGraw-Hill Press, New York.

Martellotti, M.E. (1941). An analysis of the milling process. *Trans. of the ASME*, **63**, 677-700.

Sutherland, J.W. and DeVor, R.E. (1986). An improved method for cutting force and surface error prediction in flexible end milling systems. *ASME J. of eng. for industry*, **108**, 269-279.

CASE BASED APPROACH FOR MESH SPECIFICATION IN ADAPTIVE FINITE ELEMENT ANALYSIS

Abid Ali Khan, Xu Yuan Ming and Shen Zhang

Beijing University of Aeronautics and Astronautics, Beijing 100083, China

Abstract: This paper describes the application of an Artificial Intelligence paradigm called CBR (Case Based Reasoning) to the initial mesh selection in the adaptive finite element analysis system. It gives a brief overview of the system modules which perform the analysis on the structures by accessing the main module of the system called Case Base where previously solved cases are stored for future use. The system matches the current problem with similar stored case. The mesh distribution in the form of numerical values is represented by the object-oriented technique, which can be easily adapted by the topology of new problem in the same structural domain. As building of Case Based Reasoner, even in its simplest form involves a number of issues, such as case representation, case indexing, case matching and retrieval. All these issues are briefly discussed in this paper. It is believed that the use of developed knowledge representation scheme in the CBR systems will be a step ahead for Case Based Reasoning to become a powerful problem solving tool in the structural analysis domain.

Keywords: Finite element analysis, Object oriented programming, Knowledge representation, Case based reasoning, Error estimation, Mesh specification.

1. INTRODUCTION

The generating of finite element model involves great experience by the experts in the related analysis domain. Such tasks can be time consuming and error prone if done manually. Many approaches have been proposed to generate finite element model. A number of expert systems such as EXPERT (Babuska and Rank, 1987), IQFEM (Breitkopf and Kleiber, 1987), SACON (Bennett and Englemore, 1983), SESCON (Fjellheim and Syversen, 1983), FEMOD (Chen and Hajela, 1988) and FEMES (Gong and Xu, 1995) have been developed over the period of time to resolve the problems involved in model generation by using conventional rule based reasoning techniques. Other approaches (Shephard and Finnigan, 1988; Finnigan, et al., 1989; Unruh and Anderson, 1992) involve the incorporation of Computer aided designing (CAD) techniques, Intelligent decision making and advanced data management system into finite element analysis, so as to form the hybrid integrated intelligent structure analysis system. Although all these approaches have shown some success in analyzing certain type of structures but still there are a lot of issues that need to be addressed and resolved to formulate a more universal and highly automated approach in the finite element analysis domain. One of the difficult tasks for the FEM users is to set the spatial mesh over the geometrical domain. Specifying mesh densities that capture the behavior of local phenomena, can typically become a process of trial and error. Intelligent users of the FEM for numerical discretization have always endeavored to assess the accuracy of their results before using them in practice. Unfortunately, the most obvious procedure of doing this by study of refinement convergence is generally too expensive to be used frequently.

This paper proposes an intelligent-based reasoning

approach called CBR (case-based reasoning), to resolve the issue of mesh specification in adaptive finite element analysis. As it is well known that, the good results of adaptive finite element analysis and computational expenses highly depend on the cycles of iterations, error estimation and rate of convergence. These features of adaptive finite element analysis are directly effected by the selection of initial mesh, hence the selection of this mesh is highly important for a specific problem. To provide a truly robust adaptive finite element analysis system the specification of the most appropriate initial mesh is substantial. Therefore, the basic idea of the devised approach is to use the specific knowledge, in terms of mesh patterns of previous analysis episodes to generate an initial mesh for the new problems. Mesh patterns of already solved problems serve as cases to be adapted for the similar structure analysis problems. Once the mesh pattern is adapted, a consequent adaptive analysis is performed, in which the approximation is successively refined to achieve a predetermined standard of accuracy.

2. ADAPTIVE FEM PROCESS

The objective of an adaptive finite element analysis is to control the discretization error by increasing the number of degree of freedom in regions where the previous finite element model is not adequate and achieve the maximum rate of convergence of the finite element approximation economically (Zhu, *et al.*, 1991). It is therefore essential to have quantitative assessment of the quality of approximate solution and the capabilities of the refinement. The complete process of adaptive analysis is shown in Figure 1, where error measures for a given mesh are based on numerical results obtained during the solution process. These error estimations provide information about accuracy and reliability of the computed solution, they also guide to refine the mesh in an efficient manner to achieve a certain accuracy. The Z-Z posteriori error estimator (Zienkiewicz and Zhu, 1987) is being used during the adaptive process of the proposed analysis system.

The mesh refinement methods for adaptive finite element analysis can be classified into two main categories as; methods based on experimental criteria and methods based on the finite element error estimations. In order to achieve the optimal results researchers are using the following four means of adaptive remeshing in different ways, these methods are based on the error estimates.

h- version; It achieves the accuracy by refining the mesh using a given type of finite element

p- version; It keeps the mesh fixed and accuracy is

achieved by increasing hierarchically the order of the approximation used.

r- version; It achieves the accuracy by only relocating the nodes, without increasing the total number of nodes and elements.

hp- version; It is a proper combination of h- and p-versions.

The selection and application of the refinement methods depend on the type of element and the required engineering accuracy for a specific problem domain. The most common method is the h-version, however, the use of the p-version or the hp-version of adaptive analysis is considered when a very high accuracy is required.

The essential steps for adaptive algorithm can be summarized as follows:

Step 1; Select the initial mesh.

Step 2; Perform the finite element analysis.

Step 3; Estimate the error for each element by using the initial finite element solution computed in step 2.

Step 4; If the errors are acceptable produce the output otherwise proceed to step 5.

Step 5; Subdivide the elements according to error measure of each element in order to refine the mesh and repeat step 2 to 4.

In this paper the adaptive analysis process has been applied to the two-dimensional plate problem with a posteriori error estimator and h- method for refining the meshes to achieve the required accuracy.

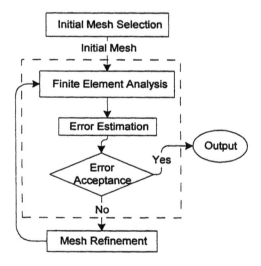

Fig. 1. Adaptive finite element Analysis.

3. CASE BASED REASONING

The focus of the research presented here is the issue of mesh specification in the adaptive finite element analysis. Case based reasoning uses a representation of specific episodes of problem solving to learn to solve a new problem. A CBR provides the potential for developing knowledge-based systems more easily than generalized knowledge-based approaches. The case base is not tightly bound to the compiled knowledge of rules; rather, it is free to postulate solutions based on less restrictive matching routines that make it more flexible thus it permits solutions of problems to be proposed even when domain knowledge is lacking.

Case base is one of the most important modules of such systems. It includes a representation of a set of previously solved problems. The content of each case is basically a description of the previously solved problem. The two major considerations in case based reasoning are, identifying a similar previous case and determining what changes and what stay the same. These two are referred to as Recall and Adapt as shown in the model of case based reasoning in the Figure 2. Given a new problem to solve, the recall module matches, retrieves and selects the similar case from the case base. The adapt module includes a recognition of the differences between the selected case and the new problem and decisions regarding what aspects of the case are to be changed to fit the new problem. On completion of the analysis and successful solution of the problem it is added to the case base, for use in future problem solving episodes. In this way the case base is continuously being updated through the learning module of the CBR system.

4. SYSTEM ARCHITECTURE

An integrated finite element analysis system proposed by author is under development (Khan, 1996). The system is based on the principles of case based reasoning, and uses the adaptive finite element analysis techniques to compute the results thus the realization of the complete system comprises two major parts, case based reasoner and adaptive finite element analysis process. Building of Case Based Reasoner, even in its simplest form involves the resolution of the issues of case representation, case indexing, case retrieval and ranking of cases. Whereas, adaptive finite element analysis involves development of mesh generator, error estimator and selection of refinement strategies. The following sub-sections explain the key issues in the case based system devised for the problem solving strategy.

Figure 3 shows the framework for the integrated and

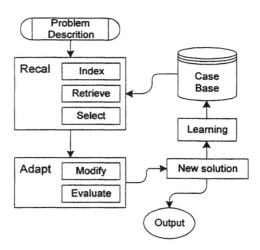

Fig. 2. An elaborated model of Case based reasoning.

intelligent system. When a new problem is given to the system the selection module matches a set of features and retrieves similar case from the case memory (case base). The mesh of the retrieved case is corrected at the critical points by using numerical adaptation. On achieving a specified accuracy the results are fed to results analysis module which extracts the features from the solution to store as a new case in the case base and display the output.

In the first phase of the system development, two dimensional plate problems have been considered for analysis to evaluate the system accuracy and observe the system behavior.

4.1 Case representation

The representation of the case, in the problem domain is vital for the entire integrated system. Case representation is a process of determining the contents of cases and their organization. Case contents for finite element modeling problem, includes problem description and solution algorithm.

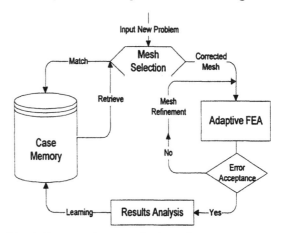

Fig. 3. Integrated system framework

Problem description to formulate a complete case includes: geometrical attributes of the concerned structure, its load type and position, boundary condition information, as well as material and physical data of the structure necessary for analysis. The parameters essential to define the case are stored in a set of four files known as node resource file, node connectivity file, structural geometry attributes file, and node coordination file. Each case consists of a case name and a particular set of parameters defining the attributes in these four files.

A typical case for a structure composed of three geometrical shapes is shown in Figure 4. This case is named as "C1" where C stands for a case based on a structure with three geometrical shapes and 1 means it is the first case in this category. All the four files of the case are named after the case name and parameters are saved as attribute-value pairs. The cases are indexed by their names in such a way that just by looking at the name of the case it can found that the case belongs to which category, i.e., The structure is composed of how many boundaries and sharp edges. This simple classification and organization of cases is not only realizing the indexing but also shortening the matching process. The matching or similarity analysis as discussed below is only performed in the similar category of structures.

4.2 Class hierarchy

A class hierarchy has been designed to realize and implement the case representation strategy as shown in Figure 4. It incorporates all the information essential for finite element analysis. The parameters present in the case are structure geometry attributes, load type and position of its application, boundary constraints, structure material data and physical data such as, cross sectional area and thickness of the plate. The class identification is done on the basis of a detailed analysis about the relationship among the

```
Case Name: c1
        (Geometrical structural is formed by
         three boundaries)
Node Point resource file: _c1.rsc
        Node #        resource type
          1              1
          2              1
          ...
Node Connectivity file: _c1.cnc
Structure Geometry file: _c1.str
        No of boundaries, Load, bc : 3, 0, 0
        No of boundary points      : 4, 1, 4
        ...
Node Coordination file: _c1.xyz
        (Formed from *.str file for new mesh)
```

Fig. 4. The representation of a complete case

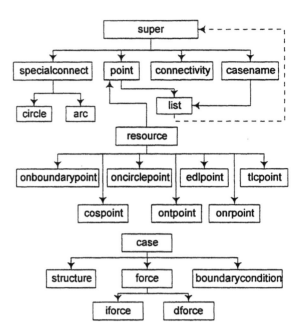

Fig. 5. Class hierarchy for case and node points

functions that are required for case representation in case based reasoning. The candidate classes are identified as follows and arranged into hierarchies as shown in Figure 5.

Class super. It is the most general class in the class hierarchy.

Class specialconnect. It defines special connections, such as connections on circle and arc boundaries.

Class point. It is handling all the resources for points in the structure geometry of a case.

Class connectivity. Node connectivity for the mesh is defined in this class.

Class onboundarypoint. It handles points on the outer boundaries.

Class oncirclepoint. It handles points on the circle boundaries.

Class edlpoint. It handles equidistant points on the line.

Class tlcpoint. It handles points generated by the intersection of two entities.

Class cospoint. It handles construct point (a point that can divide a complex geometry into two basic geometry shapes).

Class onrpoint. It handles point in rectangular elements that subdivides the element into smaller triangles.

Class ontpoint. It handles point in triangular elements that subdivides it into smaller triangles.

Class structure. It is a virtual class with several members, for sending and receiving messages from the most general class named as super.

4.3 Matching & retrieval processes

Matching each case index against the target case becomes time consuming and expensive as the case base grows in size. Therefore in the case retrieval a filtering technique is applied by the case name, which can identify the cases with same number of boundaries for detailed matching. A weighted feature scheme also known as a nearest neighbor approach is applied for the detailed matching of the similar cases. The weight assignment is done during the match process in the structure geometry attribute file. As all the parameters in the case are numerical values which can be easily compared for a complete or partial match in order to determine the total weight W_{total} for a case. The procedure for the attribute comparison between the target attribute (a) and new case attribute (a') is explained in the following few equations.

$$\text{If} \quad value(a) = value(a') \quad (1)$$

$$\text{then} \quad W_i(a) = 1 \quad (2)$$

which means match is complete and attribute a is weighted as 1 or 100%.

$$\text{If} \quad value(a) \neq value(a') \quad (3)$$

$$\text{then} \quad W_i(a) = value(a) - value(a') \quad (4)$$

This clearly indicates that the weight W_i for the attribute will be less than 1. This process is followed in complete cycle unless all the attribute values are compared between the target and new case. After every comparison the weight assigned is added in the previous weight $W_{previous}$ in order to compute the W_{total} as shown in equation 5.

$$W_{total} = W_{previous} + W_i(a') \quad (5)$$

When W_{total} for all the similar cases is computed in so called similarity analysis, the case with the highest weight is considered the most applicable. The case with highest weight is then retrieved so as to use its mesh on the new problem structure topology. In order to adapt the selected mesh for the new problem its validity is checked by looking if all the nodes in

the mesh reside with in the outer boundary of the new structure topology. In case the nodes are found in an invalid region the mesh is modified and all such nodes are moved into the valid region of the structure geometry.

5. A CASE STUDY

An application of the above mentioned reasoning technique on a plate with hole is being shown below in Figure 6. When a two dimensional problem of a plate with hole in the center and an angular edge is given to the system see Figure 6(c), it accessed the case base in order to match the similar target case. As the problem consists of an inner and outer boundaries therefore all the cases from the case base with two boundaries were retrieved for matching. In the similarity analysis all the indices are matched and the case of a rectangular plate with the hole was weighted highest and thus selected for adaptation of the mesh pattern to the problem structure topology. All the nodes from the target case are adapted to the structure topology one by one depending on their resource types i.e., all the nodes located on the inner and outer boundary of the target case are placed on the problem structure inner and outer boundaries respectively. The elements are then generated on the basis of connectivity information of the target case. In this way mesh pattern of the selected case is completely adapted by the new problem as an initial mesh for further adaptive refinement process.

6. CONCLUSION

A CBR approach to mesh specification for adaptive finite element analysis with object oriented representation of the cases has been proposed. The

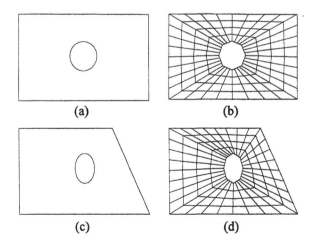

Fig. 6. (a) Target case structure geometry. (b) Target case mesh. (c) New problem structure geometry. (d) Adapted mesh from the target case.

proposed approach can be beneficial for the fast formulation of initial mesh for a particular structure through the retrieval and adaptation of successful cases analyzed in the past. It is believed that the use of such techniques and adaptation of less expensive meshes will reduce the speed of adaptive process by reduced refinement steps. Moreover, the use of already optimized mesh on a similar structure will also boost the efficiency of the analysis. In future it is expected that further investigation and working on the existing principles will evolve a much universal methodology that can be used for aeronautical structures.

REFERENCES

Babuska, I., and E. Rank (1987). An expert system for the optimal mesh design in the hp-version of the finite element method. *International Journal of numerical methods in engineering.* **24**, 2087 - 2106.

Bennett, J. S., and R. S. Englemore (1979). SACON : a knowledge based consultant for structural analysis. *6th International joint conference of artificial intelligence Tokyo.* 47 - 49.

Breitkopf, J., and M. Kleiber (1987). Knowledge engineering enhancement of finite element analysis. *Communication application of numerical methods.* **3**, 359 - 366.

Chen, J. L., and P. Hajela. (1988). *FEMOD : A consultative expert system for finite element modeling. Computer and structures*, **29**, 1, 99 - 109.

Fjellheim, R., and P. Syversen (1983). An expert system for Sesam-69, structural analysis program selection. *Report # CR-83-6010.* Computas, Norway.

Gong, Y. N., and Y. M. Xu (1995). FEMES : Finite element modeling expert system. *Journal of mechanical strength (in Chinese).* **17**, 2, 14 – 19.

Finnigan, P. M., A. Kela, and J. E. Davis (1989). Geometry as a basis for finite element automation. *Engineering computations*, **5**, 147 – 1160.

Khan, A. A., (1996). Towards integrated and intelligent adaptive finite element analysis, *MSc Thesis.* Beijing University of Aeronautics & Astronautics, Beijing, China.

Shephard, M. S. and P. M. Finnigan (1988). Integration of Geometric Modeling and advanced finite element preprocessing. *Finite element analysis design*, **4**, 147 – 238.

Unruh, V. and D. C. Anderson (1992). Feature-based modeling for automatic mesh generation. *Engineering with computers*, **8**, 1 - 12.

Zhu, J. Z., E. Hinton and O. C. Zienkiewicz (1991). Adaptive finite element analysis with quadrilaterals. *Computers and Structures,* **40**, 5, 1097 - 1104.

Zienkiewicz, O. C., and J. Z. Zhu (1987). A simple error estimator and adaptive procedure for practical engineering analysis. *International Journal of numerical methods in engineering.* **24**, 337 - 357.

Computer Simulation of Continuous Manufacturing from Forging to Machining for the DFM

J. H. Liou Ph.D.

D. Y. Jang Associate Professor

Aircraft Certification Institute
Civil Aeronautics Administration
Taiwan, Republic of China

Industrial Engineering Dept.
Seoul National Polytechnic University
Seoul, Korea
(Former MAE Dept.
University of Missouri-Columbia
U.S.A.)

Abstract: Design for manufacturing (DFM) is a way to provide design engineers with complete information on the characteristics of product quality. Designers can then decide remedial manufacturing procedures, if necessary, in the design stage without requiring actual manufacturing operation. As a way to develop a computing tool for the DFM and stress information in products, a three-dimensional finite element model to simulate a continuous manufacturing process from forging to machining was developed using a nonlinear finite element program. In the simulation, MILS 11595 was used as a workpiece material and stress developments in each stage of operation were calculated.

Keywords: DFM, Forging, Maching, Stress, Computer simulation

Introduction

A recent trend in manufacturing is to fabricate a product to certain dimensions through forging or casting, after which machining processes such as milling and turning are applied to control the dimensions and accuracy of the final product. The process is called "hard turning". Today, hard turning is one of the important process in order to eliminate a separate finishing operation such as grinding. Taking a final finishing pass on a lathe eliminates the labor required for a second grinding setup and time on the grinder. Hard turning also consumes less energy per metal removed than grinding does. It is more environmentally friendly. Hollow shafts used mainly in the automobiles and other industries, are fabricated through radial forging and machining to obtain the desired surface finish and to adjust dimensions. The advantages of this process are a smooth surface finish, tight tolerance, considerable material or weight saving, preferred fiber structure and increased material strength.

Forged components with high residual stresses generated by the radial forging process will receive more stresses and undergo some deflections after machining due to the resultant high residual stresses, thus losing the dimensional accuracy. Residual stresses can also cause stress corrosion for certain materials when used in a corrosive environment. The residual stress distribution in high strength metal is of great importance since its presence, above a certain level, can shorten the service life of critical components used in the severe service conditions. In order to decrease the amount of tensile residual stress or change it into compressive stress, post processes such as heat treatment and shot peening are used frequently after each manufacturing process. However, these processes are complicated and may cause the undesirable effect on the dimensions of the product. They are also time consuming and expensive, and can affect cost and delay of delivery. By introducing computer simulations in the stage of process design, it is possible to verify the design for manufacturing, and bring the process design closer to completion before testing it on the actual machine. That process simultaneously reduces the test frequency [1]. Furthermore, through stress analysis in a product, we can estimate the post processing conditions to relieve residual stresses and decide the optimum operational conditions for forging and machining which will yield favorable stress conditions in a product.

With reliable computer simulations to integrate the manufacturing process from forging to machining, it is possible to produce products with surface integrity and reliable performance at low production costs and high rates. The developed simulation models can be applied to the unified optimization of manufacturing operations to realize the adaptive control of optimization for maximum production and surface integrity. The tool which simulates the manufacturing process can provide design engineers with complete information on the characteristics of product quality. Designers can then ascertain whether or not a particular manufacturing process adversely affects the quality of products. They can then decide remedial manufacturing procedure, if necessary, in the design stage without requiring actual manufacturing operation. The integrated manufacturing system can also be used for product/process optimization of precision manufacturing, which will produce products with the desired quality and maximum production rate. Hence, the objective of this research was to develop nonlinear finite element (FE) computer models to simulate the manufacturing process on an axi-symmetric product in order to find the stress developments during forging and machining. This research also provided an analytical tool and theoretical backgrounds to the analysis of the hard turning.

Development of a Model to Simulate Continuous Manufacturing using FEM Theory

Application of FEM to manufacturing problems began as an extension of the structural analysis technique to the plastic deformation regime. Early applications of FEM to manufacturing problems were based on the rigid-plastic formulation developed from the Prandtl-Reuss equations. Since elastic strains are neglected, the stresses below the yield stress are not known. This implies that effects such as residual stresses and spring-back which are critical to the analysis of metal forming cannot be predicted. Hence, elasto-plastic material was assumed in this study and the material was assumed to obey the Von-Mises yield criterion and its associate flow-rule. The deformation process for elasto-plastic materials is associated with the boundary value problem, where the stress and strain field solutions satisfy the equilibrium equations and the constitutive equations in the domain and the prescribed boundary values.

The weak form of equilibrium equation neglecting body force is expressed by

$$\int_{v} \sigma_{ij,j} \delta v_i dV = 0 \qquad (1)$$

where σ_{ij} is the component of stress tensor, δv_i is an arbitrary variation of velocity, and a comma denotes

partial differentiation. Using the divergence theorem and the symmetry of the stress tensor and imposing essential boundary conditions, $\delta v = 0$ on S_v, where S_v is specified velocity field boundary and v_i is prescribed on S_v. Eq. (1) becomes

$$\frac{1}{2}\int_{v} \sigma_{ij}(\delta v_{i,j} + \delta v_{j,i})dV - \int_{S_F} \sigma_{ij}v_i n_j dS = 0 \qquad (2)$$

where n_j is the unit normal to the surface and S_F is the surface where traction is prescribed, namely $\sigma_{ij}n_j = T_i$ on S_F. Decomposing the stress tensor into the deviatoric component σ'_{ij} and the hydrostatic component σ_m, Eq. (2) becomes

$$\int_{v} \sigma'_{ij}\delta\dot{\varepsilon}_{ij}dV + \int_{v} \sigma_m\delta\dot{\varepsilon}_v dV - \int_{S_F} T_i\delta v_i dS = 0 \qquad (3)$$

where $\dot{\varepsilon}_{ij}$ is the strain-rate and $\dot{\varepsilon}_v$ is the volumetric strain-rate. The final equation is obtained by replacing the integrating of the first term by the effective stress $\bar{\sigma}$ and the effective strain-rate $\dot{\bar{\varepsilon}}$. The incompressibility constraint on the admissible velocity field of Eq. (3) can be removed by a penalty function constant C [2], as

$$\delta\pi = \int_{v} \bar{\sigma}\delta\dot{\bar{\varepsilon}}dV + K\int_{v} \dot{\varepsilon}_v \delta\dot{\varepsilon}_v dV - \int_{S_F} T_i\delta v_i dS = 0 \qquad (4)$$

where $\bar{\sigma} = \sqrt{(3/2)\sigma'_{ij}\sigma'_{ij}}$, $\dot{\bar{\varepsilon}} = \sqrt{(2/3)\dot{\varepsilon}_{ij}\dot{\varepsilon}_{ij}}$, and C is a large positive constant. The effective stress for a specific material is determined by uniaxial tension or compression tests as a function of effective strain and effective strain-rate.

Eq. (4) is the basic equation for the finite element discretization. Once the solution for the velocity field that satisfies the basic equation is obtained, then the corresponding stresses can be calculated using the flow rule and the known mean stress distribution.

In the analysis, the solution is obtained by using the Newton-Raphson method. The method consists of linearization and application of convergence criteria to obtain the final solution. Linearization is achieved by Taylor expansion near an assumed solution point $v = v_o$, namely,

$$\left[\frac{\partial\pi}{\partial v_i}\right]_{v=v_o} + \left[\frac{\partial^2\pi}{\partial v_i \partial v_j}\right]_{v=v_o}\Delta v_j = 0 \qquad (5)$$

where Δv_j is the first-order correction of velocity v_o, and the higher-order terms are neglected. Eq. (5) can be written in the form

$$K\Delta v = f \qquad (6)$$

where K is called the stiffness matrix corresponding to the tangential direction, and f is the residual of the nodal point force vector. Since Eq. (6) represents only a Taylor series approximation, the displacement increment correction, Δv is used to obtain the next displacement approximation

$$v^{i+1} = v^i + \Delta v \qquad (7)$$

The iteration is continued until appropriate convergence criteria are satisfied.

There are some basic assumptions made in this study:
a. The deforming material is considered to be a continuum.
b. Uniaxial tensile or compression test data are correlated with flow stress in multiaxial deformation conditions.
c. Anisotropy and Baushinger effect [3] are neglected.
d. Volume remains constant [2].
e. The yielding criterion is assumed to follow the Von Mises assumption [3].

In order to handle the contact and friction between the workpiece and the die, and the workpiece and the mandrel, a stiffness relationship between two contact areas must be established. Otherwise, the two areas will pass through each other and assembled finite element stiffness matrix will be singular. This relationship is established through a spring that is put between the contacting areas when contact occurs as shown in Fig. 1. The rigid Coulomb model was applied since in the radial forging process, the workpiece slides continuously between the die and the mandrel.

FEM Model

A hollow shaft as shown in Fig. 2 will be used as an example of an axi-symmetric shaft and is assumed to be fabricated by radial forging and machining. Using the ANSYS program, deformation shapes and stress distributions at various stages of each bite-operation were calculated. The residual stress information on the forged products was used to calculate the stress distribution due to the machining operation. Due to the axisymmetric characteristics of radial forging and machining, only a section of full three-dimensional workpiece was analyzed.

To model manufacturing process, it was necessary to develop a suitable representation of the forging dies, the mandrel, the contact areas, and the workpiece. Three different types of elements were used in modeling forging: (1) SOLID45 with 8 node 3-D solid element for the die and the mandrel; (2) VISCO107 with 8 node 3-D large strain element for the workpiece; and (3) CONTACT49 3-D general contact element for the contact surfaces between the workpiece and the die, and between the workpiece and the mandrel. Additional assumptions were made to obtain reasonably accurate results: (i) rotational feed was neglected; (ii) the material was assumed to have elasto-plastic behavior during manufacturing process; (iii) the deformation process was assumed as isothermal; and (iv) the deformation of forging dies and mandrel was neglected.

The detailed information of modeling and boundary condition are given in Table 1. The die and mandrel were assumed to be rigid bodies with higher elastic modulus. The workpiece was assumed to be symmetric on both sides which were composed of r-z planes.

The simulating strategy of the forging operation used in this study was to model the manufacturing operation on a "stroke-by-stroke" (increment by increment) basis [4]. In each stroke, the workpiece was punched by a small increment in the axial direction. Then, a systematic iteration method (Newton-Raphson method) was utilized to obtain the new equilibrium condition which was the initial condition for the next stroke. The solution was incrementally obtained by using an automatic-load-adjustment option. The frictional force between the die and the workpiece, and between the mandrel and the workpiece was simulated through the 3-D general contact element, CONTACT49. The stroke was continuous until the workpiece passed through the die. In the simulation of forging process, the whole process was divided by five stages and each stage included hundreds of strokes. The numbers of stroke depended on the contact area. This procedure is also called as the update Lagrangian method [5]. After the forging simulation, the pre-stresses produced by forging would be considered as the initial stresses in the machining simulation.

After the residual stresses in the workpiece generated by the forging operation were obtained, the "loading cycle" method [6,7], was utilized to calculate residual stress distributions in the final product processed by the machining. The workpiece was assumed to be fixed at both ends. The loading process was divided into three stages and the unloading process was divided into two stages. The workpiece used in the simulation was alloy steel MILS 11595 [8,9]. Fig.3 shows an example of the meshed model of the die, mandrel, and workpiece. The forming geometry and process variables used in the forging operation are listed in Table 2. The machining condition used in the hard turning operation on the forged part is given in Table 3.

The reason for applying slow and light machining was due to heavy cutting resistance in hard turning [10]. The coordinate of the finite element model of machining operation on the forged part is shown in Fig. 4. Three components of cutting force were assumed to act on the forged part.

Calculations were performed on the CRAY90 in the Pittsburgh Super Computer Center (PSC). The ANSYS51 (maximum wavefront is 1500) FEM code was used in this study. The files of results were transferred through File Transfer Protocol (FTP). The post-processing was done on the VAX computer.

Results and Analysis

The effectiveness of each process using the developed model has been verified by comparing the results with experimental data and published work [11]. Calculation using Table 2 and Table 3 were performed for continuous manufacturing from forging to machining.

Fig. 5 show the effective-strain distributions in the forged product at stage 2. The deformation can be divided into three zones [4]; sinking, forging, and sizing zones. In the sinking zone, the workpiece was fed into the die and the mandrel with only small change in diameter. The effective-strain was approximately 10% in the sinking zone as shown in Fig. 5. In the forging zone, the workpiece was forged completely to the die land with the most of deformation taking place in this zone. The effective-strain reached its maximum value 30%. In the sizing zone, the workpiece passed the die land and the strain spring-back due to the elastic or elasto-plastic deformation. The results also showed that effective-strain was greater at the outer and the inner surfaces of workpiece than at its sub-surface. This might be due to the workpiece being subjected to intense shear between the contact surfaces of die and mandrel.

The effective-strain after the forging process is shown in Fig. 6. The maximum strain was found to occur at the outer and the inner surfaces at the start of deformation, when the die initially contacted the workpiece and decreased radially through the wall to its minimum value at the sub-surface.

The stress development in the axial direction for stages 3 is shown in Fig. 7. At the beginning of forging, the outer surface of the sizing zone showed a small tensile stress region. However, the sinking and the forging zones were dominated by compressive stresses. As the workpiece continued to feed into the die land, the tensile stress region increased and the compressive region decreased. An interesting result of the simulation was that although the tensile stress region increased during the forging process, the magnitude of compressive stresses in the forging zone remained constant at each stage. The stresses in the radial direction at stages 3 of the forging process is shown in Fig. 8. It showed the similar pattern as the axial stresses, but values of tensile stresses in the radial direction were smaller than those in axial direction. The maximum tensile stress was located at the beginning of deformation, when the die initially contacted the workpiece. It was found that radial stresses in the sizing zone were smaller than those when workpiece passed the sizing zone. This was due to the effect of spring-back (elastic recovery) after the sizing zone and caused the existence of the higher tensile stresses [12]. The radial stresses exhibited radial variations from compressive stresses at the inner surface to tensile stresses at the outer region. Fig. 9 represents the stress distribution of the hoop direction at forging stages 3. It also showed the tensile stresses at the outer surface and compressive stresses at the inner surface, but the maximum tensile stress occurred at the middle part of the outer surface.

The residual stress distribution in the axial direction is shown in Fig. 10. It can be seen that the residual stresses were tensile at the outer surface with the maximum 317 MPa and compressive stresses at the inner surface. The radial residual stresses of the forged product are shown in Fig. 11. It showed that the tensile stresses were at the outer surface and the compressive stresses were at the inner surface. The maximum tensile stress occurred at the both ends of the workpiece with a maximum value of 173 MPa. The hoop residual stress distribution of forged product is shown in Fig. 12. The tensile stresses were at the outer surface of workpiece with the maximum value of 232 MPa. Compared with the axial, the radial, and the hoop residual stresses, maximum tensile and compressive stresses were found in the axial direction. The tensile residual stresses in the axial and hoop directions might have been caused by the frictional force which was always opposite to the moving direction. That is, when the workpiece was fed in the die land, the frictional force tended to pull the workpiece in the opposite direction, thus causing the tensile stresses at the outer surface of the workpiece. The tensile radial stresses might have been generated due to the spring-back effect of the workpiece in the radial direction.

The residual stress distributions of the outer and the inner surfaces along the axial direction showed that the outer surface was dominated by tensile residual stresses with the maximum value at the rear part of workpiece. The axial stresses had the minimum stress (-48 MPa) at the front part of the workpiece and increased along with the axial direction to the maximum stress (300 MPa) at the rear part of the workpiece. However, the inner surface presented most of compressive residual stresses with only small tensile stresses in the front part of the workpiece. The axial stresses showed large variations between the inner and outer surfaces, but the radial and hoop stresses did not show significant change with the thickness. The maximum tensile stress occurred at the outer surface of the workpiece along the axial direction.

After the forging simulation, the workpiece was assumed to be fixed on the both ends. By considering the pre-residual stresses produced by the forging process, the cutting, thrust, and axial forces [13] in the hoop, radial and axial directions were then applied on the surface according to the loading cycle method. The instantaneous stress distribution at loading stage for axial direction is showed in Fig. 13. It can be seen that the axial direction showed compressive stresses on the outer surface. Same stress distributions were found in the radial direction. The hoop direction showed compressive stress in front of the tool tip and a large tensile stress behind the tool tip. That is, during the machining process, a large compressive plastic deformation zone is formed ahead of the tool edge and a tensile plastic deformation zone behind it.

The final residual stress distribution due to the forging and machining process is shown in Fig. 14. It can be seen that the hoop and axial stresses had maximum tensile stresses about 300 MPa at the outer surface while radial stresses presented small tensile stresses of 10 MPa at the outer surface. As the thickness increased from the outer surface, the hoop and the axial stresses decreased and reached their minimum values at the inner surface. However, the radial stresses showed very small variations between the inner and the outer surfaces. Compared to the residual stresses due to the forging process, it showed that axial and radial stresses did not have much variation but the hoop residual stresses increased from 110 MPa to 300 MPa at the outer surface.

From the FEM simulation, it can be found that the largest residual stress was found in the axial direction when the workpiece was manufactured by forging process only. By considering the pre-residual stresses produced by forging process, the machining simulation showed that the axial and radial stresses did not have significant change after the machining process but the hoop residual stresses became the most important components on the products. The radial residual stresses could be neglected due to its small value when compared to the hoop and the axial stresses.

Conclusion

In the present study, a three dimensional finite element model was developed to simulate the stress development from the radial forging to machining process. The workpiece was assumed to have elasto-plastic behavior during the cold forging and machining process. The results from this research could be summarized as follows:

1. The outer surface of forged products was dominated by tensile residual stresses with maximum values at the rear part of the forged products.
2. The inner surface of forged products showed compressive residual stresses and only small tensile residual stresses in the front part of workpiece.
3. The front part of forged products was dominated by the axial residual stresses, but the rear part of workpiece showed significant hoop stress.
4. Surface residual stresses due to forging and machining operations were tensile in the axial and hoop directions.
5. The machining process did not cause significant change for the axial stress.

Acknowledgments

The authors acknowledge the support from the Pittsburgh Super Computer Center (PSC) under Grant No. DDM9400029. Thanks are also due to reviewers

References

[1] Yano, H. and T. Akshi (1994). Application of analytical simulation to forged part design. *TOYOTA Technical Review*, **Vol. 43, No. 2**, pp. 20.
[2] Kobayashi, S., S. Oh, and T. Altan (1989). Metal forming and finite element method. *New York Oxfod*.
[3] Malvern, L. E. (1969). Introduction to the mechanics of a continuous medium. In: *Prentice-Hall*, N.J.
[4] Domblesky, J. P., R. Shivpuri and B. Painter (1995). Application of the finite element method to the radial forging of large diameter tubes. *J. of Materials Processing Technology*, **Vol. 49**, pp. 57
[5] Swanson Analysis System, Inc. (1993). *ANSYS user's manual*, **Vol. IV, Revision 5.0**, Houston, PA.
[6] Liu, C. R. and M. M. Barash (1976). The mechanical state of the sublayer of a surface generated by chip removal process, part 1 & 2. *J. of Engr. for Ind., ASME*, **Vol. 98**, Nov., pp. 1192
[7] Liu, C. R. and C. Z. Lin (1985). Effects shear plane boundary condition on shear loading in orthogonal machining. *Int. J. Mech. Sci.*, **Vol. 27**, pp. 281.
[8] Hoffmanner, A.L. and K.R. Iyer (1978). Residual stress control in precision swaged rifle barrels. *NAMRC VI*, pp. 180.
[9] Harvey, P. D. (1982). Engineering properties of steel. *American Society for Metals*, Metal Park, Ohio.
[10] Stovicek, D. (1992). Harding part turning. *Tooling and production*, **Vol. 57**, pp. 25.
[11] Liou, J. H. (1996). Study of stress developments in axi-symmetric products fabricated by forging and machining, Ph.D. dissertation, U. of Missouri-Columbia.
[12] Sawamiphakdi, K., P. K. Kropp and G. D. Lahoti (1990). Investigation of residual stresses in drawn wire by the finite element method. *J. of Engr. Mats. and Tech. ASME*, **Vol. 112**, pp. 231.
[13] Nakayama, K. and M. Arai (1976). On the storage of data on metal cutting forces. *Annals of the CIRP*, **Vol. 25**, pp. 13.
[14] Jang, D. Y., J.H. Liou, T.R. Watkins, K.J. Kozaczek and C.R. Hubbard (1995). Characterization of Surface integrity in Machined Austenitic Stainless Steel. *Manufacturing Science and Engineering, ASME*, **MED-Vol. 2-1/MH Vol.3-1** pp. 399.

Items	No. of Element	Type of Element	Boundary Condition
Die	42	SOLID 45	Rigid
Workpiece	300	VISCO 107	Symmetry on both sides
Mandrel	10	SOLID 45	Rigid
Contact areas	2418	CONTACT 49	With frictional force

Table 1 Elements and boundary conditions

Geometry of workpiece	
Outside radius of workpiece	= 33 (mm)
Inside radius of workpiece	= 7.94 (mm)
Mandrel radius	= 7.94 (mm)
Outside radius of forged tube	= 30 (mm)
Reduction in area (percent)	= 18.20 %
Die configuration	
Length of die land	= 12.7 (mm)
Die inlet angle (α)	= 6°
Process condition	
Friction	= rigid Coulomb friction
Friction coefficient	= 0.05
Contact stiffness	= 4.0E6 (N/mm)

Table 2 Forging geometry and process condition [8]

Cutting Speed (rpm)	Feed rate (mm/rev)	Depth of cut (mm)	Tool radius (mm)
310	0.104	0.762	0.397

Table 3 Machining conditions [14]

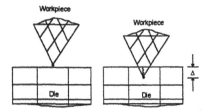

Fig. 1 Schematic of contact problem [5]

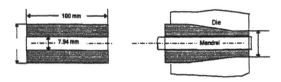

(a). Mil 11595 alloy ingot   (b). Radial forging process

Fig. 2 Dimension of specimen

Fig. 3 Finite element model for radial forging simulation

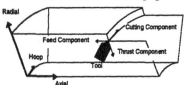

Fig. 4 Schematic view of machining on the forged part

Fig. 5 the instantaneous effective strain distribution at stage 2

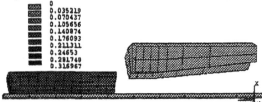

Fig. 6 The effective strain distribution after forging process

Fig. 7 The instantaneous axial stress distribution at stage 3

Fig. 8 The instantaneous radial stress distribution at stage 3

Fig. 9 The instantaneous hoop stress distribution at stage 3

Fig. 10 The instantaneous axial stress distribution at stage 5

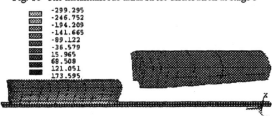

Fig. 11 The instantaneous radial stress distribution at stage 5

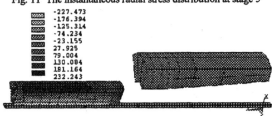

Fig. 12 The instantaneous hoop stress distribution at stage 5

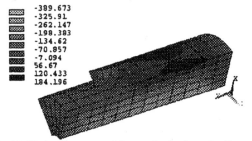

Fig. 13 The instantaneous axial stress distribution at loading stage

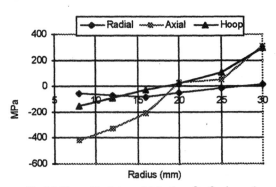

Fig. 14 The residual stress distribution after forging and machining simulation

PLACING SMD PARTS ON ELECTRONIC CIRCUIT BOARDS WITH PARALLEL ROBOTS - CONCEPTS AND CALCULATORY DESIGN

Jürgen Hesselbach, Hanfried Kerle

Institute of Production Automation and Handling Technology
Technical University Braunschweig
P. O. Box 3329, D - 38023 Braunschweig, Germany

Abstract: The problem of placing SMD parts on electronic or printed circuit boards is mostly solved today by using automatically and rapidly working special machines - which is adequate for mass production with large run quantities - or by using conventional serial robots with big moved masses - which should be changed, especially in the case of small or medium run quantities. Two alternative types of robots are presented here: parallel robots that can meet all technical requirements.

Keywords: Robotic manipulators - Assembly robots - Robot kinematics - Positioning systems - Mechanisms

1. INTRODUCTION

The manufacture of electronic circuit boards has increased remarkably during the last years. This is due to the more and more smaller becoming dimensions of electronic parts (resistances, capacitors, etc.) in the light of reducing weight and size. The demand for place units for wireless SMD parts (SMD = Surface Mounted Devices) has grown on an international scale by 12%, in Germany by around 20% (N. N., 1995).

The place technology used for SMD parts comprises two steps: In the first step the SMD part is put down on the board - the metallic ends of the SMD part match at the place foreseen with some printed metal foil paths on the board - and is fixed by an adhesive substance; in the second step the SMD part is fixed permanently and electrically in a soldering process, e. g. flow soldering, Fig. 1 (Wies, 1996).

The two steps are normally done automatically by special handling units (pick & place devices) or by robots, e. g. of the SCARA type. Besides, there have to be feeders for the SMD parts in the periphery of the robot.

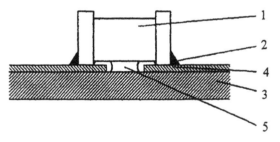

Fig. 1   Placing a SMD part 1 on a circuit board 3
(2: solder, 4: metal foil path, 5: adhesive)

The task of handling SMD parts of a few millimetres in size imposes special requirements on the performance of the robot:

- High precision when giving position and orientation to the working platform (WP) with gripper or tool,
- high velocity of the WP within the workspace,
- small dimensions of the robot and its links compared with those of the SMD parts to be handled.

In total these points come close already to the characteristic specifications of a robot for micro-assembly purposes (Hesselbach, et al., 1996; Hesselbach, et al., 1997). The question arises which robot is the best to meet all these requirements. Apart from single and expensive solutions we nowadays must realize that conventional serial robots consisting of a series of links or arms have reached the limits of their workabilities, whereas parallel robots with several arms between a fixed platform (base platform (BP)) and the WP become a more and more interesting alternative. The structure of parallel robots is based on closed-loop kinematic chains which provide the following technical advantages:

- All the main drives can be located in the BP and therefore need not be moved.
- The structural stiffness is high.
- The mechanical construction is simple and compact, many parts of it are of the same kind and can easily be exchanged.
- Backlash in joints and gears is not cumulative.

These advantages recommend the use of parallel robots for handling and placing tasks of high precision and velocity. Concepts of design must include the structure, namely the type of links and joints and the number of degrees of freedom (DOF) dependent on these elements. After having chosen some dimensions it is possible to evaluate geometric and kinematic relations by simulation procedures in order to adapt the results to the demands of the task of placing SMD parts on a circuit board with prescribed dimensions.

Taking as an example a positioning space of a rectangular prism form with the dimensions 750 mm × 450 mm in an X-Y-plane and 150 mm in Z-direction perpendicular to that plane a feasibility study based on parallel robots shall be presented on the following pages. The bottom area of the prism limits the region for arranging electronic circuit boards to be placed with SMD parts.

2. STRUCTURAL SYNTHESIS

At the beginning it is worthwhile thinking about how many DOF F of the parallel structure will be necessary. In case there are no redundancies the number F (number of independent drives) coincides with the number of global coordinates that describe the position and orientation in space of the WP or its Tool Centre Point (TCP) P. We need F = 3 for placing a SMD part in the given positioning space: X_P, Y_P, Z_P. However, it will be necessary to turn the part by an angle φ around the Z-axis when it is taken out of the feeder and must be placed and/or

corrected in its orientation on the circuit board. This means that the number of DOF is at least F = 4[1].

There are several possibilities to choose 4 global coordinates of the WP; it depends further on the type of the parallel robot: planar or spatial, fully-parallel or hybrid (parallel-serial), Table 1.

Table 1 Distributed global coordinates of the WP

No.	Basic	Additional	Robot type
1	2: X_P, Y_P	2: Z_P, φ	planar/hybrid
2	3: X_P, Y_P, φ_1	1: φ_2	planar/hybrid
3	3: X_P, Y_P, Z_P	1: φ	spatial/hybrid
4	4: X_P, Y_P, Z_P, φ	0: -	spatial/fully-parallel

Case No. 2 in Table 1 points out the fact that a first rotation φ_1 of the WP around the Z-axis is achievable with a planar robot with turning and sliding joints only and all links parallel to the X-Y-plane. Such a planar robot of the type DELTA (Gosselin, et al., 1988) has three DOF (rotary or translatory) but the angle φ_1 is very limited and therefore most be supplemented by an additional rotation unit giving φ_2. The additional coordinates automatically lead to hybrid types.

Two structures are chosen from Table 1: case No. 1 (planar) and case No. 3 (spatial). Both structures are based on parallelograms thus providing that the WP is always led parallel with constant orientation in the X-Y-plane or parallel to it.

3. STRUCTURAL ANALYSIS

The global coordinates - X_P, Y_P or X_P, Y_P, Z_P respectively - of the WP or TCP are united in the vector **x** (x_1 = X_P, etc.) whereas the rotational or translational coordinates of the drives - q_1, q_2 or q_1, q_2, q_3 respectively - are components of the vector **q**. The components of both vectors are related by closed-loop equations (constraint conditions) expressed by a one-column matrix **F**:

$$\mathbf{F(x, q)} = \mathbf{0} \qquad (1)$$

(Gosselin, et al., 1990). The normally nonlinear equations resulting from the constraint conditions are solved in two directions, i. e.

- Direct Kinematic Problem (DKP):
  q is given or known, **x** is unknown;

---

[1] With micro-assembly tasks it is nevertheless recommended to use robots having 6 DOF because of the need of fine-adjustment of the parts to be assembled.

* Inverse Kinematic Problem (IKP):
  x is given or known, **q** is unknown.

If Eq. (1) is differentiated with respect to time t, it is easily verified that

$$dF/dt = \mathbf{A} \cdot (d\mathbf{x}/dt) + \mathbf{B} \cdot (d\mathbf{q}/dt) = 0 \qquad (2)$$

is valid, where

$$\mathbf{A} = \partial F/\partial \mathbf{x} \equiv \mathbf{J}_{DKP}, \quad \mathbf{B} = \partial F/\partial \mathbf{q} \equiv \mathbf{J}_{IKP} \qquad (3)$$

are two JACOBIAN matrices which are needed to solve the DKP or IKP correspondingly.

The Eqs. (2) and (3) are not only the basis of velocity calculations but also most important to get an insight into the structural sensitivity of a parallel robot (Kerle, et al., 1997):

> * If **A** is singular, say det(**A**) = 0, it is not possible to solve the DKP or to calculate velocities dx_i/dt (i = 1, 2, ...);
> * If **B** is singular, say det(**B**) = 0, it is not possible to solve the IKP or to calculate velocities dq_i/dt (i = 1, 2, ...).

Singular configurations or already positions of the WP and/or arms in the vicinity of singular configurations must be avoided for mathematical and safety operation reasons (Hesselbach, et al., 1996).

Moreover, if Eq. (2) is raised onto a level of finite differences Δx_i and Δq_i - i. e. as components of the vectors $\Delta \mathbf{x}$ and $\Delta \mathbf{q}$ - we have $\Delta F \approx 0$ and

$$\Delta \mathbf{x} \approx - \mathbf{A}^{-1} \cdot \mathbf{B} \cdot \Delta \mathbf{q} \qquad (4)$$

as well as

$$\Delta \mathbf{q} \approx - \mathbf{B}^{-1} \cdot \mathbf{A} \cdot \Delta \mathbf{x} \qquad (5)$$

with the two structural sensitivity matrices

$$\mathbf{J}_{S(DKP)} = - \mathbf{A}^{-1} \cdot \mathbf{B} \qquad (4a)$$

and

$$\mathbf{J}_{S(IKP)} = - \mathbf{B}^{-1} \cdot \mathbf{A} = [\mathbf{J}_{S(DKP)}]^{-1} \qquad (5a)$$

which help to study the influences of error deviations or resolution differences (increments) of the input side on the output side or vice versa.

4. EXAMPLES

4.1 Concept of a planar parallel robot with 2 DOF

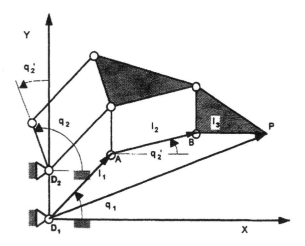

Fig. 2   Planar parallel robot with two DOF

Fig. 2 shows the sketch of a planar parallel robot with two rotary drives characterized by the angles q_1 and q_2. The robot itself lies in the X-Y-plane (or in a plane parallel to it) with the motor axes in D_1 or D_2 respectively. Because of the use of three parallelograms there is no rotation of the WP and there are only three significant lengths of links: l_1, l_2, l_3. The point P belongs to the WP and represents the TCP at the same time or the place where to fix the additional rotational and translational axis for the tool (gripper). The input and output vectors are $\mathbf{q} = (q_1, q_2)^T$ and $\mathbf{x} = (X_P, Y_P)^T$.

Taking D_1 as the origin O of the coordinate axes X and Y the vectors marked by arrows must close, i. e.

$$\underline{OA} + \underline{AB} + \underline{BP} - \underline{OP} = \mathbf{0} \qquad (6)$$

which leads to the components of **F** in Eq. (1):

$$F_1 = l_1 \cdot \cos(q_1) + l_2 \cdot \cos(q_2') + l_3 - X_P = 0, \qquad (7)$$
$$F_2 = l_1 \cdot \sin(q_1) + l_2 \cdot \sin(q_2') - Y_P = 0, \qquad (8)$$

where $q_2' = q_2 - \pi/2$. Because the next steps as regards the evaluation of Eqs. (3) - (5) are straightforward, they are omitted here.

Workspace calculations. At first glance the following quantities may be varied for workspace calculations and investigations:

* Length l_3,
* lengths l_1 and l_2, assumed that $l_1 < l_2$,
* lengths l_1 and l_2, assumed that $l_1 > l_2$,
* lengths l_1 and l_2, assumed that $l_1 = l_2$,
* q_1 and q_2 to achieve symmetrical or asymmetrical forms.

At a second glance it becomes clear that varying l_3 only shifts the point P along the X-axis, therefore l_3 is kept constant at 100 mm. At a third glance (supported by calculations) only the absolute differences $|l_1 - l_2| \neq 0$ and $= 0$ make sense and give different results.

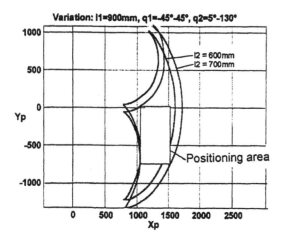

Variation: l1=900mm, q1=-45°-45°, q2=5°-130°

Fig. 3   Two workspace bottom areas of the planar parallel robot of Fig. 2

Fig. 3 as a representative figure for many calculations of similar kind displays two asymmetrical solutions with $l_1 = $ const. $= 900$ mm: One is useful when $l_2 = 700$ mm is chosen, the other solution with $l_2 = 600$ mm is not. To demonstrate this the given positioning area is placed into the region limited by the corresponding curves.

Resolution of the drives. Because of the simple construction of the planar robot it is recommended to estimate the required resolution of the two drives according to the following considerations: The crucial position of the point P is the one where it has the largest distance from the origin O, i. e. $\Sigma l_i = l_1 + l_2 + l_3$. Given Δs as the prescribed position accuracy on the periphery of a circle with the radius Σl_i, the necessary minimum resolution of the drives will be

$$\Delta q_2 \leq \Delta q_1 = \arctan(\Delta s / \Sigma l_i). \qquad (9)$$

From that it follows for instance that the two rotary drives must have a resolution better than $3.4 \cdot 10^{-4}$ degrees if $\Delta s = 10$ μm is prescribed and $\Sigma l_i \geq 1700$ mm is given.

4.2 Concept of a spatial parallel robot with 3 DOF

The idea is to use three translational drives each of them moving a twin set of parallel bars according to Fig. 4. The parallelograms keep the WP always parallel to the BP or X-Y-plane. In addition translational (direct) drives are more and more the basis of innovative solutions for assembly tasks of high precision.

The well-known DELTA robot (Clavel, 1988) also has three parallelograms, but driven by rotary drives. The first who forwarded a patent with translational drives was Hervé (1991). All three translational directions pass through the origin O of the spatial

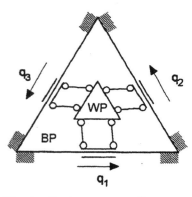

Fig. 4   Basic structure of a spatial parallel robot with 3 DOF (translational drives)

X-Y-Z coordinate system fixed in the BP. For descriptive and calculatory purposes it is sufficient to deal with one branch and a running index i = 1, 2, 3, Fig. 5. The input and output vectors are $\mathbf{q} = (q_1, q_2, q_3)^T$ and $\mathbf{x} = (X_P, Y_P, Z_P)^T$, the latter being coordinates of the TCP.

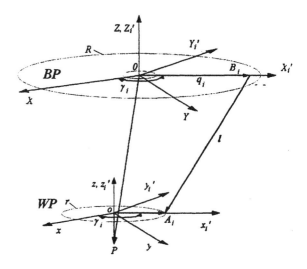

Fig. 5   Vector system represents a spatial parallel robot with 3 DOF and translational drives

The point B_i on the BP represents a slider with the variable distance q_i from O, the point A_i lies on the WP with a constant distance (radius) r from the origin o of the local x-y-z coordinate system of the WP. Both points as the ends of a bar of length l are centres of CARDAN or universal joints having f = 2 joint DOF (angles).

Each component of the one-column matrix $\mathbf{F}$ in Eq. (1) may be written as

$$
\begin{aligned}
F_i = F_i(X_P, Y_P, Z_P; q_i) = \\
= [X_P + (r - q_i) \cdot \cos(\gamma_i)]^2 + \\
+ [Y_P + (r - q_i) \cdot \sin(\gamma_i)]^2 + \\
+ (Z_P + p)^2 - l^2 = 0,
\end{aligned}
\qquad (10)
$$

where $i = 1, 2, 3$ and $-p$ is the local z-coordinate of the point P on the WP. If symmetry is assumed, γ_i comes from $\gamma_i = (i - 1) \cdot (2\pi/3)$.

Workspace calculations. Apart from singularity and structural sensitivity aspects the size and shape of the workspace settles the question about the workability and usability of this robot for the given task of placing SMD parts. Because of the fact that the WP always remains parallel to the BP it is adequate to divide the workspace into several parallel „slices" on the negative Z-axis between a lower limit $-Z_{pmax}$ and an upper limit $-Z_{Pmin}$. In general the following quantities may now be varied:

- Radius r of the WP,
- length l of the arms,
- angles γ_i ($i = 1, 2, 3$),
- symmetric or asymmetric arrangement of γ_i.

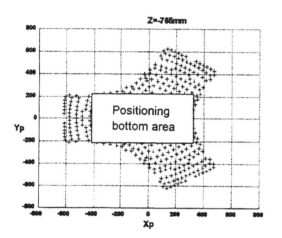

Fig. 6   Workspace slice at $Z = -765$ mm of the spatial parallel robot of Fig. 4 and Fig. 5

It is impossible to give all the details here; but, one result of a tedious row of calculations is shown by Fig. 6: $r = 100$ mm, $l = 1000$ mm, $\gamma_i = (i - 1) \cdot (2\pi/3)$. The workspace slices at the upper limit $Z = -Z_{Pmin} = -765$ mm as well as at the lower limit $Z = -Z_{pmax} = -915$ mm (not shown) both allow to place the given positioning area into the slice areas marked by crosses.

Resolution of the drives. Eq. (5) - the IKP - enables calculating the Δq_i-differences when the Δx_i-components are given. If we assume $\Delta X_P = \Delta Y_P = 0$ and $\Delta Z_P = 10$ μm and 20 μm alternatively on the negative Z-axis, we can determine $\Delta q_1 = \Delta q_2 = \Delta q_3 = \Delta q$ at the most unfavourable position $-Z_{Pmin}$ of the WP. Fig. 7 shows for instance that with $l = A_iB_i = 1000$ mm and a position accuracy of $\Delta Z_P = \Delta s = 10$ μm we need a resolution of the drives of less or better than 1.5 μm! The accuracy of the robot now depends on the accuracy of the measuring system,

on the transmission of motion between machine elements and on the magnitude of backlash in joints - generally on the quality of manufacturing all the single component parts of the robot.

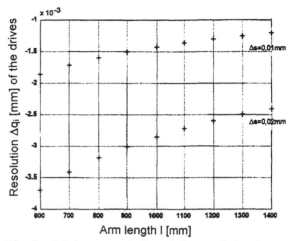

Fig. 7   Minimum resolution of the translational drives for given global accuracy $\Delta Z_P = \Delta s$

CONCLUSIONS

Summing up it is recommended to build the next generation of robots for pick & place operations of SMD parts on the basis of parallel structures. The advantages are eye-catching: Parallel structures combine high stiffness with low moved masses, which is fundamental for tasks of high precision and high velocity in production engineering.

The mathematical tools for computer-aided simulation as regards geometry and workspace, singularity control and sensitivity analysis have been discussed. The tools may be used for selecting the proper drives together with the measuring systems, too.

REFERENCES

N. N. (1995). Drei neue SMD-Bestückautomaten. *Elektronik*, **24** (11), 20.

Wies, D. (1996). Aufbau und Untersuchung von Handhabungsgeräten mit Parallelstruktur für eine SMD-Bestückung. Student's project at the Technical University Braunschweig (not published).

Hesselbach, J., R. Pittschellis, R. Thoben and H.-S. Oh (1996). Handhabungsgeräte für die Mikromontage. *Zeitschrift f. wirtsch. Fertigung*, **91** (9), 437-440.

Hesselbach, J., R. Pittschellis, E. Hornbogen and M. Mertmann (1997). Shape memory alloys for use in miniature grippers. In: *Proc. Conf. on Shape Memory and Superelastic Technology*, Monterey (CA), USA (in print).

Gosselin, C. M. and J. Angeles (1988). The optimum kinematic design of a planar three-degree-of-freedom parallel manipulator. *Jl. of Mech., Transmission, and Autom. in Design,* **110** (1), 35-41.

Gosselin, C. M. and J. Angeles (1990). Singularity analysis of closed-loop kinematic chains. *IEEE-Trans. on Robotics and Automation,* **6** (3), 281-290.

Kerle, H., J. Hesselbach and N. Plitea (1997). Structural sensitivity analysis of parallel robots. In: *Proc. 6th Int. Workshop on Robotics in Alpe-Adria-Danube-Region (RAAD '97),* Cassino, Italy (in print).

Hesselbach, J., H. Kerle and N. Plitea (1996). On some aspects of parallel robots control. In: *Proc. 27th ISIR Symposium,* pp. 683-687, Milan (Italy).

Clavel, R. (1988). DELTA, a fast robot with parallel geometry. In: *Proc. 18th ISIR Symposium,* pp. 91-100.

Hervé, J. M. (1991). Dispositif pour le déplacement en translation spatiale ... . French patent No. 9100286.

DEVELOPMENT OF A TASK-LEVEL ASSEMBLY PLANNING SYSTEM USING DUAL ARMS

H. W. Kim*, B. H. Hwang*, J. S. Park*, I. H. Suh*, B. J. Yi**

*Hanyang Univ. Dept. of Electronics Eng.
**Hanyang Univ. Dept. of Control & Measurement Eng.

Abstract: In this paper, it is proposed that a task-level assembly planning system that consists of Object Data Base(ODB) generator, Target Data Base(TDB) generator, and Task-Level Command Interpreter(TCI). ODB generator provides the information on the dimensions of each part to be assembled. Considering the information of ODB and the assembly sequence of the target object given by 3D GUI environment, TDB generator finds an optimal path of each assembling part while avoiding obstacles. TCI generates a set of primitive robot control commands for the robot to follow the path planned by TDB, where a collision avoidance scheme for dual arms is also considered. Then, these primitive commands are downloaded to our dual-arm control system to perform the given assembly task. To show the validity of our proposed planning system, an assembly work building a house with 13 Lego blocks is demonstrated.

Keywords: Assembly language, Path planning, Obstacle avoidance, Automation, Robot

1. INTRODUCTION

Automation technology for manufacturing processes using robots has been developed in industry. Specifically, automatic assembly becomes a demanding technology due to the increase of labor cost and the complexity of assembling processes. Until now, most of robot languages have been developed for robot operation itself. However, they are not adequate to a novice who is not familiar with the given robot system. Therefore, development of a task-level language being able to describe and program assembly tasks easily would facilitate robot operation by field users(Halperin and Wilson, 1995).

The task-level language is a new type of language that the user only sets start and end position of the part to be assembled, with no knowledge of robot kinematics. Starting from AUTOPASS(Lieberman and Wesley, 1977), several researches on the task-level language have been accomplished. In general, development of general-purpose task-level robot language is very difficult because of different style and noncompatibility of robot languages(Maniere, et al., 1992). In this paper, several task-level commands for assembly tasks are defined and a task-level assembly planning system, which automatically converts the task-level commands to robot commands for control of a SCARA-type dual arms, is proposed.

This paper is organized as follow: in section 2, the task-level commands are defined, and our assembly planning system is briefly explained. In section 3, algorithms for ODB, TDB, and TCI are described. Section 4 explains the collision avoidance method for dual arms. Experimental results and the discussion of future works follow in section 5 and 6, respectively.

2. TASK-LEVEL ASSEMBLY PLANNING SYSTEM

2.1 Definition of Task-Level Command Set

Task-level language should consider the connection relation among parts. Ko and Lee(Ko, et al., 1993) classified the relation of parts into four types such as *Against, Fits, Tight-fits,* and *Contact.* Homem de Mello, Arthur C. Sanderson (Homem de Mello, et al., 1991), J. P. Thoms and P.N Nissanke (Thoms and Nissanke, 1995) introduced a "*relation model graph*" that consists of information on parts and connection.

In this paper, task-level commands are defined with consideration of relation of parts. They include four basic motion commands ; *PICK, PLACE* for position information and *ATTACH, INSERT* for position or sensor information(Cervera, *et al.*, 1995).

These commands can be utilized for several real tasks using SCARA-type dual arms as follows.

PICK robot# <part>

: Pick up the part located at <part> by robot# and move up the end-effector maximally. In case of SCARA-type dual arms, it is assumed that the robot avoids collision with the partially assembled object.

PLACE robot# <part> TO <location>

: Move to <location> maintaining current Z position, then lower end-effector to <location> perpendicularly.

ATTACH robot# <part1> TO <part2>

: Move toward contact location of <part2> while maintaining the contact force in a fixed boundary. If the robot arrives at the goal position, stop and put <part2> to the goal position.

INSERT robot# <part1> INTO <part2>

: Insert <part1> into <part2> while adjusting position and angle errors. If the robot reaches the goal, stop inserting and release <part1>.

The commands, PLACE, ATTACH, and INSERT, have an assumption that the robot picks a part before assembly. Basically, the robot moves along the taught path to assemble the picked part on the object, and moves backward along the same path to pick another part. Usually, any assembly task can be described by combination of PICK and one of the other three task commands.

2.2 The Task-Level Assembly Planning System

The structure of our task-level assembly planning system is divided by two parts. In the first part, the task-level command interpreter interprets the given task-level program, translates the program into robot commands, and downloads it to the robot system for assembly.

In the second part, the target data base generator let the user select assembly sequence by using graphic environment, and then task-level command for a specified target is automatically generated. Then, the information is saved in TDB file. In general, for a specific target, assembly sequences are not unique.

The structure of assembly planning system is shown in Fig. 1 and the role of each component is described as follows:

① TDB (Target Data Base)
   data file that includes task-level commands and robot commands composing of task-level command.
② TCI (Task-level Command Interpreter)
   translates task-level commands into robot commands by using TDB.
③ EDB (Element Data Base)
   data file that includes the geometric information of elementary objects.
④ ODB (Object Data Base)
   data file that includes the 3-D geometric information of each assembling part and the connection information among elementary objects.
⑤ Obstacle Modeling
   converts obstacles described in the task space into the configuration space by using information of ODB.
⑥ Path Planning
   generates a path to move from the start position to the goal position

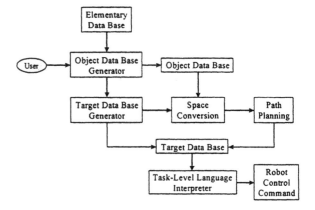

Fig. 1. System Block Diagram

3. THE DATA BASE

3.1 Parts Modeling and Object Data Base

A mechanical assembly is a composition of interconnected parts forming a stable unit. Each part is a solid rigid object, that is, its shape remains unchanged. For assembly planning, modeling of parts is needed. So the information of parts such as length, width, height or grasp point, and grasp configuration is saved in a file.

In this paper, parameters and connection information of elementary objects are saved in EDB to represent geometric information of parts as shown in Fig. 2. Information on the connection among elementary objects, grasp points, postures of robot are saved in ODB. The structure of ODB is given by

```
struct data_base {
    char *name;          // part's name
    char *cur_loc;       // part's feeding location
```

```
double grasp_point[3];      // grasp point (x, y, z)
double grasp_angle;         // 4th joint angle
struct item {
    int cur_obj;            // kinds elementary object
    double length;          // object's length
    double width;           // object's width
    double height;          // object's height
    double radius;          // object's radius
    double dx, dy, dz;      // offset from origin of
                            // part's coordinate
    int connect;            // connection method
                            // (OR, XOR, XNOR)
} feature[ ];
} object[ ];
```

where "cur_loc[]" denotes the current position of the feeder (e.g., conveyor system) which supplies parts, and "feature[]" represents the information of elementary objects and connection relation(OR, XOR). "grasp_point" and "grasp _angle" denote the center position of the end-effector and 4th joint angle of the robot, respectively.

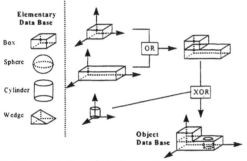

Fig. 2. Representation of 3D Connection Relation

3.2 Constructing Target Data Base

Once ODB is constructed, then construction of TDB follows. TDB is a data base including assembly sequences, assembly path, connection point of parts, and task-level commands and corresponding robot commands. By using TDB generator, users select assembly sequence and the assembled location of parts in graphical method, and generates the path for assembly. The information is saved in TDB according to the following data structure:

```
struct target_data {
    char *name;            // target name
    int connect_num;       // number of parts constructing
                           // a target
    struct con {
        char *task;        // task command's name
        int obj_num;       // part's number to be assembled
        int robot_num;     // which robot assembles a part?
        int mate_part;     // part's number to be mated
        char *grasped;     // part's name to be assembled
    } connect[ ];
} target;
```

4. PATH PLANNING

4.1 Construction of Obstacle Space

In general, the target object is constructed by assembling a number of parts. Then, at each step of assembly, a partially constructed object becomes an object to the moving robot. For successful assembly, collision between the moving robot and the partially constructed object should be avoided. Therefore, the path planning of the robot should be performed in the configuration space that is formed by using 3D information in ODB. In the configuration space, a safe path means a continuous path that does not intersects with any obstacles (Latomb, 1990).

The direction of Z axis of SCARA type robot is constrained vertically and the position of end effector is determined by 1st and 2nd joint angle θ_1, θ_2 respectively. So, obstacle space can be constructed without searching all joint angles as the following algorithm.

< Algorithm for constructing configuration space >

1. Determine the 4th joint angle θ_4 according to the orientation of the part to be assembled.

2. Obtain the smallest polyhedron that encompasses a part by using geometric information of each part.

3. Determine the bound of the joint displacement θ_3 according to the size of the polyhedron calculated in step 2.

4. Project the polyhedron to X-Y plane.

5. Check whether the projected polygon of moving part intersects with other parts previously assembled while increasing 1st and 2nd joint angle by unit degree. If polygons intersect each other, find the location of the collision by searching the bound calculated in step 3. Collision points are saved in data base of the configuration space.

4.2 Path Planning

Initially, set the joint angles as the initial position when the robot picks up a part and then set the joint angles as the goal position when the assembly is completed. The path that connects the two positions must not collide with other parts already assembled, and the assembly should be made as quick as possible.

In this paper, it is proposed that a path planning algorithm that takes into account of minimum assembly time and minimum distance with relatively little computation.

1. Check if the line connecting initial position (P_s) and the goal position (P_g) intersects obstacles.

2. If the line intersects obstacles, search the direction that does not collide with obstacles along the fundamental directions.

3. Select the direction that has the closest distance to the goal position among the fundamental directions having no collision.

4. Set the next via point as the closest position to the goal position.

5. Move to the via point.

6. Repeat step 1 to step 6 until the goal position is reached.

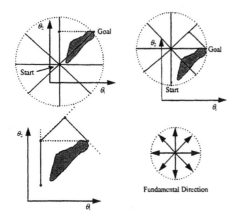

Fig. 3. Determination of Via Points

According to the above algorithm, a path with two via points is obtained in Fig. 3. However, it is observed that the path is not the shortest one. Now, to optimize the distance, the path is searched backward from the goal position to the initial position.

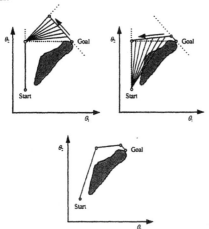

Fig. 4. Distance Optimization

First, let the goal position be a current position, the previous via point be **a**, the via point previous to **a** be **b**, and the point that moves along the line connecting the goal position to the point **a** be **c**. Then, move the point **c** by the unit magnitude until the line connecting **c** and **b** does not collide with the obstacle. Fig. 4 shows the resulting path.

4.3 Collision Avoidance

Robots working in the same workspace should avoid collision each other. In general, semaphore(in other word, flag) is used to control the sequence of multiple processors. In this paper, collision avoidance is achieved by using the resources supported by the robot control system.

Samsung FARA SRC421 robot controller supports number of registers. Two robot arms share information through these registers. When the assembly task starts, each robot program sets three registers with certain values and check the registers whenever execution of each task-level command is completed. To clarify it, an example of robot-level program is introduced :

Robot1	*Robot2*
WHILE(@R3 != 99)	WHILE(@R3 != 99)
DELAY 0.1	DELAY 0.1
ENDWHILE	ENDWHILE
READY	READY
@R1 = 0	@R2 = 0
(other robot commands)	*(other robot commands)*
WHILE(@R2 == 10)	WHILE(@R1 == 10)
DELAY 0.1	DELAY 0.1
ENDWHILE	ENDWHILE
@R1 = 10	@R2 = 10
(other robot commands)	*(other robot commands)*

In above robot program, "@" means shared register. If robot 2 initially assembles a part in the common workspace, the program sets the register R1 as zero and the register R2 as 10, and then robot 1 waits until robot 2 finishes its job, which can be monitored by the state of register 2. When the value of register 2 becomes 0, robot 2 stops while robot 1 starts its job.

5. EXPERIMENTAL RESULTS

In this paper, the proposed automatic assembly planning system is constructed using Motif program, which is run in Lynx OS mounted on IBM-PC. Initially, the user assembles the parts graphically on a screen, then the planning system generates task-level language and converts it to robot-level commands automatically. These converted commands are downloaded to the robot controller for assembly. Fig. 5 is a front view of dual-arm robot system used in experiment.

Fig. 5. Dual-Arm Robot System

To show the validity of our proposed planning system, a practical assembly job using 13 pieces of Lego block is performed. Fig. 6 is an example of target object.

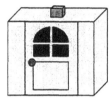

Fig. 6. House Model Example

The following three figures show the three steps of assembly job ; generating the object date base by using ODB generator(Fig. 7), assembly on a screen and path planning in TDB generator(Fig. 8), and conversion the task-level program into the robot program in TCI(Fig. 9). Finally, the resulting programs are downloaded to robot controller.

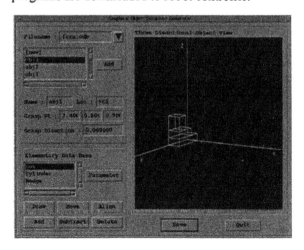

Fig. 7. Graphical ODB Generator

When the TDB is constructed using TDB generator, the sequences are selected by users graphically and the task-level program is generated automatically as shown in the left-side of Fig. 9, and then the robot program for each robot is automatically generated, as shown in the right-side of Fig. 9, without any aid from the users.

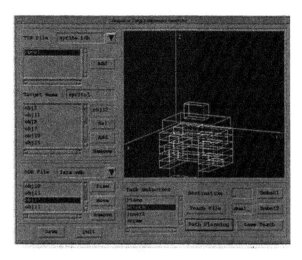

Fig. 8. Graphical TDB Generator

Fig. 9. Task-level Command Interpreter

Fig.10 shows the conversion of task-level commands to robot commands by using task-level command interpreter(TCI).

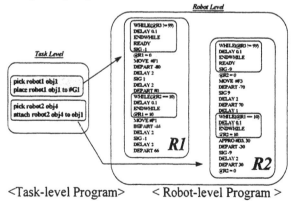

<Task-level Program>   < Robot-level Program >

Fig. 10. Converted Robot Command

In the robot-level programs, the codes surrounded by round boxes are used for collision avoidance. The generated robot programs are downloaded to robot controller for assembly. Fig. 11 demonstrates the given assembling job for house model.

Fig. 11. Robot that assembles parts

6. CONCLUSION AND FUTURE WORK

In this paper, an assembly planning system is proposed. A task-level language was developed on the purpose of facilitating robot operation by field users. The proposed system consists of Object Data Base(ODB) generator, Target Data Base(TDB) generator, and Task-Level Command Interpreter (TCI). The proposed scheme can be run real-time and its practicality was proven experimentally by constructing a house model using 13 Lego blocks.

The future works involve complex parts modeling, optimal sequence generation automatically among the possible assembly sequences, and sensor-based task-level language using information from vision and force sensors.

REFERENCES

E. Cervera, Angel P. del Pobil, E. Marta and M. A. Serna(May, 1995). A Sensor-Based Approach for Motion in Contact in Task Planning. *Proc. of the IEEE Int. Conf. on Robotics and Automation*, pp. 468-473.

D. Helperin and R. H. Wilson(May, 1995). Assembly Partitioning along Simple Path : the Cost of Multiple Translations. *Proc. of the IEEE International Conference on Robotics and Automation*, pp.1585-1592.

N. Y. Ko, B. H. Lee and M. S. Ko(July, 1993). An Approach to Robot Motion Planning for Time-Varying Obstacle Avoidance Using View-Time Concept. *ROBOTICA*, **Vol.11**, pp.315-327.

L. S. Homem de Mello and A. C. Sanderson(April, 1991). Representations of Mechanical Assembly Sequences. *IEEE Trans. On Robotics and Automation*, **Vol. 7**. No. 2. pp.211-227.

J. C. Latomb(1990). Robot Motion Planning.

L. I. Lieberman and M. A. Wesley(1977). AUTO-PASS : An Automatic Programming System for Computer Controlled Mechanical Assembly. *IBM J. Res. Develop*, **Vol. 21**, No. 4, pp.321-333.

E. C. Maniere, B. Espiau and E. Rutten(July, 1992). A Task-Level Robot Programming Language and its Reactive Execution. *Proc. of the IEEE/RSJ Int. Conf. on Intelligent Robots and Systems*, pp. 2751-2756.

J. P. Thoms and P. N. Nissanke(May, 1995). A Graph-based formalism for Modelling Assembly Tasks. *Proc. of the IEEE Int. Conf. on Robotics and Automation*, pp.1296-1301.

ON THE ISOTROPIC CONFIGURATIONS OF SIX DEGREES-OF-FREEDOM PARALLEL MANIPULATORS

Ronen Ben-Horin, Moshe Shoham and Joshua Dayan

Department of Mechanical Engineering

Technion - Israel Institute of Technology

Technion City, Haifa 32000, Israel

E-mail: shoham@hitech.technion.ac.il

Abstract: Isotropy is considered as an important feature in the design of all robotic manipulators, whether serial or parallel. When focusing on kinematics considerations, rather than dynamics, the condition number of the Jacobian matrix can be used to quantify the isotropy of the manipulator. For a parallel manipulator who's links forms a circular platform, an isotropic configurations is shown to exist and an appropriate design will make it realizable.

Keywords: Parallel, Robot, Kinematics, Manipulator, Manoeuvrability

1. INTRODUCTION

The end-effector motion of a robot manipulator is usually obtained as a result of a coordinated motion of several motors. Along different directions - different coordination of the motors actions is required. It is desirable that theses end-effector motions are performed by similar speeds of the motors, since large variations of the speeds reduce the end-effector accuracy and its ability to exert the same velocity and force in all directions. This feature is known as the manipulator's "manipulability" and it has been investigated by several investigators (see, for example, Yoshikawa, 1985a,b). This feature is a local one and it differs from the kinematically related concept of "dexterity", which means the ability to reach a point from different directions (Gupta and Roth, 1982).

To define the manipulability, the Jacobian matrix is used. It maps the end-effector velocity into the joint rates. The most commonly used measures of local manipulability are the Jacobian determinant, the condition number and the minimum singular value of the Jacobian matrix (Klein and Blaho, 1987). From many aspects of view the condition number appears to be most suitable for expressing the local dexterity of parallel manipulators. Manipulators whose Jacobian matrix is capable of attaining a condition number of unity, being called isotropic (Salisbury and Craig, 1982). The Concept of

kinematic isotropy has been used as a criterion in the design of planar and spherical parallel manipulators (Gosselin and Angeles, 1988; Gosselin and Lavoie, 1993; Mohammadi-Daniali and Zsombor-Murray, 1994).

In this paper, the isotropic condition of a six degree-of-freedom parallel manipulator, often referred to as the generalized Stewart Platform (Innocenti and Parenti-Castelli, 1993), is derived.

Fig. 1: A six degrees-of-freedom parallel manipulator.

2. AN ARCHITECTURE OF A SIX DEGREES-OF-FREEDOM PARALLEL MANIPULATOR

Figure 1 depicts a six degree-of-freedom parallel manipulator, comprises six extensible links. Each link has a prismatic actuator and is connected between a moveable and a stationary platforms. The link ends are jointed one to the stationary platform and one to the moveable platform by a universal and a spherical bearings, respectively, although, two spherical joints are sometimes used. This mechanism is referred to as The Stewart Platform, even though earlier versions of it were known (Merlet, 1994).

3. PARALLEL ROBOT VELOCITY TRANSFORMATION

The Jacobian matrix of the manipulator is the linear transformation relating the joints rates of the manipulator and its end-effector velocity. For parallel manipulators the Jacobian matrix was given by Merlet (1989) and by Pitten and Podhorodeski, (1993) and, for the sake of presenting a complete case, it is summarized here, too.

Let vpi be the velocity of each of the link attachment points on the moveable platform, then, the following relation exists:

$$\mathbf{v}_{pi} = \mathbf{v} + \omega \times \mathbf{p}_i \qquad (1)$$

where, $\mathbf{v}$ is the linear velocity of a point on the moveable platform, which, instantaneously, coincides with the origin of the frame of reference; $\mathbf{w}$ is the angular velocity of the moveable platform; $\mathbf{p}_i$ is the position vector locating link i attachment point on the moveable platform, with respect to the desired frame of reference.

The joint rates $\dot{\mathbf{q}}_i$ are the projections of the velocities $\mathbf{p}_i$ onto a unit vector $\mathbf{e}_{si}$ along the sliding direction of the corresponding prismatic joint:

$$\dot{\mathbf{q}}_i = \mathbf{e}_{si} \cdot \left(\mathbf{v} + \omega \times \mathbf{p}_i\right) = \mathbf{e}_{si} \cdot \mathbf{v} + \mathbf{e}_{si} \cdot \omega \times \mathbf{p}_i$$
$$(2)$$

or, using properties of scalar triple products:

$$\dot{\mathbf{q}}_i = \mathbf{e}_{si} \cdot \mathbf{v} + \mathbf{p}_i \times \mathbf{e}_{si} \cdot \omega \qquad (3)$$

Arranging the relations of (3) into a single matrix equation, the joints rates of a six degrees-of-freedom parallel manipulator are:

$$\dot{q} = \begin{bmatrix} \dot{q}_1 \\ \vdots \\ \dot{q}_6 \end{bmatrix} = \begin{bmatrix} e_{s1}^T & \left(p_1 \times e_{s1}\right)^T \\ \vdots & \vdots \\ e_{s6}^T & \left(p_6 \times e_{s6}\right)^T \end{bmatrix} \begin{bmatrix} v \\ w \end{bmatrix} = \mathbf{J} \begin{bmatrix} v \\ w \end{bmatrix}$$
$$(4)$$

where $(\cdot)^T$ denotes the transpose operation. The matrix $\mathbf{J}$ represents the Jacobian matrix of a parallel manipulator. Note that the transformation (4) represents a mapping from the end-effector velocity to the manipulator's joint rates, whereas the transformation for a serial-chain manipulator is a mapping in the opposite direction (joint rates to end-effector velocity). Also, note that the matrix $\mathbf{J}^T$ represents the plucker line coordinates (Hunt, 1978) of the line of action of the corresponding actuated joint.

4. ISOTROPY CONDITIONS

Let the singular value decomposition (Klema and Laub, 1980) of J be:

$$\mathbf{J} = \mathbf{U}\Sigma\mathbf{V}^T \qquad (5)$$

where, $\mathbf{U}$ and $\mathbf{V}$ are orthonormal matrices, whose columns are the so called left and right singular vectors of $\mathbf{J}$, respectively, and $\mathbf{S} = \mathrm{Diag}(s_{max},...,s_{min})$ is a diagonal matrix containing the real non-negative singular values of $\mathbf{J}$.

If $\mathbf{J}$ is invertible then:

$$\mathbf{J}^{-1} = \mathbf{V}\Sigma^{-1}\mathbf{U}^T \qquad (6)$$

Since $\mathbf{U}$ and $\mathbf{V}$ are orthonormal, the singular values of $\mathbf{J}^{-1}$ are the reciprocals of the singular values of $\mathbf{J}$. Physically, the maximum singular value represents the maximum gain of the matrix, in terms of the Euclidean distance, that is:

$$\|\mathbf{J}\|_2 = \max_{\|\mathbf{x}\|_2=1}\|\mathbf{J}\mathbf{x}\|_2 = \sigma_{max} \qquad (7)$$

Similarly, the minimum singular value represents the minimum gain of the matrix. In other words, over the entire domain of unit input vectors x, the magnitude of Jx must lie between the minimum and the maximum singular values, inclusive.

For a square, nonsingular matrix $\mathbf{J}$, the condition number is defined as:

$$\mathrm{Cond}(\mathbf{J}) = \|\mathbf{J}\|\|\mathbf{J}^{-1}\| \qquad (8)$$

and thus the condition number with respect to the spectral norm, is simply the ratio of maximum to minimum singular values. Geometrically, the space of unit input vectors x can be represented as a unit hyper sphere, and the space of image vectors Jx can be represented as a hyper ellipsoid. The lengths of the axes of the hyper ellipsoid are given by the singular values of the matrix $\mathbf{J}$. Note that because $\mathrm{Cond}(\mathbf{J})=\mathrm{Cond}(\mathbf{J}^{-1})$ it is unimportant in which direction the Jacobian was defined.

For good dexterity, it is desirable to have the Jacobian hyper ellipsoid as close to a hyper sphere as possible. At that point (when the Jacobian is represented by a hyper sphere), the manipulator's ability to generate forces and velocities is independent of direction and, therefore, Salisbury and Craig (1982) called it the "isotropic point". Hence, a Jacobian matrix is isotropic if all its singular values are identical and non-zero. This is equivalent to saying that, if $\mathbf{J}$ is isotropic then:

$$\mathbf{J}^T\mathbf{J} = \sigma^2\mathbf{I}_6 \qquad (9)$$

where, s is the common singular value and $\mathbf{I}_6$ is the identity matrix of order six.

Because the manipulator describes for both positioning and orienting tasks, its Jacobian entries are dimensionally inhomogeneous. Indeed, the ith column of $\mathbf{J}^T$ is composed of the plucker coordinates of the line of action of the corresponding actuated joint. Three of these coordinates, associated with the unit vector along the actuated joint, are dimensionless, whereas the remaining three, corresponded to the moment of these axes around the origin of the frame of reference, have units of length. This dimensional inhomogeneity leads to dimensional inconsistency when evaluating the manipulator condition number. In fact, in this case three singular values of the Jacobian are dimensionless, whereas the remaining three have units of length. Therefore, it is impossible to order the singular values from largest to smallest. In order to cope with this problem Angeles (1992) and Pitten and Podhorodeski (1993) suggested to write the point-velocity equations of the manipulator in a nondimensional form, by suitably dividing both sides of the velocity equation by a natural length of the manipulator. To be effective, the measure of length for the translational velocities must reflect the perceived cost of rotational vs. translational velocity. Furthermore, it is critical to note that the translational velocity refers to the velocity of the point, which instantaneously coincides with the reference origin. The isotropic configurations, resulting from the dimensionless Jacobian, will provide optimal dexterity at the end-effector point which coincides with the reference origin. These optimal configurations will be functions of the choice of rotational vs. translational weights and of the choice of the reference frame location. Choosing a shorter natural length emphasizes translational terms, and conversely, a longer natural length emphasizes rotational terms.

We consider the joints on the platform to be distributed in circles. This symmetry distribution is better from work-space point of view. Then, if the radius of the moveable platform is chosen as the desired natural length, the cost associated with translational velocity will be equivalent to that associated with rotational velocity, as seen from equation (4). The following dimensionless Jacobian is obtained when the radius of the moveable platform is used as the natural length :

$$
\mathbf{J} = \begin{bmatrix} \mathbf{e}_{s1}^T & \left(\mathbf{e}_{p1} \times \mathbf{e}_{s1}\right)^T \\ \vdots & \vdots \\ \mathbf{e}_{s6}^T & \left(\mathbf{e}_{p6} \times \mathbf{e}_{s6}\right)^T \end{bmatrix} \tag{10}
$$

Where, $\mathbf{e}_{pi}$ is a unit vector from joint i of the movable platform to the origin of the platform-attached coordinate system.

Substitute the dimensionless Jacobian $\mathbf{J}$ from equation (10) into equation (9) yields:

$$
\begin{bmatrix} \sum_{i=1}^{6} \mathbf{e}_{si}\mathbf{e}_{si}^T & \sum_{i=1}^{6} \mathbf{e}_{si}\left(\mathbf{e}_{pi} \times \mathbf{e}_{si}\right)^T \\ \sum_{i=1}^{6} \left(\mathbf{e}_{pi} \times \mathbf{e}_{si}\right)\mathbf{e}_{si}^T & \sum_{i=1}^{6} \left(\mathbf{e}_{pi} \times \mathbf{e}_{si}\right)\left(\mathbf{e}_{pi} \times \mathbf{e}_{si}\right)^T \end{bmatrix}
$$
$$
= \sigma^2 \mathbf{I}_6 \tag{11}
$$

If equation (11) is to hold, then the first diagonal block of the foregoing matrix equals to $s^2\mathbf{I}_3$, where $\mathbf{I}_3$ is the identity matrix of order three, and the off-diagonal block must vanish; i.e.:

$$
\sum_{i=1}^{6} \mathbf{e}_{si}\mathbf{e}_{si}^T = \sigma^2 \mathbf{I}_3 \tag{12a}
$$

$$
\sum_{i=1}^{6} \left(\mathbf{e}_{pi} \times \mathbf{e}_{si}\right)\mathbf{e}_{si}^T = \mathbf{0}_3 \tag{12b}
$$

where $\mathbf{0}_3$ is the null matrix of order three. Taking the trace of both side of equation (12a) yields:

$$
\sigma = \sqrt{2} \tag{13}
$$

Equation (13) shows that in every isotropic configurations of the manipulator, where the links are distributed in a circle, all the singular values are identical and equals to $\sqrt{2}$.

The isotropy of the lower diagonal block of equation (11) leads to:

$$
\sum_{i=1}^{6} \left(\mathbf{e}_{pi} \times \mathbf{e}_{si}\right)\left(\mathbf{e}_{pi} \times \mathbf{e}_{si}\right)^T = \sigma^2 \mathbf{I}_3 \tag{14a}
$$

while the upper off-diagonal block must vanish; i.e.:

$$
\sum_{i=1}^{6} \mathbf{e}_{si}\left(\mathbf{e}_{pi} \times \mathbf{e}_{si}\right)^T = \mathbf{0}_3 \tag{14b}
$$

Equations (12b), (14a) and (14b) are satisfied .if and only if:

$$
\left\{\forall i, i = 1,2,...,6 \mid \mathbf{e}_{pi} \perp \mathbf{e}_{si}\right\} \tag{15}
$$

From equation (15) one can see, that any isotropic configuration of the manipulator demands that the sliding directions of the prismatic joints are tangent to the platform circle, when projected onto a horizontal plane, e.g., the stationary plane. Therefore, it leads to a general result that a properly designed six degrees-of-freedom parallel manipulator can have isotropic configurations. This result is compatible with the numerical analysis of Pitten and Podhorodeski (1993) and of Zanganeh and Angeles (1995). Two symmetrical structures of a parallel manipulator with six degrees-of-freedom, at isotropic configuration, are shown in figure 2.

CONCLUSIONS

The main conclusion drawn from the above analysis shows that any parallel manipulators can be designed in such a way that the isotropy property is exhibited. At the required point of isotropy the vector connecting this point to the articulation point and the vector along the link are orthogonal.

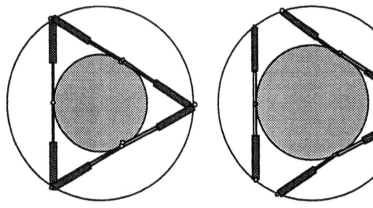

Fig. 2: Isotropic configuration of two six degrees-of-freedom parallel manipulators

REFERENCES

Angeles J. (1992). The Design of Isotropic Manipulator Architectures in the presence of Redundancies. *the Int. J. Robotics Res.*, 11 (3), 196-201.

Gosselin C. and Angeles J. (1988). The Optimum Design of a Planar Three-Degree-of-Freedom Parallel Manipulator. *ASME J. mech., Trans., Auto. in Design*, 110, 35-41.

Gosselin C. and Lavoie E. (1993). On the Kinematic Design of Spherical Three-Degree-of-Freedom Parallel Manipulators. *the Int. J. Robotics Res.*, 12 (4), 394-402.

Gupta K. C. and Roth B. (1982). Design Considerations for Manipulator Workspace. *ASME J. Mech. Design*, 104 (4), 704-712.

Hunt K. H. (1978). *Kinematic Geometry of Mechanisms*. Clarendon Press, Oxford.

Innocenti C. and Parenti-Castelli V. (1993). Forward Kinematics of the General 6-6 Fully Parallel Mechanism: An Exhaustive Numerical Approach Via a Mono-Dimensional-Search Algorithm. *ASME J. Mech. Design*, 115, 932-937.

Klema V. C. and Laub A. T. (1980). The Singular Value Decomposition: Its Commutation and Some Applications. *IEEE Trans. Automatic Contr.*, 25 (2),164-176.

Klein C. and Blaho B. (1987). Dexterity Measures for the Design and Control of Kinematically Redundant Manipulators. *the Int. J. Robotics Res.*, 6 (2), 72-83.

Merlet, J. P. (1994). Parallel Manipulators: State of the Art and Perspectives. *Adv. Robotics*, 8, 586-594.

Mohammadi-Daniali H. R. and Zsombor-Murray P. J. (1994). The Design of Isotropic Planar Parallel Manipulators. *Proc. 1st World Automation Congress, Maui*, 2, 273-280.

Pitten K. H. and Podhorodeski R. P. (1993). A Family of Stewart Platforms with Optimal Dexterity. *J. Robotic Sys.*, 10 (4), 463-479.

Salisbury J. K. and Craig J. J. (1982). Articulated Hands: Force Control and Kinematic Issues. *the Int. J. Robotics Res.*, 1 (1), 4-12.

Yoshikawa T. (1985a). Dynamic Manipulability of Robot Manipulators. *J. Robotic Sys.*, 2, (1), 113-124.

Yoshikawa T. (1985b). Manipulability of Robotic Mechanisms. *the Int. J. Robotics Res.*, 4 (2), 3-9.

Zanganeh K. E. and Angeles J. (1995). On the Isotropic Design of General Six-Degree-of-Freedom Parallel Manipulators. In: *Computational Kinematics* (Merlet J. P. and Ravani B. eds.), pp. 213-220.

INTEGRATED ROBOT NAVIGATION IN CAD ENVIRONMENT

Gernot Kronreif, Man-Wook Han

Institute for Handling Devices and Robotics
Vienna University of Technology
Floragasse 7a, A-1040 Vienna, AUSTRIA
Tel: +43-1-5041835, Fax: +43-1-5041835-9
E-mail: {kronreif, han}@ihrt1.ihrt.tuwien.ac.at

Abstract: The proposed navigation system for mobile robot platforms is designed for both application of mobile systems in unknown environment as well as for pre-defined surroundings. A special Graphic User Interface (GUI) should provide a convenient access to the real as well as to the simulated robot, and to the representation of the environment. Through this GUI, the user can send commands to the robot, monitor command execution by seeing the robot actually moving on the screen, visualize instantaneous and cumulated sensor data. The user should also be able to create and modify a simulated environment by means of standard CAD functions, and use it to test robot programs.

Keywords: Mobile Robots, Navigation, Path Planning, Control, Simulation

1. INTRODUCTION

Mobile robot systems are gaining more and more importance in both, service sectors (with one special emphasis on applications in healthcare) and in manufacturing areas. These applications have all in common the tele-operation or semi-autonomous operation of robot platforms in various scenarios, among them:

- Factory automation projects, where robot vehicles are used to transport components between distant machining and (dis-) assembly sites;

- Operation in hazardous environments, including the deployment of mobile robots in mine excavation, and the use of autonomously navigation robots inside nuclear facilities for inspection purposes;

- Planetary and space exploration, using autonomous rovers and probes, and the employment of telerobotic systems in space construction;

- The use of robots for deep-sea surveying and prospecting;

- Assistance for the handicapped, application of „service robots" in healthcare; or

- „Service robots" for personal use - e.g. cleaning robots.

These applications address many problems in robot-component technologies, including sensor interpretation and integration, real-world modeling, actuator and sensor control, path planning and navigation, task-level planning and plan execution, and global monitoring and control of the robot system as a whole - all which is combined in the catch-all term „control" in the following.

The approach for a CAD-control system „combination" introduced in this paper is strongly influenced by the special needs of small and medium sized companies. As a consequence, one of the most important features always to keep in mind for the program development is to find a real „low-cost" solution - without having the need of expensive hard- and software and/or costly operator training for running these software tools.

2. COMPONENTS OF AN INTEGRATED NAVIGATION SYSTEM

One can identify numerous problem-solving activities as being integral parts of „control" of mobile robot system and can group them in three major areas:

Perception:
⇨ Sensor interpretation:
On this level the acquisition of sensor data and its interpretation is done.

⇨ Sensor integration, Recognition:
Due to the intrinsic limitations of any sensory device, it is essential to integrate information coming from qualitatively different sensors, such as stereo vision systems, sonar devices, laser range sensors, etc. On this level, information is aggregated and assertions about specific portions of the environment can be made.

⇨ Real-world modeling:
To achieve any substantial degree of autonomy, a robot system must have an understanding of its surroundings, by acquiring and manipulating a rich model of its environment of operation. This model is either based on assertions integrated from the various sensors - on the other hand, some of the data might be available from previous planning steps (e.g. previously stored maps).

Planning and Control:
⇨ Global planning:
To achieve a global goal proposed to the robot, these activities provide task-level planning for autonomous generation of sequences of actuator, sensor and processing actions. Other activities needed include simulation, error detection, diagnosis and recovery, and re-planning in the case of unexpected situations or failures.

⇨ Scheduling:
This activity is responsible for appropriate scheduling of different activities in order to execute the task-level plan given, while adapting to changing real-world conditions as detected by the sensors.

⇨ Monitoring of overall robot activity, Supervisoring:
On this level a supervisory module oversees the various activities and provides an interface to a human user.

Actuation:
⇨ Navigational activities:
For autonomous locomotion, a variety of problem-solving activities are necessary, such as short-term and long-term path planning, obstacle avoidance, detection of emergencies, etc.

⇨ Actuator control:
These activities take care of the physical control of the different actuators available to the robot. A set of primitives is provided to free the higher levels of the entire system from low-level details.

To meet the requirements of a hierarchical modular system, there is a separation between (low-level) functions realized with the robot deamon and (high-level) functions integrated in the introduced CAD-navigation system.

3. INTEGRATED ROBOT NAVIGATION IN CAD-ENVIRONMENT (IRN)

The key features of this CAD-navigation tool for mobile robots are as follows:
☑ Definition of the environment
• Definition by the user, using CAD functions
• Automatic mapping by means of sensor information

☑ Programming the robot on task-level
As a first step towards this feature, a „Meta-Language" for definition of the robots task has to be designed. This command structure should be able to combine the definition of the task together with the needed mode of surveillance in an appropriate way.

☑ Path planning and collision avoidance
For navigation, generation of the shortest free path and collision avoidance are two important activities. For generation of the free path Kohonen's self-organizing map shall be applied for the control system. For collision avoidance „Fuzzy Control" is one of the suitable methods because of inaccurate sensor information. For application of fuzzy control success is depending on finding the optimal fuzzy structure.

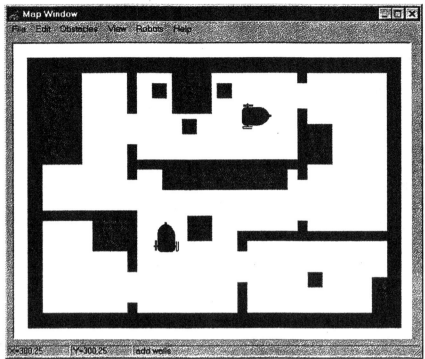

Fig. 1: Map Window

☑ Simulation of the robot programs

Beside the definition of a simulated robot environment, this simulator should serve as a testbed for the development of robot programs to have them validated before downloading to the robot daemon. Because of offering a big amount of training data, this simulation environment is very useful - even for the Neural Network path planning algorithms. To provide an appropriate testbed for program validation, there must be a suitable simulation of sensory data as well. Therefore, the simulation module shall include representation of encoder position, bumper, and sonar sensory (including reflections and cross-talks).

☑ Monitoring

The first attempt was to use AutoCAD as graphical interface together with an appropriate AutoCAD ADS application. Due to the graphical overhead of the CAD-program the application would be slowed down in a massive way. As a consequence the software is now planned to be programmed in Delphi 2.0 respectively C++ builder. For combination to CAD systems an interface for exchange of graphical data will be added to the IRN.

3.1 System components

The main components of the IRN are described in the following.

Managing of the robots world:

The „Map" window (Fig. 1) represents the particular environment of the automation. In the global „Map" window all the robots are represented simultaneously.

Here, one can define the environment and create different robots working in this environment.

Definition of the environment: The environment of the robot can be modeled as follows:

- Automatic mapping:
  The robot follows the walls in a certain distance. Via radio ethernet / modem connection the coordinates of the route are sent to the stationary station. After this an appropriate algorithm will be applied to the above mapping data to simplify the obtained curves to a combination of simple shapes (rectangle, circle, ...) and to get the (almost) real working space of the robot.

- Loading a graphical representation of the environment from a CAD system:
  There is an interface to import of *.dxf* (resp. other graphical standards) files which can be obtained from various CAD systems.

- Built-in CAD functions:
  One can create his own obstacles and walls using basic built-in functions to create simple shapes.

- Combinations of the features above:
  Combination of the three steps above to define the environmental map.

Managing of various robots:

To get a higher degree of flexibility for the application one can define several different robots for each environment. The software manages the special information of each robot which are:

- different sensors,
- dimension of the robot,
- properties like gripping devices, speed, minimum driving circle and others,
- data transfer,
- interpretation module for the motion "Meta - Language".

All these information for any particular robot can be stored in a separate configuration file.

Control of one robot:

The software allows interactive control of each robot. For comfortable use the IRN provides a „Control" window (Fig. 2). The sensor information of the chosen robot are displayed in a special panel (short and long distance sensors). An actual program file can be displayed and the window also displays information about the actual state of the considered robot. To save room on the screen one can hide parts of the control window.

Using the joy-stick panel one can move the robot by means of mouse, keyboard or hardware joy-stick events. A much smarter version is to „talk" to the robot using a special „meta - language" - a task-oriented language (running on the stationary system) developed especially for mobile robots and an interpreter (running on the certain mobile platform) which includes all the necessary features for the control automation of the robot. This language provides primitives which make the use of robot's low-level software transparent to the user. By this way the robot programmer is able to debug or develop software for any type of mobile robot which

has installed the interpreter module.

System architecture of the „meta-language": The system consists of a kernel and several tasks (Fig. 4). The kernel manages the tasks and their succession, according to their priority and conditions. The kernel also includes an event processor, which deals with asynchronous events and reacts according to the programmer's specifications. Usually a robot will have a main task and several satellite tasks which are triggered as responses to certain events and providing a regulator mechanism intended to assure the good working of the robot system. Along with these tasks the system automatically creates special tasks which can be divided in two classes: security tasks and functional tasks.

The first mainly deal with providing the robot with certain protection, independently from the custom programmed tasks. These tasks will have the highest priority as they take care of the physical existence and state of the robot. Functional tasks provide the robot with very useful and necessary instruments, such as path planning, collision avoidance, environment model acquisition, robot position localization and correction.

The kernel: The kernel is the module which is created on system start-up and is active as long as the robot is power-supplied. The kernel is composed of a task scheduler, a language interpreter and an event processor.

The **task scheduler** decides which task is going to be given processing time, i.e. which task will be passed to the language interpreter to be executed. The task scheduler has a task priority list used to decide which task is going to get processing time. The tasks which have the same priority level are dealt with using a circular list. The scheduler always chooses for running the task in the head of the list and takes care of moving that task at the end of the

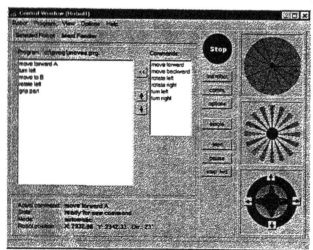

Fig. 2: Control Window

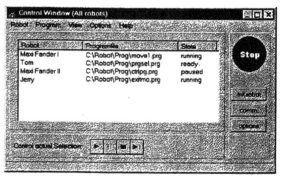

Fig. 3: Control window for all robots

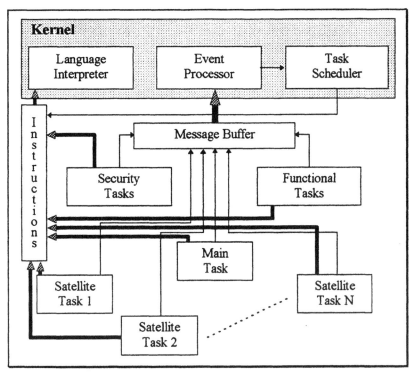

Fig. 4: System architecture „Meta-Language"

circular list, in such a way that all the tasks which have the same priority get an advanced position in the list.

One of the most important issues of the scheduler is to assure that before the next task is switched to, the stopped task has its state saved, in order to be able to correctly continue its execution next time it will get processing time.

The **language interpreter** reads the program's instructions, checks the correctness of the syntax and executes the actions if syntax was correct. In case of a syntax error, the interpreter reports the error but doesn't stop the program, leaving the task of stopping an action to the security or satellite tasks.

The **event processor** manages messages put by the tasks in a message buffer. A message is an integer value associated with a string which represents the message name. The association is made automatically by the interpreter at the beginning of the program, by specifying each message name after the MESSAGE keyword, in the following syntax format:

```
MESSAGE <message1> [<message2>
    ... <messageN>]
```

The message allows the event processor to find the appropriate action that is to be done. There are two types of messages: predefined messages and user-defined messages. The predefined messages are related to events which are expected to happen to a mobile robot, such as collision, encounter an

obstacle, low battery power. Of course, such messages trigger predefined actions, but can also be dealt with by the programmer. However, the predefined action cannot be bypassed and it is the reflex and first action taken by the system. User-defined messages are those specified by the programmer and they trigger user-defined actions. The actions triggered by a message, either predefined or user-defined, may be specified after the MESSAGE declarations, in the following format:

```
ON <message-name>:<action>
```

The action may be either a simple command, a procedure or summoning of a task. The message buffer is implemented as a pipe, following the principle of First-In-First-Out. The tasks put the messages in the pipe and the event processor reads these messages, decodes them and runs the specified action. Of course, when the pipe is empty, the event processor doesn't do anything. Also the tasks must detect the situation when the pipe is full. In such a case, the task stops itself and reports this to the task scheduler, in order to allow another task to be run, avoiding time waste. The stopped task can send the message next time it gets processing time.

Control of all robots simultaneously:
Considering that there are more than one robots in the environment and that the robots affect each other there is a need for a control window for all robots (see Fig. 3). By that it is possible to select some or all robots and to start the assigned program file. Since the system provides with the information about the actual position of each robot, routes can be

planned in the way that there is no handicap between them. As one other example, one can set particular synchronization points in the program files if there are conditions between two robots.

Fig. 5: Mobile Platform „Maxifander" by DBI

Simulation, Supervision:
A main part of the IRN is off-line simulation of the developed programs for each robot. For all activities one can switch between real and simulated robots. For the latter ones control commands affect only the representation on the display. A special simulator calculates the sensor information - the actual environment is being replaced by that way.

After simulation of the robot's program, the instructions (coded in meta-language) can be transferred to the real robots and executed. Beside interpreting the particular statements, the interpreter module running on the mobile platform is also responsible for sending back raw sensory information. Same as for simulated sensor data, all the measurements are displayed on the control screen in order to enable supervision of the entire robot state.

4. IMPLEMENTATION

The proposed navigation system for mobile robot platforms is being implemented on mobile robot platform MaxiFander, produced by Denning Branch International (Fig. 5). The option of mounting heavy equipment (up to 25 kg) on the robot such as robot arms or even a PC chassis as well as possible handling at outdoor conditions, like rough, uneven surfaces, gravel, etc., together with the very simple structure of the robot makes this system to an

excellent tool for education in robotics. Together with the wide array of sensors provided by the robot (rotating sonar transducer, infrared proximity detectors, touch sensors, microphones („ears"), optical line followers), MaxiFander allows a comprehensive view of mobile robot control techniques. The robot can be programmed using language C or C++ - download of the application programs is possible by means of floppy disk or serial interface - for further implementation of the IRN system the robot will be equipped with radio modem connection.

REFERENCES

1. Iyengar S.S. and Elfes A. (editors), „Autonomous Mobile Robots: Perception, Mapping and Navigation", IEEE Computer Society Press, 1991.

2. Everett H.R., „Sensors for Mobile Robots: Theory and Application", AK Peters Ltd, 1995.

3. Dillmann R., Rembold U. and Lüth T. (editors), „Autonome Mobile Systeme 1995", Springer Verlag, 1995.

4. Han M.-W., Kopacek P., „Neuro-Fuzzy Concept for Mobile Robot", in Proceedings of the Workshop on Advanced Control Systems, 1996, Vienna, pp. 129-136.

5. Han, M.-W., Kolejka, T., „Artificial Neural Networks for Control of Autonomous Mobile Robot", in Preprints of the IFAC Workshop on Intelligent Manufacturing System (IMS 94), 1994, Vienna, pp. 163-168.

REAL-TIME SCHEDULING SYSTEM FOR DISTRIBUTED PRODUCTION MANAGEMENT SYSTEMS

Itsuo Hatono * Tohikazu Nishiyama * Hiroyuki Tamura *

* *Graduate School of Engineering Science, Osaka University*
Toyonaka, Osaka 560, JAPAN
Internet: hatono@sys.es.osaka-u.ac.jp

Abstract. This paper deals with a real-time scheduling system for distributed production management systems. To develop a real-time scheduling system, we must take the architecture of the production management system into account, because in the real-time scheduling system. In this paper, we discuss the architecture of real-time scheduling systems and the cooperative scheduling protocols, and the dispatching algorithms for implementing a real-time scheduling system on the distributed production management systems.

Keywords. Real-time scheduling, Distributed production management system, Cooperative scheduling protocol, Distributed simulation

1. INTRODUCTION

Recently, many architectures of next generation of production management systems are proposed(Okino 1992, Ueda 1992). Each production system are based on different concepts such as *holonic*(Okino 1992), *fractal(Sihn 1996)* and *bionic*(Ueda 1992), and so on., but almost all the production management system is developed as a distributed system to cope with the uncertain economic environment and diversified customer's needs. Furthermore, in the manufacturing systems under such uncertain environment, we often come across the sudden change of production plans, emergency jobs, reconfiguration of manufacturing systems, and so on. In this case, real-time scheduling(Harmonosky and Robohn 1991) is more effective than conventional off-line scheduling(Baker 1974).

To develop a real-time scheduling system, we must take the architecture of the production management system into account, because in the real-time scheduling system, we must handle a huge amount of data in the production system in real-time(Parunak 1990). For each next generation of production management system mentioned previously, some real-time

scheduling systems are proposed(Arai *et al.* 1996, Sugimura *et al.* 1996), but it is insufficient to apply these scheduling methods to real manufacturing systems. In this paper, we discuss the architecture of the real-time scheduling systems, the cooperative scheduling protocols, and the dispatching algorithms for real-time scheduling systems on distributed production management systems.

2. DISTRIBUTED PRODUCTION MANAGEMENT SYSTEMS

In flexible manufacturing, since the production environment varies dynamically, the production management system must have functions to deal with failures of components, and reconfiguration or expansion of production facilities. To cope with these difficulties, it seems that it is more effective to construct a production management system as a distributed system. In the centralized production management system, the whole production system will stop, if the host CPU breaks down. However, in the distributed production system, the whole system will not stop, even if some components break down. In this paper, the production management system is assumed to be distributed to each component such as machining center, AGV and so on.

However, in many cases, the distributed production systems with no hierarchies are not always practical, because communication traffic between each scheduling process is too large if the number of machine tools in the production system is large. In those cases, the manufacturing systems are often constructed as hierarchical distributed systems. In this paper, we evaluate the performance of the real-time scheduling systems on not only the distributed production management systems but also the hierarchical ones.

3. REAL-TIME SCHEDULING SYSTEMS

To generate a schedule in real-time, it is necessary to resolve three kinds of conflicts as follows:

(1) Selections of alternative machine tools (This is called routing of job),
(2) Job selections for transporting,
(3) Job selections from an input buffer.

In the distributed real-time scheduling system, these conflicts are resolved locally in the scheduling processes concerned with the conflicts in real-time. In this paper, we assumed the following conditions:

(1) In the distributed real-time scheduling system, we consider each machine tool as a subsystem. A scheduling process in each subsystem resolves the above conflicts locally.
(2) In the hierarchical real-time scheduling system, we assume that a scheduler manages the lower level schedulers.
(3) Each scheduling process has only local information concerned with the equipments or the controllers managed by the scheduling process. By using communication facilities, the scheduling process can obtain the status of the other machine tools, and the whole system of that.
(4) In flexible manufacturing, in general, it is difficult to manage the information concerned with each job, such as processing information, due dates, and so on, when the number of jobs to be processed becomes large. In this paper, the information is recorded in "tag," which is attached to each job. The information is read by each scheduling process and is used to resolve conflicts.

In this paper, the information contained in tags consists of name of job, lot number, processing information of jobs, history of processing, due date, sum of processing time up to the current time, processing start time, degree of importance of due dates and so on.

To implement the real-time scheduling system on a production management system, we must develop the architecture of the real-time scheduling system that conforms to various kind of that of the production management systems. In this paper, we propose an architecture of the real-time scheduling system that can be applied to the various kind of architectures.

4. REAL-TIME SCHEDULING SYSTEM ON DISTRIBUTED PRODUCTION MANAGEMENT SYSTEMS

In this paper, we assume the scheduling functions implemented in the production management system consists of three kinds of schedulers: (1) factory level scheduler, (2) shop level scheduler, and (3) equipment level scheduler. In the followings, we describe the functions of each scheduler.

4.1 Factory level scheduler

We assumed that the factory level scheduler controls the whole production system. Therefore, there is only one factory level scheduler in the production system. In this paper, the factory level scheduler consists of the two kinds of schedulers as follows:

Factory level job scheduler

A factory level job scheduler controls the flows of jobs between the shops using the global information such as status of input and output buffers of each shop, utilization of each production equipment, and so on. When the factory level scheduler receives a scheduling requirement of a job from a shop level scheduler or a human manager of this factory, the factory level scheduler selects a shop for the next operation of the job, and returns the scheduling results the shop level scheduler or the manager for the scheduling requirement.

Factory level AGV scheduler

The factory level AGV scheduler controls the AGVs that convey jobs between the shops. The factory level AGV scheduler selects the alternative AGV to reduce the waiting times of jobs just after the selection of a shop for the next operation. **Fig. 1** shows the overview of communications for scheduling in a factory level scheduler.

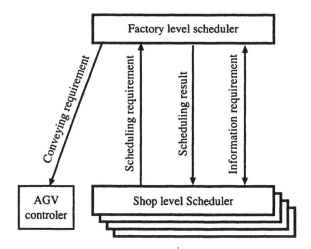

Fig. 1. Overview of communications for scheduling in a factory level scheduler.

4.2 *Shop level scheduler*

A shop level scheduler controls the flows of jobs in a shop, and controls the AGVs which convey jobs between the equipments based on the local information obtained in the shop. Furthermore, the shop level scheduler controls the flows of jobs among shops using contract net based cooperative scheduling algorithms(Hatono *et al.* 1994). In this case, it is not necessary to develop the factory level scheduler. The shop level scheduler consists of three scheduling modules as follows:

Shop level cooperative scheduler

A cooperative scheduler has scheduling functions as follows:

- The cooperative scheduler receives the processing requirement from the other scheduler such as the factory level scheduler and shop level schedulers, and send the scheduling requirement to the shop level job scheduler.
- The cooperative scheduler selects a shop for the next operation of jobs using the cooperative scheduling algorithm, if there exists no factory level scheduler.
- The cooperative scheduler generates the schedules for recovering from machine breakdowns and the emergency jobs.

Shop level job scheduler

A shop scheduler resolve the three kinds of conflicts in the shop: selections of alternative machine tool, job selections for conveying, and job selection from an input buffer of each machine tool in the shop. In this paper, we assume that appropriate dispatching rules are used to resolve the conflicts.

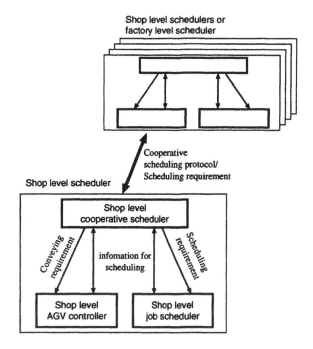

Fig. 2. Overview of communications for scheduling in a shop level scheduler.

Shop level AGV controller

The shop level AGV controller controls the AGVs that convey jobs between the equipments. The shop level AGV scheduler selects the alternative AGV to reduce the waiting times of jobs just after the selection of the equipment for the next operation. **Fig. 2** shows the overview of communication for scheduling in a shop level scheduler.

4.3 *Equipment level scheduler*

A equipment level scheduler generates the processing orders of jobs in the input buffer. Furthermore, the equipment level scheduler require the shop level scheduler to select the equipment of the next operation of each job.

We can develop the various kinds of hybrid systems of distributed and centralized architecture by omitting the appropriate level schedulers. **Fig. 3**(a)~(d) shows the examples of hybrid real-time scheduling systems.

5. COOPERATIVE SCHEDULING PROTOCOLS

We need to develop cooperative scheduling protocols and dispatching algorithms to resolve the three kinds of conflicts, if there are no schedulers in the upper layer. In this paper, we develop the cooperative scheduling protocols and dispatching algorithms supported the functions as follows:

SS: Shop scheduler   ES: Equipment scheduler

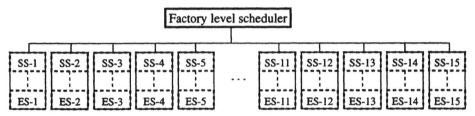

(a) Centralized real-time scheduling system.

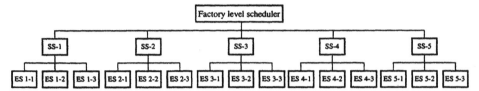

(b) Hierarchical Real-time scheduling system which consists of factory, shop, and equipment level schedulers.

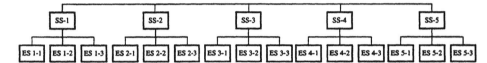

(c) Distributed real-time scheduling system which consists of shop and equipment level schedulers.

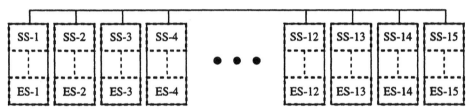

(d) Distributed real-time scheduling system which consists of shop and equipment level schedulers, which each shop level controller manages only one equipment level scheduler.

Fig. 3. Examples of hybrid real-time scheduling system.

- Cooperative decision making mechanism using Contract net protocol to select an alternative equipment or scheduler,
- Dispatching algorithms to select a job from an input buffer,
- Dispatching algorithms to select a job to be convey next,
- Dispatching algorithm for emergency jobs that are given the highest priority,
- Recovery mechanism from breakdowns of equipments,
- Simple deadlock avoidance mechanism.

system, the scheduling process of each machine tool manages the status of buffers of each machine. Therefore, a buffer that was not filled when the machine tool was selected may be filled when a job is actually conveyed to the buffer. To avoid this phenomena, in this paper, the scheduling process reserves a vacancy of the input buffer of the selected machine tool. We apply the contract net based cooperative scheduling protocol selecting an alternative machine tool for a normal job to be processed next.

5.1 Cooperative Scheduling Protocol for Normal Jobs

In the cooperative scheduling protocol for normal jobs, first, a machine tool that a job arrives at the output buffer notifies the information of the arrived job to all the scheduling process of alternative machines, and then selects a machine tool for the next process of the job. However, in this real-time scheduling

5.2 Cooperative Scheduling Protocol for Emergency Jobs

When an emergency job arrives at the input buffer of the current machine tool, the cooperative scheduling protocol for emergency jobs is applied to select the next machine tool from alternatives. The outline of the cooperative scheduling protocol is as follows:

(1) The current machine tool calculates the estimated operational end time when an emergency job arrives at the input buffer of the current machine tool, and sends the conveying requirement to the current cell AGV that the emergency job must convey to the next machine tool at the estimated operations end time .

(2) The cell AGV that received the conveying requirement calculates the earliest conveying start time by using the status of the job queue, and returns the conveying start time to the current machine tool.

(3) The current machine tool calculates the estimate arrival time to each alternative machine tool of the next process based on the earliest conveying start time, and sends the arrival time and the other information about the emergency job to each alternative machine tool. The current machine tool waits to be returned the earliest operational start time and the estimated processing time from each alternative machine tool.

(4) The current machine tool sends the information gathered in (3) to the cell AGV in the current cell.

(5) The cell AGV received the information selects an appropriate alternative machine tool and sends the results to the current machine tool. Furthermore, the cell AGV registers the emergency job and the selected alternative machine tool to the own emergency job list.

(6) The selected alternative machine tool obtains status of the input buffer. When the input buffer is nearly filled, the machine tool sends the message that any job must not convey to the machine tool until the emergency job arrives at the input buffer. If the selected alternative machine tool belongs to the other cell, the cell AGV in the current cell sends the time when the job arrives at the unloading machine of the current cell, to the inter-cell AGV that conveys the job from the unloading machine to the loading machine of the next cell. The inter-cell AGV sends the the time when the job arrives at the loading machine, to the cell AGV of the next cell.

5.3 *Dispatching Algorithms*

To keep the due date of the emergency jobs, we need to develop a dispatching algorithm taken emergency jobs into account. The outline of the dispatching algorithm is as follows:

Step 1. Select a normal job from jobs in the input buffer using an appropriate dispatching rule, when no emergency jobs are registered.

Step 2. Select jobs whose operations are finished until the earliest arrival time of emergency jobs. If no jobs are selected, wait until an emergency job are arrived.

Step 3. Select a job using an appropriate dispatching rule from the jobs selected in (2).

Detail algorithm is omitted because of page limits.

5.4 *Recovery Mechanism from Breakdowns of Equipments*

In the case that machine breakdowns occur, it is necessary to convey the jobs in the input buffer of the failed equipment to other alternative equipments and to cancel all the reservations concerned with the failed equipments. In this paper, we develop the outline of a recovery mechanism from equipment breakdowns as follows:

Step 1. The scheduler control the failed equipment notify the other schedulers of the failure.

Step 2. If there exist emergency jobs in the production system, cancel all the reservations concerned with the emergency jobs.

Step 3. Send a scheduling requirement for each emergency job to the upper level scheduler. The upper level scheduler selects an alternative equipment and reserves a vacancy of the input buffer if possible. If there exit no upper level schedulers, selects an alternative equipment and reserves using the cooperative scheduling protocol for emergency jobs.

Step 4. Send a scheduling requirement for each normal job to the upper level scheduler. The upper level scheduler selects an alternative equipment and reserves a vacancy of the input buffer if possible. If there exit no upper level schedulers, selects an alternative equipment and reserves using the cooperative scheduling protocol for normal jobs.

6. DISTRIBUTED PRODUCTION SYSTEM SIMULATOR

In real-time scheduling systems, however, it is rather difficult to optimize the schedule and to estimate the performance of control computers and communication facilities theoretically, because the behavior of the production management systems are too complex. Therefore, to develop the scheduling algorithm and protocols that are suitable to the production management systems, we need to evaluate the algorithm and protocols by computer simulations. However, it is difficult to develop the simulation system to verify and to evaluate the protocols by using conventional discrete events simulation technique, because it is too complex to develop the simulation model of the real-time scheduling system. To cope the difficulty, we developed a distributed simulation systems to evaluate a real-time scheduling system. The simulation system proposed in this paper can simulate the real-time scheduling system on not only a perfectly distributed production management systems but also hierarchical ones. Since we developed the simulation system as a virtual production system, we can

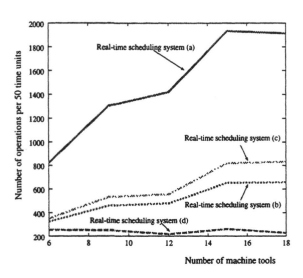

Fig. 4. Average number of operations on machine tools in every 20 time units.

simulate the detail behavior of the scheduling algorithms and protocols.

7. NUMERICAL EXAMPLES

In this paper, we apply the distributed simulator to real-time scheduling systems which have the architecture shown in Fig. 3 (a)~(d). Consider that the production system consists of 5 shops, each of which consists of 3 machine tools. In the production systems, we assume that 10 kinds of jobs are processed and the number of processes of each job and the number of each job are 30 and 3, respectively.

Fig. 4 shows the average number of operation on machine tool in every 20 time units when the number machine tools are varied. In Fig. 4, real-time scheduling system (a), which is developed as a centralized system, the number of operation increases in proportion to the increase of the number of machine tools. However, in real-time scheduling system (d), which is developed as a distributed system, the number of operation is not varied when the number of machine tools increase. Therefore, the overhead for scheduling in real-time scheduling system (d) is the least in the real-time scheduling system (a)~(d). We can estimate that the performance of the computer of the factory level scheduler in real-time scheduling system (a) must be 10 times as much as that of real-time scheduling system (d).

8. CONCLUSION

In this paper, we discussed the architecture of the real-time scheduling system, and proposed the cooperative scheduling protocols and dispatching algorithm for emergency jobs

for real-time scheduling systems on the distributed production management systems. Furthermore, we developed a distributed production system simulator to evaluate the efficiency and the performance of the distributed real-time scheduling system.

Further research might be focused on evaluating the efficiency of the cooperative scheduling protocols by implementing for larger scale production systems, and revising the protocols. Moreover, we might be develop the virtual production systems to evaluate the various kinds of cooperative protocols for real-time scheduling systems on the distributed production management systems.

REFERENCES

Arai, E., L. Jihong and S. Amnuay (1996). Distributed production system to realize flexible scheduling. In: *Proceedings of Japan/USA Symposium on Flexible Automation*. Boston. pp. 1365–1372.

Baker, K.R. (1974). *Introduction to Sequencing and Scheduling*. John Wiley and Sons. New York.

Harmonosky, C.M. and S.F. Robohn (1991). Real-time scheduling in computer integrated manufacturing: Review of recent research. *International Journal of Computer Integrated Manufacturing* 4(6), 331–340.

Hatono, I., K. Tachibana, M. Umano and H. Tamura (1994). Distributed real-time scheduling for flexible manufacturing. In: *Proceedings of JAPAN-U.S.A. Symposium on Flexible Automation – A Pacific Rim Conference –*. Kobe. pp. 803–810.

Okino, N. (1992). A prototyping of bionic manufacturing system. In: *Proceedings of the ICOOMS '92*. pp. 297–302.

Parunak, H.V.D. (1990). Distributed AI and manufacturing control: Some isuues and insights. In: *Decentralized A.I.* (Y. Demazeau et al., Eds.). pp. 81–101. North-Holland. Amsterdam.

Sihn, W. (1996). Paradigm shift the cooperation: The fractal factory. In: *Proceedings of the 6th IFIP TC5/WG5.7 International Conference on Advances in Production Management Systems – APMS'96*. Kyoto. pp. 305–308.

Sugimura, N., M. Hiroi, T. Moriwaki and K. Hozumi (1996). A study on holonic scheduling for manufacturing system of composite parts. In: *Proceedings of Japan/USA Symposium on Flexible Automation*. Boston. pp. 1407–1410.

Ueda, K. (1992). A approach to bionic manufacturing systems based on DNC-Type information. In: *Proceedings of the ICOOMS '92*. pp. 303–308.

ARCHITECTURAL DESIGN
FOR OPEN INTELLIGENT MANUFACTURING SYSTEM

Hyoung Joong Kim*, Gi Taek Kim*, and Byung-Wook Choi**

*Department of Control and Instrumentation Engineering
Kangwon National University
Chunchon 200-701, Korea
** Advanced Manufacturing System R&D Office
Korea Institute of Industrial Technology
Chonan 330-820, Korea

Abstract: The next generation manufacturing systems will exploit new features from distributed computing, object-oriented software engineering, and Internetworking. In this respect, the role of agents for intelligent manufacturing systems is addressed. This paper mainly surveys the state-of-the-art of the distributed object and agent technologies, which are the core of the intelligent manufacturing systems. Open architecture for interoperability is stressed. Standardization activities are also addressed.

Keywords: Intelligent manufacturing systems, Intelligent agents, Open Architecture, Distributed computing, Interoperability

1. INTRODUCTION

During the last decade many concepts regarding manufacturing systems have been emerged. They include numerical control, computerized numerical control, direct numerical control, flexible manufacturing systems (FMS), and computer-integrated manufacturing (CIM) systems. Reliable and robust networking techniques, microprocessors, artificial intelligence, computer vision, and database technologies have enabled the breakthrough in modern manufacturing systems. Modern manufacturing systems have exploited information technology.

However, past results have mainly focused on the cell level or below. Manufacturing cell control has been extensively researched and has resulted in practical cell control architectures. Thus the next step is to integrate the independent cell control systems into the so-called intelligent manufacturing system (IMS). Intelligent systems have been generally attributed to the system controlled or managed by introducing artificial intelligence techniques such as genetic algorithm, fuzzy logic or neural networks (Kusiak, 1990; Parsaei and Jamshidi, 1995). Moreover, they can adaptively pursue goals varying over time. They can extract knowledge and information from unstructured data.

However, future IMS must have more features than those artificial intelligence techniques. IMS models may, for example, include biological manufacturing system (BMS) (Kanji, 1994) holonic manufacturing system (HMS) (Van Brussel, *et al.*, 1995), and virtual manufacturing system (VMS) (O'Leary, *et al.*, 1997). HMS is a highly decentralized system consisting of cooperative, autonomous, and intelligent agents, called holons. They yield an agile and self-organizing manufacturing systems and support global optimization. Similarly, the main idea of BMS is to model manufacturing systems by imitating biological metamorphosis and symbiosis. In other words it can be taken as having similar capabilities and composition to HMS with an additional ability to organize, repair, grow and evolve by itself at all levels within its domain in much the same way as biological organism does.

Manufacturing systems can have both the hardware agents and software agents. Robot is a typical example consisting of hardware agents. Software agents are now being studied and implemented. Intelligence will be added to the software agents such that they are extensible, flexible, performance tunable, fixable, and especially interoperable.

It is evident that next generation manufacturing systems will become more decentralized and be geographically spread. They will be connected to either Fieldbus, Ethernet, ATM, or possibly wireless LAN. They may be connected even to the Internet (Erkes, 1996). The distributed cells will be operated under different operating systems. Diverse communication protocols will be introduced. They will be connected to heterogeneous networks. Mobile and multimedia computing capabilities will also be utilized. The problem is therefore how to keep those different systems to be interoperable. Interoperability becomes a hot issue due to the great diversity of existing standards and proprietary product spectrums from different vendors, and large number of special purpose devices.

Architectural design of the IMS is an important problem (Van Brussel et al., 1995). Open architecture is one of important objectives for implementing the IMS. The agent is a major component of the open IMS, and it must be open and interoperable. It must also be collaborative with others. The focus of this paper will be placed on the interoperability of agents. This paper mainly surveys the state-of-the-art of the distributed object and agent technologies which are the core of the intelligent manufacturing systems. To achieve better interoperability among agents, open architectural issues and standardization activities are stressed. Thought any new results are presented in this paper, it is intended to highlight the main ideas behind each of concepts of intelligent manufacturing systems and standardization activities for interoperable, object-oriented systems, and intelligent agents.

2. INTEROPERABILITY

One problem of the current manufacturing systems is to cope with fluctuating demands. FMS has been introduced to offer flexibility and responsiveness to the manufacturing floor. On the other hand, FMS changes purchaser's attitudes in just the same way it needs a changed attitude to introduce it (Luggen, 1991). As a consequence, manufacturing systems must be more flexible than ever. The far-reaching influence of Internet over the manufacturing systems will be apparent. Internet will be connected to the manufacturing systems and it will force them to be more business-driven. That is to say, manufacturing systems must give a solution to small volume but wide variety of items. For example, Internet shopping will more frequently demand small volume of goods. The orders will be subject to the fad.

Virtual manufacturing can adaptively solve the peak-and-valley demands (O'Leary, et al., 1997). A typical example of suffering peak-and-valley demands is the defence industry. As demand for military goods decreases, these firms need to use unused resources for alternative production. One form of virtual manufacturing comes from the Internet shopping. When a user places an order, an intelligent agent collects the orders and assigns them to factories or warehouses, for example, nearest to the client placed the order. An intelligent agent can negotiate with other agents to get optimized results. Of course, each agent has its objectives and constraints to optimize system performance.

By the introduction of the ARPA Agile Infrastructure for Manufacturing Systems (AIMS) (Park, 1993) the AIMS project will create a national infrastructure for agile manufacturing. The project focuses on the rapid prototyping and fast-turnaround production of small lots (Park, 1993). The manufacturing systems have been evolved from an isolated one to the national infrastructure, and will evolve to the world-wide infrastructure. Internet has played an important role for speeding up interconnecting systems through the network over the world. It facilitates the virtual manufacturing systems to be come true. Moreover, Internet introduces the new technologies including hypertext, Web browser, home shopping, electronic commerce, multicasting, and virtual manufacturing. Agent technologies have gained impetus when they are applied to the Internet.

Certainly Internet will also play an important role in the future. More systems will be interconnected to the networks. Then how to make the different platforms to be interoperable? CORBA (common object request broker architecture) can be a solution. Of course, so can be DCOM (distributed component object model). A benefit of object or object-oriented programming is reusability of software. Traditional software has kept method (sequences of computer instructions) and data (information which the instructions operates on) apart However, in object-oriented programming, they are merged into a single indivisible thing: an object. Now distributed object technologies, such as CORBA or DCOM, break the rule that every object must reside on the local computer; now objects can post services through a broker, and the objects themselves can even be written in different programming languages, as far as they use the same interface definition language or, in short, IDL (Cohen et al., 1994). Distributed object technology is a new computing paradigm that allows objects to be distributed across a heterogeneous network, and allows each of the components to interoperate as a unified whole. One feature of the distributed objects is that they can message each other transparently anywhere on the network (Orfali et al., 1996). When a software is written in Java, it can be said that the software is platform-independent. Interoperability is far more than platform-independence. For example, Java with HTTP-CGI is inferior to Java with CORBA (Orfali et al., 1996). CORBA objects are location-independent, which allows objects to be physically moved around on the network (e.g., to reduce network traffic). Java is even more extreme in this aspect, since the objects can be moved to any machine on the network on the fly (Fayad and Cline, 1996). As is mentioned above, CORBA provides the mechanisms by which objects transparently make requests and receive responses. The CORBA ORB is an application framework that provides interoperability between objects, built in (possibly) different languages, running on (possibly) different machines in heterogeneous distributed environments.

The kernel of distributed objects is independent software components. The smart components can play in different networks, operating systems, languages, and tools. On the other hand, components is also an object in the sense that they support inheritance, polymorphism, and encapsulation. A componet is a software IC (analogous to hardware integrated chips), where Cox spoke of a software revolution that would result from the use of software ICs (Cox and Novobilski, 1991). Therefore, component technology can improve maintainability, software reusability, and platform-independence, and reduce application complexity as hardware ICs did. The distributed object infrastructure is by definition component infrastructure. At the lowest level, a component infrastructure provides an object bus, the object request broker (ORB). ORB ties the distributed objects together. ORB also provides means for objects to locate and activate other objects on a network, regardless of the processor or programming language used to develop objects. ORB makes these things happen transparently to the developer. Therefore, the ORB is the middleware which allows interoperability in heterogeneous networks of objects. A significant benifit of ORBs is that they provide a means for locating, activating and communicating with other objects while hiding their implementation details. The hiding of implementation details within objects is one of the key features of object-oriented technology to managing complexity in distributed computing. Objects only know what services other objects provide (their interfaces), and not how they provide those services (their implementation).

It is almost impossible to have interoperable networks of objects without standards. Object Management Group (OMG) is a consortium to define the standards for distributed object systems in heterogeneous environments. The objective of OMG is the definition of the Object Management Architecture (OMA). It includes four sets of standards; one of them is CORBA and the others are Common Object Services Specification, Common Facilities, and Application Objects. In 1992, OMG approved a standard architecture called CORBA that defined the services to be provided by an ORB. Since then, several vendors have been working on their own distributed object computing products, primarily ORBs. To date there are about a dozen ORB implementations. Some are commercially available now; the rest are expected to be released soon. Most, if not all, of these claim to be CORBA-compliant or plan to be in the near future.

3. INTELLIGENT AGENTS

Agents are not new in manufacturing systems. For example, AARIA (Autonomous Agents at Rock Island Arsenal) (Parunak, et al., 1997) and above-mentioned AIMS (Park, 1993) are manufacturing systems based on agents. Now, many agents exist: task agents, resource agents, inventory agents, tool agents, mobile agents, learning agents, and so on. However, agent implementation is just at the start line.

An agent is an autonomous software that encapsulates a service (e.g., database, applications, and software tool). Agents communicate with each other by exchanging messages in a common language, an agent communication language (ACL). Agents can be executing on a different hardware platform, can be written in different programming languages, and can execute under different operating systems. In this sense, open architecture of interoperable agent is required. A common agent interface allows the exchange of information.

The fundamental attributes of agents include persistence (maintaining a consistent state over time which is unchanged capriciously), autonomy (exercising exclusive control over their internal state and behavior), reactivity (reacting to changes in their environment), mobility (moving from one location to another while preserving their internal state), learning (accumulating knowledge based on past experience), and so on. Of course, not all agents can have those attributes. Distributed agents collaborate each other by passing messages. The open agent architecture (OAA) is a framework for integrating a community of heterogeneous software agents in a distributed environment.

The openness states that agents can be created in multiple programming languages and interface with existing legacy systems. It is desirable to have a single programming language for creating agents. However, no language has been satisfactory. Thus, open architecture can support interoperability among agents created in different programming languages and interface.

Agents must be added or replaced individually at runtime in a transparent manner to other agents. Otherwise, system will be interrupted and readjusted whenever a new agent is added or replaced. An agent can become faulty. It must be replaced without disturbing the system. Otherwise, system may not be fault-tolerant. Mobile agent can move from one location with or without notifying its behavior and register at another location. Other agents must be able to communicate with the mobile agent wherever it is located. Future systems must be able to hand ubiquitous agents due to the commercialization of mobile components.

Intelligent agents must have ability to communicate with each other using an expressive communication language, work together cooperatively to accomplish complex goals, act on their own initiative; and use local information and knowledge to manage local resources and handle requests from peer agents. ACL is a language that is designed to support interactions among intelligent software agents. It has been successfully used to implement a variety of information systems using different software architectures. ACL can be thought of consisting of three parts: vocabulary, KIF (Knowledge Interchange Format), and KQML (Knowledge Query and Manipulation Language). An ACL message is a KQML expression in which the arguments are terms or sentences in KIF formed from words in the ACL vocabulary. An ontology is a vocabulary appropriate to a specific application area. Shared ontologies are under development in many important application domains such as planning and scheduling. In the agile manufacturing cell shared ontologies for PDES (Product Data Exchange using STEP) standard are under construction.

The Foundation for Intelligent Physical Agents (FIPA) is an international non-profit association of companies and organisations which agree to share efforts to produce specifications of generic agent technologies. Other organizations for standardization include the Agent Society, OMG, W3C, and so on. The purpose of FIPA shall be pursued by identifying, selecting, augmenting and developing in a timely fashion specifications of generic agent technologies that are usable across a large number of IPAs (Intelligent Physical Agents) and provide a high level of interoperability with other agents-based applications. The target of FIPA-specified agent technologies are IPAs. They are devices intended for the mass market, capable of executing actions to accomplish goals imparted by or in collaboration with human beings or other IPAs, with a high degree of intelligence. As a rule FIPA selects and adapts existing technologies and only occasionally develops its own technologies. Thus, FIPA keeps close contact with formal standardization bodies, industry consortia and government agencies, such as ARPA, CEC, DAVIC, IETF, ITU, MPEG, OMG, TINA, W3C, and so on.

4. CONCLUSION

Intelligent manufacturing systems will be based on the distributed object and intelligent agent technology to achieve better overall performance. Goals of the intelligent agents and the intelligent manufacturing systems are eventually the same. In fact the core of the intelligent manufacturing systems is the intelligent agents. Future manufacturing systems will be more complex and distributed. Thus, interoperability will be a hot issue for implementation. Open architectures are needed in this sense. Standardization effort of distributed object brokers and intelligent agents are addressed. The works of OMG and FIPA are sketched. Standardization of STEP, VRML, and so on, is also important for implementation of the intelligent manufacturing system though it is not covered in this paper.

REFERENCES

Cohen, P., A. Cheyer, M. Wang, and S. Baeg (1994), An open agent architecture, *Working Notes on AAAI Spring Symposium on Software Agents*, 1-8.

Cox, B., and A.J. Novobilski (1991), *Object-Oriented Programming : An Evolutionary Approach*, 2nd ed., Addison-Wesley

Erkes, J.W., K.B. Kenny, J.W. Lewis, B.D. Sarachan, M.W. Sobolewski, and R.N. Sum, Jr. (1996), Implementaing shared manufacturing services on the world-wide web, *Communications of the ACM*, **39**, 34-45.

Hardwick, M., D. Spooner, T. Rando, and K. Morris (1996), Sharing manufacturing information in virtual enterprises, *Communications of the ACM*, **39**, 46-54.

Kanji, U. (1994), *Biological Manufacturing Systems*, Kogyochosakai Publication Co., Tokyo.

Kusiak, A (1990), *Intelligent Manufacturing Systems*, Prentice-Hall.

O'Leary, D.E., D. Kuokka, and R. Plant (1997), Artificial intelligence and virtual organizations, *Communications of the ACM*, **40**, 52-59.

Orfali, R., D. Harkey, and J. Edwards (1996), *The Essential Client/Server Survival Guide*, 2nd ed., John Wiley & Sons

Park. H., J. Tenenbaum, and R. Dove (1993), Agile infrastructure for manufacturing systems (AIMS): A vision for transforming the US manufacturing base, http://web.eit.com/creations/papers/DMC93/DMC93.html.

Parsaei H.R., and M. Jamshidi, Ed. (1995), *Design and Implementation of Intelligent Manufacturing Systems*, Prentice-Hall.

Parunak, H.V.D., A.D. Baker, and S.J. Clark (1997), The AARIA agent architecture: An example of requirements-driven agent-based system design, *Proceeding of the First International Conference on Autonomous Agents* (ICAA'97), Marina del Rey, CA, February 6-8, 1997.

Van Brussel, H., P. Valckenaers, L. Bongaerts, and J. Wyns (1995), Architectural and system design issues in holonic manufacturing systems, *Proceedings of the 3rd IFAC Workshop on Intelligent Manufacturing Systems*, Bucharest.

MAINTENANCE OF MANUFACTURING DATABASES USED IN CAPM SYSTEMS

[1]Eric Bonjour, [1,2]Pr. Pierre Baptiste

[1]Laboratoire d'Automatique de Besançon / UMR 6596 CNRS
ENSMM / Université de Franche-Comté
Institut de Productique, 25 Rue Alain Savary, 25000 Besançon - France
Tel : +33.81.40.27.94 / Fax : +33.81.40.28.09
[1]IUT Belfort-Montbéliard Dept GTR, 4, Place Tharradin BP 427
25211 MONTBELIARD Cédex
[2]KAIST Dept Computer Science   373-1, Kusong-Dong, Yusong-Gu
TAEJON, 305-701 Coree du sud
e-mail : ebonjour@ens2m.fr ; baptiste@ens2m.fr

Abstract: Data describing shop floor activities should be of good quality to ensure efficient Production Planning and Control. In this purpose, this paper proposes some guidelines to achieve efficient modelling and maintenance of Manufacturing DataBases (MDB) used in Computer Aided Production Management (CAPM) systems. Short industrial examples and simulation results illustrate these guidelines and prove that database simplifications are possible without degrading CAPM results significantly.

Key words : Database maintenance, Manufacturing systems modelling, Production Scheduling, database quality, sensitivity analysis

1. INTRODUCTION

Many companies have difficulties in using their CAPM results (Computer Aided Production Management) because of poor manufacturing data quality. They have not fully used the potential of their CAPM software since data is either poorly structured (poor modelling), or poorly used, or poorly maintained. They have been compelled to introduce improved structures to correct proposals made by their CAPM systems. Due to a lack of a real maintenance strategy for MDB, significant differences may appear between what is actually happening at the shop floor level and what is modelled in MDB.

The analysis of previous problems points out that maintenance costs have rarely been taken into account in the MDB design phase. Numerous research works have dealt with database modelling (tools and methods like (Berio et al., 1995), works like (Companys et al., 1990)). An implicit assumption of MDB model designers is that data should be of high quality when it is used. In practice, even if CAPM experts advise users to

ensure good data quality to have efficient CAPM work (Browne et al., 1988 p135), (Giard, 1988 p494-501)...), MDB suffer from poor data quality ((Plossl, 1990), (Clement et al., 1992], ...).

This paper intends to propose guidelines to achieve efficient modelling and maintenance of MDB. It will focus on the links between dimensions of database quality, the required precision of CAPM processing (Production Planning, Scheduling...) and maintenance costs.

This work is divided into 3 main parts.

First, the main dimensions of database quality are defined. Second, data maintenance is defined and research works concerning data maintenance and structuring in the field of CAPM are described. Then, the main causes of value non-quality are explained. Third, this paper proposes important guidelines to maintain MDB and to simplify data structures according to the quality dimensions already described. It takes simulation experiments in scheduling and industrial cases into account.

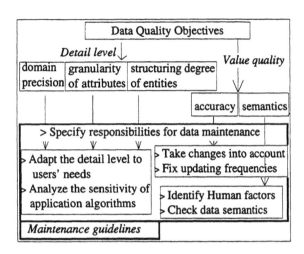

Fig. 1. Preventive actions for data maintenance

2. DATABASE QUALITY

To identify problems of data quality and the adequate maintenance actions (see fig. 1. and § 4) , it is necessary to specify data quality dimensions. According to T.Redman's classification (1992), they can be divided into two main groups concerning value quality and model quality. In this paper, two other dimensions are introduced : structuring degree and semantics.

2.1. Model quality

Precision of domains, granularity of attributes and structuring degree of entities are features of model quality. These three dimensions represent the level of detail used to model the real world.

The precision of domains. According to T. Redman (1992), precision "refers to the level of detail in the measurement or classification scheme that defines the domain". The greater the number of values in the domain, the greater the precision. Precision is often badly used instead of accuracy.

The granularity of attributes. It refers to the number of attributes that are used to represent a single concept (Redman, 1992). For instance, to model the "time" concept in MRP (Material Requirements Planning), it is possible to use either only the "lead time" attribute in the item master file or several attributes in the routing file (processing time, set-up time, transport time) and in the work centre file (waiting time). (see (Ranky, 1990) for more details) The second set of attributes provides greater granularity (with more attributes) than the first one.

The structuring degree of entities. This dimension represents the level of detail used to model entities of the manufacturing process (product, routing, work centre). The structuring degree is computed as the ratio between the number of stored entities in database and the number of entities which actually exist.

If all the entities in the manufacturing system are identified, this ratio equals 1. The more the

manufacturing system is modelled in MDB, the greater this ratio is. On the contrary, the more aggregate the entities are, the more reduced the structuring degree is.

In a CAPM system, the structuring degree of entities could be applied to the following entities :
- routings by counting the operations used to describe the manufacturing process,
- work centres by identifying the actual resources at the shop floor level and the work centre which is their counterpart in database,
- products by counting the parent/component links included in the Bill Of Materials (BOM) and used to describe the product composition. The lowest degree of structuring is reached for the mono-level BOM, i.e., all components are listed at the same level.

A. Kusiak (1990 p12) notes that the weakest volume of handled information takes place at the management level (aggregate data) and the biggest at the shop floor level (detailed data). Consequently, in a CAPM system, different structuring degrees should be used.

2.2. Value quality

Accuracy and semantics are the main dimensions of value quality, as far as CAPM systems are concerned.

Accuracy. Accuracy is commonly accepted as being the most fundamental quality dimension. It is usually admitted that a datum is accurate if the recorded value is within an acceptable, predetermined domain of value, a tolerance margin.

Data values change with time. According to T. Redman (1992), "a datum value is up-to-date if it is correct in spite of a possible discrepancy caused by time-related changes to the correct value ; a datum is outdated at time t if it is incorrect at t but was correct at some time preceding t." In MDB, data inaccuracy is due to changes that occur in the manufacturing system and that are not taken into account to update the modified data. It is not due to random, unexpected disturbance in the manufacturing process. For quantitative data, an inaccuracy ratio can be defined as follows :

$$\text{inaccuracy ratio} = \frac{\text{actual value} - \text{stored value}}{\text{stored value}} \quad (1)$$

The stored value is fixed in the database and is used to measure the actual difference. The greater the ratio inaccuracy is, the more inaccurate data is.

Semantics. Many authors deal with conceptual (model) semantics in terms of constraints only (see (Pels and Wortmann, 1990)). However, it is important to separate model semantics and actual data semantics. The design semantics of a data model may be changed (intentionally or not) when it is implemented or used. In addition, data can have several meanings according to their users (semantics consistency). A "correct value" depends on data semantics, i.e., on the choice of the process which has created it (algorithm, collection method, monitoring, copy from an other database...).

3. ON THE MAINTENANCE OF MDB

First, this part defines what is called data maintenance and its limits. Second, it presents existing works related to data maintenance in the field of CAPM. Third, the main human factors of non-quality are explained. Finally, algorithm sensitivity analyses are described to specify the impact of data inaccuracy on algorithm results.

3.1. Definition

Traditionally, the maintenance of database does match with an evolution of the conceptual schema of the database (constraints, relations, entity types) and hence, of the application programs that use this database (see (Berio et al., 1995), (Companys, et al., 1990) ...).

In this paper, data maintenance refers to actions on the database that do not affect the conceptual schema, that is, the creation or deletion of entity, the checking of data accuracy and if necessary, the updating, the checking of control parameters, the checking of semantics consistency, of database completeness but also the change in the level of detail required to describe the manufacturing system properly (in a CAPM system, the conceptual schema is imposed as soon as the software is chosen).

This kind of preventive maintenance is rarely dealt with in literature. However, it should allow a continuous change of MDB to adapt them with the modelled environment that is steadily changing. It is a non-value-adding activity but necessary to assure good results of CAPM processing.

3.2. Existing research works

This part presents existing works that provide guidelines to structure and maintain MDB in the field of production management (Production Planning and Control, Scheduling ...). Despite its importance, this topic has called little attention.

How to structure MDB. J. Clement et al. (1992) describe an interesting industrial case concerning the relevant level of data structuring. They show that data structuring has to find a global trade-off between work control, the volume of data to be maintained and the users' satisfaction. To streamline data maintenance, reporting, and any non-value-adding activity, the authors describe steps to question database structures.

A level is created in the bill of material when parent/component relationship is defined. In a similar way, a routing step is created when an operation/work centre relationship is defined for a parent item or part. Levels are determined by interruptions in factory flow or by stocking materials at intermediate points in the production process. Routings steps are assigned to specify activities and their sequences, to set standard times for work and set-ups to perform detail capacity planning, to schedule operations in detail and to track actual work precisely.

But the goal of companies today is to eliminate Work-In-Progress stocks, reduce set-up times, end costly reporting, and not create much data maintenance. The goal is to reach flatter bills of materials (BOM) because production has to flow with fewer interruptions and flatter routings with production flows based on cells and kanban scheduling.

How to share MDB maintenance responsibilities. In the ABCD classification proposed by O.Wight to evaluate CAPM performance (see (Lanvater and Gray, 1989 p335)), questions refer to the necessity of a decentralised data maintenance. Two points of the questionnaire (I.3 and I.4) highlight that each CAPM user is responsible for the quality and maintenance of data that he (she) uses. J. Clement et al. (1992) describe the two ways of maintaining data : in a centralised approach, the data files are maintained by a given group ; in a decentralised approach, the data files are maintained at the source. Anyway, the point is, errors need to be quickly corrected when they are identified and data quality objectives have to be clearly defined.

3.3. Causes of value non-quality

In this part, main causes of value non-quality are discussed. They highlight a lack of regular updating due to human factors.

Inaccuracy. According to A. Hatchuel et al. (1988), CAPM users have a natural behaviour that consists in correcting information provided by the system before making their decisions but they often neglect to update this information in the database when it is necessary. A lack of maintenance policy to decide when updating is useful results in a lack of updating !

P. Dubois et al (Dubois, 1992) note that the work of CAPM users is evaluated in the short term whereas MDB updating is considered as an additional task in the short term but has repercussions in the long term only. Users tend to neglect maintenance.

Concerning the updating cost, it can be evaluated between 1 and 20 dollars. Data collection and capture are not the only time-consuming tasks to update data. Human factors can arise : when and how the decision has been made to update data, how long it takes to go and collect data, how long the operator speaks with other people during this task... Moreover, this task is laborious and is generally achieved after other works, with bottom priority.

Data semantics. Semantics bias are often used to adapt the use of a software program to a given company. It is then necessary to check at once that the results of application programs are not false because of these changes and that all the users give the same meaning to attributes and to entity types.

3.4. Simulation results on algorithm sensitivity

In case of flow shop scheduling (Bonjour, 1996), simulation experiments have shown that 20% of inaccuracy on processing times imply less than 5% of loss on results (sensitivity according to the makespan (Cmax) or the mean flowtime). Flowshop heuristics that are used in this simulation, are well-known and largely discussed in the flowshop literature. Figure 2 only intends to present global trends of heuristics in terms of increasing data inaccuracy. It does not intend to study these heuristics in detail.

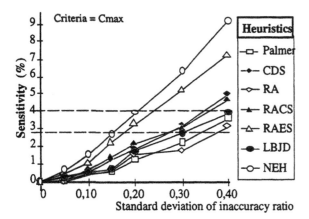

Fig. 2. Evolution of sensitivity in terms of data inaccuracy

R.W. Conway et al. (1967, p39-43) note that in practice, processing times are inaccurate and scheduling decisions are based on estimated times. They study the sensitivity of the SPT (*Shortest Processing Time*) rule that gives an optimal solution for the scheduling problem with one machine. They conclude that SPT is very little sensitive to the inaccuracy of processing times.

These results show that the study of algorithm sensitivity is important to determine a required level of accuracy. Data maintenance results in a trade-off between the profit due to better results and the cost of data maintenance. An inaccuracy ratio between 0,1 and 0,2 seems to be reasonable and consequently, scheduling results are not degraded significantly.

4. GUIDELINES TO MAINTAIN MDB

In most companies, there is a lack of "preventive" maintenance to check data value and a lack of "perfective" maintenance to continuously improve

the suitability of the modelling to the reality according to the detail level dimensions.
This part proposes guidelines to maintain MDB and to simplify models according to the detail level dimensions. Industrial examples illustrate the main guidelines. They can be summarised as follows (see Fig.1.) :

1- To specify responsibilities for data maintenance

2- To adapt the detail level to the changing manufacturing process and to new users'needs
   - Global or detailed data (set-up times...)
   - What detail level is required ?
   - How many levels are required in the BOM or in the routings ?
   - What data precision is required ?

3- To analyse the sensitivity of algorithms
   - To adapt data inaccuracy and
   - To adapt the detail level

4 To take into account changes in the manufacturing process
   - Attitude towards data modification due to SMED
   - Updating processing times after an improvement on the process

5- To fix value updating frequencies

6- To identify Human factors disturbing data quality

7- To check data semantics

These guidelines are now explained and illustrated.

4.1. To specify responsibilities for data maintenance

The part 3.2. has pointed out that maintenance could be either centralised or decentralised. The point is, responsibilities for data maintenance must be clearly defined.

4.2. To adapt the detail level to the changing manufacturing process

The detail level should be suitable to the decision level, to the decision-making horizon (long term, medium term, short term) that depends on the production lead times, to the adequate level of work control that depends on the production organisation and to the adequate precision of CAPM results. For instance, MRP (medium term) requires fewer details than scheduling (short term) ; flowshop routings are shorter than jobshop routings.

The initial choice of the detail level is made when the CAPM system is implemented but this decision should be questioned periodically or when special events occurs (for instance, users express new needs in data, the organisation of the manufacturing process has changed...). The problem is that generally, the initial implementation of software has been fixed for a long time.

However, experience has been evidenced that it is useless to structure data excessively. A trade-off should be found between the level of detail, data inaccuracy due to changes and the quality of CAPM results, i.e., users' satisfaction. The MDB is only a of reality and can't be 100% accurate. A difference is also logical. Maintenance costs can be reduced by modelling only the relevant and important data. Anyway, other data of low importance will not be updated after changes.

Global or detailed data. Material Requirements Planning (MRP), Capacity Requirements Planning (CRP) and scheduling do not require the same granularity of data. For instance, sequence-depending setup times may be defined in scheduling but they are useless for Capacity Planning.

Simulation results have shown that, up to a relative ratio of 0,1, an attribute could be deleted as regards others and however, results could be about the same. This implies that times lower than about 0,1 of the processing times could (or should) be neglected when the production system is modelled (even in the aim of a scheduling application). For instance, setup times, cleaning-up times... lower than 10% of the processing times could be neglected at the detailed level but they could be taken into consideration in *global* data like the actual availability of the work centre. An other example. Scrap factors lower than about 0,05 of the usage quantity could be eliminated and managed in a global "usage quantity" with a particular semantics : 1,05 times the normal usage quantity. The precision will certainly be sufficient. The aim is to assure efficient modelling and easy maintenance of databases whereas data interactions would become easier to understand.

What detail level is required ? When using MRP systems in a particular factory, there is no guideline regarding the detail level (the number of levels in the bill of material, the number of operations in the routing...). CAPM software do not impose any constraints.

Changes in the manufacturing process could be taken into account to simplify data maintenance. For instance, in a company, an analysis of granularity has led to delete some attributes (set-up times) which are no more important as regards others (processing times and actual availability of the work centre in the shop). They represent less than 5% of the others. Results are not degraded. The granularity of the "time" concept is modified to reduce maintenance costs.

A new detail level can be required to meet new users'needs in CAPM results precision. For instance, a manual assembly shop is divided into 4 cells according to product families and with a cell manager. The application of CRP allows the production manager to globally plan the required number of workers in the shop. The objective of Senior management is to make each cell manager responsible for the required number of workers in his cell. The structuring degree of the shop will increase since 4 new work centres will be identified in the database to plan the labor load in each cell.

How many levels are required in the BOM or in the routings ? Part 3.2. has pointed out how to structure MDB. To structure BOM and routings, all the concerned departments have to find a global trade-off between work control, the volume of data to be maintained and the users' satisfaction.

What data precision is required ? When data is inaccurate enough, it is useless to quantify some concepts precisely (for instance, time in routings or quantity of usage in Bills of Materials). According to G.W. Plossl (1992), "systems in the past have buried managers in data of high precision but dubious accuracy and little relevance."

4.3. To analyse the sensitivity of algorithms

To adapt data inaccuracy. Part 3.4. has shown that the best heuristics (NEH, RAES) are the most sensitive to data inaccuracy. Hence, the required level of data accuracy depends on the algorithm sensitivity and on the required precision of results.

To adapt the detail level. For instance, simulation results in flowshop have shown that a work centre can be neglected in the modelling when it is followed by a work centre with more than twice its load. Scheduling is very little sensitive to this simplification.

4.4. To take into account changes in the manufacturing process

ISO 9000 standards demand that companies update data after each improvement actions but they do not take the maintenance costs into account. In addition, it is rarely obvious to determine when significant changes on the manufacturing process occur. There is always a delay before updating.

Attitude towards data modification due to SMED and Updating processing times after an improvement on the process. The following conclusions come from flowshop scheduling simulations. When SMED actions or process improvement actions are carried out on a bottleneck machine, it is profitable to update the setup times after reductions greater than 15% of the average processing time because actual gains are only 50% of potential gains that would be obtained if data would be really updated. If an improvement of 0,2 happens on the three most loaded machines, it triggers the same degradation than an homogenous deterioration of 0,2. This means that data updating is very important when progress actions are performed otherwise, heuristic performance could be greatly degraded.

4.5. To fix value updating frequencies

According to an ABC classification of the entities, data have not the same importance in the manufacturing process and require various accuracy and then, different updating frequencies.

In a shop, reasonable inaccuracy levels on processing times were fixed for each class of operation and according to the precision required by the processing of the Capacity Requirements Planning. Then, appropriate updating frequencies (per year) were determined to reach these inaccuracy levels and to limit global maintenance costs. Updating was costed at about 3$. A maintenance policy was specified to check data according to these frequencies and by a "cycle counting" approach. The global maintenance of these times was costed at about 260$ a year.

Table 1 Costs of maintenance according a ABC classification

operation	class-A	class-B	class-C
reasonable inaccuracy	10%	15%	20%
updating frequency of a value	3 /year	1/year	0,5/year
number of operations per class	15	20	45
estimated costs of maintenance per class	135$	60$	67$

4.6. To identify Human factors disturbing data quality

In many companies, monitoring data are used to reward operators. Their interest is to ignore poor performance and to respect stored data as an objective. Data semantics is biased. Anyway, data quality depends on the Human behaviour.

4.7. To check data semantics

Once a year, the CAPM manager checks the semantics consistencies of important attributes by specifying the attribute meaning during a meeting with the different departments. Semantics is maintained in a data dictionary.

5. CONCLUSION

Up to now, data maintenance has been neglected. Companies have a rather intuitive approach of maintenance and do not know how much it costs. This paper has highlighted some important guidelines for data maintenance. In practice, it seems reasonable to expect an inaccuracy ratio in the range of 10 to 20%. Concerning the design of a database (for a specific application : planning, scheduling...), designers and users should be aware of inaccuracy that is likely to affect data quality and to take it into account to select relevant attributes only. The entities in the same entity type have not the same importance in the manufacturing process and then, do not require the same level of detail. Thus, it should be possible to simplify numerous current data structures. In perspective, this work will be broaden according to 2 axes : (i) definition of maintenance policies to fix optimal period of data updating, (ii) impact of structuring degree of entities on scheduling performance.

6. REFERENCES

Berio, G., Di Leva, A., Giolito, P., Vernadat, F. (1995). "The M*-object methodology for Information System design in CIM environments", IEEE Transactions on systems, man and cybernetics, Vol 25, N° 1, pp 68-85

Bonjour, E. (1996). "La qualité et la mise à jour des Bases de Données Techniques utilisées en GPAO", Ph.D, Université de Franche-Comté, Besançon

Browne, J., Hahren, J., Shivnan, J. (1988). "Production management system, a CIM perspective", Addison Wesley

Clement, J., Coldrick, A., Sari, J., (1992). "Manufacturing Data Structures: building foundations for excellence with bills of materials and process information ", the Oliver Wight Companies

Companys, R., Falster, P., Burbidge, J.L. (1990). "Databases for Production Management", Amsterdam, North-Holland - proceedings of the IFIP TC5/WG5.7 Working Conference on design, implementation and operations of databases for Production Management, Barcelona, Spain, 10-12 may 1989

Conway, R.W., Maxwell, W.L., Miller, L.W. (1967). "Theory of scheduling", Addison Wesley

Dubois, P., Montagne-Vilette, S. (1992). "De la conception des systèmes d'informatisation de la gestion de production : une question de temps ?", In: "Les nouvelles rationalisations de la production", Cépaduès-Editions, Toulouse, 1992, pp 87-106

Hatchuel, A., Sardas, J.C., Weil, B. (1988). "La mise en oeuvre et le pilotage d'une GPAO : à chaque étape ses difficultés", Revue française de Gestion Industrielle, N°3, pp 49-63

Giard, V. (1988). "Gestion de production", 2è édition, Economica, Paris

Landvater, D.V., Gray, C.D. (1989). "MRP II Standard System, a handbook for Manufacturing Software Survival", the Oliver Wight Companies,

Pels, H.J., Wortmann, J.C. (1990). "Modular design of integrated databases in Production Management Systems", Production Planning and Control, Vol 1, N°3, pp 132-146

Plossl, G.W. (1990). "Cost accounting in manufacturing: dawn of a new area", Production Planning and Control, Vol 1, N° 1, pp 61-68

Redman, T. C. (1992). "Data quality: management and technology", New York, NY: Bantam Books

THE PRODUCTION CONTROL WITH THE JIT-BUFFER

Dipl.Ing.Dr.Soonhee KANG
Add.: Hoerlgasse 14-2-2-11, 1090 Vienna, Austria
e-mail: Soonhee_Kang@blackbox.at

Institute for Betriebswissenschaften, Arbeitswissenschaft and Betriebswirtschaftslehre,
Technical University of Vienna, Austria

Abstract: Although the philosophy of the JIT production is wellknown in many manufacturing companies as an ideal solution to shorten throughput time and to cut back on work-in-progress inventory, it cannot always be easily realized. The reason is, that there is no compact concept for the JIT production besides the KANBAN system above all in job-shop manufacturing.
And it's true that the KANBAN system can fulfill the most important requirement for the JIT production, the rule circle with the pull principle, but it can be used only on a line manufacturing system because of the physical KANBAN box. Therefore the KANBAN system is an inacceptable solution to realize the JIT production in a different manufacturing system, for example in job-shop manufacturing system.
The essential development of the new concept **production control with the JIT-Buffer** was to realize the pull principle on the rule circle with the JIT-Buffer. The JIT-Buffer is virtual and filled with work hours for manufacturing orders. In order to manage the JIT-Buffer, the simulation is put in. The production control that is carried out in the simulation will be transferred in the real manufacturing system.
The advantage of the production control is, that it can be used on all manufacturing systems from line manufacturing system to flexible manufacturing system.

Keywords: JIT, simulation, job-shop manufacturing, production control, KANBAN

1. INTRODUCTION

1.1 Change of the manufacturing factories

In the last fifteen to twenty years, a strong structural change has occurred in the production program of most manufacturing factories. Until the 1960's, demands were not sufficiently satisfied because of the poorness of production programs, of mass production of similar products with few variants, and of manufacturing for an anonymous market. Increasingly saturated markets because of mass production led strongly to an order producing job-shop manufacturing, to the development of immense numbers of complete solutions for the customers' desires, and to the complex development of manufacturing structures.

The result of these developments was an increase of throughput time, which led to an increase of work-in-progress inventory; therefore, the manufacturing factories required a more coordinating system for production planning and control. The goal of these systems was to organize the production systems in order to cut back the work-in-progress inventory, to reduce the throughput time, and to keep delivery dates with high efficiency of capacities.

The classic systems of production planning and control (PPC) could not reach the goals despite massive use of computer systems. They could not manage the two important real manufacturing processes splitting and overlapping at the manufacturing; therefore, the attempt to manage the complex processes in job-shop manufacturing through the ppc-systems was a failure.

In addition, many of the job-shop manufacturing companies searched for a solution in the JIT-Production in order to adapt themselves to the structural change from the mass production for an anonymous market to the job-shop manufacturing with the goals of good delivery faith, short throughput time, low work-in-progress inventory, and high efficiency of capacities.

But a ppc-system with a concept of the JIT-Production is not widely known in the job-shop manufacturing. It is handicapped by the lack of a concrete concept for the conversion of the philosophy of the JIT-Production into practice. Specifically, if it is about the job-shop manufacturing, the conversion into practice is more difficult, because of the dynamic complex of manufacturing structure. The KANBAN-System is an excellent concept for the JIT-Production, but it imposes on the job-shop manufacturing a few restraints that make the use difficult. Thus, a new concept for the JIT-Production in the job-shop manufacturing had to be developed.

1.2 Simulation of PPC (production planning and control)

1.2.1 Simulation

According to the VDI guidelines 3633, the simulation is a reproduction of dynamic processes in a model to come to a realization that is transferable to reality. And Komarnicki defines: „Simulation is first technology that includes generation of a model, that is a copy of the real situation, and after it the experiments will be started by the model.

The generation of a model means the building of a model that should reflect a real system with dynamic processes. The simulation means the built model to experiment in order to know, how the model reacts to a certain input. In fig. 1 a graphic presentation and a mathematical model of a simulation model are shown.

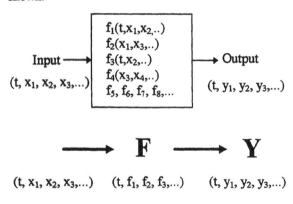

Fig. 1 simulation model

For the functions of the model f_1, f_2, f_3, f_4, ... in fig. 1 one can find, that some of them are dependent on time and others independent of time. The dependent functions are the functions whose results or values change in time. The weather situation, work-in-progress inventory and number of visitors in a stadium, etc. are belonging to those. The independent functions are the functions whose results or values don't change in time: number of machines, number of seat places in a stadium, etc. The mathematical simulation model F is then a combination of all the functions and again a dependent function in time $F(t, f_1, f_2, f_3, ...)$.

The values of the input variables $X(t, x_1, x_2, x_3, ...)$ can be given in the function $F(t, f_1, f_2, f_3, ...)$. After the processing of the values in the model the output values $Y(t, y_1, y_2, y_3, ...)$ can be get. By this process the model was experimented once. If the experiment is run successively for a time, then it is a simulation.

In fig. 2 circal of the simulation of the model is shown. After the circle the output variables $Y(t, y_1, y_2, y_3, ...)$ can be controlled and regulated for the input variables $X(t, x_1, x_2, x_3, ...)$.

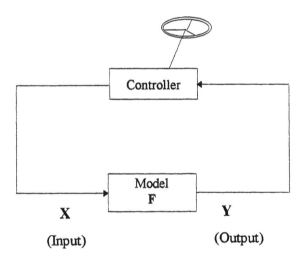

Fig. 2 Simulation

1.2.2 Simulation of PPC

Experts of the classic ppc-systems believe that only simulation can offer a solution to the combination of the dynamic processes in the job-shop manufacturing. Schmidt and Ortmann suggest for the ppc-system the use of simulation that shows in the computers the realistic material flow at processing and the waiting for orders. The goal of it's use in ppc-system is that it should bring dynamic processes of a production system and its material flow under control, but the present ppc-systems could not control it.

The effort of organizing dynamic processes realistically in the ppc-systems was often tried in a simulation. But there was no remarkable success in the job-shop manufacturing up to now. Lack of a leading concept is responsible for it. Actually the simulation itself does not mean a solution or a optimizing method, which leads automatically to the generation of a solution, but it imitates only the running process through the model technique. It implies that a concept must be developed and by means of it a model must be built to simulate the dynamic processes realistically.

If a model is built, it should be integrated in a ppc-system as seen in fig. 3.

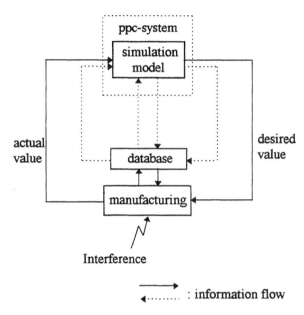

Fig. 3 simulation model in ppc-system

1.3 Purpose of the paper

The classic ppc-systems cannot reflect the dynamic processes realistically enough for production control. And the increase of job-shop manufacturing has made the development of a new concept necessary, which can organize a production system to the JIT production. In this paper is developed the production control with the JIT-Buffer in order to perform in the two conditions, to reflect the dynamic processes and to perform the JIT production. The production control with the JIT-Buffer should organize a production system of the job-shop manufacturing realistically for the JIT production, where different products are produced, many machines are set up, and stocks are often filled with surplus.

The simulation is used in the new concept and a simulation model will be built to reflect the dynamic processes of the production system. The simulation model should be integrated in a ppc system and control the whole production system for the JIT production. With that good delivery faith, short throughput time and low work-in-progress inventory can be reached.

2. PRODUCTION CONTROL WITH THE JIT-BUFFER

2.1 Conditions for the JIT production

Just-in-Time is a production strategy with the requirement to stand by or produce materials at the right time, with the right quality and quantity on the right place. The next figure shows the structure of the conditions for the JIT production and means: the more the conditions are met, the better the JIT production can be realized. In fig. 4 the conditions for the JIT production are presented.

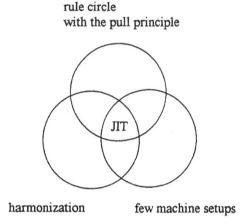

Fig. 4 conditions for the JIT production

2.1.1 Rule circle with the pull principle

The basic condition for the JIT production is the rule circle with the pull principle for the material flow to offer materials at the right time as seen in fig. 5.

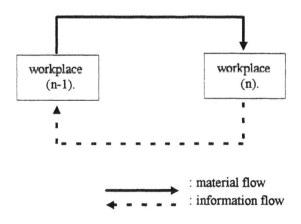

Fig. 5 rule circle with the pull principle

There must be designed an impluse rule (information flow) for the rule circle according to at what times a workplace pull material from the last workplace (material flow). The rule circle connects then the two workplaces with the information and material flow. If the rule circle is expanded from the two workplaces to all workplaces, the whole material flow can be included in the rule circle and controlled for the JIT production as described in fig. 6.

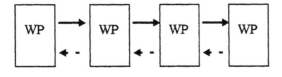

WP: workplace

———————————→ : material flow

◄ - - - - - - - - · : information flow

Fig. 6 material and information flow in the rule circle with the pull principle

2.1.2 Harmonization

Before a ppc-system tries the JIT production in manufacturing, already the manufacturing system should be harmonized or synchronized. The reason is, that the JIT production is possible only in a stable manufacturing system. The stable manufacturing system can be reached only by low variation of all production factors: piece-work-time, machine setup time, capacities, demands, etc. And low variation can be balanced by harmonization.

The harmonization means alignment of the working rhythm in the manufacturing system. The less products there are in the manufacturing, the easier the harmonization will be, for example in mass production, where there are only a few production factors to harmonize. The more different the products, the more difficult it is, for example in a special order manufacturing, because the different products lead to high variation of the production factors. And the harmonization in a job-shop manufacturing is not easy because of the variety of the production processes.

2.1.3 Few machine setups

Machine setup is rearrangement of a workplace for work on a new manufacturing order. During the machine setup the workplace will be blocked. This causes unstable material flow that can exclude the workplace from the JIT production. If there are many machine setups, the workplace will not realize

the JIT production. And if the whole production system needs many machine setups, then the whole JIT production will be failed to perform. This problem characterizes the special order manufacturing. Only where there are few machine setups, e.g. line production, series production, the JIT production can be set in and run successfully.

2.2 Production control with the JIT-Buffer

2.2.1 JIT-Buffer

The beginning of the new concept was how the JIT production could be realized in job-shop manufacturing, where the KANBAN system could not be used because of some restraints. A fundamental hindrance for the job-shop manufacturing was the KANBAN box in which definite products must be taken.

Another instrument was needed instead of a box in order to take in many products from different orders of the job-shop manufacturing. For this purpose the JIT-Buffer was developed.

The JIT-Buffer is a virtual container that exists only in a simulation model instead of a real buffer in front of a workplace. The size of a JIT-Buffer will be measured not by volume but by time. In the JIT-Buffer many different waiting orders will be taken in together with the conversion to their work hours. Then the inventory of the JIT-Buffer (actual height) corresponds with the sum of their work hours.

$$H_{act} = \sum_{k=1}^{n} WH_k$$

H_{act}: actual height of the JIT-Buffer
WH_k: work hours for next process of the k-th wating order in buffer

In this way the buffer is not regarded any more as a physical stock, but as a stock with hours. This conversion of the buffer is clearer and more useful for a dynamic material flow in the job-shop manufacturing, because the inventory is directly related to the capacity of workplaces through the JIT-Buffer. And by the actual height H_{act} of the JIT-Buffer it can be easily recognized whether a workplace is overloaded or not.

2.2.2 Control by the JIT-Buffer

The new concept has to perform the most important condition for the JIT production, the rule circle with the pull principle. The JIT-Buffer is set in the rule circle as seen in the fig. 7.

The impulse to the rule circle, by which the products

are pulled from the last workplace, is given through a limit of the JIT-Buffer. This limit is called „limit height". The principle of the new production control is: the actual height H_{act} of the JIT-Buffer should be continually kept at the limit height H_{lim} as near as possible. Owing to the principle the actual height H_{act} can be held constantly.

- → : information flow

→ : material flow

.... : limit height H_{lim}
        (desired value)

——— : actual height H_{act}
        (actual value of
        the JIT-Buffer)

JB: JIT-Buffer on a workplace

WP: workplace

a, b: sequence of the works

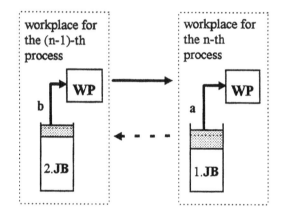

Fig. 7 rule circle with the pull principle in the production control with the JIT-Buffer

As soon as the workplace of the n-th process takes an order from the 1. JIT-Buffer (a in the fig. 7) and the actual height H_{act} goes under the limit height H_{lim}, a new order from the 2. JIT-Buffer will be taken on the workplace of the (n-1)-th process (b in the fig. 7) in order to reach again the limit height H_{lim} of the 1. JIT-Buffer. If the actual height H_{act} does not fall the limit height H_{lim} of the 1. JIT-Buffer in spite of taking a order, then a new order will be not taken from the 2. JIT-Buffer on the workplace.

In case there are many workplaces for the (n-1)-th process, one workplace must be chosen to reach again the limit height H_{lim} of the 1. JIT-Buffer. And if through the chosen order the actual height H_{act} of the 1. JIT-Buffer tops the limit height H_{lim}, then no other workplace can take any new orders for working.

2.2.3 Limit height H_{lim} of the JIT-Buffer

The JIT-Buffer is a buffer like a stock before a workplace. A definite limit is given to all JIT-Buffers. The sum of the work hours for waiting orders in the JIT-Buffer shouldn't be over this limit. It will hold the work-in-progress inventory low and realize the JIT production.

The higher the limit height H_{lim} is, the more the work-in-progress inventory and the worse the chance for success become. The lower it is, the less the work-in-progress inventory and the better the chance for the JIT production become. Naturally, if the limit height H_{lim} is too low, there can be a risk of loosing efficiency of capacities. Therefore it is very important to set the limit height H_{lim} exactly.

The task of the limit height H_{lim} is to take in the unstable material flow that is caused by inharmonization of some production factors, and to guarantee a stable material flow. The limit height H_{lim} will be set from parts of some production factors that couldn't be completely harmonized for the JIT production and that can lead to unstable material flow. The typical production factors that can't be completely harmonized in a job-shop manufacturing are piece-work-time, setup time and capacities.

$$H_{lim} = h_w + h_c + h_m + h_s$$

H_{lim}: limit height of the JIT-Buffer
h_w: part of the piece-work-time
h_c: part of the capacity
h_m: part of the machine setup
h_s: safety part

The limit height H_{lim} can be determined through the simulation of the built model. The separate determination of the parts will be wrong, because instability of a production factor can influence the instability of another production factor.

2.2.4 Conditions for the production control with the JIT-Buffer

Before this concept is set in a production system, the following conditions should be met.

a. Harmonization
The harmonization is an important condition for this concept, because different values of a production factor through various manufacturing processes in job-shop manufacturing should be aligned and harmonized. By means of the harmonization the basis for the stable material flow can be created. Without the harmonization this concept can't be realized in job-shop manufacturing. The high variation of the values of production factors can prevent the job-shop manufacturing from the JIT production. It's of particular importance to harmonize the piece-work-time and capacities.

b. few machine setups

The machine setup is an interruption of the manufacturing process and causes an instable material flow. Too many machine setups can remove the advantages of the JIT production that is obtained by the production control with the JIT-Buffer. Therefore it will be pointless if this concept is set in a production system with many machine setups.

c. Database

The simulation model needs actual data before every manufacturing period. The job of the database is to supply the simulation model with actual, reliable, correct and safe data before every manufacturing period in order to inform the model about actual situation of the material flow and to get reliable results after the simulation.

The application field of the new concept extends from mass production to job-shop manufacturing. The success of this concept will be dependent on whether the simulation model is built like a real production system and the material flow in the model is run realistically, e.g. splitting and overlapping.

It is still to hope that the production control with the JIT-Buffer will be applied in many manufacturing companies, above all in job-shop manufacturing companies. And the new concept will be important in the future for the automatical manufacturing that will take over more and more by the developing of the automation technique.

3. CONCLUSION

As seen in the KANBAN system the rule circle with the pull principle is the most important condition for the JIT production. The new concept can perform it by the JIT-Buffer. The essential point is that the JIT-Buffer is not a real but a virtual object and is located only in a simulation model that is built according to a production system. The JIT production through the JIT-Buffer can be shown in the simulation model before every manufacturing period. After the simulation the production system will be organized so that the real material flow is runned like in the simulation.

4. REFERENCES

Adam, D. (1992). *Fertigungssteuerung 38/39*.

Komarnicki, J. (1980). *Simulationstechnik*, p.14

Schmidt-Weinmar, G. and L. Ortmann (1991). *Zeitdynamische Simulation bei PPS-Systemen*.

Kapoun, J. (1987). Modellsimulation in der Logistik, eine grundsätzliche Betrachtung, *Fördertechnik* 56, p.12.

Zäpfel, G. (1989). *Neuere Konzepte der Produktionsplanung und -steuerung*, p.142.

Wildemann H. (1984). *Flexible Werkstattfertigung durch Integration von Kanban-Prinzipien*, p.39.

A GA BASED APPROACH TO ROUTING PROBLEM:
A CASE STUDY OF CONSTRAINTS GRAPH DRAWING PROBLEM

Katsumi Hama

Mechanical Engineering, Hakodate College of Technology
Tokuracho 14-1, Hakodate, Hokkaido, 042 Japan

Masaaki Minagawa

Sapporo Gakuin University, Bunkyodai-11, Ebetsu, Hokkaido, 069 Japan

Yukinori Kakazu

Hokkaido University, Kita 13, Nishi 8, Kita-Ku, Sapporo, 060 Japan

Abstract: This paper describes an attempt to solve routing problem of electricity cables using GA algorithms. Determining the shortest routes of cables and pipes is a common optimization problem in manufacturing, and the problem increases its complexity depending on the problem size and given geometrical and technical constraints. In our problem setting, it is reasonable to use a rectilinear Steiner tree taking practical situations into account and it is attempted to solve a routing problem to find a minimum cost spanning tree. Based on the proposed methodology, computational experiments are carried out under different conditions and some experimental results are shown.

Keywords: Graphs, Drawings, Tree structure, Combinatorial, Optimization problem Routing algorithms, Genetic algorithms

1. INTRODUCTION

A Graph is a convenient concept to represent abstractly concrete information and relational structures between system elements. Important tasks for a graph-based application are how to generate the layouts of the graphs and how they can be customized. Examples of such tasks include networks, entity-relationship diagrams, data-flow graphs, electric circuits and so on (Biedl, 1996), (Papakostas and Tollis, 1996). Thus graph is a method for visualizing information and system structures. To let visualization of graph function as an effective means for concept transmission, it is necessary to create good diagram, but there is a limit in manual work. So, an automatic method using computer is indispensable for realization, in that case, it is required to solve a problem concerned with human recognition and computational complexity.

Various graph drawing algorithms are proposed until now. Those are classified theory oriented algorithm, skillful heuristic algorithm and algorithm

based on simulation using mechanical model and thermodynamic model (Sugiyama, 1993). In layout of such general automatic graph drawing, aesthetic elements, for example the number of edge crossings and symmetrical property, are used as evaluation criterion in many cases.

On the other hand, in routing problem which is handled in this study, rather minimizing the number of bends and the total length of edges on a routing than making it beautiful is important. In this way, determining the shortest routes of cable and pipe is a common optimization problem in manufacturing, and the problem increases its complexity depending on the problem size and given geometrical and technical constraints. The problem of finding shortest route and minimal spanning tree in graphs is special cases of the Steiner problem in graph theory (Agarwal and Shing, 1990). Given a graph and a designated set of vertices, it is to find a minimum cost graph spanning the designated vertices. Considering real situation, rectilinear Steiner tree which connects a set of vertices with horizontal and vertical line segments is effective and its concept can be also applied to this study (Ho *et al.*, 1990). Since a basic spanning tree can be generated in deterministic way, when all the degrees of vertices are given, the problem will be finding optimal layouts for the edges of the tree. In the case of no spatial constraints, for suitable starting point for deriving the optimal tree, a minimum spanning tree can be obtained. When constraints are given, however, a minimum spanning tree may not necessarily be an adequate initial structure for optimal routing.

So, for this reason, we apply GA both to find layouts for edges of the determined spanning tree and to select different structures of the tree itself (Julstrom, 1993), (Esbensen, 1995). In this study, the routing space geometry is represented using cellular representation (two dimensional space grid) ; terminal points (e.g. power supplies) and obstacles (e.g. electrical equipments) are given as a cell and as collections of cells, respectively. The problem is to connect all these points avoiding obstacles. GA based approach uses two kinds of string. One is a fixed length string for deciding the spanning tree and another is variable length for representing layout for each edge of the tree. Each individual in population corresponds to a spanning tree and is evaluated according to the route length, the number of bends, the length which doesn't pass along equipment and wall and the length of overlaps between the layouts. The structure of tree and edge layouts are altered appropriately using two fundamental GA operations.

Through formulating the proposed methodology and carrying out computational experiments according to it, we inspect its applicability and learning performance.

2. THE PROBLEM DESCRIPTION

The purpose of this section is to suggest a fundamental problem setting. So we describe the problem P using the following 6-tuple.

$$P = (S, Q, U, \Phi, \Lambda, C) \tag{1}$$

where
S : the problem space for placement
Q : the set of terminal points
U : the set of attribute values of terminal points
Φ : the mapping which gives routing
Λ : the route of routing
C : evaluation to a routing

Suppose that a two dimensional shape in the problem space is represented as the union of convex set as shown in Fig.1. That is,

$$\Sigma = \bigcup_i \sigma_i \tag{2}$$

where
$$\sigma_i = \bigcap_{i=1}^{M} X_i \quad , \quad X_i = \{p \mid F_i(p) \geq 0\} \quad (p \in R^2)$$

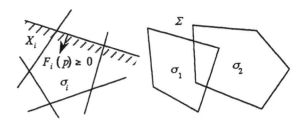

Fig.1 A two dimensional shape in problem space

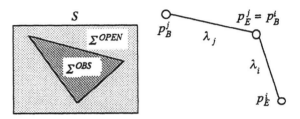

Fig.2 Two different areas   Fig.3 A routing connected
in problem space   two subroutes

Denote the whole space by S and locatable area, obstacle area by Σ^{OPEN}, Σ^{OBS}, respectively as shown in Fig.2. Then the relation of them is described as

$$S = (\Sigma^{OPEN}, \Sigma^{OBS}) \quad or \quad \Sigma^{OPEN} = S - \Sigma^{OBS} \tag{3}$$

The set of terminal points Q and the set of attributes of them U are expressed as

$$Q = \{q_i \mid i = 1, 2, \cdots, N_{TERM}\} \quad (q_i \in \Sigma^{OPEN}) \tag{4}$$

$$U = \{u_i \mid i = 1, 2, \cdots, N_{TERM}\}$$

The route of routing is described as a set of subroutes as follows:

$$\Lambda = \{\lambda_i \mid i = 1, 2, \cdots, N_{TERM}\} \tag{5}$$

where
$$\lambda_i = (p_B^i, p_E^i) \quad , \quad p_B^i, p_E^i \in \Sigma^{OPEN}$$

Suppose that the route which connects p_B^i to p_E^i is represented with straight line. Then if $p_B^i = p_E^j$ or $p_B^j = p_E^i$, it implies that two subroutes, λ_i and λ_j, are connected as shown in Fig.3. Denote a set of points which exists on the line segment connected such two points by t_{ij}. And let Γ be the set of terminal points on the route of routing, it described as

$$\Gamma = \{\gamma_i \mid i = 1, 2, \cdots, N_{NODE}\} \tag{6}$$

That is, this permits to generate a new set of via points B besides the set of terminal points given by Q. Therefore the relation is represented as follows:

$$\Gamma \supseteq Q \quad , \quad B = \Gamma - Q \tag{7}$$

Represent the route of routing with graph expression such as

$$G = (V, E) \tag{8}$$

where
$$V = \{v_i \mid i = 1, 2, \cdots, N_{NODE}\}$$
$$E = \{e_i \mid i = 1, 2, \cdots, N_{SEG}\}$$

In case that a routing from one source of power supply to all destinations becomes a tree structure, the relation of the number of vertices and that of edges is $N_{SEG} = N_{NODE} - 1$. Denote the length of each subroute by λ_i, the total length is obtained as

$$L = \sum_i |\lambda_i| \qquad (9)$$

The mapping which gives routing is expressed as

$$\Lambda = \Phi(S, Q, U) \qquad (10)$$

where this map also includes generating via points. Using above descriptions, therefore, the routing problem can be represented as

$$min \quad L + \Psi \qquad (11)$$

where Ψ is penalty term and is used in case such as conditions except an absolute constraint are disturbed. As its condition, it is required that a routing does not enter in the constraint area as shown in Fig.3 and it is described as

$$for \ \forall \ i, j \quad t_{ij} \not\subset \Sigma^{OBS} \qquad (12)$$

3. THE GENERATION OF SPANNING TREE

In the beginning of this section we explain the Steiner problem briefly. This problem is that given a set of n points on the plane, consider generating the shortest network which connects them all with segments. On this occasion, it is possible to add to an arbitrary number of points at any position. These are called Steiner points, in case of $n = 3$, an additional point is put on the position where three lines connected its point to each of given three points meet at an angle of 120 degrees each other. Fig.4 (a) shows a spanning tree on $n = 8$. Fig.4 (b) is a tree added Steiner points and it is called Steiner tree. This is an shortest network. In practical routing, however, it is necessary to consider space constraints and appearance problem and it is difficult to correspond in this form. So, to keep generality, the structure which makes Steiner tree orthogonal as shown in Fig.4 (c) is thought about.

The route of routing is equivalent to a tree structure as a connected graph without loop in graph theory. There is degree as a means to represent the tree structure. This is equal to the number of edges connected with each point. The degree of each point is determined by generating a sequences of $n - 2$ digits. Adopting case that $n = 5$ and the generated sequence is "323", we indicate a procedure to generate spanning tree which is starting point for routing. Number surrounded by circle in Fig.5 represents the degree of the point.

(1) Number from 1 to n in each of n points and initialize degree of each point to 1.
(2) Make a sequence of $n - 2$ digits by generating randomly number from 1 to n and increment the degree of corresponding number one by one.

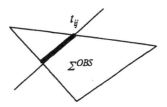

Fig.3 An example of absolute constraint

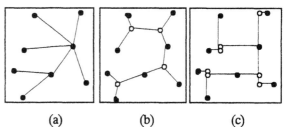

Fig. 4 (a) Spanning tree, (b) Steiner tree
(c) Rectilinear Steiner tree

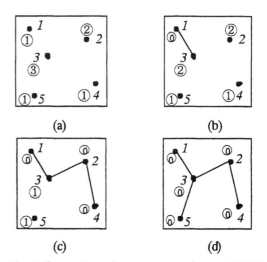

Fig. 5 Generation of spanning tree (string "323")

(3) Check the degree in order from the point of number 1 and connect the point whose degree is 1 with the point which corresponds to first number of the generated sequence. Decrement the degree of each point one by one.
(4) Repeat step (3) until the degree of every point is 0, except for two points whose degree are 1.
(5) The last edge joins these two points.

4. THE APPROACH METHOD

In addition to representing the routing space which includes terminal points and obstacles using cellular representation, we prepare the grid for layout $G_r(Q)$ obtained by drawing horizontal and vertical lines through each of terminal points as shown in Fig.6 (a). During the search of the routes, the cables will run in four orthogonal direction (forward, back, left and right) avoiding obstacles. Fig.6(b) shows an example of edge layout. The candidate set of via points B is expressed same as Eq.(7) as follows:

$$B = \Gamma - Q \qquad (13)$$

269

where
$$\Gamma = Q \times Q$$

Then, a subroute (edge) is described as

$$\lambda_i = \Phi_1\left(p_b^i, p_e^i\right) \tag{14}$$
$$= \left(p_b^i, b_1^i, b_2^i, \ldots, b_{Ni}^i, p_e^i\right), \quad b_j^i \in B$$

where Φ_1 is the mapping which gives a sequence of orthogonal segments between two given points.

The algorithm based on GA uses two kinds of strings. One is a sequence of digits for deciding the spanning tree and another is a pattern sequence for representing layout for each edge of the tree. The latter is managed every each edge and movement of four orthogonal direction is represented by integers from 0 to 3. Therefore, the collection of variable length strings which are constructed by these digits give a solution of the routing problem.

Assigning a maximum length of string according to the size of objective space, initial population is randomly generated considering the following constraints.

(1) Each cable is permitted to change its extension direction only at intersection points of the layout grid.

(2) The weight in selecting next direction is set up such as straight routing can be selected in a high probability.

To evaluate performance of each individual, we define an evaluation function and adopt the individual having the maximum value as a solution. The following values are included as function parameters:

1) The length of straight part and the number of bend.
2) The length passing along equipment and wall.
3) The length of overlap among layouts.

Considering the above parameters, we define the evaluation function of each individual F_i as follows:

$$F_i = a_1 \sum_{j=1}^{n-1} f_j + a_2 G, \quad f_j = \frac{1}{\alpha L_j + \beta C_j + \gamma W_j} \tag{15}$$

where n is the number of given points, L_j is the length of edge, C_j is the number of bends, W_j is the length which the route is not passing along wall and equipment, G is the length of the overlaps among the layouts for all edges of the tree and a_1, a_2, α, β, γ are coefficients, respectively.

By this, the strings with high performance are reproduced in high probability. As selection operation, proportional strategy and elite strategy are used together.

Next, we set necessary electric capacity to each given point, when two points are connected, we consider such routing as the difference of electric capacity between them is as small as possible. That is, in case as Fig.7 (a), suppose that the routing of Fig.7 (c) is better than that of Fig.7 (b). In accordance with this conception, f_j in Eq.(15) is redefined as

$$f_j = \frac{1}{\alpha L_j + \beta C_j + \gamma W_j} + \delta \frac{R_1 + R_2}{R_{max}} \tag{16}$$

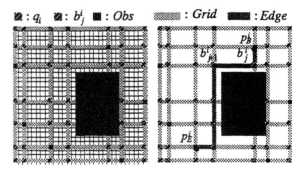

$\boxtimes: q_i$ $\boxtimes: b_j^i$ $\blacksquare: Obs$ $\text{▨}: Grid$ $\blacksquare: Edge$

(a) Grid for layout $Gr(Q)$ (b) An edge layout

Fig.6 Orthogonal representation of layout

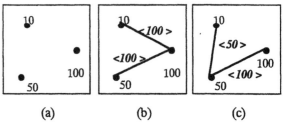

(a)   (b)   (c)

Fig.7 An example of routing considered electric capacity

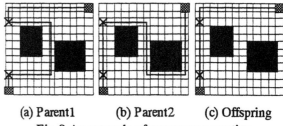

(a) Parent1   (b) Parent2   (c) Offspring

Fig.8 An example of crossover operation

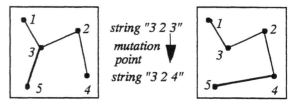

(a) Before operation   (b) After operation

Fig.9 An example of mutation operation

where R_1, R_2 are each electric capacity of two connected points, R_{max} is the either big one and δ is coefficient.

The tree structure and edge layouts are altered appropriately using the following two fundamental GA operation.

The crossover is performed locally in edge unit. First, two other edges having two same intersections (crossover points) from two parents is selected. The layout between those intersections is estimated partially and the better portion is inherited. An example of crossover operation is shown in Fig. 8.

Mutation changes the structure of the spanning tree by changing degrees of some points. An example of mutation operation is shown in Fig. 9.

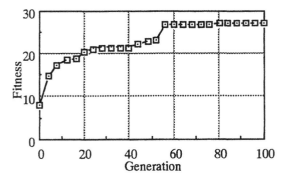

Fig. 10 Transition of the fitness (without capacity)

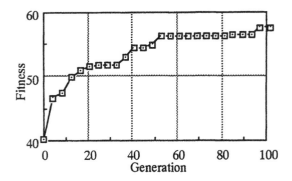

Fig. 12 Transition of the fitness (with capacity)

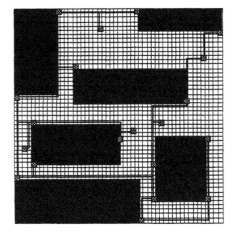

Fig. 11 The result of routing (without capacity)

5. COMPUTATIONAL EXPERIMENTS

Based on the above problem setting, we carried out some computational experiments under different conditins to inspect usefulness of solution by GA. Foudamental experimental conditions for small size problem are as follows:

· The size of space : $X = 50$, $Y = 50$
· The number of given points : $n = 20$
· The maximum length of gene : 100
· The population size : $Popsize = 40$
· The probability of mutation : 0.5
· $a_1 = 2000$, $a_2 = 0.005$, $a_3 = 0.1$, $b = 0.8$, $c = 0.3$

Each value of coefficients was decided from data obtained under situation that there are not obstacles. Here, as result of computational experiments, the change of fitness in each generation and the route of routing captured actually are shown. In the figure of routing, small box and big one indicate terminal point and obstacle, respectively. Fig.10 and Fig.11 are the results on normal connection. The fitness is changing in steps. Those portions are regarded as the change of the time when the solution got away from a local one. The captured routing result is satisfied with criterion of evaluation in point of that passing along wall and equipment as well as minimizing total length.

Fig.12 and Fig.13 show the results of case that electrical capacity is considered under the same experimental conditions. The tendency of change of

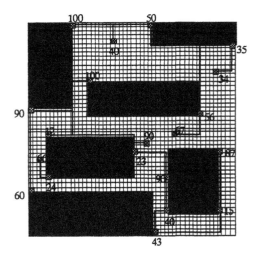

Fig. 13 The result of routing (with capacity)

fitness is similar to that of the normal case. From only the result of routing, it is difficult to judge whether the routing is good or bad. However, with respect to connecting between two points having comparatively near value of capacity, the result is contented with to a certain degree.

Next, we performed a little larger size problem increased the size of space and the number of given points. Here, the result of spanning tree used as the initial state for routing is also indicated. The results of case that $X = 70$, $Y = 70$ and $n = 40$, are shown in from Fig.14 to Fig.16. And Fig.17 to Fig.19 show the results of case that $X = 100$, $Y = 100$ and $n = 100$. Fitness is gradually rising in both. In the case of spanning tree and routing, there is considerable room for improvement in each case. But from the viewpoint of satisfying absolute condition and making no loop over all space, it is thought that both routing results are good comparatively.

6. CONCLUSION

1) We took up the routing problem as one of graph drawing problems and proposed a GA based approach to solve it.
2) The route of routing is captured by combining a deterministic generation of spanning tree from a set of given points with a stochastic layout for each edge of the tree.

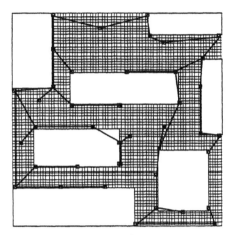

Fig. 14 The result of spanning tree

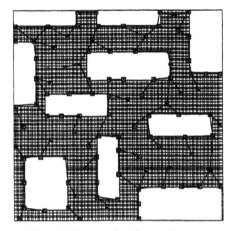

Fig. 17 The result of spanning tree

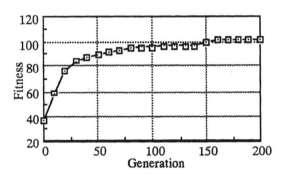

Fig. 15 Transition of the fitness

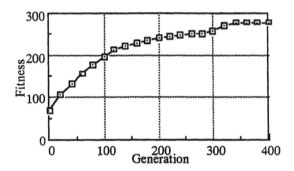

Fig. 18 Transition of the fitness

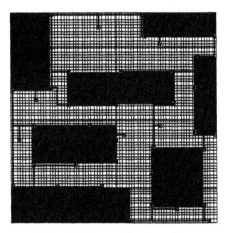

Fig. 16 The result of routing

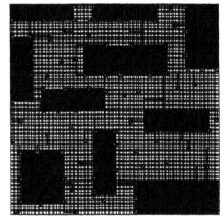

Fig. 19 The result of routing

3) Based on the proposed methodology, we carried out experiments and showed a usefulness of our approach.

4) Corresponding to a practical problem included increase of given points will be discussed in our future work.

REFERENCE

Sugiyama, K. (1993). Automatic Grapf Drawing and its application, *Library of SICE*, 1-43.

Agarwal, P. K. and Shing, M.T. (1990). Algorithms for Special Cases of Rectilinear Steiner Trees : 1. Points on the Boundary of a Rectilinear Rectangle, *Network*, vol.20, 453-485.

Ho. J. M.. Vijayan, G. and Wong, K. (1990). New algorithms for the Rectilinear Steiner TreeProblem, IEEE Trans. on Computer-Aided Design, vol.19, no.2, 185-193.

Julstrom, B. A. (1993). A Genetic Algorithm for the Rectilinear Steiner Problem, Proc. of the Fifth Internatonal Conference on GAs, 474-480.

Esbensen, H. (1995). Finding Optimal Steiner Tree in Large Graphs, Proc. of the Sixth International Conference on GAs, 485-491.

Biedl, T. C. (1996). New Lower Bounds for Orthogonal Graph Drawings, Graph Drawing, Springer, 28-39.

Papakostas, A. and Tollis, I. G. (1996). Issues in Interactive Orthogonal Graph Drawing, Grap Drawing, Springer, 419-430.

A NEW FAMILY OF FUZZY SET ALBEBRAS DERIVED FROM INFORMATION THEORY

Guy Jumarie

Department of Mathematics, University of Quebec at Montreal,
P.O. Box 8888, Downtown St; Montreal, Qc;
H3C 3P8; Canada
jumarie.guy@uqam.ca

Abstract. One of main problems which arise in implementing intelligent systems via approximate reasoning based on fuzzy sets is the suitable selection of the corresponding fuzzy set algebra, which depends upon the kind of problems under consideration. In this way, it may be of interest to have several algebras at hand, and the purpose of the present paper is to provide such other alternatives. This new approach works via information theory, and proposes new definitions for the union, the intersection and the complement of fuzzy sets. The relation with the min-max modelling is exhibited.

Keywords: Fuzzy set algebras, information theory, approximate reasoning, union, intersection, complement of fuzzy sets, informational algebra

1. INTRODUCTION

One of the problems which occur with approximate reasoning is that there is not one possible model only, but that, on the contrary, several approaches are available, in such a manner that, it is not always easy to decide which one is the best in a given framework. So, in some instances, in the implementation of intelligent systems involving approximate reasoning, it may be wise to use several different models at the same time, and then to compare their results in order to avoid wrong making decision.

Approximate reasoning can be considered in two different points of view: probabilistic reasoning and fuzzy reasoning. In probabilistic reasoning, we deal with exact concepts, and it is the rationale which involves probabilistic arguments; whilst in contrast, in fuzzy reasoning, we are manipulating data which involve some vagueness in their definition. In the latter case, it is customary to expand approximate reasoning by using fuzzy algebras, so that various schemes will yield various reasoning models, therefore various ways to test a given result before any making decision.

In the present paper, we propose a new model of fuzzy set algebra, say an *informational fuzzy set algebra* , which is mainly based upon Shannon information theory.

2. OBSERVATION AND INFORMATION THEORY

2.1 Observation with informational invariance

Several years ago (Jumarie, 1978, 1986) we suggested that any observation would be ruled by a so-called "principle of observation with informational invariance" which can be stated as follows:

Principle of observation with informational invariance. Assume that an observer R is observing an object S. Assume further that S is characterized by an amount of information I. Clearly, I is the total amount of information involved in the definition of S, which furthermore is needed to completely identify the latter. Then, the observation process will be performed in such a way that the amount of information measured by R is constant and equal to I. ∎

In words, in order to observe an object, one will enlight it, one will rotate it, but in any case, one neither destroys nor creates information about this object.

2.2 Observation with invariance of information loss

There are many practical situations where one cannot meaningfully assume that there is no loss of information when one observes something. In order to take account of this feature, one will assume that the observation process as the whole is characterized by the grade, the degree of this decrease of information, and one wll assume that the following assumption holds.

Principle of observation with invariance of information loss. Assume that an observer R is observing an object S which is characterized by an amount of information I(S). Assume further that the observation process involves some defects in such a manner that the amount of information seized by R is not I(S) but I(S/R) which is lower than I(S). Then this observation process will run in such a way that the amount of information I(S/R) is constant.∎

All the problem now is to find a suitable modelling for this information loss, and in the next section, we shall see that the so-call Renyi entropy is an excellent candidate to this end.

3. COMPLEMENT OF FUZZY SETS VIA RENYI ENTROPY

3.1 Information theoretic framework

Renyi entropy Let X denote a 1-D real valued random variable defined by the (complete) probability distribution $(p_1,..,p_n)$; then its Renyi entropy of order α, $\alpha > 0$, $\alpha \neq 1$, is defined by the expression

$$H_{R,\alpha}(X) = \frac{1}{1-\alpha} \sum_{i=1}^{n} p_i^{\alpha}, \quad (1)$$

The Renyi entropy is lower than the shannon entropy when α is larger than the unity, and as a result, in an observation with information loss, one can assume that the amount of information which is seized by the observer is not H(X), but rather is $H_{R,\alpha}(X)$ with $\alpha > 1$.

3.2 Complement of fuzzy sets revisited

We refer to the probabilistic meaning of membership functions, and we introduce the Reny entropy

$$H_{R,\alpha}(f,x) = (1-\alpha)^{-1} \ln(f^{\alpha}(x,A) + \\ + f^{\alpha}(x,A')\ (2)$$

where f(x,A) and f(x,A') are the membership functions of the fuzzy set A and of its complement A' respectively.

In the observation with invariance of information loss, one will postulate that

$$H_{R,\alpha}(f,x) = C(\alpha) = constant, \quad (3)$$

As a result, one would have the equality

$$f^{\alpha}(x,A) + f^{\alpha}(x,A') = \\ \exp\left\{(1-\alpha)C(\alpha)\right\}. \quad (4)$$

3.3 Information theoretic meaning of Zadeh's definition of fuzzy set complements

To some extent, Zadeh's fuzzy set complement can be associated with an observation process with informational invariance. But this is true to some extent only; since strictly speaking, in the case of informational invariance, one would have rather

$$-f(x,A)\ln f(x,A) - (1-f(x,A)) \\ \ln(1-f(x,A)) = constant. \quad (5)$$

4. AN INFORMATION THEORETIC APPROACH TO INTERSECTION OF FUZZY SETS

4.1 Application of Shannon entropy

Proposition 4.1 A definition for the intersection of the two fuzzy sets A and B, which is fully consistent with Shannon entropy of random variables, is given by the equation

$$f^2(x,A \cap B) = f(x,A)f(x,B)∎ \quad (6)$$

Proof By using various arguments on the practical modelling of fuzzy sets, one can write

$$H(\ (x \in A) \cap (x \in B) = \\ H(x \in A) + H(x \in B) = \\ h(f,x) + h(g,x) = \\ -\ln f(x,A) - \ln g(x,B). \quad (7)$$

4.2 Application of Renyi entropy

The arguments can be duplicated step by step, and it will provide the same result. The points of importance lies in the following remarks:
(i) The Renyi entropies of the events (x element of A) and (x element of B) are still

$$h_{R,\alpha}(f,x) = -\ln f(x), \quad (8)$$
$$h_{R,\alpha}(g,x) = -\ln g(x), \quad (9)$$

(ii) Next, for two independent random variables, one still has the equality

$$H_{R,\alpha}(E_1, E_2) = H_{R,\alpha}(E_1) + H_{R,\alpha}(E_2) \quad (10)$$

5. AN INTERFORMATION THEORETIC APPROACH TO THE UNION OF TWO FUZZY SETS

5.1 Preliminary remark

Let E denote a random experiment which provides the results e1, e2,..,en with the respective probabilities p1, p2,..,pn. Its Shannon entropy and its Renyi entropy can be meaningfully considered as the entropy of the union of these events, clearly

$$H(E) = H(e_1 \cup e_2 \cup .. \cup e_n), \quad (11)$$

and

$$H_{R,\alpha}(E) = H_{R,\alpha}(e_1 \cup .. \cup e_n), \quad (12)$$

As a result, for the events e1,..ej only, we shall refer to the entropy of incomplete probability distributions which involve the partial sum of the probabilities as normalizing coefficients.

5.2 Application to union of fuzzy sets

Definition 5.1 A definition of the union of fuzzy sets, which is fully consistent with Shannon entropy of random variables on the one hand, and which furthermore satisfies the condition $f(x, A \text{ union } A) = f(x,A)$ on the other hand, is provided by the expression

$$\ln f(x, A \cup B) = f(x, A) \ln f(x, A) + \\ + f(x, B) \ln f(x, B) \,/\, \\ f(x, A) + f(x, B) \blacksquare \quad (13)$$

Definition 5.2 A definition of the union of fuzzy sets, which is fully consistent with entropy of random variables on the other hand, and which furthermore satisfies the condition $f(x, A \text{ union } A) = f(x,A)$, is provided by the expression

$$\ln f(x, A \cup B) = \\ -\frac{1}{1-\alpha} \ln \frac{f^\alpha(x, A) + g^\alpha(x, B)}{f(x, A) + g(x, B)} \blacksquare (14)$$

These definitions can be supported by noticing that if we assume they do not hold on the one hand, and if we consider the special case when the membership functions are characteristic functions of crisp sets, then one would get a contraciction.

6. EXTENSIONS TO n FUZY SETS

6.1 Intersection and union of n fuzzy sets

In order to simplify the reading of the paper (a student would say for pedagogical reasons), in the preceding section, we restricted ourselves to the union and the intersection of two fuzzy sets only, but the reader will have understood easily that the arguments which we have used can be easily applied to n sets.

The intersection of n fuzzy sets Ai, i=1,..,n, defined by the membership functions f(x,Ai), i = 1,..,n, is yielded by the equation

$$f_1 \cap .. \cap f_n = (f_1 .. f_n)^{1/n}, \quad (15)$$

and the union is such that

$$\ln(f_1 \cup f_2 \cup ... \cup f_n) = \\ -\frac{1}{1-\alpha} \ln \frac{f_1^\alpha + f_2^\alpha + ... + f_n^\alpha}{f_1 + f_2 + ... + f_n} \quad (16)$$

In the special case when $\alpha = 1$, this equation turns to be

$$\ln(f_1 \cup f_2 \cup .. \cup f_n) = \\ -\sum_{i=1}^{n} f_i \ln f_i \Big/ \sum_{i=1}^{n} f_i \quad (17)$$

6.2 On Associativity of Union and Intersection of fuzzy sets

It is clear that these definitions do not satisfy the associativity requirement. So, the question which then arises, is whether this property is a compulsory prerequisite or on the contrary whether can be dropped.

We think that, at this stage of the discussion, we must carefully make the difference between two cases, depending upon the purpose of the modelling. We may want to build up a theoretical scheme of approximate reasoning which is as simple as possible, and in this way, it may be recommended to assume that associativity should hold. But we may want also to describe human approximate reasoning, and then, it is not sure at all that associativity is meaningful.

So, as far as independency of fuzzy variables could be sufficient to disqualify assiociativity, or at least to make it doubtful in fuzzy algebra, not only human reasoning modelling, but also artificial engineering devices should reflect this feature.

Moreover, we point out that the non associative fuzzy set algebra so obtained is quite consistent with our purpose. Indeed, as it is mentionned in the introduction, our main goal is to have several concurrent models implemented in the same device, and in this way we shall compare the results of assiociative algebras with those of non associative algebras.

7. INCLUSION OF FUZZY SETS

7.1 On the implication "If A then B"

Production systems (in artificial intelligence) are based on the repetitive application of the implication

If A Then B

In the fuziness framework, this rule is extended in the form

If A (to degree f) Then B (to degree g).

It is customary to associate this expression with the definition

$f(x, \text{If A Then B}) = \min(f(x,A), f(x,B))$

which is the direct application of the min-definition of fuzzy set inclusion.

In our approach to fuzzy set algebra via information theory, we have to follow another way for implementing the implication, and it is the purpose of the next section.

7.2 Inclusion of fuzzy sets revisited

In the information theoretic approach, we shall choose to extend the usual definition of crisp set inclusion (A is contained in B if x element of A impleas x element of B), and we shall do it as follows.

Definition 7.1 Let A denote a fuzzy set defined by the membership function $f(x,A)$ on the universe U. We shall say that an element x of U is in A whenever $f(x,A)$ is strictly positve.■

Definition 7.2 Let A and B denote two fuzzy sets defined by the membership functions $f(x,A)$ and $f(x,B)$ respectively. We shall say that A is contained in B whenever

$$f(x,A) > 0 \Rightarrow f(x, A \cap B) > 0 \ ■$$

The basic difference with the min-definition is that, here, in the information theoretic approach, one can have A included in B even when $f(x,A)$ is greater than $f(x,B)$ for some x. In other words, the present definition picture the fact that, in practice, inclusion would be working on the set A as the whole, intead of locally, at each point.

8.COMPARISON WITH OTHER RESULTS

8.1 Informational approach and min-max rule

Lemma 8.1 The following inequalities hold

$$\min (f,g) \leq f \cap g \leq f \cup g \leq \max (f,g) \tag{18}$$

Lemma 8.2 Consider the expression

$$f \cup g = \frac{f^2 + g^2}{f + g} \tag{19}.$$

With the definition of fuzzy set inclusion as in Definition 7.2, if $f1(x,A1)$ is included in $f(x,A)$ and $g1(x,B1)$ is included in $g(x,B)$, then the union of f1 and g1 is included in the union of f and g.

8.2 About the De Morgan's laws

It is clear that the new definitions so obtained for the union and the intersection of fuzzy sets, together with the definition of the complement do not satify the De Morgan's laws. Here, they are only more or less satisfied, what is, moreover, quite in the wake of approximate reasoning.

If we take as a compulsory requirement that they should be satisfied, then we shall firstly define the complement with either the union or the intersection, and then we shall use the De Morgan's laws to derive the remaining composition rule. By this way, one can obtain the following ensemble of fuzzy set algebras.

9. SUMARY OF FUZZY SET ALGEBRAS

As a synthesis, in the present section, we summarize the basic equations of the various fuzzy set algebras so obtained by using information theory. For the sake of completeness, we shall begin with the min-max rule and the probabilistic one.

Model 1 (Zadeh)

$$
\begin{aligned}
f + f' &= 1 \\
f \cap g &= \min (f,g) \\
f \cup g &= \max (f,g)
\end{aligned}
$$

Model 2 (probabilistic like)

$$
\begin{aligned}
f + f' &= 1 \\
f \cap g &= fg \\
f \cup g &= f + g - fg
\end{aligned}
$$

Model 3 (via de Morgan's laws)

$$
\begin{aligned}
f + f' &= 1 \\
f \cap g &= (fg)^{1/2} \\
f \cup g &= 1 - \left[(1-f)(1-g) \right]^{1/2}
\end{aligned}
$$

Model 4 (via De Morgan's laws)

$$ f^2 + f'^2 = 1 $$

$$f \cap g = (fg)^{1/2}$$

$$(f \cup g)^2 = 1 - \left[\left(1 - f^2\right)\left(1 - g^2\right)\right]^{1/2}$$

Model 5 (via De Morgan's laws)

$$f + f' = 1$$

$$f \cup g = \exp\left\{\frac{f \ln f + g \ln g}{f + g}\right\}$$

$$f \cap g = 1 - \exp$$

$$\left\{\frac{(1 - f)\ln(1 - f) + (1 - g)\ln(1 - g)}{(1 - f) + (1 - g)}\right\}$$

Model 6 (via De Morgan's laws)

$$f^2 + f'^2 = 1$$

$$f \cup g = \frac{f^2 + g^2}{f + g}$$

$$(f \cap g)^2 = 1 -$$

$$-\left[\frac{(1 - f^2) + (1 - g^2)}{(1 - f^2)^{1/2} + (1 - g^2)^{1/2}}\right]^2$$

Model 7 (via De Morgan's laws)

$$f^\alpha + f'^\alpha = 1 \quad , \quad \alpha > 0 \; , \alpha \neq 1$$

$$f \cap g = (fg)^{1/2}$$

$$(f \cup g)^\alpha = 1 -$$

$$-\left[(1 - f^\alpha)(1 - g^\alpha)\right]^{1/2}$$

Model 8

$$f^\alpha + f'^\alpha = 1 \quad , \quad \alpha > 0 \; , \alpha \neq 1$$

$$f \cup g = \left[\frac{f^\alpha + g^\alpha}{f + g}\right]^{-1/(1-\alpha)}$$

$$(f \cap g)^\alpha = 1 -$$

$$-\left[\frac{(1 - f^\alpha) + (1 - g^\alpha)}{(1 - f^\alpha)^{1/\alpha} + (1 - g^\alpha)^{1/\alpha}}\right]^{-\alpha/(1-\alpha)}$$

In the next section, we shall summarize some possible desiderata for the operations on fuzzy sets, which could be of help to select the best model for a given kind of problem.

10. REVIEW OF SOME DESIDERATA FOR OPERATIONS ON FUZZY SETS

On the surface, some possible requirements for a "good" definition of intersection of fuzzy sets could be as follows.

(i) Symmetry. For every pair of fuzzy sets, one should have

$$f(x, A \cap B) \quad f(x, B \cap A) \quad (20)$$

(ii) Invariance of the intersection with the universe U,

$$f(x, A \cap U) = f(x, A) \quad (21)$$

(iii) Idempotence,

$$f(x, A \cap A) = f(x, A) \quad (22)$$

(iv) The intersection of the fuzzy sets (A,f) and (B,g) should yield the intersection of crisp sets when f and g are the characteristic functions of two crisp sets.
(v) The following inplication should hold,

$$A_1 \subset A \; , B_1 \subset B \Rightarrow A_1 \cap B_1 \subset A \cap B \quad (23$$

(vi) The following implication should hold

$$A \subset B \Rightarrow f(x, A \cap B) = f(x, A) \quad (24)$$

(vii) Associativity,

$$f(x, A \cap (B \cap C)) = f(x, (A \cap B) \cap C) \quad (25)$$

(viii) The fuzziness of the intersection of A and its complement A' should be small.
(ix) The fuzziness of the intersection of A and B should be a fuzzy number.
(x) For any λ, $0 < \lambda < 1$, one should have the equality

$$\lambda f(x, A) \cap \lambda f(x, B) =$$
$$= \lambda (f(x, A) \cap f(x, B)) \quad (26)$$

It seems that the requirements (i) to (vi) should be compulsory. As we pointed out, associativity can be or may not be satisfied. Assumption (viii) is quite understandable, and there remains of course to define whay do we mean by "small". The requirement (ix) refers to the fact that an observed membership function is only an element of the family of membership functions associated with the same fuzzy set. The requirement (x) would emphasize the improtance of membership magnitude in the operations on fuzzy sets.
In the above statements, we have considered the intersection of fuzzy sets, but the reader will state easily the equivalent for union of fuzzy sets.

11. CONCLUDING REMARKS

There is not only one model of approximate reasoning, but several ones; and this feature reflects what happens, in practice, in human rationale. And Zadeh himself, on several times in various meetings, agreed with this remark, when he admitted that other alternatives to the min-max rules, would be quite acceptable at first glance.

Once again, we do not claim that the min-max approach is not good or is not satisfactory!! And even on the contrary we showed how information theory can bring some support to Zadeh's definition of fuzzy set complement! All we say is that, since the rationale of approximate reasoning is not free of mistakes, it may be wise to have several different schemes which will work simultaneously, so that we can compare their results before making a decision of some importance.

Our main point of departure has been to consider membership as a subjective probability. Fuzzy set scientists have repeatedly pointed out that membership cannot be thought of as probability, and that is correct; but the pai (j1-f) can be dealt with as such, at leats for rough preliminary manipulations. And it exactly what we did here.

There remains to make an exhaustive comparison of all these algebras, to exhibit their main characteristics and differences.

ACKNOWLEDGEMENT

This paper was written whilst the author, who is a French citizen from West Indies (Guadeloupe Island) has been living in Canada. All the help and the support came from outside the country.

REFERENCES

Bellman, R. and Giertz, M.; On the analytic formalism of the fuzzy set theory, *Information Sciences,* vol 5, pp 149-156, 1973

Goguen, J.A.; Concept representation in natural and artificial lanhuages: axioms, extensions and applications for fuzzy sets, *International Journal of Man-Machine Studies,* vol 6, pp 513-561, 1974

Haack, S.; Do we need fuzzy logic?, *International Journal of Man-Machine Studies,* pp 537-445, 1979

Hersh, H.M. and Caramazza, A.; A fuzzy set approach to modifiers and vagueness in natural language, *Journal of Experimental Psychology: General,* vol 105, no 3, pp 254-276, 1976

Jumarie, G.; Theory of relative information and communication (in french), *Annals of Telecommunication,* vol 33, Nos 1-2, pp 13-27, 1978

Jumarie, G.; *Subjectivity, Information, Systems. Introduction to a Theory of Relativistic Cybernetics,* Gordon and Breach, London, 1986

Jumarie, G.; New concepts of fuzzy-subjective probabilities and their applications to soft-making decision, *Proceedings of the Second IFSA Congress,* pp 568-571, 1987

Jumarie, G.; *Relative Information, Theories and Applications.* Springer Verlag, Berlin, 1990

Jumarie, G.; Further results on the mathematical relations between probability, possibility and fuzzy logic. The rationale of subjectivity via invariance of information loss, *Systems Analysis, Modelling and Simulation,* (to appear 1997)

Klir, G.J.; Probability-possibility conversion, *Proceedings of the 3rd IFSA Congress,* Seattle, pp 408-411, 1989

Koksalan, M.M. and Daagli, H.C.; A fuzzy programming approach to departemental planning, *Technical Report,* Middle East Technical University, Ankara, Turkey, 1981 (also presented at the CORS/TIMS/ORSA Joint National Meeting, Toronto, Canada, 3-6 May, 1981)

DESIGN OF A FUZZY LOGIC CONTROL
USING A SINGLE VARIABLE

*Byung-Jae Choi, **Seong-Woo Kwak, and **Byung Kook Kim

* : Dept. of Electronic Engineering,
Joong Kyoung Technical College,
155-3, Jayang-dong, Dong-gu, Taejon, 300-100, Korea
E-mail : bjchoi@gaia.kaist.ac.kr

** : Dept. of Electrical Engineering,
Korea Advanced Institute of Science and Technology,
373-1, Kusong-dong, Yusong-gu, Taejon, 305-701, Korea

Abstract: In this paper, we design a new FLC (Fuzzy Logic Control) using a single fuzzy input variable. Since most of conventional FLCs commonly use error and its derivative as fuzzy input variables. Then the fuzzy rule table is established on a two-dimensional space of error and its derivative. From properties of the table, a new variable called a signed distance is derived, which is used as a sole fuzzy input variable. Thus, the number of total rules is greatly reduced. And almost equal control performances are obtained. In order to compare the control performances of above cases, we will perform computer simulations using a nonlinear plant.

Keywords: fuzzy logic control, PD-type FLC, sliding mode control, signed distance.

1. INTRODUCTION

There has been growing interest in using fuzzy set theory for control systems. Since L. Zadeh had introduced the fuzzy set theory and E. H. Mamdani had applied to the areas of automatic control, the FLC has emerged as one of the most active areas for research in the application of fuzzy set theory (Lee, 1990). In particular, it appears useful control method for the plants with the difficulty in derivation of a mathematical model or with the limitation in performances using conventional linear control schemes. The majority of works in the field of the fuzzy control theory use error and its derivative as fuzzy input variables (Lee, 1993; Driankov et al., 1993). Also, either control input or incremental control input is used as the variable representing the rule-consequent (then-part of a rule).

Tang and Mulholland (1987) developed the relation between the scaling factors of the fuzzy controller and the control gains of the equivalent linear PI controller. And they showed that the FLC can be used as the multiband control. Lee and Tu (1993) proposed a FLC that is based on the SMC to formulate the linguistic rules. Kawaji and Matsunaga (1991) explained a design method of the FLC based on the SMC to guarantee the stability of the system. Palm (1992 and 1994) proposed a sliding mode fuzzy control which generates the absolute value of the switching magnitude in the sliding mode control law using error and its derivative. Since most of them use two fuzzy input variables in the rule-antecedent, the rule tables are constructed on a two-dimensional space.

In this paper, we design a new FLC using a sole fuzzy input variable. Most of the conventional FLCs use error and its derivative as variables representing the contents of the rule-antecedent. And they generate either a control input or an incremental control input. Then, the fuzzy rule table is established on a two-dimensional space of the phase plane, that is, it has two dimension which is composed of error and its derivative. Generally, the two-dimensional rule table has the skew symmetric property and it appears that the absolute magnitude of a control input is proportional to the distance from its diagonal line. This property allows us to derive a new variable which is called a signed distance. This variable is the distance with a sign to a current state from a straight line through the origin. Thus, a signed distance can be negative. Now, we use this variable as a single fuzzy input variable. The control scheme of a new FLC is less fuzzy due to using a single variable instead of two as input variables of the FLC. Also the number of fuzzy rules is greatly reduced compared to the case of the conventional FLC. Furthermore, control performances are nearly the same as those of the conventional FLC. And this control algorithm easily produces the SMC (Sliding Mode Control) with BL (Boundary Layer). That is, it is simply showed that the SMC with BL is merely a kind of the FLC. Hence, the stability of the FLC is implicitly guaranteed from the inherent robustness property of the SMC. In order to compare the control performances of above cases, we perform computer simulations using a nonlinear plant and show that the results of the proposed FLC are nearly the same as those of the conventional FLC.

In Section 2, we give a short review of the conventional FLC. And then, we describe a new design method for the FLC and the equivalence between a new FLC and the SMC with BL in Section 3. Continuously, we discuss the control performances of the conventional and the proposed FLC via computer simulations in Section 4. And concluding remarks will be offered in Section 5.

2. THE CONVENTIONAL FUZZY LOGIC CONTROL

The FLC appears useful control method for the plants with the difficulty in derivation of the mathematical model or with the limitation in performances using conventional linear control schemes. Most of the fuzzy controllers use error and its derivative as the process state variables representing the contents of the fuzzy rule-antecedent (if-part of a rule). Also, either a control input or an

incremental control input is used as the variable representing the contents of the rule-consequent (then-part of a rule). Namely, in the conventional FLC, a control input or an incremental control input is generated by error and its derivative. This results from the linear PD or PI control scheme.

Let the plant to be controlled be a system of nth order (linear or nonlinear) state equation:

$$
\begin{aligned}
x^{(n)} &= f(x, \dot{x}, \cdots, x^{(n-1)}, t, u) \\
y &= g(x) \\
e &= x_d - x
\end{aligned}
\quad (1)
$$

with

$$
\begin{aligned}
\mathbf{x} &= (x_1, x_2, \cdots, x_n)^T \\
&= (x, \dot{x}, \cdots, x^{(n-1)})^T,
\end{aligned}
\quad (2)
$$

where $\mathbf{x}$ is the process state vector, and y, u, e, x_d and t are process output, control input, error, desired value, and time parameter, respectively. And where f and g are the linear or nonlinear functions. In the following, we do not write the time parameter t as an argument of states or functions of states for simplicity.

For a large class of systems, FLCs are mostly designed with respect to the phase plane of error and its derivative. That is, a fuzzy value for a control action is determined according to fuzzy values of error and its derivative.

The following rule is the typical form of the FLC which generates a control input using error and its derivative.

$R_{old}^{(i)}$:

If e is $LE^{(i)}$ and $\dot{e}$ is $LDE^{(i)}$ then u is $LU^{(i)}$,

where $LE^{(i)}$, $LDE^{(i)}$, and $LU^{(i)}$ are the linguistic values taken by the process state variables e and $\dot{e}$, and the control input u in the ith rule, respectively. And

$$
i = 1, 2, \cdots, I,
\quad (3)
$$

where I is the total number of rules. Above rule shows that the control input is directly obtained from error and its derivative. Hence, this is called a PD-type FLC or a position-type FLC.

In the case of using five linguistic values, the rule table for the FLC is mostly given by Table 1.

For example, one rule of Table 1 is as

follows :

R^1_{old} : *If e is LE_{-2} and $\dot{e}$ is LDE_2 then u is LU_0.*
As shown in Table 1, the linguistic values for a control input are obtained in two-dimensional space of error and its derivative, also we know that the total number of rules are 25 in this case. Subscripts -2, -1, 0, 1, and 2 imply fuzzy linguistic values of Negative Big (NB), Negative Small (NS), ZeRo (ZR), Positive Small (PS), and Positive Big (PB), respectively.

Table 1. Rule table for the FLC.

$\dot{e}$＼e	LE_{-2}	LE_{-1}	LE_0	LE_1	LE_2
LDE_2	LU_0	LU_1	LU_1	LU_2	LU_2
LDE_1	LU_{-1}	LU_0	LU_1	LU_1	LU_2
LDE_0	LU_{-1}	LU_{-1}	LU_0	LU_1	LU_1
LDE_{-1}	LU_{-2}	LU_{-1}	LU_{-1}	LU_0	LU_1
LDE_{-2}	LU_{-2}	LU_{-2}	LU_{-1}	LU_{-1}	LU_0

Above rules can be expressed as the following equation:

$$u_{old} = h_1(e, \dot{e}), \qquad (4)$$

where $h_1(\cdot, \cdot)$ is commonly an arbitrary nonlinear function. Also, the function can be linear by using product-sum and center-average methods for inferencing and defuzzification, respectively. Then following equation holds (Mizumoto, 1995).

$$u_{old} = K_1 \cdot e + K_2 \cdot \dot{e}, \qquad (5)$$

where K_1 and K_2 are arbitrary positive constants. That is, we can design a FLC which has the same form as the linear PD controller.

3. DESIGN OF A NEW FLC USING A SINGLE VARIABLE

From Table 1, we can see that the rules represent the skew symmetric property with respect to the main diagonal. If the quantization levels of the independent variables are halved, then the boundaries of the control regions are less fuzzy. Furthermore, as they are infinitesimal, the boundaries are no longer fuzzy. That is, as they approach almost zeros, the boundaries become the straight lines rather than the step types (Fig. 1). Then the control law describes the multilevel relay controller with five bands (Tang and Mulholland, 1987).

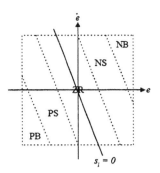

Fig. 1. Rule table with infinite quantization levels.

Consider the line which pass the origin and is parallel to these straight lines.

$$s_l : \dot{e} + \lambda e = 0. \qquad (6)$$

Then the sign of the control input is positive for $s_l > 0$ and negative for $s_l < 0$. Thus, this line is called a switching line because the control inputs have the opposite sign at above and below regions of the line. Also, the magnitude of the control input is proportional to the distance from the switching line. That is, the more the distance increase the more the absolute magnitude increase.

Now, we introduce a new single variable. It is the distance with the sign to the operating point from the switching line and is derived by the followings. Let $H(e, \dot{e})$ be the intersection point of the switching line and the line perpendicular to the switching line from a particular operating point $P(e_1, \dot{e}_1)$. These are illustrated in Fig. 2.

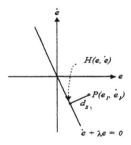

Fig. 2. Derivation of a signed distance.

Then d_1, the distance between $H(e, \dot{e})$ and $P(e_1, \dot{e}_1)$, is expressed by the following equation :

$$d_1 = [(e - e_1)^2 + (\dot{e} - \dot{e}_1)^2]^{1/2}. \qquad (7)$$

Eq. (7) is reduced by the following equation.

$$d_1 = \frac{|\dot{e}_1 + \lambda e_1|}{\sqrt{1+\lambda^2}}. \tag{8}$$

From Fig. 1, we can see that the absolute magnitude of the control action u_1 at a particular operating point is linearly proportional to the distance, d_1. Namely, the following relation holds.

$$|u_1| \propto d_1 = \frac{|\dot{e}_1 + \lambda e_1|}{\sqrt{1+\lambda^2}}. \tag{9}$$

Without loss of generality, Eq. (9) can be rewritten as the following equation.

$$|u| \propto d = \frac{|\dot{e} + \lambda e|}{\sqrt{1+\lambda^2}}. \tag{10}$$

Also, we can rewrite Eq. (10) as follows:

$$\begin{aligned} u \propto d_s &= sgn(s_l) \cdot \frac{|\dot{e} + \lambda e|}{\sqrt{1+\lambda^2}} \\ &= \frac{\dot{e} + \lambda e}{\sqrt{1+\lambda^2}}, \end{aligned} \tag{11}$$

where d_s is called a signed distance which is the distance with a sign and

$$sgn(s_l) = \begin{cases} 1 & \text{for } s_l > 0 \\ -1 & \text{for } s_l < 0 \end{cases}. \tag{12}$$

From the relationship (11), a fuzzy rule table can be established on a one-dimensional space of d_s instead of two-dimensional space of e and $\dot{e}$. In the majority of conventional FLCs, the rule table was established on a two-dimensional space of error and its derivative, but now the table can be constructed on a one-dimensional space using only a single variable, d_s. That is, the control actions can be determined by only d_s. Then the rule form of a new FLC is changed to as follows :

$$R_{new}^{(i)} \ : \ \textit{If } d_s \textit{ is } LDL^{(i)} \textit{ then } u \textit{ is } LU^{(i)},$$

where $LDL^{(i)}$ is the linguistic value of a signed distance in the ith rule. Then the rule table can be established on a one-dimensional space like Table 2.

Table 2. Rule table for a new PD-type FLC.

d_{sl}	LDL_{-2}	LDL_{-1}	LDL_0	LDL_1	LDL_2
u	LU_{-2}	LU_{-1}	LU_0	LU_1	LU_2

From Table 2 we can see that the total number of rules is greatly decreased compared to the case of the conventional FLC. Therefore, we can easily increase the number of rules for the purpose of a fine control.

Proposition: If we assume a linear FLC, then the following equation holds :

$$u = K_3 \cdot d_s, \tag{13}$$

where K_3 is a constant that is obtained by a linear FLC.

Proof: This means that the relationship between input and output of a fuzzy logic-based system is linear. For this, we use a particular type of membership functions, inferencing methods, and defuzzification methods. Membership functions are used triangular types that intersect at the height of 0.5 (Fig. 3). In Fig. 3, $\widetilde{d_{si}}$ represents the fuzzy set of d_{si}. Let d_{s1} and d_{sN} be the minimal and maximal values of a signed distance.

$$d_{s1} \le d_s \le d_{sN}, \tag{14}$$

where N is the number of fuzzy sets for a signed distance.

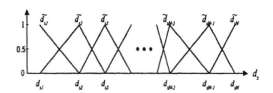

Fig. 3. Fuzzy sets for d_s.

By using the product-sum inferencing method, we can obtain the following final consequence $\widetilde{u'}$.

$$\begin{aligned} \mu_{\widetilde{u'}}(u) &= \sum_{i=1}^{N} \mu_{\widetilde{u_i'}}(u) \\ &= \sum_{i=1}^{N} \mu_{\widetilde{d_{si}}}(d_s) \cdot \mu_{\widetilde{u_i}}(u), \end{aligned} \tag{15}$$

where $\widetilde{u_i}$ represents the fuzzy set of u_i. Since only two rules are fired for given any fact d_s, (15) is reduced by the following equation.

$$\begin{aligned} \mu_{\widetilde{u'}}(u) &= \mu_{\widetilde{u_n'}}(u) + \mu_{\widetilde{u_{n+1}'}}(u) \\ &= \mu_{\widetilde{d_{sn}}}(d_s) \cdot \mu_{\widetilde{u_n}}(u) + \mu_{\widetilde{d_{sn+1}}}(d_s) \cdot \mu_{\widetilde{u_{n+1}}}(u), \end{aligned} \tag{16}$$

where

$$\mu_{\tilde{d}_{sn}}(d_s) = \frac{d_{sn+1} - d_s}{d_{sn+1} - d_{sn}} \qquad (17)$$

$$1 \leq n \leq N-1 \qquad (18)$$

$$\mu_{\widetilde{dsn+1}}(d_s) = 1 - \mu_{\tilde{d}_{sn}}(d_s). \qquad (19)$$

We consider the height defuzzification method and use any triangular type for the membership functions of u which intersect at the height of 0.5. Let u_j be the center or peak value of $\widetilde{u_j'}$ $(j=1,2,\cdots,N)$. Then the fuzzy rule $R_{new}^{(j)}$ can be expressed as follows (Mizumoto, 1995):

$$u_j = K_3 \cdot d_{sj}. \qquad (20)$$

And the following equation holds.

$$\mu_{\tilde{u}_i}(u_i) = 1. \qquad (21)$$

Now, we can get the representative point for the resulting fuzzy set $\tilde{u}'$ as follows:

$$u = \frac{\sum_{i=n}^{n+1} \mu_{\tilde{u}_i'}(u_i) \cdot u_i}{\sum_{i=n}^{n+1} \mu_{\tilde{u}_i'}(u_i)}$$
$$= \frac{\mu_{\tilde{d}_{sn}}(d_s) \cdot u_n + \mu_{\widetilde{d_{sn+1}}}(d_s) \cdot u_{n+1}}{\mu_{\tilde{u}_n'}(u_n) + \mu_{\widetilde{u_{n+1}'}}(u_{n+1})}. \qquad (22)$$

Substituting (17), (19), (20), and (21) into (22), we can obtain (13). ☐

Now, we show that the control law for a new FLC is similar to that for the ordinary SMC with BL. If we assume that $K > 0$ is the maximum value that can be obtained by the fuzzy logic-based controller, (13) can be rewritten as follows:

$$u = \begin{cases} K , & \text{for } u > K \\ -K , & \text{for } u < -K \\ K_3 \cdot d_s , & \text{for } -K \leq u \leq K . \end{cases} \qquad (23)$$

By some manipulations, Eq. (23) is rewritten as follows:

$$u = \begin{cases} K , & \text{for } s_l > \varPhi \\ -K , & \text{for } s_l < -\varPhi \\ \dfrac{K}{\varPhi} \cdot s_l , & \text{for } -\varPhi \leq s_l \leq \varPhi \end{cases} \qquad (24)$$
$$= K \, sat(\frac{s_l}{\varPhi}) ,$$

where

$$\varPhi = \frac{K\sqrt{1+\lambda^2}}{K_3} . \qquad (25)$$

The final equation of (24) is nearly the same as the control law of the SMC with BL. That is, the control law of the SMC with BL is derived from a FLC. Thus, we may say that a FLC is an extension of the ordinary SMC with BL.

4. SIMULATION EXAMPLE

As a simulation example, we consider an inverted pendulum system. Fig. 4 shows the plant that consists of a pole and a cart. The cart moves on the rail tracks to the horizontal direction left or right. The control objective is to balance the pole starting from arbitrary conditions by supplying a suitable force to the cart. For simplicity, we do not consider the position of the cart. Then the plant dynamics is expressed as:

$$\ddot{\theta} = \frac{g\sin\theta + a\cos\theta - \mu_p w^2 l\cos\theta\sin\theta}{l(4/3 - \mu_p\cos^2\theta)} ,$$

$$\mu_p = \frac{m_p}{m_p + m_c} ,$$

$$a = \frac{F}{m_p + m_c} ,$$

where g is an acceleration due to gravity($=9.8\ m/\sec^2$), and F is the applied force. $m_c(=1.0\text{kg})$ and $m_p(=0.1\text{kg})$ are masses and $l(=0.5\text{m})$ is the pole length.

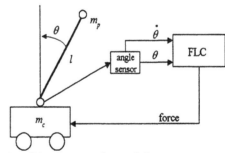

Fig. 4.   The inverted pendulum system.

Fig. 5 represents the fuzzy sets for error, change-of-error, and signed distance. And we use product-sum and center-average method for rule inferencing and defuzzification, respectively.

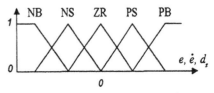

Fig.   5.   The   fuzzy   sets   for   error, change-of-error, and signed distance.

Figures 6 and 7 show the simulation results of tracking performances and control inputs, respectively. Here (a) and (b) represent the cases of the conventional and proposed FLCs, respectively. As shown in figures, the control performances are nearly the same. Furthermore, since our FLC has only 5 control rules, it is much simpler, requires less computation, and has nearly equal performance to the conventional 2-input FLC.

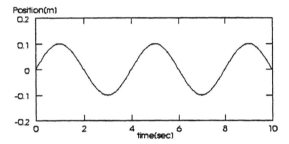

(a) Conventional FLC.

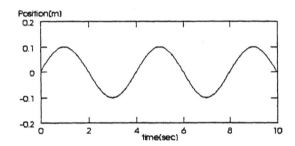

(b) Proposed FLC.

Fig. 6. Comparison of tracking performances.

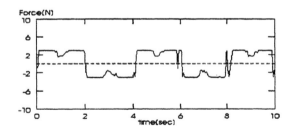

(a) Conventional FLC.

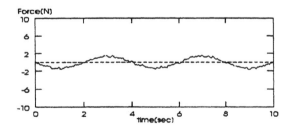

(b) Proposed SMC-type FLC.

Fig. 7. Comparison of control inputs.

5. CONCLUSIONS

We have described two design methods for the FLC. Since the proposed design method requires a sole fuzzy input variable, it has some advantages. One of them is the considerable reduction of the total number of control rules required to obtain the desired performance. So, the computational complexity has been greatly decreased and generation of rules becomes very easy. These advantages has been confirmed by computer simulations.

We also demonstrated that a FLC is almost equivalent to the control law of the SMC with BL. Thus, the stability of the FLC is implicitly guaranteed from the inherent robustness property of the SMC.

REFERENCES

Lee, C.C. (1990). Fuzzy Logic in Control Systems : Fuzzy Logic Controller – Part I , II, *IEEE Trans. Syst., Man, Cybern.,* **20(2)**, pp. 404–435.

Lee, J. (1993). On methods for improving performance of PI-type fuzzy logic controllers, *IEEE trans. Fuzzy Syst,* **1(4)**, pp. 298–301.

Driankov, D., H. Hellendoorn, and M. Rainfrank (1993). *An Introduction to Fuzzy Control.* Springer–Verlag Berlin Heidelberg.

Tang, K.L. and R.J. Mulholland (1987). Comparing fuzzy logic with classical controller design, *IEEE Trans. Syst., Man, Cybern.,* **17(6)**, pp. 1085–1087.

Lee, T.-T. and K.-Y. Tu (1993). Fuzzy Logic Controller Design based on Variable Structure Control, *Proceedings of the 1993 IEEE/RSJ Int. Conf. on Intelligent Robots and Systems,* pp. 958–964.

Kawaji, S. and N. Matsunaga (1991). Fuzzy Control of VSS Type and Its Robustness, *IFSA '91, Brussels,* pp. 81–84.

Palm, R. (1992). Sliding Mode Fuzzy Control, *IEEE Int. Con. on Fuz. Syst.,* pp. 519–526.

Palm, R. (1994). Robust Control by Fuzzy Sliding Mode, *Automatica,* **30(9)**, pp. 1429–1437.

Galichet, S. and L. Foulloy (1995). Fuzzy controllers : Synthesis and Equivalences, *IEEE Trans. Fuzzy Syst.,* **3(2)**, pp. 140–148.

Mizumoto, M. (1995). Realization of PID controls by fuzzy control methods," *Fuzzy Sets and Systems,* **70**, pp. 171–182.

CONTRIBUTION OF FUZZY LOGIC TO MODELLING EXPERTISE IN INTELLIGENT MANUFACTURING PROCESS PLANNING SYSTEMS

Cédric Derras, Muriel Lombard, Patrick Martin

Centre de Recherche en Automatique de Nancy (CNRS URA DO 821)
Equipe Ingénierie de Conception et de Fabrication
Faculté des sciences - BP 239
54506 Vandoeuvre Cedex - France
Phone : (33) 3.83.91.21.50- Fax (33) 3.83.91.23.90 - e-mail : derras@cran.u-nancy.fr

Abstract : This paper aims at describing how fuzzy logic can be used to model process planning expertise in an integrated design context. The presented approach is AI-based : process planning rules do not have the same importance in every case, and this importance depends on the context related to the given problem. This context is expressed through rules that are often imprecise, as they are strongly related to know-how, experience..... It is this context modelling that can handle fuzzy logic concepts through expert rules dynamic weighting.

Key-words: Fuzzy expert systems, Intelligent knowledge-based systems, Decision support systems, Integration, CAM, Mechanical engineering, Knowledge engineering.

1. INTRODUCTION

The process planning activity in manufacturing engineering stands as the interface between the design and the machining activity. Therefore, as a validation of the design production in terms of feasibility and as a planning of the machining activity, it constitutes the key of an efficient integration, the key of integrated design.

Then, in order to develop intelligent decision support tools for the process planning activity, one of the problems a manufacturing engineer encounters is the identification and the semantic comprehension of the knowledge involved in this engineering process. The problem is that this knowledge is often imprecise. Actually, it is issued from know-how, experience, personal or enterprise uses.

In the particular case of process planning, there already exist several general rules that have been implemented in several Computer Aided Process Planning (CAPP) systems.

But one has to conclude that these systems suffer from a lack of flexibility concerning the implemented expertise, when they are confronted with the various problems that arise from the manufacturing production (Ong and Nee 1996). In fact, the manufacturing production needs intelligent CAPP systems (Kusiak, 1990) that should be able, in the authors' view, to qualify the importance of general rules, depending on the context of the problem. It is therefore necessary to represent this context, that is deeply related to the profession. As the context of a problem is a dynamic one through the evolution of criteria defining it, the aim of the presented work is therefore to dynamically weight the importance of an expert process planning rule in relation with this context.

After having detailed this problem, this paper presents on the basis of an example, how fuzzy logic concepts enable the modelling of this imprecise knowledge to perform a dynamic weighting, and how the method has been implemented.

Finally, perspectives of future work deal with the potential of fuzzy logic to model the design information, so that the process planning reasoning may participate in the design process through its expert constraints. This might lead from the development of intelligent CAPP systems towards the development of intelligent integrated design systems.

2. INTELLIGENT PROCESS PLANNING PROBLEM

The process planning problem has evolved these last years from preparing machining activity towards integrating machining constraints in the design process. Actually, up to now, existing systems aim at producing a process plan able to manufacture a product that is totally defined before reasoning. Now, from a partial definition of the part to design, it is important to identify the knowledge enabling to begin process planning reasoning, in order to produce :

- several macro-process plans enabling a more efficient reactivity of the workshop controlling activity,
- but also modifications and/or precision of the part to be machined definition.

Regarding the IMS program (Kurihara, *et al.*, 1996), this paper therefore corresponds to the Process issue, as it deals with :

- a more flexible processing module in a CAPP system,
- better interaction and harmony among various functions namely the manufacturing design and process planning.

Thus, this new problem aims at designing intelligent CAPP systems. However, the process planning activity is known to be an under constraints planning. These constraints are implemented in declarative CAPP systems through "true or false" rules, whatever the problem is. But in manufacturing production, encountered problems are so different and especially the context in which they have to be solved, that these expert constraints should be regarded more or less important (Ong and Nee, 1996). This is what is already tackled in the PROPEL (Tsang, 1988) process planning system in which rules are weighted. But the weighting coefficients are static ones, while it is necessary to make them evolve, in order that the importance given to rules should totally be adapted to the context (part characteristics, production ones...) in which the problem has to be solved.

This might make the system more flexible, more adapted to a particular case, intelligent, through associating to each process planning rule its exact importance during the decision process. The context in this case supports product design information as well as those related to its machining, through expert rules manipulating different types of criteria :

- geometrical ones (tolerances, dimensions...),
- topological ones (emerging surfaces...),
- economical and temporal ones (time allowed, size of series, manufacturing costs...),
- technological ones (surface state, material, tooling...).

The context modelling (expert rules) emerges from know-how which expression is as imprecise as uncertain, on two different points of view :

- information modelling describing each criterion (data : characteristics, attributes...),
- the way each rule, manipulating each type of criterion, influences the importance of a given process planning rule (computation : expression of the rule, links between rules...).

3. FUZZY REASONING APPLIED TO DYNAMIC WEIGHTING OF A PROCESS PLANNING RULE

The fuzzy set concept has been proposed by Zadeh (1965). It aims at allowing an element to belong more or less strongly to a class. As this concept enables the modelling of imprecise information and reasoning with imprecise knowledge, it constitutes the fundamental substrate of fuzzy logic.

In the literature, few applications of fuzzy logic to the process planning problem exist. Zhao (1995), Singh and Mohanty (1991) deal with optimal selection of process plans among several solutions, based on a fuzzy definition of criteria and constraints related to the scheduling activity. More related to the manufacturing activity, Ong and Nee (1996) propose a fuzzy procedural computation of matrices, in which dependence degrees between manufacturing features (tolerances, dimensions...) are expressed. This contribution consists in the way of computing the coefficients standing in matrices and obtained empirically. It does not deal with the expertise, deeply related to the profession, leading to obtaining these coefficients.

This paper presents the method used to handle the modelling of this expertise, which is imprecise (information and computation), based on an example.

3.1. The example process planning rule.

During reasoning to obtain the process plan of the part presented in Fig. 1, Brissaud (1992) writes the following process planning rule :

"when two bores are secant, one has to machine the little one before the big one",

in order to avoid cylindricity problems related to the little bore. This rule is then systematically applied while there are some cases when this is not necessary (when the little bore is a little bit smaller than the big one for instance).

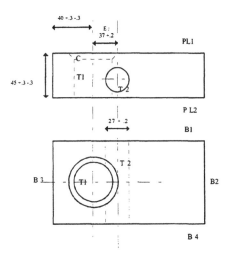

Fig. 1. Part used to present the method.

The importance of this rule depends on a context defined by several criteria and rules manipulating them. Among the different types of criteria the economical and temporal ones are not presented. Fig. 2 shows information enabling the calculation of technological ($\alpha = \arccos[2E/D1]$), topological ($\delta = D2/D1$) and geometrical (tolerance quality θ of the little bore) ones.

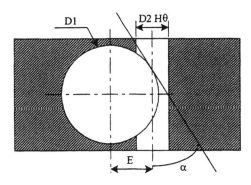

Fig. 2. Part information defining criteria.

These criteria are imprecise regarding their definition (α is more or less large, θ is more or less tight...), as well as regarding their computation (rules manipulating them). Actually :

- IF [(α is large AND δ is big) OR
  (α is small AND δ is big AND θ is large) OR
  (α is large AND δ is little AND θ is large)]
  THEN ρ is little.

- IF [(α is small AND δ is little) OR
  (α is small AND δ is big AND θ is tight) OR
  (α is large AND δ is big AND θ is tight)]
  THEN ρ is important,

with ρ being the weighting coefficient of the process planning rule.

The fuzzy method enabling the representation of these criteria as well as rules manipulating them is composed of the five following steps.

3.2. Fuzzification.

This step enables to characterise a given variable (numerical, linguistic...) into a set of linguistic terms (gradual categories) taken from the human language (Fodor and Roubens, 1994).

For instance, expertise associated to the α criterion is so that α is small between 0 and 30 degrees, and large between 60 and 90 degrees. The universe of discourse related to α is then characterised by two fuzzy sets which membership functions are presented in Fig. 3.

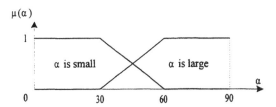

Fig. 3. Terms characterising the angle α.

The same kind of modelling is used for other criteria and for ρ. This step constitutes the modelling of the informational aspect of knowledge.

3.3. Input variables conjunction.

This step constitutes the modelling of rules premises. A general fuzzy proposition is composed of elementary propositions related together with logical connectors (AND, OR).

It was decided to use the minimum t-norm operator to model the AND connector, and the maximum t-conorm operator to model the OR connector.

The fuzzy vector resulting from this conjunction is the composition of all membership grades for each term of all variables.

3.4. Inference.

The inference in fuzzy reasoning is performed using the generalised modus ponens, after having chosen an implication operator, which is the Lukasiewicz one in the present case. The min operator is used as the t-norm of the generalised modus ponens. With f_{MPGi} the membership function of the generalised modus ponens, f_{RLi} the one of the implication and $f_{Ri}(x)$ the one of the characteristic fact :

$$\forall \rho \in [0;1], f_{MPGi}(\rho) = \sup_{x \in X} \min(f_{Ri}(x), f_{RLi}(x,\rho)) \quad (1)$$

Relatively to the fact : $\alpha=50$, $\delta=0.4$ and $\theta=11$ (Fig. 4), the results of this step are presented through the membership functions of rules conclusions on Fig. 5.

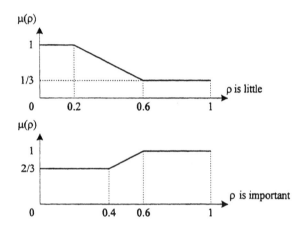

Fig. 5. Results of inference.

3.5. Aggregation of partial conclusions (Fig. 6).

The inference leads to a different conclusion for each inferred rule. These conclusions have to be aggregated in order to produce a single membership function for the ρ variable. In this aim the minimum operator is used as the t-norm involved in this calculation :

$$\forall \rho \in [0;1], f(\rho) = \min(f_{\rho \text{ is little}}(\rho); f_{\rho \text{ is important}}(\rho)) \quad (2)$$

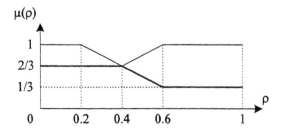

Fig. 6. Result of conclusions aggregation.

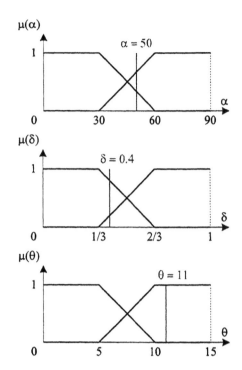

Fig. 4. Representation of the characteristic fact.

3.6. Defuzzification.

This last step aims at translating a membership function into a numerical information, exact value of the weighting coefficient related to the process planning rule. Among the existing methods (Yager and Filev, 1994), the centroid one which is the most used in the authors' knowledge was chosen to perform this task.

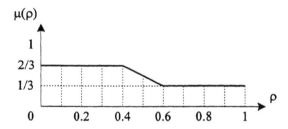

Fig. 7. Conclusion defuzzification.

Then, from a division into ten parts of the universe of discourse related to ρ (Fig. 7), the result is :

$\rho = 0.417$.

Relatively to the observed fact, the result corresponds to the experts' sensibility : a rather little importance of the rule. Actually, even if both diameters are very different (which would classically make one decide that the rule has to be applied), this result is explained by the fact that α is a bit large, and that θ is specified large.

4. IMPLEMENTATION OF THE METHOD

In order to develop a computer platform supporting the whole method, the CAD-X1[1] knowledge-based systems generator was chosen. It is totally adapted to planning under constraints (process planning principle), as it is based on a constraint propagation technology. It also enables to implement information in the form of a logical model, independently from the implementation of the expert rules. Actually, this system is an application development environment containing an object-oriented data management system, a constraint propagation based inference engine, as well as a man/machine interface generator.

Several languages describe the application :

- The ODL (Object Description Language) is used to describe objects (class instances) with which the inference engine works,
- The CDL (Class Description Language) enables to describe object classes that shall then be manipulated by RDL and ODL languages,
- The RDL (Rule Description Language) enables to describe constraints supporting experts reasoning. This is a declarative language that uses a syntax and words closely related to the human language.

4.1. Fuzzy rule implementation.

All concepts used in fuzzy rules (rules, premises, conclusions, variables, fuzzy sets...) are implemented as object classes. As a fuzzy rule is manipulating fuzzy sets that describe a linguistic variable, premises and conclusions of fuzzy rules are fuzzy sets of variables. All these classes are implemented with the CDL, and their instances with the ODL. Fig. 8 shows the "LINGUISTIC_VARIABLE" class as well as one of its instances.

4.2. Fuzzy reasoning implementation.

All logical calculations (AND, OR...), as well as other ones (implication operations, inference operations...) are expressed in the form of constraints. The calculations of implication operations, etc., are based on an equation modelling of fuzzy sets. These constraints are grouped into rules in CAD-X1, and are implemented using the RDL. In the actual application, one rule file is used for each step of the method. Fig. 9 shows a rule that associates its important weight to each process planning rule, which is the value of the "weight" object.

[1] CAD-X1 : KADETECH Society, 40 av. Guy de Collongue, 69130, Ecully, FRANCE

```
{ LINGUISTIC_VARIABLE
  - CHARACTERISTICS
    # Definition ("class of linguistic variables")
    # Position   (0,451,255,300)
  - COMPONENTS
    . List_fuzzy_sets   FUZZY_SETS        compose
    #valuator           (operator)
    . Value             REAL              fundamental
    #valuator           (inference engine)
    . Def_min           REAL              fundamental
    #valuator           (operator)
    . Def_max           REAL              fundamental
    #valuator           (operator)
}
        ┌──▶ { weight : LINGUISTIC_VARIABLE
                . Def_min        = 0.000
                . Def_max        = 90.000
                . List_fuzzy_sets = ( weight_is_little,
                                      weight_is_important )
                . Value          = 0.417
             }
```

Fig. 8. An implemented class, one of its instances.

```
{ CAPP_RULE_WEIGHT_ALLOCATION

  - OBJECTS
    whatever_is  "capp_rule"  a CAPP_RULE
            with "weight"      the Weight
                 "l_var"       the List_linguistic_variable ;

    whatever_is  "var"        a LINGUISTIC_VARIABLE
            in       "l_var"
            with     "value"   the Value
            so_that  "var" = &weight;

  - CONSTRAINTS
                    "weight" = "value";
}
```

Fig. 9. Example of implemented reasoning.

4.3. Results and conclusion.

Table 1 presents the results related to different values of α, δ and θ. The tinted case in Table 1 gives 0.373 as a value of ρ, for $\alpha = 10$, $\delta = 0.8$ and $\theta = 10$. Actually, the little bore is a bit smaller than the large one, and the little bore's tolerance quality is quite large. Therefore, even if α is very little, it is of a rather small importance to machine the little bore before the big one.

Table 1. Different ρ values for different α, δ and θ ones.

	$\alpha = 10$	$\alpha = 50$	$\alpha = 70$	
$\theta = 5$	$\rho = 0.803$	$\rho = 0.627$	$\rho = 0.549$	$\delta = 1/3$
$\theta = 10$	$\rho = 0.617$	$\rho = 0.503$	$\rho = 0.416$	$\delta = 1/3$
$\theta = 5$	$\rho = 0.617$	$\rho = 0.503$	$\rho = 0.416$	$\delta = 0.8$
$\theta = 10$	$\rho = 0.373$	$\rho = 0.245$	$\rho = 0.197$	$\delta = 0.8$

Regarding the implementation, as CAD-X1 does not offer a real data structuring, it seems that an external database is needed. Another point is that the fuzzy method is more an algorithmic than a declarative one; it is then more easily implementable using procedural languages than with declarative ones like the CAD-X1 description languages. Therefore, instead of having all the development supported by CAD-X1, the ideal architecture should be (Fig. 10) :

- all information stocked in an external database,
- the fuzzy method implemented by a procedural language, producing CAPP rules' weights,
- weighted CAPP rules implemented in CAD-X1.

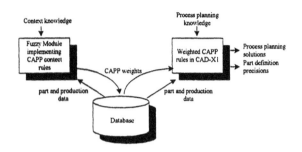

Fig. 10. Ideal architecture of the system.

5. CONCLUSION AND FUTURE WORK

Finally, fuzzy logic answers the initial problem, which is to model the context in which the process planning activity is performed, through the dynamic weighting of CAPP rules.

Moreover, fuzzy logic seems to be a potential tool enabling the integration of the process planning reasoning in the design process. As fuzzy logic enables to directly manipulate linguistic and symbolic variables, it can enable the design function to give imprecise information to the system. A downstream activity like process planning could then benefit of more degrees of freedom concerning its decisions regarding the product. Therefore, the precision of this information could be "brought up" to the design function by this activity, taking into account its own constraints, and would finally allow to produce a completely precise definition plan that would respect constraints of both design and downstream activities (process planning,...).

For instance, toleranced dimensions are intrinsically imprecise information of the plan definition that can be modelled with fuzzy logic. The fuzzy number notion (Yager and Filev, 1994) enables the modelling of this information in order to manipulate it like it really is (an imprecise dimensional value) and not to simultaneously manipulate several information (nominal value, maximum tolerance and minimum tolerance).

Globally, in an integrated design context, fuzzy logic brings a larger modelling flexibility in terms of parts being computed to the process planning problem. During the decisional process, fuzzy logic also enables to keep all possible solutions, whatever their quality degree. Through its semantic power, fuzzy logic allows the modelling of a strong semantic, near the one standing in experts' language and sensibility, as it represents and manipulates information like experts syntactically do.

Finally it is foreseen to precisely identify the design/process planning integration knowledge, to model its imprecise aspect (design information) with fuzzy logic, so that the process planning reasoning may participate in the design process. This might lead from the development of intelligent CAPP systems towards the development of intelligent integrated design systems.

REFERENCES

Brissaud, D. (1992). *Système de conception automatique de gammes d'usinage pour les industries manufacturières*. PhD. Dissertation, Université Joseph Fourier, Grenoble.

Fodor, J. and M. Roubens (1994). *Fuzzy Preference Modelling and Multicriteria Decision Support*. Kluwer Academic Publishers.

Kurihara, T., P. Bunce and J. Jordan (1996). Next Generation Manufacturing Systems (NGMS) in the IMS Program. In: *Proceedings of the 2nd International Conference on the Design of Information Infrastructure Systems for Manufacturing (DIISM'96)* (J. Goossenaerts, F. Kimura, H. Wortann, Ed.), pp. ims3-ims9. Kaatsheuvel The Netherlands.

Kusiak, A. (1990). *Intelligent Manufacturing Systems*. Prentice Hall, Englewood Cliffs, New Jersey.

Ong, S.K. and A.Y.C. Nee (1996). Fuzzy-set-based approach for concurrent constraint set-up planning. *J. Intelligent Manuf.*, 7, 107-120.

Singh, N. and B.K. Mohanty (1991). A fuzzy approach to multi-objective routing problems with application to process planning in manufacturing systems. *J. Prod. Research*, **29.6**, 1161-1170.

Tsang, J.P. (1988). The PROPEL process planner. In: *Proceedings of the 19th CIRP Seminar on Manufacturing Systems*, Vol. 17-2, pp. 115-123.

Yager, R.R. and D.P. Filev (1994). *Essentials of Fuzzy Modelling and Control*. Wiley-Interscience.

Zadeh, L.A. (1965). Fuzzy sets. *J. Information and Control*, **8**, 338-353.

Zhao, Z. (1995). Process planning with multi-level fuzzy decision-making. *J. C.I.M. Systems*, **8.4**, 245-254.

NEURAL NETWORKS AND FUZZY ROBOT CONTROL

Man-Wook Han and Peter Kopacek

Institute for Handling Devices and Robotics, Vienna University of Technology, Austria
Floragasse 7A, A-1040 Vienna, Austria
Tel.: +43-1-504 18 35, Fax: +43-1-504 18 35 9
E-mail: {han, kopacek}@ihrt1.ihrt.tuwien.ac.at

Abstract: Artificial neural networks (ANN) and fuzzy control are two advanced control techniques. Neural networks are applied in many fields because of their learning and parallel processing capabilities. Fuzzy system shows the advantages for easy implementation of system. The conventional control techniques are not sufficient to control mobile robots, because of their dynamic and non-linear behavior. Because of their internal computing mechanism ANN and fuzzy controllers are able to solve these problems well. In this contribution the application of neural networks and fuzzy control as well as integration of both techniques for the navigation is presented.

Keywords: Neural networks; fuzzy control; mobile robot; robot navigation; obstacle avoidance.

1. INTRODUCTION

In the recent year the need on the mobile manipulators consisting of a mobile platform and robot manipulators is increasing for the handling and transportation tasks in many fields, for examples in the manufacturing as well as in service area. For the mobile manipulator the position control and motion control are more complex than that of the conventional industrial robots. The mobile platform works in a partially or fully unknown environment. Such robots differ from industrial robots because they will be individually designed for the execution of given tasks in a specific environment following predefined organizational scheme. Intelligent mobility and intelligent handling are two key features for the intelligent behavior of a mobile manipulator. Fig. 1 shows the components supporting the intelligent mobility.

Two basic requirements for the realization of intelligent mobility are the intelligent sensors and intelligent control algorithms. The industry is waiting for the intelligent robot equipped with external sensors. As already known the development of such external sensors like visual, auditive, force torque is going on very well in research institutes and laboratories. But until now only few of these sensor concepts are available for industrial applications at a reasonable price. Intelligent control algorithm is developed with the combination of artificial intelligence and control system. The progress in the artificial intelligence and sensors serve as the basis for the further development of intelligent robot control.

Two representatives in the advanced control techniques are artificial neural networks (ANN) and fuzzy control. ANN are applied because of their learning capability and parallel processing nature. Two techniques are suitably applied to solve the nonlinear dynamic problem because of their internal processing nature.

While ANN and fuzzy systems are applied in practice, the drawbacks of both techniques are reported, such as the time consuming learning process of neural networks or the existence of an efficient knowledge base for the fuzzy system. Each system can be applied in combination with other methods, such as conventional techniques, expert systems and genetic algorithms. In order to compensate the weakness through reciprocal application of advantages of each method ANN are integrated in fuzzy systems and vice versa. Several methods for the integration of ANN with fuzzy system are already reported, for examples, neural networks model the one of parts of a fuzzy system, and ANN are also applied for the rule extraction of fuzzy system, and others. "Neurofuzzy" is emerged with the aim to combine the learning capability of artificial neural networks with plausibility of fuzzy system. The integration ways are depending on the application, available data and learning algorithms. Neurofuzzy is applied in following fields: the inverted pendulum problem (Bouslama, 93), electric load forecasting (Dash, 95), flight control of a helicopter (Ushida, 94) and others. But there are few publications for the navigation of a mobile robot using neurofuzzy (Sulzberger, 93). The neurofuzzy can be called when one of the system parts gives influences to adjust the parameters of other system.

The aims of this work are to show detail problems by the realization of intelligent mobility for the mobile manipulator and to investigate how neural networks and fuzzy control as well as neurofuzzy methods can be effectively applied.

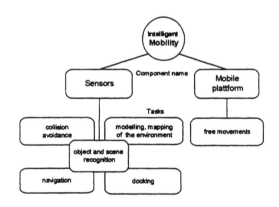

Fig. 1 Components supporting intelligent mobility

2. ROBOT NAVIGATION

For the navigation of a mobile robot, there are several problems to solve, such as modeling of the environment, collision avoidance, navigation, docking, and sensor information processing. The control systems applied for navigation of a mobile robot are conventional (linear) control, adaptive control, ANN, and fuzzy control. The main goal of robot navigation system is the development of real

time, sensor based intelligent respectively autonomous navigation system. The main task of control system is the to generate, execute actions and adapt the behavior based on the environmental changes.

Fig. 2 Nomad200 („Tom")

At the Institute for Handling Devices and Robotics two mobile platforms are currently working. The mobile platform „Tom" (Nomad200 by Nomadic Technologies) is equipped with 16 Ultrasonic sensor, CCD-camera, tactile sensors, and a lift mechanism to perform mobile manipulation tasks.

The other mobile platform „Jerry" (Maxifander by DBI) has a sonar system with 4 rotating heads, infrared sensors, tactile sensor, and other optical and auditive sensor. It can be controlled by clap.

Both robots are tested for the wall following, man-following, collision avoidance, and others. Based on the work in simulation base as well as with real mobile robots followings are found out;

⇨ there are tolerances between planned and actual path internal as well as external influences, such as the unevenness of the surfaces.

⇨ robots do not move smoothly around the corner.

⇨ it is difficult to get out from the small space with small entrance/exit, because of the capability of sensors.

⇨ if two robots simultaneously works with ultrasonic sensors in a room, there are sensing problems.

⇨ the sensor system can not detect the obstacles which are lower than the sensing height.

Therefore it is necessary to develop the more effective control concept and to implement other sensing system, such as vision system, improving the above described weakness.

Fig. 3 Maxifander („Jerry")

3. SENSOR

As being autonomously or (partly autonomously) guided systems, they typically need to move in an unstructured environment without having a-priori knowledge. Regarding to economical reasons, one is interested to increase the autonomy of these systems so that they can take over some tasks of the operator, who again can spend his time for other, more special tasks. To provide this autonomy, there is a need for sensor based guidance - thinking in terms of „low-cost" solutions - operating on the base of ultrasonic sensors.

An ultrasonic sensor system used for the autonomous guidance is required to reliably detect a vast variety of objects. The degree of reflection of the ultrasonic energy differs from almost 100% for a huge plane wall perpendicular to the acoustical axis of the sensor down to 20% for the legs of a human being or even 1% for a tiny object such as the leg of a chair. „Smart Sensors", the combination between sensor element and behavior control, are designed for the reliable and robust execution of sensor based functions in the real world. Requirements for ultrasonic sensor systems are listed systematically in Fig. 4.

The performance and robustness of a behavior control is strongly related to the characteristics of the sensor technology. Therefore approaches using ultrasonic sensors for guidance of mobile robots concentrate on:

* the appropriate sensor design by beam forming techniques

* optimization of the physical sensor configuration

* the sensor management

* the behavior control

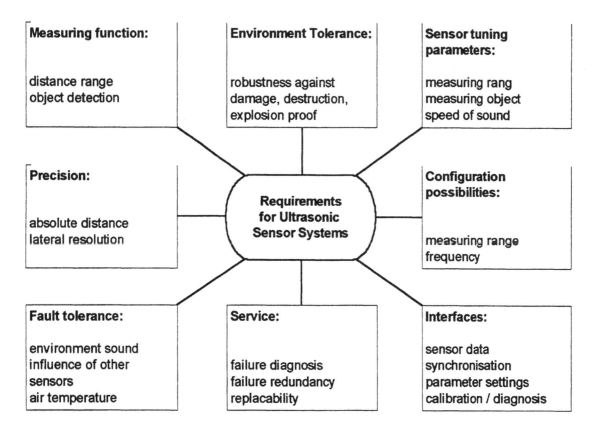

Fig. 4 Requirements for ultra sonic system

4. ARTIFICIAL NEURAL NETWORKS AND FUZZY CONTROL

In the area of artificial intelligence artificial neural networks (ANN) is widely applied from signal processing to control. The trends in the application of neural networks are to develop new learning algorithms and to apply known ANN structures and learning algorithms in practice.

A great advantage of fuzzy methods compared with conventional data processing is that fuzzy systems implement expert knowledge in three parts of a fuzzy algorithm: fuzzy rule set, linguistic form, and membership functions. For application of a fuzzy algorithm, expert knowledge must be transformable into rules. Generally fuzzy rules can be easily formulated. Otherwise, if the design of rule set is impossible, the fuzzy element would be unusable. If the rule set is formulated correctly, the fuzzy element is a very powerful tool, which is capable of making decisions based on even fuzzy input data.

The collision avoidance is a non-linear and time-varying process which can be modeled difficult in mathematical form. ANN and fuzzy systems are suitable to solve this problem. For collision avoidance fuzzy systems can be easily applied because most collision avoidance procedures will be carried out on rule-base. Fuzzy systems can be applied with a small number of rules. Fuzzy system is robust, but not adaptive. ANN are adaptive, but they need extensive learning process. In case of application of ANN for the collision avoidance the all possible situations should be investigated and trained in the network. It is a very complicated and extensive process. For the collision avoiding problem the fuzzy system can solve this problem easier than ANN. Generally the collision avoiding procedure should be carried out in real-time. For real-time operation there are requirements in following fields, such as the sensory data processing and information processing of a controller. In real-time operation ANN can be hardly applied because of the duration of the learning. The speed of information processing of a controller plays an important role. Following approaches are investigated for the integration of neural networks with fuzzy control to speed up the information processing and to increase the reliability of system. One is the optimization of fuzzy controller using ANN and other is the training of ANN with fuzzy system. Additionally the expert knowledge is fed into the neural networks reducing the duration of learning process.

The drive system of Nomad200 consists of three independent axes. The first axis is the translational axis, which drives three wheels of the robot. The steering axis changes the direction of the robot's wheels. The translational axis and steering axis

move the robot in the plane in a non-holonomic fashion. The third axis is responsible to move turret with respect to the base. All three axes can be controlled in two different control modes: velocity and position control. In this work the velocity control mode is chosen. As input one bumper signal, 16 infrared signals, and 16 ultra sonic sensor signals are available. Based on the robot controller and kinematic structure the control commands are different. For Nomad200 the motion commands are translational velocity mode, steering velocity mode, and turret velocity mode. For Maxifander and Khepera motion commands is the angular velocities of each wheel.

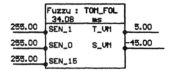

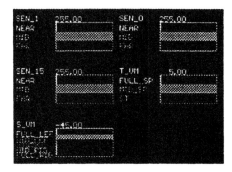

Fig. 5 An example for fuzzy control using 3 sensor signals (with Ilmenauer Fuzzy Tool)

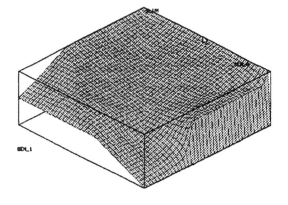

Fig. 6 3-D Characteristics for the steering direction (s_vm) of the robot

4.1 Optimization of a fuzzy controller using ANN

Fig. 5 shows an example of a fuzzy control which generate the motion commands using three sensor signals. Fig. 6 shows the 3D characteristics of fuzzy control for the steering direction of Nomad200.

By the application of fuzzy system the optimization of fuzzy system is necessary. In practice it is not recommendable to apply fuzzy system without any optimization. The simulation results between a fuzzy system with optimization and other without optimization show differences. There are two common optimization of fuzzy control:

- Optimization of membership function of inputs and outputs - in this case there is no changes of rules - the membership functions will be shifted.

- Optimization of rules - In this case each rule has „weight" and weights are optimized.

Fig. 7 shows the structure for the optimization of a fuzzy control using ANN. ANN can be applied to optimize the fuzzy system and parts of it, such as the fuzzification, defuzzification process and rule-based. The optimization of the fuzzy algorithm is accomplished by variation of membership function parameters and fuzzy rule base.

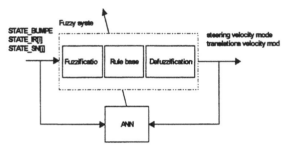

Fig. 7 The optimization of a fuzzy controller using ANN

4.2 Training of ANN with fuzzy system

Fig. 9 shows the robot approaches to the wall. Fig. 10 shows the three layer neural networks, which generate the velocity control commands based on the sensor signals (Fig. 8). The Backpropagation network with one hidden layer is applied. Input layer consists of 33 neurons and information is fed from the sensor on-line (one bumper signal, 16 infrared signal, and 16 ultra sonic signals). The outputs of network are the motion commands, such as translational as well as steering commands.

The success of neural networks application depends on the learning, choice of suitable network structure, and the preparing the input and output patterns and others. For the application of the neural networks the preparation of patterns is one of important tasks. In most cases the training of networks can be carried out by means of simulator. Without sufficient numbers of patterns the learning can not be executed successfully. Fuzzy control serves in first of all to find the optimal structure of the networks and to feed into the expert knowledge.

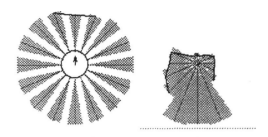

Fig. 8 Sensor views

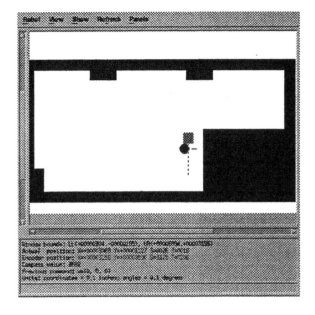

Fig. 9 Approaching to the wall

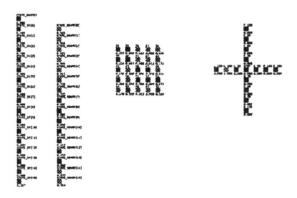

Fig. 10 Three-layer neural networks for collision avoidance based on the sonar data

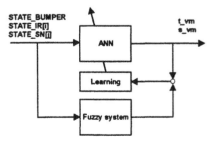

Fig. 11 Training of ANN with fuzzy system

Fig. 12 shows a simulation result of wall following with Nomad200. The working space is 5,4m wide and 11,5 meter long.

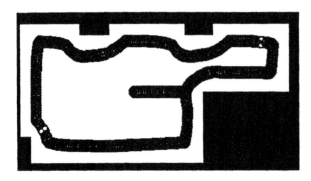

Fig. 12 Simulation result of mobile platform
(Nomad200) for wall following

5. CONCLUSION AND FUTURE WORKS

The integration of neural networks and with fuzzy control is presented. The test and comparison of effectiveness of each algorithm using the mobile robots (Nomad200 and Maxifander) are going now.

One of on going works is the application of a vision system. When the two robots are working simultaneously, there is the sensing problem because of the mutual sonar influences. On-Board ultra sonic sensors and infrared sensors can not detect some objects, such as tables which are little lower then the sensing height. On the other side development of intelligent control concepts for the robot manipulator using neurofuzzy is going now. In near future the manipulator will be built and equipped on the mobile robot to enable the handling the material. A new research headline at the institute is the application of robots in the entertainment.

REFERENCES

Bouslama, F, and Ichikawa, A. (1993). Application of neural networks to fuzzy control, *Neural Networks*, Vol. 6, pp. 791-199,

Dash, P.K., Liew, A.C., Rahman, S. and Dash, S., (1995) Fuzzy and Neuro-fuzzy Computing Models for electric Load Forecasting, *Engineering Application Artificial Intelligence*, Vol. 8, No. 4, pp. 423-433

Ushida, H., Yamaguchi, T., Goto, K. and Takagi, T., (1994). Fuzzy-Neuro Control using Associative Memories, and its Applications, *Control Eng. Practice*, Vol. 2, No. 1, pp. 129-145

Sulzberger, S. M., Tschichold-Gürman, N.N. and Vestli, S., (1993). FUN: Optimization of Fuzzy Rule Based Systems Using Neural Networks, *Proceedings of IEEE International. Conference on Neural Networks.*

Glasius, R., Komoda, A. and Gielen, S. (1995). Neural Network Dynamics for Path Planning and Obstacle Avoidance, *Neural Networks*, Vol. 8, No. 1, pp. 125-133

Han, M.-W. and Kopacek, P. (1995). Neuro-Fuzzy and Mobile Robots, *Preprints of 4th Int'l Workshop Robotics in Alpe-Adria Region (RAA '95)*, July 6-8 1995, Pörtschach, Austria, Vol. 1, pp. 155-158

Brown, M. and Harris, C., (1994). *Neurofuzzy adaptive modelling and control*, Prentice Hall International

Kosko, B., (1992). *Neural networks and fuzzy systems*, Prentice Hall International

Preuß, H.-P. ,Tresp, V., (1994). Neuro-Fuzzy, *atp-Automatisierungstechnische Praxis* 36, pp. 10-24

Probst, R. and Kopacek, P. (1996). Service robots:: Present Situation and Future Trends, *Proceedings The Second ECPD International Conference on Advanced Robotics, Intelligent Automation and Active Systems*, Sep.26-29, Vienna, Austria, pp. 45-52

Wolf, T., (1993). Die Synergie in Neuro-Fuzzy, *Design & Elektronik, 9/1993*, pp. 4-8. (in German)

Zazala, A.M.S, and Morris, A.S., (1996). *Neural networks for robotic control: theory and application*, Ellis Horwood

ROBOTICS, SURGERY AND SAFETY

Enzo Gentili *, Alberto Rovetta †, Stefano Reschigg *, Giovanni Valentini *.

* Universita' di Brescia - Dipartimento di Ingegneria Meccanica
Via Branze, 38 - 25123 Brescia - Italy
† Politecnico di Milano - Dipartimento di Meccanica
Piazza Leonardo da Vinci, 32 - 20133 Milano - Italy

Abstract: This paper deals with problems related to safety in medical robotics and computer assisted surgery. It is focused in particular on a medical system for liver biopsy which the Telerobotics Department of the "Politecnico di Milano" has been developing and that we are presently experimenting and testing at Niguarda Hospital, Milan. Here there is a description of the system and of the procedures that have been followed in order to obtain the CE certification. The CE declaration of conformity is necessary for the use of the robotical device and its free circulation inside EEC.

Keywords: Medical systems, Medical applications, Robotics, Safety.

1. INTRODUCTION

Robotics, medicine and surgery have been developing together in a fast way during recent years: there are many advantages in this union, both for the patient and the operator.

In performing a biopsy with a robot we aim at higher precision, greater flexibility, faster operations and a lower probability of the patient catching an infection, since a direct contact with the medical staff is not required. Certainly, on the other hand, the introduction of robotics has increased the systems complexity and the risk of system failure. That is why safety plays a fundamental role in design, use and maintenance.

In April 1995 Rovetta's staff performed a prostatic biopsy with a SCARA robot on a human patient in a clinical context. This time there is an additional reason of attention, because the liver is a very delicate organ, since there are many blood vessels in the surrounding area. That is why a new robot with six degrees of freedom is needed, reaching a greater ability to pinpoint trajectories.

The aim of this paper is to demonstrate that robotics and surgery can work together with great results, and to describe the medical system as well as the procedures to be adopted for reaching high safety in normal working conditions.

2. SYSTEM DESCRIPTION

The system is composed of the following macrocomponents:
- a robot with six degrees of freedom
- a control box
- a PC
- an echograph
- stereoscopic cameras.

To perform a biopsy, firstly the doctor localizes the

Fig.1. Prostatic robotical biopsy in April 1995, Policlinico Hospital, Milan.

the biopsy target using the echograph probe. In oder to reach the target correctly, its coordinates has to be known in the robot reference system.

The echograph has got its own reference system, which is able to localize an object in the probe plane. Probe position, referred to the robot reference system, is provided by two stereoscopic cameras, which detect three leds present on the probe according to this procedure.

1) At the beginning leds and lights over workspace are turned off.

2) A check is executed so that no lighting point is present, so that there is nothing which could interfere with led identification.

3) Leds are turned on and the scanning of the image provided by the left camera begins, in order to analyze single pixels brightness.

4) When a pixel with a characteristic brightness is found, a scanning of the surrounding area is executed with the aim of verifying that the first pixel detected is really a led.

5) If the pixel number is superior to a specified limit, then it is sure that it is not an undesiderable reflection.

6) The centroid of the detected pixels is calculated by the telemeter.

7) This procedure is repeated three times.

8) Having found the left image of the leds, the procedure is performed once again from point 4 with the right hand camera.

In the end a PC elaborates the three coordinates

couple obtained with the scanning and gives back the 3-dimensional probe coordinates.

The system operator chooses the biopsy target on the PC video, using the mouse on the echograph returned image. The PC, knowing the probe coordinates in the robot reference system and the target ones, is able to determine the target coordinates in the robot reference system.

A confirmation is asked and then the biopsy is performed.

3. CERTIFICATION

We define certification the act in which a third and independent part, a recognized organization, declares that a particular product, process or service is in conformity with a specific standard or with an other prescriptive document. The certification is carried out with the emission of a conformity declaration and the affixing of a conformity mark, that is, identification badges which belong to the organization appointed for the certification.

The reasons which can induce a producer or a supplier to ask an organization a conformity declaration for his own product are:

- the necessity of demonstrating to an interested part, for example a client, that a product has been really manufactured in accordance with a particular standard, and therefore it respects the

Fig. 2. Biopsy needle.

specific requirements;
- the necessity of demonstrating to an interested part that a product distinguishes itself for its quality, reliability and life;
- the necessity of obtaining for a product free circulation inside EEC, avoiding preventive controls in foreign member countries. In particular, the European Community has imposed to affix the CE mark on the products whose characteristics could not respect safety and health requirements in any way.
The CE mark vouches for the obtaining of European certification, that is, the recognition of conformity with the requirements specified in communitary directives.
The Directive we must refer to, in our case, is 93/42/EEC. It follows the so called "new approach", adopted by the European Community in 1985, which is based on these fundamental principles:
- the community legislative harmonization confines itself to the adoption, by means of Directives which shall be incorporated in every member country, of necessary safety qualifications, which have to be satisfied by "regulated" products, that is those which can threaten danger or damage to citizens and environment safety and integrity, in order to freely circulate inside EEC;
- the competent legislative organizations, CEN and CENELEC, have to elaborate technical standards, taking into account the momentary technological level. These standards are not compulsory, but they have a voluntary character;
- Public Administrations are obliged to recognize to the products manufactured in accordance with harmonized standards, the presumed conformity with the necessary safety qualifications fixed by Directives. The producer has consequently the faculty to make products which do not conform to the standards, but in this case he must directly demonstrate that his products answer the necessary applicable Directive qualifications.
This new approach is a consequence not only of

the birth of the common market, but it tries to follow the actual situation which is characterized by a really fast technological development and a considerable delay in standards production.
As we have already said, the Directive we have to refer to for our system is 93/42/EEC, a Directive for the purpose of medical devices. It provides for three different procedures in order to get the certification of conformity. These procedures correspond to three levels of risk in using bio-medical technologies.

4. SAFETY AND RISKS

Risk is the combination of probability of occurrence of a particular hazardous situation and damage entity correspondent. The safety task is to reduce risks as much as possible, so as to avoid that an unfavourable event creates a danger, which can possibly produce a damage or an injury. It will act on the damage entity through safeguards and it will lower the probability by means of prevention.
After setting machine limits (space limits, use limits and time limits), the first step is to identify hazards and make a risk assessment. Risk reduction must start from machine design, and only if it is not possible to act in this way,with guards and/or safety devices. However, operators will have to be informed about any persistent hazard or risk.
In order to detect any possible hazard and to prevent them, it is useful to follow the scheme given by specific standards. EN 60601-1 is a European standard about general requirements for safety of medical electrical equipment. As Directive 93/42/EEC, it identifies a list of general hazard types. The most relevant are electrical and mechanical ones.
You can achieve electrical protection through:
- prevention of a contact between the patient or the operator and a machine part which could be in tension because of an insulation failure using wrappers or assembling dangerous parts in inaccessible places;
- current and voltage restriction in the machine parts which can be or are voluntarily touched by the operators and/or the patient, during the normal working conditions or in case of first failure.
Mainly it is important to safe guard the patient, because he is in an extremely weak condition, he has not been trained and you can not foresee his reactions. To this end the needle has been carefully insulated and an electrodispersive plaque is positioned on the patient, next to the biopsy target. This expedient minimises the path of dispersion current through the patient.
Obviously the system has been properly earthed and a list of countermeasures has been taken in

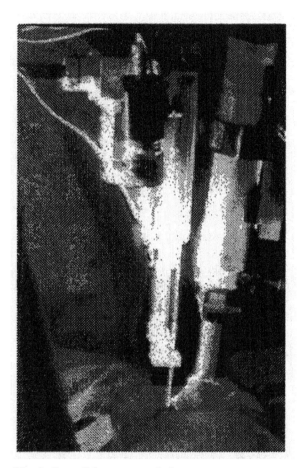

Fig. 3. One of the tests carried out.

accordance with the standards. A high number of tests have been carried out, concerning in particular insulation, limitation of tension and energy, leak and auxiliary currents and protection earthing.

Working continuity is guaranteed by an autonomous electrical energy source present in hospital, which allows the continuation of the operation even in case of a black-out.

Mechanical hazards can be conditioned by:
- shape: sharp elements, sharp corners
- relative position: can give birth to crushing or entangling zones while they are moving
- mass and stability: potential energy of elements which can move under the effect of gravity
- mass and speed: kinetic energy of elements in controlled or uncontrolled movement
- aceleration
- insufficient mechanical resistence.

In order to prevent these hazards, robot workspace has been limited, at the moment by software but in the future hopefully there will be physical limitations. The robot sensorization has been managed placing a force sensor on the end effector. If a threshold is exceeded, the biopsy is automatically stopped. This event occurs even in the case of the needle penetrating the body too deeply. There are also software and hardware safety stop functions. The biopsy operation is performed only if a specific button is continuously

pressed, giving a "hold-to-run" control. If this button is released the robot stops and comes back. A "pause" button has been provided, too. There are also two special electromechanical devices which can interrupt the power supply. Hazards generated by radiations or vibrations and thermal character hazards have also been taken into account, observing prescriptive measures contained in the standards.

5. CONCLUSIONS

A liver biopsy system is about to be used normally in an out patients department: robotical applications in medicine are not only possible, but are beneficial.

What matters is that devices are designed so not to endanger neither the safety and health of the patient nor the operator's, when devices are used within the conditions and the purposes foreseen.

Risks should be acceptable, in ratio to the benefits given to the patient, and the safeguards have to be adhered to the present technological level, considering also the economic aspect.

Besides this, medical systems have to guarantee the performances which are attributed by the "builder". In order to be as precise as possible, many functional tests have been performed and will be carried out on dummies. However, present results are surely encouraging.

In the near future we shall embark on a new project with the aim of performing a tele-robotic biopsy, that is a robot that will not be controlled locally. It is an important and difficult step, which will be both demanding and fascinating.

"Team working" and co-operation are fundamental, so that every personal skill is optimised in order to reach a beneficial common aim.

6. ACKNOWLEDGEMENTS

The authors would like to thank Eng. Sala and Eng. Bressanelli of Telerobotics Laboratory, Politecnico of Milan, who made this system a reality; Eng. Leone of ABB Robotica (Milan) because of the kind use of the robot; Eng. Badi of Niguarda hospital (Milan) for his support and the Telecom Italia.

Special thanks to Mrs. Mary Flynn who checked the manuscript.

REFERENCES

Costi, G., E. Gallo, G. Garibotto et. al. (1995). Force feedback control of robotic arm in neurosurgical applications. *Ninth World*

Congress on the Theory of Machines and Mechanism, **3,** 2161-2165,

De Baere, T., A. Roche, J.M. Amenabar, C. Lagrange, et al. (1996) Liver Abscess Formation After Local Treatment of Liver Tumors. *Hepatology,* **23,** 1436-1440.

Rovetta, A. and R. Sala (1994). Sensorization of surgeon robot for prostate biopsy operation. *First Internat. Symposyum on Medical Robotics and Computer Assisted Surgery,* Pittsburgh.

Rovetta, A. and R. Sala (1995). Robotized biopsy. *Second Symposyum on Medical Robotics and Computer assisted Surgery,* Baltimore.

Rovetta, A. and R. Sala (1995). Robotics and telerobotics applied to a prostatic biopsy on a human patient. *MRCAS,* Pittsburgh.

Rovetta, A. and R. Sala (1996). Execution of robot assisted biopsies within the clinical context. *Image Guided Surgery,* **1, n. 5,** Wiley-Liss.

Sala, R. (1995). Construction of a new automatic telemeter for medical applications and robotic telesurgery. *Ninth World Congress on the Theory of Machines and Mechanism,* Milan.

ROBOT INTELLIGENCE AND FLEXIBILITY
FOR SMOOTH FINISHING OF CAR BODY

Byung-Hoon Kang, Jin-Dae Kim, Jin-Yang Yoo, Jong-Oh Park
Kwang-Se Lee*, Hyun-Oh Shin*

Advanced Robotics Research Center, Korea Institute of Science and Technology
* Automation Research Lab., Hyundai Motor Co.
jop@kistmail.kist.re.kr

Abstract : Robotic finishing such as car body grinding is one of critical automation processes because of direct influence of robot system capability on final surface quality. Difference of each car surface even in the same car model is also one another critical problem. Here one suggestion on robotic finishing system with intelligence and flexibility. Laser vision scans the local surface of each car body, from sensing data after signal processing surface and bead are modeled, after that exact robot path is generated. All these processes are integrated in one robot finishing system, which is applied in actual production line. Such intelligence and flexibility of robot system shows adaptability even in production line as well as further interesting research topics. Finally technological extension of such system to more complicated workpiece is explained.

Keywords : Robotic finishing, Laser vision, surface modeling, Bead modeling, Robot path generation, Robot intelligence, Robot flexibility.

1. INTRODUCTION

Robotic finishing remains still as one of the most exciting automation process. The main reasons for that are firstly direct and serious influence of robot movement on final process quality of workpiece, reaction force behaviour of robot caused by contact during finishing, and various specific process parameters in each finishing process. Car body finishing might be the typical example for such robotic finishing. Many different ways to automate the car body finishing can be considered, for example, simple grinding without any compliance nor adaptive devices, with passive compliance devices, with adaptive control using contact force recognition, or with 3D shift using reference position recognition. In this research all possible unreliable factors have been counted, for example, jig and fixture error, car body positioning error, positional and orientational difference of local target surface between each car body even in the same car model.

2. RECOGNITION OF CAR BODY SURFACE

2.1 Environment

Removal of weld bead on car body surface has been being executed manually. The grinding work for such bead removal requires highly skilled experience and sputter and dust generated by grinding has very bad influence on the worker's health. Such environment induces the automation of manual grinding process using robot. But the conventional method is to teach-in the robot path into robot controller and to repeat the preprogrammed path for grinding, but such method is not adaptable to the unexpected change in car body positioning and body surface (Whitney, *et al.*, 1990; Park, *et al.*, 1993). The reasons in detail are firstly, the incorrect positioning of car body on conveyor and conveyor itself. The second reason is the ununiformity of position and volume of welded bead. And the another important reason is the incorrectness of the local surface to be finished even in the same car model. Finally flexibility caused by model change in the same production line requires effective recognition of car body surface and automatic generation of robot path.

2.2. Suggestion

As mentioned above, to achieve the required surface finishing quality the effect of incorrectness caused by various influence factors should be minimized and optimal robot path should be generated. In this paper automatic and flexible generation method of robot finishing path for weld bead removal in production line will be explained.

At the present in production lines in other countries 2D CCD camera and linear displacement sensor are used for car body recognition (Sekine, *et al.*, 1990). But the achieved information will be merely used for 3D shift of total preprogrammed path, not for correction of fine change of robot path. Such algorithm cannot solve the surface difference of each car shape.

This suggested method for improvement of finishing robot system is summarized as follows : laser sensor which is attached at robot wrist scans each car surface, which arrives at finishing station, measurement data on the local surface will be derived and from measurement data local surface to be finished and also weld bead will be surface modeled and finally optimal robot path will be generated.

Because the robot path is newly generated for each car body, the positioning and orientational error of car body and simultaneously fine local surface error of each car, shape and positional error of bead can be solved in our suggested system..

The robot system requires laser scanning vision sensor, vision frame grabber, CPU for sensing data processing and robot path generation, appropriate articulated robot and some peripheral devices. The total system configuration is like as Fig. 1 and system process flow is like as Fig. 2.

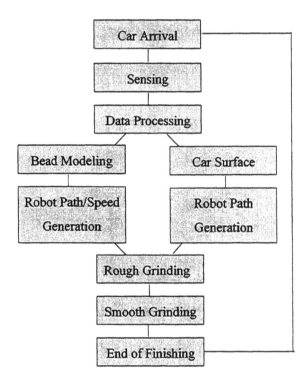

Fig. 2. System process flow

3. SENSOR SIGNAL PROCESSING AND ANALYSIS

3.1. Body surface to be recognized

One example for robotic finishing is as in Fig. 3,

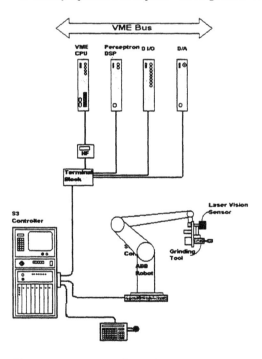

Fig. 1. System configuration

Fig. 3. Local surface to be finished

which is located in rear part of car body and brazed line between roof and side panel, and has curved line in 3D space with continuously varied curvature. The length of the bead is not over 200 mm and both ends are bent about up to 90 °.

For measurement of welded bead on car surface, laser scanning sensor attached at robot wrist moves according to the following position and sequence like as in Fig. 4. For exact calculation of start point of bead sensor moves from S_1 to S_2. For measurement of body curvature and the position of welded bead sensor moves along M_1, M_2, ... M_n. Robot positions and sensing times are synchronized each other using D I/O channel, from each sensing data surface data and welded bead data will be derived separately and each derived data will be converted into input data for surface modeling and bead modeling in each. Finally the sensor moves to the position E_1 and then to E_2, from such measurement the end position of grinding is calculated.

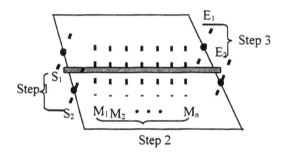

Fig. 4. Measurement sequence and positions for car body surface

3.2. Consideration on measurement

If finishing robot is applied in existent production line, the available cycle time will be on of dominant factors for system configuration and component selection, because sensing, processing, modeling, robot path generation and finishing processes should be finished in the constrained time. Since the time required for rough and fine grinding is process-oriented and directly quality influent, the time reduction might be almostly impossible. But the measurement time can be reduced through various experiments. On the assumption of allowable car body tolerance up to ± 10 mm in 3D space, sensing process was optimized. At the start and end position of finishing the finishing path changes abruptly, therefore the finishing edge has to be ground very sharp. For that the sensing should be also very exact. For exacter sensing the sensor head will be located perpendicular to the edge line of car surface. The number of measuring line is reduced to 7 in consideration of enough recognition of sculptured

surface and acceptable minimum measuring line. In each measuring the center position of bead cut is calculated and with reference of center position the local surface is modeled. For reduction of consumed time sensing process runs during constant movement of robot and it takes not over 9 sec.

3.3. Filtering and Feature Extraction.

With each sensing line about 800 3D measurement position data are generated. These data contain noise as well as surface data, therefore should be filtered as the preprocessing stage. Noise tends to a high-frequency zone, therefore only data within $\mu \pm 3$ σ will be further processed (Gonzales, *et al.*, 1992). After achievement of surface data without noise influence, the center line of bead will be acquired and then feature extraction process for separation of bead and adjacent surface data will be followed. Feature extraction is executed by converting each measurement data into a group data like Eq. (1) and then by iterative polygon approximation (Pavlidis, *et al.*, 1974).

$$G^i = \{(x^i_1, y^i_1, z^i_1), \cdots, (x^i_n, y^i_n, z^i_n)\} \quad -(1)$$

where ,i = number of sensing iteration

n = number of sensing data

In this method, the straight line is acquired by initial point (x^i_1, y^i_1, z^i_1) and end point (x^i_n, y^i_n, z^i_n) of group data like as Eq. (2) and maximum value(d_m) can be derived by calculation of every points within group and its perpendicular distance (d_k).

$$l = \{(x,y,z) | \alpha * y + \beta * z + \gamma = 0, \alpha^2 + \beta^2 = 1\}$$

$$d_k = |\alpha * y^i_k + \beta * z^i_k + \gamma| \quad -(2)$$

$$d_m = (d_k)_{MAX}$$

In case of the value over allowable distance, new group data is generated and this method iterates for feature extraction till there are no more separable group. Fig. 5 shows the result of application of iterative polygon approximation to car surface measurement data.

Among the extracted feature values the farest feature value from main axis of measurement data will be the center position of welded bead. The width of welded bead lies generally within the same zone, therefore by using bead center line surface data and bead data can be separated. In each sensing line, 2 points in upper and lower side of center line in each and 7 times of such line, that is, totally 28 points configure surface data and also are used as input data for surface modeling.

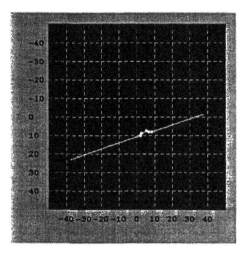

Fig. 5.   Result of feature extraction

The finishing process is divided into rough and fine finishing, especially in rough finishing without recognition of exact bead position severe damage on surface caused by incorrect contact of grinding disk on bead occurs. Therefore bead modeling is required from separated bead data. In each line, 9 points on the bead are measured. With total 7 line, 63 measuring data of bead will be used for input to bead modeling.

4.   SURFACE MODELING

4.1   Realization of Ferguson modeling

With the above mentioned method of measurement, although it is possible to generate robot path from measuring data of car body surface , it is problematic that all the car body surface must be measured in detail to make a entire surface correct. If the measurement is carried out to entire surface, it takes many times to take sensing. Processing and saving data are also another problem. At this point, It is necessary to use Ferguson surface with a view to acquiring smooth surface from measurement data and generating robot tool path to a region of excluded from sensing. In method of Ferguson surface modeling, a unit patch included in entire composition surface is appointed and positional, tangential and twisting vector of its four edge is inputted. In the last step, the point on the patch to come within the value of given parameter, $(0 \leq u, v \leq 1)$, is calculated by using Ferguson equation.

Therefore, the entire surface can be taken from each connecting unit patch with another acquired by this method. Positional vector to come within the value of given parameter is calculated from Eq. (3) on the unit surface patch. Where r is positional vector in

measurement point, t and s represents the tangential vector in u and v direction, x is twisting vector, u and v represent coordinate of unit surface patch.

$$
\begin{aligned}
&r(u,v) \\
&= (x(u,v), y(u,v), z(u,v)) \\
&= UCQC'V'
\end{aligned}
$$

$$
= \begin{bmatrix} 1 & u & u^2 & u^3 \end{bmatrix}
\begin{bmatrix} 1 & 0 & 0 & 0 \\ 0 & 0 & 1 & 0 \\ -3 & 3 & -2 & -1 \\ 2 & -2 & 1 & 1 \end{bmatrix}
\begin{bmatrix} r_{00} & r_{01} & t_{00} & t_{01} \\ r_{10} & r_{11} & t_{10} & t_{11} \\ s_{00} & s_{01} & x_{00} & x_{01} \\ s_{10} & s_{11} & x_{10} & x_{11} \end{bmatrix}
\begin{bmatrix} 1 & 0 & -3 & 2 \\ 0 & 0 & 3 & -2 \\ 0 & 1 & -2 & 1 \\ 0 & 0 & -1 & 1 \end{bmatrix}
\begin{bmatrix} 1 \\ v \\ v^2 \\ v^3 \end{bmatrix}
$$

-(3)

Tangential vectors in u-direction and v-direction can be obtained by differentiating r(u,v) which is on the unit patch for u and v, and also can take normal vector by cross product of these acquired two vectors. Generally, the surface of object to be measured has complicated shape which is not defined by Ferguson surface equation. For the purpose of representing surface of this object, it is used to connect unit patch to neighbor patch with smoothness. Smooth surface can be acquired by putting tangential and curvature value on equal terms. .

$$
\mathbf{s}_{i-1,j} + 4\mathbf{s}_{i,j} + \mathbf{s}_{i+1,j} = 3(\mathbf{r}_{i+1,j} - \mathbf{r}_{i-1,j}); \quad i = 1,2,\cdots,M-1; \quad j = 0,1,\cdots,N
$$
$$
\mathbf{t}_{i,j-1} + 4\mathbf{t}_{i,j} + \mathbf{t}_{i,j+1} = 3(\mathbf{r}_{i,j+1} - \mathbf{r}_{i,j-1}); \quad i = 1,2,\cdots,M; \quad j = 0,1,\cdots,N-1
$$
$$
\mathbf{x}_{i-1,j} + 4\mathbf{x}_{i,j} + \mathbf{x}_{i+1,j} = 3(\mathbf{t}_{i+1,j} - \mathbf{t}_{i-1,j}); \quad i = 1,2,\cdots,M-1; \quad j = 0,1,\cdots,N
$$

-(4)

If modeling is completed to unit surface patch, it is possible to get positional and tangential vector at arbitrary position of car body.

4.2   Bead and surface modeling

Modeling of bead as discussed in chapter 3.3 forms sixty-three Ferguson unit patches by using nine points per measured line. The shape of smooth bead can be modeling as tangential vector is acquired by using chord length ratio and ferguson composite surface is formed in each unit patch.

The shape of car body surface can be modeled when each unit patch is compounded as the same method of bead modeling.

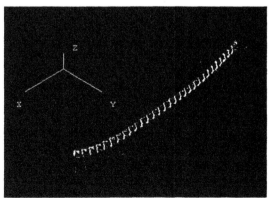

Fig. 6.1.   Bead Modeling

306

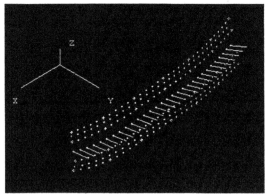

Fig 6.2.   Surface Modeling

5.   ROBOT PATH GENERATION

5.1 Decision of start and end point for tool path

If there is a satisfied modeling result to a car body, it is possible to calculate positional, tangential, and normal vector in all the inner point of car body. Optimal robot tool path can be generated by using modeling data which are perfect.

It is necessary to have interpolation of bead position to make an accurate tool path because modeling does not include the parts of start and end point of bead. Interpolation is carried out by using information of bead modeling and edge line which is measured in sensing process. Tangential vector is calculated to longitudinal direction of the bead modeling and center position of bead is extended, and the cross product between extended line and edge line of car body is calculated, and then, start and end point are acquired. Bead point is interpolated with constant interval between these points and points of modeled bead. Directional vector of bead is also calculated by using points of modeled bead.

5.2 Robot position and orientation

The removal of welded bead is fulfilled with two processes, rough grinding and fine grinding. In each case, the robot path is different each other. Rough grinding is the process to remove welded bead on the car body with grinder disk. In rough grinding process, grinder disk is positioned accurately to center of the bead at first. And then removing work is performed by constant force pressed on the bead surface. At this time, grinder stone contacts to car body surface, critical damage might occur. For this reason, it is necessary to get information of accurate bead position, bead feature, and bead height. It is important to have exact robot position and orientation while grinding is in progress. For the purpose of prohibiting the contact between grinder stone and car body surface, robot must be hold the position to the normal direction of car surface which

is neighbor of the bead. Robot position can be calculated by information of bead modeling and start and end points of bead which are taken by linear interpolation. Robot posture can be calculated by normal vector of surface modeling.

Fine grinding is performed not only to remove remaining bead after the rough grinding but also to maintain smoothness of car body surface. Flexible type disk which has higher mesh number than that of rough grinding is normally used. Thus, to grind welded bead and car surfaces, grinder disk should have contacted to much portion of car surface plane. Important thing in fine grinding is to maintain surface smoothness to be one free surface. The posture information of robot is very important for this reason. Tool direction of robot using normal vector from car surface modeling can be calculated. Ferguson modeling shows the orientation of vertical direction. as follows Eq. (5).

$$\frac{\partial r(u,v)}{\partial u} = (\frac{\partial x(u,v)}{\partial u}, \frac{\partial y(u,v)}{\partial u}, \frac{\partial z(u,v)}{\partial u}): \begin{array}{l}\text{Tangential vector}\\ \text{in u direction}\end{array}$$

$$\frac{\partial r(u,v)}{\partial v} = (\frac{\partial x(u,v)}{\partial v}, \frac{\partial y(u,v)}{\partial v}, \frac{\partial z(u,v)}{\partial v}): \begin{array}{l}\text{Tangential vector}\\ \text{in v direction}\end{array}$$

$$\frac{\partial r(u,v)}{\partial u} \times \frac{\partial r(u,v)}{\partial v}: \begin{array}{l}\text{Tangential vector in vertical}\\ \text{direction}\end{array}$$

-(5)

When grinding process proceeds, grinding position is not only to follow the center of bead but also to follow upper and lower parts of the bead.

6.   EXTENSION TO SEVERELY UNDEFINED SURFACE

6.1 Necessity of study on severely undefined surface

In the previous chapters of this thesis, the subjects of three-dimensional measurement data processing and surface modeling are concentrically discussed to implement smooth finishing robot system for the object of welded bead on the car body. In looking roughly into the characteristics of welded bead, the special features appear in the surface shape as follows. Bead shape by itself has some constant form for a different welding condition or working environment. When measuring a section of the bead with three-dimensional laser sensor, measurement data structure to be obtained takes the form of one to one relationship to an axis of the criteria. It is enough to acquire sectional shape of bead's surface by one and only sensing because of the morphological and geometrical characteristics of welded bead as mentioned above. But various types of burr in an industrial environment should be accomodated where appears burr's form from simple type like welding bead to complicated or not predicted type which can be seen easily in the casting. At this point, processing algorithms and course must be extended to the object whose surface is severely undefined look, that is, its measuring data structure takes the form of many to one relationship to an axis of the criteria. And it follows that development of three-dimensional range data processing algorithms is a new approach to widen adaptation and flexibility for intelligent robot system.

6.2 Sensing of crank shaft seat casting

In our laboratory, developing deburring system of the casting in automobile engine block which has discontinuous and irregular surface plane is progressing with a view to extending and enlarging robot grinding work. In this thesis, data processing is carried out to the crank shaft seat part which especially appears occlusive edge or discontinuous features in the entire engine block. Fig. 7 shows entire engine block casting, and a quadrilateral mark represents crank shaft seat part.

Fig. 7.   Car engine block casting

In sensing the crank shaft seat, the next followed items are considered chiefly. First, it is impossible to have extraction of a sectional features which are contained in crank shaft seat from obtained data by measurement of perpendicular direction to the surface plane of crank shaft seat parts, because the surface parts exist, which is not measured, caused by the facts that upper width of burr is wide and the lower is narrow. Secondly, to make occluded surface characteristics useful when measurement should be performed with the slope of forty-five degrees from perpendicular direction. It is possible to have information of occluded edge part. But in the opposite direction, the concealment region exists in performing measurement, these facts must be considered to extract exact information of surface from crank shaft seat parts. In this thesis, the measurement is carried out three times per one section. These measured data are merged together. Fig. 8. shows entire three dimensional sensing data, and that is a result of having thirty-five times sensing at an interval of 2.5 mm, changing the measurement point three times, to the 83×85 ×130(mm) crank shaft seat part .

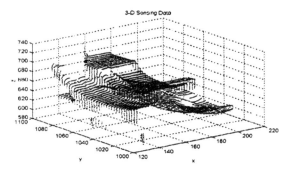

Fig. 8.   Entire 3-D sensing data of crank shaft seat

6.3 three-dimensional sensing data processing

① Characteristics of crank shaft seat surface data

Fig. 9. shows the features of cutting section to crank shaft seat which has the burr, upper plot represents a sectional shape of real burr, and lower shows how the part data appears that has existence of burr when real measurement is performed by laser sensor. Look at lower, it is difficult to divide the sensing signal into two parts, burr and noise, because the burr with noise occurs mixed.

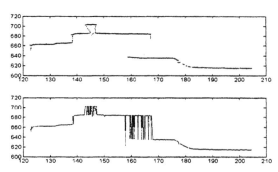

Fig. 9.   Characteristics of burr data

The fact that burr's sectional shape is discontinuous and data structure of sensing take the form of many to one relationship to an axis of the criteria. To find out accurate shape and position data of burr surface is the most important thing because robot deburring system is the work to remove the burr. As shown in lower plot of Fig. 9., because burr and noise element have high frequency features alike, noise elimination is accompanied with a removal of burr data at the same times, so, becomes only a uninteresting work. For this reason, this thesis proposes a new approach to solve this problem in three-dimensional range data processing.

② Creation of accumulated image

New approach, discussed in this thesis, is the transformation of Cartesian coordinate value which means sensing data coordinate values to image space values. This method is taken to exclude the problem of burr elimination while noise elimination, which has same meaning with removal of high frequency, is progressing. Processing method for transformed data is proposed also in the image space . So, this method has a merit to include the algorithms which are related to gray level image. Eq. (6) provides a relationship formula for transforming the Cartesian coordinate values to image space values. Eq. (7) is the inverse of Eq. (6).

$$\begin{bmatrix} x_i \\ y_i \\ z_i \end{bmatrix} = \begin{bmatrix} m_u & 0 & 0 \\ 0 & 1 & 0 \\ 0 & 0 & m_v \end{bmatrix} \begin{bmatrix} u_i \\ f_i \\ v_i \end{bmatrix} + \begin{bmatrix} s_x & s_y & s_z \end{bmatrix} \begin{bmatrix} x_{min} \\ y_{min} \\ z_{min} \end{bmatrix} \text{-(6)}$$

$$\begin{bmatrix} u_i \\ f_i \\ v_i \end{bmatrix} = \begin{bmatrix} 1/m_u & 0 & 0 \\ 0 & 1 & 0 \\ 0 & 0 & 1/m_v \end{bmatrix} \begin{bmatrix} x_i - s_x x_{min} \\ y_i - s_y y_{min} \\ z_i - s_z z_{min} \end{bmatrix} \text{-(7)}$$

where m_u : mapping grid snap for x_i,

m_v : mapping grid snap for z_i,

s_x : scale factor for x_i,

s_y : scale factor for y_i,

s_z : scale factor for z_i

③ Noise elimination and processing in image space
Measured data obtained by three dimensional laser sensor contain noise element and appear lack of information. As stated above, Cartesian coordinate values which are sensing data of crank shaft seat has the structure of many to one relationship. For this reason, data processing is difficult and it is not easy to find out some solution for settled Cartesian. So, transformation into the image space, proposed in previous chapter, is performed first, to ensure reliable flexibility and efficiency of data processing and then, cutting bottom area not interested and eliminating noise process are accomplished in the image space. The image processing methods adapted in this course are dilation, erosion, isolated point removal, thinning, etc.

Fig. 10. shows the output of these processing algorithms executed to some image frames. The processing method adapted Fig. 10. Follows. Firstly, searching 3×3 neighborhood pixels if more than one pixel has the value that is not zero, the value of pixel (i,j) is changed to the same value of its neighborhood and this process calls dilation. If more than one pixel (i,j) has zero, the pixel(i,j) is zero, erosion. In the second step, searching pixel (i,j) and its neighborhood, if pixel number which is valued lower than threshold number, the pixel (i,j) is regarded as an isolated point. The final step, thinning process is applied for the image coding and feature extraction process.

④ Image coding and feature extraction

Chain coding is the method to express order pair of vector with surface edge pixels to make scattered pixels to have a piece of information and to represent two dimensional information of image space in one dimensional space. Fig. 11. shows the chain coding of the sectional shape image of car body for the eight-directional vector and appears linear interpolation based on Eq. (8).

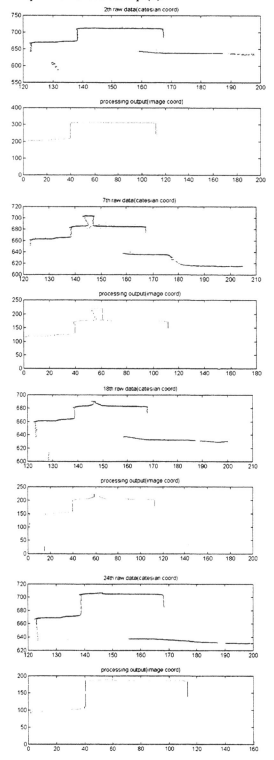

Fig. 10. Data processing result in image space

As shown in Fig. 11., each surface edge pixels can be divided into serial straight lines, therefore, feature point can also be extracted from these lines. Here, start and end point of line and cross point of two lines for feature point can be chosen.

$$ABS\left(\tan^{-1}\left(\frac{v_i - v_{i-10}}{u_i - u_{i-10}} \right) - \tan^{-1}\left(\frac{v_{i+10} - v_i}{u_{i+10} - u_i} \right) \right) \geq T \quad -(8)$$

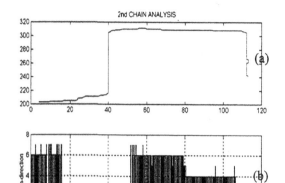

(a)

(b)

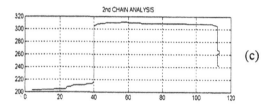

(c)

(d)

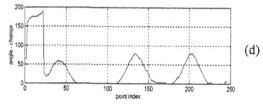

(e)

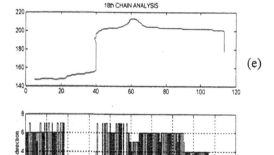

(f)

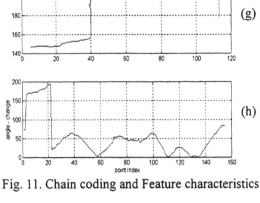

(g)

(h)

Fig. 11. Chain coding and Feature characteristics

(a) and (c) are the plot of the order pair to be carried out from chain coding in image space, (b) and (f) are the plot of chain vectors, (d) and (h) represents the output of angular value by Eq. (8).

⑤ Transformation to Cartesian coordinate for feature point

The coordinate values of image space that is obtained from feature extraction process must be transformed to Cartesian values to burr modeling and robot path generation. Table 1 shows image space values that is extracted from feature and Cartesian values are carried out transformation about eighteenth image frame. The average of error between calculated value form sensing data and measured value indirectly is $\Delta x_{ave}= -0.1260$ mm and $\Delta z_{ave}= 0.2982$ mm.

Table.1   Feature point transformation

	image u	image v	cartesian x
1	5	147	124.6240
2	22	148	131.4240
3	24	152	132.2240
4	39	157	138.2240
5	40	197	138.6240
6	54	204	144.2240
7	59	213	146.2240
8	69	204	150.2240
9	113	201	167.8240
10	113	177	167.8240

	catesian z	measure x	measure z
1	660.2188	124.4727	660.1751
2	660.6188	131.1415	661.0503
3	662.2188	132.5884	662.8009
4	664.2188	138.2315	664.3326
5	680.2188	138.6656	680.5252
6	683.0188	143.8746	683.1510
7	686.6188	146.1897	687.1532
8	683.0188	149.6624	683.5263
9	681.8188	167.6045	682.2757
10	672.2188	167.7492	672.1794

In this paper, a methodology has been presented for 3

dimensional complicated surface object processing based on its range measurement applicable to robot deburring system. And it showed the flexibility and efficiency in 3 dimensional data processing. The algorithms to solve electrical noise or data that have insufficient data structure will be the subject of further research work.

7. CONCLUSION

Robotic finishing such as car surface grinding has some critical problems for process automation, for example, jig/fixture error, car positioning error, and fine difference of car shape even in the same car model. Allowable time limit for process automation will be another critical constraint because of prefixed line cycle time.

Under such constraint here newly developed finishing robot system was explained with focus on intelligence and flexibility.

Laser vision sensing process for every each car surface was applied mainly for fine recognition of local surface difference in every possible case. Such method will be the first applied method in car surface finishing robot system at all.

System sequence is as follows : laser vision recognition of car surface, signal processing, surface modeling of local surface and surface modeling of bead, robot path generation and robotic grinding.

Noise elimination was also tried in various ways, because the weld spatter builds difficult noise component besides electrical noise. Such elimination was carried out by trial and error experiments. The result shows reliable signal preparation.

Surface modeling of local car surface and also bead shape is carried out. The finishing reference should be the adjacent surface curvature not bead by itself, because the finishing surface should be smooth with continuous curvature beyond simple bead removal. Therefore the surface was modeled with reference of existent surface curvature. For smoothness instead of exact material removal Ferguson modeler was applied and the result was also very smooth. Bead modeling was necessary for recognition of center line of bead, else the finishing quality was not acceptable. Such modeling process was optimized and integrated in in-line production line.

Every processes were integrated in one system, which runs in production line. For such integration various optimization work was carried out, for example communication reliability between different controllers.

Such developed finishing robot system shows, in one side, one possibility and further extensibility of automatic robot path generation using laser vision in other areas. In another viewpoint, such integrated finishing robot system, such as vision recognition, surface modeling, and robot path generation, which runs in production line not in office, made it possible, to apply intelligent and flexible robot system in production line as well as improvement of component technology in various corresponding fields.

REFERENCES

Besl, P.J., R.C. Jain, (1985). Three-Demensional Object Recognition. In: *ACM Computing Surveys*, vol.17, No.1, pp75-145.

Choi, (1992). CAM System and CNC Machining.

Gonzalez, R.C., *et al.* (1992). Digital Image Processing. In: Addison Wesley.

Park, J.O., *et al.* (1993). Grinding Robot System for Car Brazing Bead. In: *The Korea Association of Automatic Control*.

Pavlidis, T., *et al.* (1974). Segmentation of Plane Curve. IEEE Trans. Comput. 23, pp.860~870,

Pratt, W.K., (1991). Digital Image Processing. Second edition, Sun Microsystems, pp.472-482.

Sekine, Y., S. Koyama, and H. Imazu. (1991). Nissan's new production system : Intelligent body assembly system", In: *International Congress and Exposition*, Detroit, Michigan, SAE Technical Paper Series 910816, pp 1-12, Feb. 25-Mar.1.

Whitney, D.E., *et al.* (1990). Development and Control of an Automated Robotic Weld Bead Grinding System. Trans. Of the ASME, Vol.112, pp166~176.

ON-LINE QUALITY CONTROL FOR ENAMELLED WIRE MANUFACTURE

L. W. Bridges* and N. Mort**

* BICC Connollys, Liverpool, UK

** Department of Automatic Control & Systems Engineering, University of Sheffield, UK

Abstract: This paper addresses the important issue of on-line quality control for an enamelled wire production process. Data over a period of ten months, was gathered from a plant used for the manufacture of enamelled wire. The time indexed data was subsequently related to the off-line tested wire quality to produce a data series. This was then utilised in the derivation of models based on three different types of process control paradigms. The predicted results for wire quality from each of the three models was then compared to the off-line results to assess the suitability of each of the methods.

Keywords: Fuzzy logic, Process control, Quality control, Neural networks, Modelling

1. INTRODUCTION

BICC Connollys is the UK's leading manufacturer and supplier of enamelled and textile covered winding conductors. In an attempt to improve the enamelling plant, and the production methods within it, the company is taking radical steps to improve its ability to continuously test the product on-line. This will ultimately meet the increasingly stringent quality standards imposed by customers. Experience has shown that the most important parameter that influences product quality is enamelled wire cure. At the present time, this is measured using an established technique known as the "Tangent Delta Method" (Jorgensen, 1990). However, this is an off-line method where wire samples are tested at the end of the curing process. It is apparent that the cure of the enamelled wire is a function of the curing oven temperature, and the speed of the wire through the oven. By developing a model of the process it should be possible to predict the cure of the wire continuously throughout its manufacture thus providing an on-line quality measure.

1.1 Tangent Delta.

As we have said, test methodologies relating to the enamelling process are still performed very much off-line. As such, quality information relating to the process is acquired after a time-lag which is dependent upon the frequency of testing, and the time taken to perform the tests. This is obviously unacceptable. Reported applications of on-line quality measures are few in number (Benyo, 1969)

Recent developments in on-line test equipment have enabled surface defects to be monitored more readily, both in terms of the repeatability of the equipment and the availability with respect to cost. Trials with colour, (cure), recognition systems, have as yet proven difficult to set-up, due to the variety of enamels, which vary in colour, both by their physical nature and also by the amount of cure. However there is still a big gap between this and the most meaningful test of product quality, Tangent Delta.

2 ENAMELLING PROCESS DESCRIPTION

The process consists of an Enamelling/Curing Oven, Annealer and an In-Line Wiredrawing/Take-Up Unit.

Bare copper wire, previously drawn down, is fed into the wiredrawing unit of an enamelling oven. The wire is then drawn down to the required conductor size, whereupon it is cleaned and passed through a bath of enamel. This coats the wire, which is subsequently passed through a control die to regulate the diameter of the coating applied to the wire. The enamel coated wire is then fed through the enamelling oven where the enamel is cured onto the wire. The wire is then cooled, and the process repeated until the required enamel thickness (or build) for the application is achieved.

2.1 Enamelling Oven.

The enamelling oven is either horizontal or vertical in orientation depending on the size of wire enamelled. Horizontal ovens are commercially capable of producing wire in the range < 0.8mm diameter, whereas vertical ovens are capable of producing enamelled wire of > 0.5mm diameter. The reasons for this include wire sag, surface tension, choice of applicator and wire tension.

Once the product has been finished to the required specification, the wire is then fed back to the Wiredrawing/Take-Up unit, where it started. The wire is then wound onto a reel, which is weighed and bar-coded, prior to being put onto a pallet and sent to the warehouse.

Fig. 1. An enamelling oven at BICC Connollys, Liverpool.

2.2 Process Control.

Each oven is heated via a bank of electric catalyst heaters, controlled using 3 term on/off control. The warm air produced by the heaters is circulated around the oven by a constant speed warm air recirculation fan, via a catalyst plate. A negative airflow into the oven at both ends is produced to ensure that solvent fume does not escape.

Air is passed into the wire section of the oven which is split into two zones. The Evaporation Zone, where the solvent is separated from the enamel, producing additional heat in the oven, and hence preparing the enamel for curing. The Curing Zone, where the enamel is cured onto the wire.

The exit temperature of the oven is controlled by a second 3 term control loop on the oven, using two

variable speed fans, a cold air inlet, and a warm air exhaust.

In order to maintain the set oven temperature, the warm air recirculation is exhausted and replaced by the cold air inlet fan .The resultant airflow of both fans is set to maintain air flow equilibrium.

The temperature profile across the oven is controlled by a set of airflow dampers. These ensure that there is a decrease in temperature from the oven control set point in the curing zone to the evaporation zone at the point where the wire enters the oven.

3 TEST METHODOLOGIES

With respect to the testing of enamelled wire the most important test, that of Tangent Delta, is carried out off-line.

3.1 Tangent Delta.

The measurement of Tangent Delta is performed using modern electronic test equipment.

Wire samples are placed in a carriage with graphite contacts, this is subsequently placed into a test oven. Test information relating to the sample are entered into a computer and the test is started. The result of the test, a Tangent Delta temperature graph, (see Fig. 2), is printed out, with the dielectric loss factor, (Tan Delta value).

The appearance of the graph portrays the relationship of the dipoles inherent in the molecular structure of the enamel. Since the behaviour of these dipoles is dependent upon the internal bonds of the molecules and the temperature, the graph trends the changes in the molecular bonds as a function of temperature.

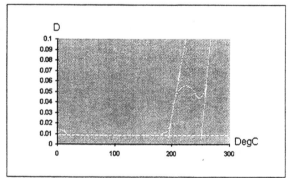

Fig. 2. Tangent Delta temperature graph, displaying dielectric loss factor for a polyesteramide enamel.

Due to the fact that the molecular structure of each enamelis different, and that the graph of each enamel varies depending upon the curing characteristics, i.e. wire speed and oven temperature, the general characteristics for each enamel and optimum tan, (cure), range can be predicted.

With reference to Fig. 2, it can be seen that the graph exhibits the characteristic valley on the steep gradient of the graph. This can be attributed to the change in oscillation mode of the dipoles, which signifies the transition of the enamel. The placement of the valley with respect to temperature indicates the cure of the enamel.

3.2 Other Off-line Tests.

The Tangent Delta method is the most popular test but it is not unique. Other tests are available such as:

Breakdown Voltage. A sample has a set of test probes attached, one to the bare copper and one to the enamelled part of the wire. A voltage is then applied across the insulation until the insulation conducts, this is thus determined to be the breakdown voltage.

Reverse Torsion. An undamaged piece of wire is clamped in a Peel tester, (a device with two chucks, one is fixed and the other is rotatable) and a force applied. The test sample is then rotated clockwise and counter clockwise, the force is removed and the material visually inspected for surface defects.

Dimension. Overall diameter, maximum and minimum gauge, bare wire diameter and enamel build are all checked using micrometers or microscopy.

Concentricity. A sample of wire is examined using a microscope following immersion in a resin and subsequent edge grinding. A microscope is then used to measure enamel concentricity.

3.3 On-line Testing..

As well as off-line tests, there exist on-line tests as follows:

High Voltage Continuity Testing. This is the traditional method of on-line testing. A high voltage is applied across the wire and test equipment detects flaws or weak spots in the enamel coating through the leakage of current via a monitoring head. The defects are counted and classed according to the degree of leakage.

Corona Discharge. A non-contact piece of equipment based upon IEC standard test requirements for winding wires. A detector head generates a high voltage, in order to generate an electric field, and subsequent corona cloud through which the wire passes. Variations in corona cloud caused by surface defects are detected as leakage currents. Faults are logged and classified according to their severity.

Laser Sensor. Diameter gauging can be performed through the utilisation of a laser scanning head. This can then be used to determine gauge.

4. QUALITY PARADIGMS

4.1 Fuzzy Logic.

Fuzzy logic, first proposed by Zadeh (Zadeh, 1965), starts with the concept of a fuzzy set. The fuzzy set by definition is not crisp or clearly defined. The very nature of the enamelling process as outlined above is thus ideal for modelling using fuzzy methods.

Application of Fuzzy Inference Systems. The process can be broken up into five distinct parts, (see Fig. 3)

1. Fuzzify Inputs,
2. Apply Fuzzy Operator,
3. Apply Implication Method,
4. Aggregate All Outputs,
5. Defuzzify.

4.2 Artificial Neural Networks.

Neural networks are composed of many simple elements operating in parallel. These elements are inspired by biological nervous systems. The functionality of the network is determined largely by the connections between elements. A network can be trained to perform a particular function by modifying or adjusting the values of the connections between elements. Neural networks have been trained to perform complex functions in various fields, including pattern recognition, identification, classification, speech, vision and control systems. Today, neural networks can be trained to solve problems that are difficult for conventional computers or human beings.

Backpropagation: One of the first and most popular neural network training algorithms is backpropagation (Rumelhart and Maclellan, 1986). For simplicity, this approach has been used in this work (see Fig 4) where a simple neural network with two input neurons, one output neuron and a single hidden layer with two neurons is tested.

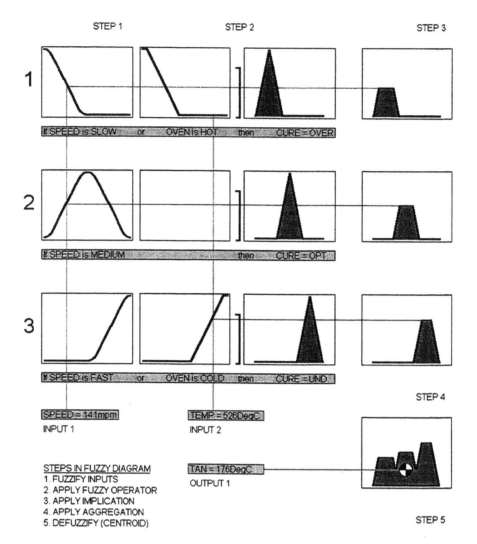

STEP 1   STEP 2   STEP 3

If SPEED is SLOW   or   OVEN is HOT   then   CURE = OVER

If SPEED is MEDIUM   then   CURE = OPT.

If SPEED is FAST   or   OVEN is COLD   then   CURE = UND.

SPEED = 141mpm

INPUT 1

STEPS IN FUZZY DIAGRAM
1. FUZZIFY INPUTS
2. APPLY FUZZY OPERATOR
3. APPLY IMPLICATION
4. APPLY AGGREGATION
5. DEFUZZIFY (CENTROID)

TEMP = 526DegC

INPUT 2

STEP 4

TAN = 176DegC

OUTPUT 1

STEP 5

Fig. 3. Fuzzy Inference Diagram for the enamelling process at BICC Connollys.

4.3 Linear modelling

The simplest of all paradigms is that of linear modelling. The system or process to be modelled is monitored and a characteristic equation is derived through graphical or mathematical representation. The derived characteristic equation is then used to model the process, around specific operating conditions.

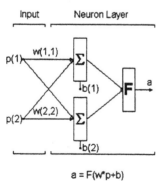

Input   Neuron Layer

$$a = F(w^*p+b)$$

Fig. 4. Simplified backpropagation network.

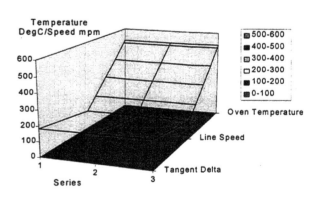

Temperature
DegC/Speed mpm

□	500-600
■	400-500
▣	300-400
□	200-300
■	100-200
▣	0-100

Oven Temperature

Line Speed

Tangent Delta

Series

Fig. 5. Surface contour for the enamelling process.

Because the model is in fact a linearisation the accuracy of it can and does vary. It is therefore best used as an indicator of a process rather than a true model. The surface contour relationship for the enamelling process may be seen in Fig. 5.

5 RESULTS

Product type, oven temperatures, fan speeds and wire speeds for each of the eight individual 'heads', of the machine were time/date keyed to the quality information. This resulted in a set of 14 inputs and 1 output for each individual product/head though a large proportion of the parameters were common.

This information was then sorted into a time series related to each specific head and surplus/spurious information was discarded. It was also decided that in order to prove the process an initial set of key parameters would be selected. These were 2 inputs, oven temperature and wire speed and 1 output, tangent delta value. The result was a workable data table, keyed to each individual take-up unit on the curing oven, and hence related to a specific product type.

The preliminary models for each of the three paradigms were tested using an initial set of test data. Subsequent data was used as test vectors, in order to prove the validity of the paradigms. It was evident during testing that a variation in oven conditions in relation to wire speeds due to different product mixes and enamel batches caused the quality model to vary.

The resultant predicted tangent delta output for the process varied in most cases and as such each of the models was tuned to better reflect the behaviour of the process.

5.1 Paradigm Testing.

Following the development of the models for each of the three control paradigms a further set of test vectors, was used to prove the tangent delta prediction accuracy of the individual models, in comparison to each other.

In order to evaluate the accuracy and facilitate comparison three measures were used:

- Arithmetic median of all predicted values,
- Standard deviation of all predicted values,
- Graphical representation of time series.

Arithmetic Median.
Because of the variance of predicted tangent delta values over a wide sample range, and the occasional spurious reading the median effectively returns a filtered average variance. The standard deviation for the product 0.355N1, may be seen in Table 1.

Table 1 Median variance of predicted tangent delta for four discrete time periods for 0.355N1

Linear	Fuzzy	Neural
10	4	3
3	-1	-3
-10	-1	-5
6	2	-1

Table 2 Standard deviation of predicted tangent delta for four discrete time periods for 0.355N1

Linear	Fuzzy	Neural
3	2	3
12	3	2
24	5	9
18	4	2

Standard Deviation.
The standard deviation of the predicted tangent delta values is a measure of the variation in the predicted values throughout the data series. The standard deviation for the product 0.355N1, can be seen in Table 2.

Time Series. The time series is a representation of predicted tangent delta results against actual tangent delta results over the three sample periods, this can be seen in Fig . 6.

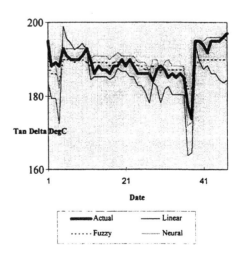

Fig. 6. Time series of predicted tangent delta for 0.355N1 enamelled wire.

6 CONCLUSIONS

It can be seen from the basic statistical analysis carried out and reported above that the fuzzy paradigm gives the most repeatable prediction of tangent delta for the process. This is true for all test vector sets. However when the time series is analysed it can be seen that the values are closely grouped around the control parameter band of the defuzzified output. It is therefore obvious that even although the model can deal with significant variations in the input fields, these cannot be translated through to the output.

One way to improve this is to increase the input and output window sizes. But care must be taken as this will result in greater fluctuations in the values predicted by the fuzzy logic model. The may result in a requirement to introduce more input variables to maintain stability

The trained back propagation neural network performs reasonably well for all of the test vectors. Again, the statistical analysis performed on the results indicates that this approach is not as good as that of the fuzzy logic model. However, the time series plot, when compared to that of the actual tangent delta maps very closely, albeit for a slight offset.

Fluctuations in the actual process quality are also followed well by the neural network, in particular for values that the network has not been trained on. It is therefore obvious that this type of paradigm has more scope to be developed at a lower level, through the careful selection of training data and the number of training epochs used.

More work should be conducted to see at what point the network is over-trained, and also the effect of different network structures in terms of the numbers of neurons in input, output and hidden layers.

The linearised model paradigm has been shown, both statistically and graphically, to be unsuitable for modelling. As might be expected, even slight variations in the process conditions give major variations in the predicted tangent delta.

The results from this preliminary study provide evidence that modelling based on fuzzy and neural architectures can be developed to assist in on-line quality control procedures for a wide range of different manufacturing processes. We have only concentrated on one in this work.

REFERENCES

Benyo, Z (1969) Enamelled Wire Manufacturing - a new method to control running production, *Journal Hungarian Heavy Industry*, **19**, 14-19, July

Jorgensen, P (1990) Tangent Delta Temperature Curves and Principles, *Dansk System Electronik A/S*

Rumelhart, D E and McClelland, J L (1986) *Parallel Distributed Processing*, 318-362, MIT Press

Zadeh, L (1965) Fuzzy Sets, Information and Control, **8**, 338-353

FUZZY-MARKOV SIMULATION TECHNIQUE FOR
PRODUCT TESTING EQUIPMENT

Valentin Arkov **,1 Timofei Breikin* Gennady Kulikov*

* *Ufa State Aviation Technical University, Russia*
** *Institute of Mechanics, Russia*

Abstract: A novel intelligent simulation technique is proposed for testing of automatic control systems. The technique combines fuzzy reasoning and Markov modelling for stochastic system simulation purposes. A learning problem is also discussed with an example of a digital test-bed.

Keywords: Simulation, fuzzy modelling, Markov models, tests.

1. INTRODUCTION

Product testing is used in manufacturing systems to ensure the quality of purchased and manufactured items (Harrington, 1985; Besterfield, 1994). Digital controllers are tested on special test-beds after assembling. It is also necessary to use test-beds before finally assembling an aircraft (Kulikov et al., 1992). A test-bed for the digital controller simulates real engine functioning, this equipment also imitates real transducers and actuators. Simulation of random disturbances allows testing of the accuracy and interference protection of the assembled unit.

Common simulation uses in manufacturing include design, redesign and process monitoring (Thomson, 1995). This paper discusses a novel simulation technique for the testing of equipment. A system combining Markov and fuzzy elements is proposed to simulate determined and random behaviour of a gas turbine engine as a whole model. Henceforth this technique will be referred to as fuzzy-Markov. Real engine data have been used to assess the viability of this novel approach. The advantage of the proposed approach to simulation is its flexibility and applicability to both linear and nonlinear systems.

---

[1] NATO Postdoctoral Fellowship in the University of Sheffield

Most fuzzy system applications are developed for control and analysis purposes (Zadeh, 1973; Larsen, 1980; Mamdani, 1977). Another group of applications is system state prediction (Jang, 1993; Frank, 1996). Fuzzy systems can also be used for adaptive signal processing (Wang and Mendel, 1993). A Markov approach to dynamic modelling was proposed in previous work (Patel et al., 1996). A Markov chain can be considered as an extreme case of a fuzzy system with a rectangular membership function. The output is a probability distribution, not a variable value. The proposed approach represents an attempt to overcome the primary difference between nonrandom fuzzy sets and the probability theory, which deals with random phenomena.

2. MARKOV MODELLING OF DYNAMIC SYSTEMS

The plant dynamic model can be described by nonlinear difference equation

$$x(t+1) = \Phi\{x(t), u(t), e(t)\}, \qquad (1)$$

where $x(t)$ is a state vector, Φ is a nonlinear function and $e(t)$ is an independent Gaussian random vector. This representation can be considered as a Markov process (Åström, 1970). The order of the process depends on the order of Eq.(1).

A Markov chain can be obtained using state quantization. The stochastic processes within the dynamical system can often be assumed to be stationary and ergodic. In this case the Markov chain is homogeneous, and its behaviour is described by the transition probability matrix P. In a high order case, the transition probability matrix represents an appropriately dimensioned hypercube. The first order model represents a three-dimensional matrix,

$$P = \{P_{ijk}\}, \tag{2}$$
$$P_{ijk} = P\{x(t+1) = X_i | x(t) = X_j, u(t) = U_k\}.$$

where P_{ijk} is the probability of transition from the state X_j to the state X_i under control U_k. The state probability is described as

$$\rho_i = \int_{X_i - \Delta/2}^{X_i + \Delta/2} p(x)dx$$
$$= P\{x \in [X_i - \Delta/2, X_i + \Delta/2]\}, \tag{3}$$

where $X = (X_1, \ldots, X_n)$ is a vector of interval centres, ρ_i is the probability of x being in i-th interval and Δ is the interval length. This Markov chain can be considered as a fuzzy system with a rectangular membership function. The terms X_i, X_j, U_k represent linguistic variables like PB, ZO, NB, etc. The fuzzy rule base contains the transition probability matrix.

Model learning becomes possible when using real data (Wang and Mendel, 1992). Hence, computing matrix P allows the creation of the rule base using real data as expert knowledge. Therefore, the model can be obtained by estimating the matrix P,

$$P_{ijk} = \frac{N_{ijk}}{\sum_j N_{ijk}}, \tag{4}$$

where N_{ijk} is the number of the corresponding transitions. Eq.(4) makes the matrix P stochastic.

Solution of the simulation problem represents the realization of a random value with a desired distribution (Kulikov et al., 1996). At every time $t + 1$ a new value $x(t + 1)$ has the probability distribution density,

$$p(x(t + 1)) = f(x(t), u(t)). \tag{5}$$

Note that $p(x(t + 1))$ represents the appropriate row of the transition probability matrix P. The generation of the demanded probability density is realized using a uniform distribution generator and functional transformation.

3. MARKOV SIMULATION

The process of the Markov simulation consists of three stages: obtaining the current Markov state, extracting the output distribution for the next step and generating a random value with the demanded distribution.

Consider the simulation of a first order Markov model with one input and one output. Note, that the transition probability matrix P is estimated before the simulation from the experimental data by Eq.(4). At the first stage, the current input $u(t)$ and output $x(t)$ are measured. Having been compared with the interval centres, the input and output are transformed to the Markov state $\{X_j, U_k\}$ by the following formulae

$$\begin{cases} X_j : j = arg \min_j |x(t) - X_j| \\ U_k : k = arg \min_k |u(t) - U_k|. \end{cases} \tag{6}$$

At the second stage, the vector p is obtained from the transition probability matrix P using the indices j and k, as shown in Fig.1a. The elements p_i of the vector p represent the state probabilities $p_i = P\{X_i\}$ for the time $t+1$. Fig.1b shows the obtained probability distribution $p(x)$. Finally, the demanded distribution is utilized for generating the random output $x(t + 1)$. The transformation method (Press et al., 1989) is used for generating a random number with a known distribution,

$$F(x) = \int_{-\infty}^{x} p(x)dx. \tag{7}$$

A uniform random number y is chosen between 0 and 1 (see Fig.1c), the desired transformation is therefore,

$$x(t + 1) = F^{-1}(y). \tag{8}$$

4. EXAMPLE

The proposed Markov simulation technique was applied to a digital test-bed. This equipment is intended for testing digital automatic control (DAC) systems after assembling. The testing of the DAC in a test-bed is also necessary before assembling the aircraft. The test-bed structure is shown in Fig.2. It consists of five elements: a digital model for engine simulation, imitators of transducers and actuators, DAC and a log device. The experimental data shown in Fig.3 were used for direct learning of the Markov model by Eq.(4). The engine dynamics at the steady state conditions can be described by the first order difference equation,

$$x(t + 1) = A(x) + Bu(t) + e(t), \tag{9}$$

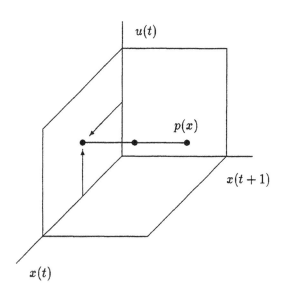

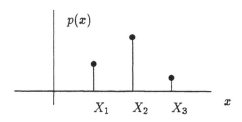

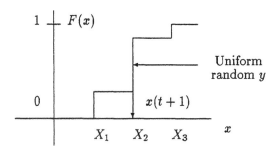

Fig. 1. Markov simulation: transition probability matrix (a), probability density (b) and simulation of demanded distribution.

where the input u is the fuel flow and the output x is the shaft speed. Having been learned, the Markov model was used for simulation. The simulated shaft speed is shown in Fig.4. The comparison of basic descriptive properties demonstrates the viability of the proposed technique, as can be seen in Fig.5.

5. FUZZY-MARKOV SIMULATION

The Markov simulation technique discussed above provides the high speed of computation because it utilizes only move and comparison operations. But, the Markov modelling approach allows a very limited number of system states. Furthermore, the transition probability matrix has a large size. For instance, the matrix P from the example shown

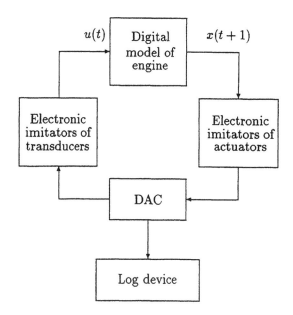

Fig. 2. Structure of test-bed for DAC.

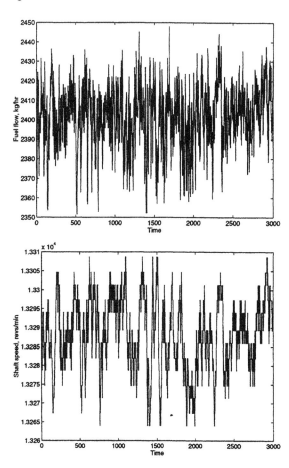

Fig. 3. The experimental data from the gas turbine engine D27: input (a) and output (b).

in Fig.4 has 13^3 elements. This disadvantage can be avoided using fuzzy logic.

Fuzzy systems have been referred to as 'universal approximator' (Castro, 1995). This can be a tool for smooth nonlinear interpolation of a multidimensional probability density function. The Markov model in this case represents a fuzzy inference system with a rectangular membership func-

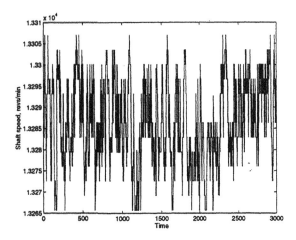

Fig. 4. Simulation of shaft speed dynamics.

tion. An adaptive-network-based fuzzy inference system (ANFIS) has been proposed for chaotic time series prediction (Jang, 1993). Stochastic time series simulation can be carried out using fuzzy inference approach combined with Markov modelling.

The structure diagram of the fuzzy-Markov simulation system is shown in Fig.6. The rule base contains information about the multidimensional transition probability matrix P. Before simulation, the rule base is learned from the experimental data. This can be done using neural networks or evolutionary computation methods. The criteria for optimization are the following: the order of the Markov model, the mean square error of the spectrum and distribution, the type of fuzzy model, the type and number of the membership functions.

The procedure for the fuzzy-Markov simulation includes four stages: fuzzification, inference, defuzzification and randomization. First of all, the input $u(t)$ and output $x(t)$ are fuzzified using the membership functions. The fuzzy values U and X with their degrees of membership are used to infer the rules from the rule base. Having been extracted, the fuzzy rules are aggregated to the fuzzy probability density function $P(X)$ of the output for the time $t+1$. After defuzzification, integration and normalization, the distribution function $F(x)$ is obtained. The normalization makes the function $F(x)$ lie in the interval $0 \leq F(x) \leq 1$,

$$F(x) = P\{x(t+1) \geq x\} \qquad (10)$$
$$= \frac{\int_{-\infty}^{x} p(x)dx}{\int_{-\infty}^{\infty} p(x)dx},$$

where $p(x)$ is the result of the defuzzification. Finally, a random value with the desired distribution $F(x)$ is generated using the transformation method from a uniformly distributed random number. Introducing the modified defuzzification

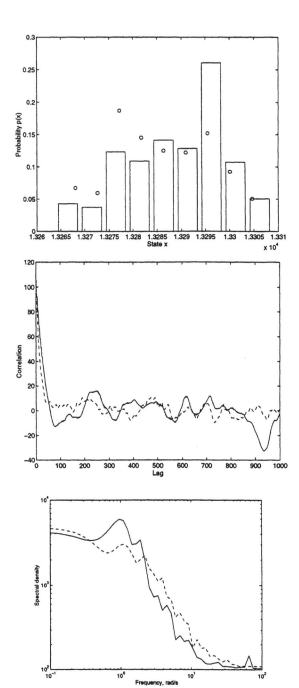

Fig. 5. Descriptive properties of the real and simulated output: histograms (a), correlation functions (b) and spectral densities (c). The real data are shown by solid line.

and randomization stages allows fuzzy simulation of a stochastic dynamic system.

6. CONCLUSIONS

The Markov modelling approach has been extended to fuzzy inference systems. The introducing of the modified defuzzification and randomization stages into a fuzzy inference system allows stochastic dynamic system simulation. The fuzzy-Markov simulation technique has been derived for product testing equipment. A numerical example

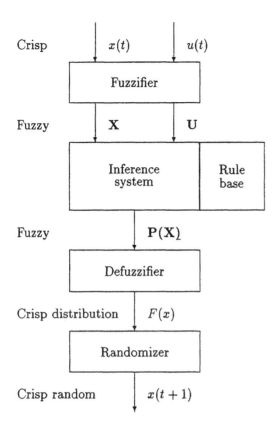

Crisp $x(t)$ $u(t)$

Fuzzifier

Fuzzy   **X**   **U**

Inference system   Rule base

Fuzzy   **P(X)**

Defuzzifier

Crisp distribution $F(x)$

Randomizer

Crisp random $x(t+1)$

Fig. 6. Fuzzy-Markov simulation system.

has been provided for a gas turbine engine controller test-bed.

7. REFERENCES

Äström, K. J. (1970). *Introduction to Stochastic Control Theory*. Academic Press. New York, London.

Besterfield, D. H. (1994). *Quality Control*. Prentice Hall. Englewood Cliffs, NJ.

Castro, J. L. (1995). Fuzzy logic controllers are universal approximators. *IEEE Transactions on Systems, Man and Cybernetics* 25(4), 629–635.

Frank, P. M. (1996). Analytical and qualitative model-based fault diagnosis and some new results. *European Journal of Control* 2(1), 6–28.

Harrington, J. (1985). *Understanding the Manufacturing Process. Key to Successful CAD/CAM Implementation*. Marcel Dekker. New York.

Jang, J.-S. R. (1993). Anfis: Adaptive-network-based fuzzy inference systems. *IEEE Transactions on Systems, Man and Cybernetics* 23, 665–685.

Kulikov, G. G., I. I. Minaev and G. I. Pogorelov (1992). Basic principles of test-bed structure and functions, with applications in testing and investigation of an aircraft engine controllers (in russian). *Aviatsionnaya Promyshlennost* 12, 31–35.

Kulikov, G. G., V. Y. Arkov and T. V. Breikin (1996). On markov model applications in aircraft gas turbine engine full authority digital controller test-beds. In: *Proceedings of UKACC International Conference on CONTROL'96*. pp. 120–124.

Larsen, P. M. (1980). Industrial application of fuzzy logic controller. *International Journal of Man-Machine Studies* 12, 3–10.

Mamdani, E. H. (1977). Application offuzzy logic to approximate reasoning using linguistic synthesis. *IEEE Transactions on Computers* C-26(12), 1182–1191.

Patel, V. C., V. Kadirkamanathan, G. G. Kulikov, V. Y. Arkov and T. V. Breikin (1996). Gas turbine engine condition monitoring using statistical and neural network methods. In: *Proceedings of IEE Colloquium on Modelling and Signal Processing for Fault Diagnosis*. pp. 1–6.

Press, W. H., B. P. Flannery, S. A. Teukolsky and W. T. Vetterling (1989). *Numerical Recipes. The Art of Scientific Computing*. Cambridge University Press. Cambridge, New York, Sidney.

Thomson, N. (1995). *Simulation in Manufacturing*. John Wiley and Sons. Exeter.

Wang, L. X. and J. M. Mendel (1992). Generating fuzzy rules by learning from examples. *IEEE Transactions on Systems, Man and Cybernetics* 22, 1414–1427.

Wang, L. X. and J. M. Mendel (1993). Fuzzy adaptive filters, with application to nonlinear channel equalization. *IEEE Transactions on Fuzzy Systems* 1(3), 161–170.

Zadeh, L. A. (1973). Outline of a new approach to the analysis of complex systems and decision processes. *IEEE Transactions on Systems, Man and Cybernetics* 3(1), 28–44.

A STUDY ON THE APPLICATION OF THE JOINT TRACKING SYSTEM TO MULTIPASS ARC WELDING USING VISION SENSOR

I. S. Chang, J. I. Lee, S. H. Rhee, and K. W. Um

Hanyang University, Department of Mechanical Engineering
17, Haengdang-dong, Sungdong-ku, Seoul, 133-791, Korea
(Phone)+82-2-290-0438, (Fax)+82-2-299-6039

Abstract: Most of the past joint tracking system using vision sensor could be used to locate joints of rather simple geometry in single pass welding operations. Therefore, in this study, a more detailed geometric description of welded joints of general shape and the vision progressing techniques for the dimensional weld joints measurements are presented. GUI software were developed and applied to joint tracking for the determination of the tracking position of the joint. The developed joint tracking system is applied to single-pass arc welding as well as multi-pass arc welding, through the process of specifying the layer procedure and tracking position.

Keywords: Vision Sensor, Autosynchronized scanner, Joint Tracking System,

1. INTRODUCTION

Welding is one of the most wide-spread joining processes used for structural fabrication. But it is traditionally a labor-intensive task depending on the skill and experience of the operator. Furthermore, manual welding is much less repeatable and coherent than automatic welding. Therefore, automation of welding is essential for the guarantee of productivity, lower cost and high quality welding products. Though many mechanical and electrical technologies have been proposed for joint tracking, machine vision can be proven the most effective for sensing the position of the material.

The process of research using vision sensor is as follows:

Agapakis (1986, 1990) demonstrated some approaches to the adjustment of welding parameters, the reliable recognition of general weld joint features, and the real-time joint tracking using machine vision. Clocksin (1985) proved some more effective applications for lab joint, T-joint and butt joint in GMA welding of thin plates. Nakata (1989a, 1989b) experimentally proposed an optimal geometrical architecture of optical device in terms of the light source, the position, the resolution, the exposure time of camera. Most of the past joint tracking systems of machine vision were restricted to locating the joints of simple geometry in single pass welding. Therefore, in this study, a more improved geometrical description of welded or unwelded joints and vision processing technique for measuring of the dimensional measurements are presented. the developed GUI software for the detection of the weld joints and the decision of tracking position of welding torch was applied to joint tracking. Using these techniques, the characteristic features of weld joints and of weld beads were extracted exactly. Through the process of specifying the layer procedure and the notification of tracking position, the developed joint tracking system can be applied not only to single pass but also to multi-pass welding.

2. VISION SENSING PRINCIPLE

Machine vision is typically employed for part finding, joint type identification, joint tracking and detection of end of seam. The vision system can be divided into two types: vision sensor using a projected pattern lighting and vision sensor using a scanning beam. The former is contouring by light sectioning, and the latter is that the object is illuminated by on single spot. Both are all based on optical triangulation. The Optical triangulation is a familiar principle in optical 3-D sensing. In industrial applications mainly vision systems using a scanning beam are used due to their robustness and the simplicity of data evaluation. The scanning beam system used in this study will be considered.

2.1 Vision Sensor using Scanning Beam

Sensors using scanning beams are used for scanning material surfaces, usually by applying a laser beam to a rotating plane mirror, polygonal mirror or pyramidal mirror, by using a precision -developed galvanometer. By using one dimensional displacement sensor to detect beams reflected off of materials by scanning, the range to the object is measured by optical triangulation. Such measured distance values are associated with information about rotation of mirror, and make 2 dimensional images. These images are called range images. In comparison with the projected pattern beam method, although it has a complicated structure and is an expensive system, the effect of the arc and the whole processing time is lessened, so usage of this system is increasing. The principle of vision sensor by scanning beam springs from optical triangulation which measures the distance from the light source to the object. Fig.1 and Fig.2 show now general optical triangulation and Scheimpflug geometry is constructed. It uses parallel lines as far apart from each other as the distance between f(focal length of the lens) and the principal plane of the lens. You can see that the position sensor is tilted towards the detection axis and this angle is founded in Eq.'s (1), (2) from Scheimpflug condition, which is the condition in which diffraction laser beams can make a clear image.

$$\beta = \tan^{-1}(\frac{f_0}{d}) \qquad (1)$$

$$f_0 = \frac{fl}{(l-f)} \qquad (2)$$

where, f is the focal length of the lens, β is the angle formed between the lens and CCD array, d is the distance between the projection axis and the center of the lens, ℓ is the distance from the object to the lens is as shown.

The advantage of such geometrical images is that any point of the projection axis is focused accurately according to the position sensor. This property allows much improvement of the depth

of view. Synchronization of the detection axis and projection axis by expending this principle to 2 dimension is the basic of synchronized scanner. For the synchronized scanner to make resolution superior, the angle between the light source and displacement had to be enlarged, which brought about the disadvantage of decrease in measurement range. The vertical displacement and horizontal displacement form a coupling, which becomes a factor in lessening the efficiency of the sensor. The autosynchronized scanner was developed to solve such problems.

As shown in Fig.3, the laser scanned on the material and the capturing light is scanned into the co-axis which is connected to the monitor. By doing this, the distance to the object across the scanning range is known, and a joint image is obtained with every scanning

3. SYSTEM CONFIGURATION

System configuration for automatic joint tracking used for sensing the joint geometry and for controlling robot is shown in Fig4. The joint system is composed of vision system, PC, motion controller, CO_2 welding machine, and 3-axis Cartesian robot. The vision system is, in general, composed of a camera and a control unit for vision processing. The vision system is composed of an industrial PC integrated with AMD 5x86 CPU (corresponding to 75MHz Pentium CPU) and a camera control unit integrated with DSP for vision processing and for laser power and scanning control. For a laser camera, this system used M-SPOT-90 from Servo Robot Inc. This camera consisted of a one dimensional CCD array, an optical profiling laser applying optical triangulation combined with autosynchronized scanning, and a band pass filter which allows only particular bandwidth to pass through.

Data obtained from the camera are processed in many steps. The first step is sensor calibration by vision system. As a result, raw profile data which consist of 256 points are transferred to industrial PC through the ISA bus. Raw profile data transferred from the vision system is vision processed, and a trajectory can be made by processing such data and by extracting the information on joint geometry.

4. VISION PROCESSING

Images generated as a result of signal transduction by vision system can be classified as intensity images or range images. Range image is an array of range values(distance) which falls under spatial points. On the other hand, intensity image is a 2 dimensional array which represents intensity value which falls under spatial points of measuring planes. In this study, range images consisting of 256 points are obtained from the vision system.

In order to apply joint tracking, raw profile data is processed in a few steps. The position of joint geometry is eventually represented as the position in the world space or as the position relative to torch tip. Raw profile data from not only range image but also intensity image may be affected by the error which acts as noise to measured profile data. Therefore, in general, (1)vision preprocessing is the first step of vision processing. By doing this, improved images having minimized error by filtering raw data profile with mean filter or median filter can be gained. But filtering has a disadvantage of increasing processing time. After vision preprocessing, it is possible to extract the features by template matching algorithm. But in this study (2)segmentation process that approximate the profile to line segments is carried instead of template matching. From such processed profiles, the feature representing the object being measured is confirmed, and then position is found out by (3)feature extraction and recognition, In case of weld joints, these features are generally classified as feature point or feature surface such as joint edge point, sloped surface and root point.

In multi-pass welding, the tracking point can be obtained from these features in accordance to layering process specifying the layer procedure.

4.1 Vision Preprocessing

Vision preprocessing is the process for improving the quality of images or for adjusting images to be suitable for specific purpose. This process includes image handling such as smoothing, sharpening and isolating of high frequency or low frequency. Raw profile data obtained as a result of the sensor calibration is represented by 256 (y, z) coordinate values. But the profiles can be expected to have noise from a variety of sources. For example, welding arc light, spatter, and multiple reflections from specular surfaces can cause the measured data to be erroneous. These introduce requirements of high operational reliability on the vision sensing for joint tracking system. Much work has been done to improve the quality of profiles. In this study, median filter, the considerably effective nonlinear processing method for discard erroneous range values, is used to smooth profiles and to remove the noise. Since rather larger windows may lose some features such as sharp V corners, it is proper to adjust a window size to 3 or 5 neighboring profile data.

4.2 Segmentation Processing

An abrupt orientation change of the profile corresponds to a feature that should be recognized. The process of partitioning the image space into meaningful regions is known as segmentation. The meaningful regions refers to the joint's features. But segmentation processing presented here is to approximate the profile to line segments in order to find such features of profiles stably. New segmentation processing was carried out by a simple algorithm suitable for range image as shown in Fig.5. Segmentation processing is also carried out symmetrically to the top, bottom, left and right of the graph. In Fig.5 m_1, m_2, d_1 and d_2 was determined by heuristic reasoning to make an optimum condition. The profile data which satisfy the above conditions are decided and saved as the breakpoint, which is the intersecting point of each line segment. The characteristic features of weld joints or weld beads can be extracted on the basis of the breakpoints obtained from such segmentation processing.

4.3 Feature Extraction and Recognition

The characteristic features of weld joints and weld beads can be induced on the basis of segmentation processing. The features of typical V groove are as shown in Fig.6. Feature extraction is finding such information on feature points and feature surfaces. By using this information, tracking points are decided according to layer procedure.

The feature extraction algorithm for V groove in this study is based on the iterative averaging technique. Of the breakpoints obtained through segmentation processing, the algorithm can be used to find the left edge point, P_2 and the right edge point, P_6 respectively in Fig.6.

First Average:

$$\bar{z_0} = \frac{1}{n} \sum_{k=1}^{n} z_k \qquad (3)$$

Find L when $Z_L > \bar{z_0}$, $Z_{L+1} < \bar{z_0}$, $1 \leq L \leq N$ (4)

Find R when $Z_R > \bar{z_0}$, $Z_{R-1} < \bar{z_0}$, $1 \leq R \leq N$ (5)

Left Edge:

$$\bar{Z} = \bar{z_0}$$

Repeat

   1. If $(Z_k < (\bar{Z} + \delta))$, $k = L \dots 1$ (6)
   Then, L = k

$$2. \quad \bar{Z} = \frac{1}{L} \sum_{k=1}^{L} Z_k \qquad (7)$$

$$\text{Until} \quad Z_k \geq (\bar{Z} - \delta), \quad \forall k \in [1, L] \qquad (8)$$

Left Edge = L

Right Edge:

$$\bar{Z} = \bar{z}_0$$

Repeat

$$1. \quad If(Z_k < (\bar{Z} + \delta)), \quad k = R \dots N \qquad (9)$$

Then, R = k

$$2. \quad \bar{Z} = \frac{1}{N - R + 1} \sum_{k=R}^{N} Z_k \qquad (10)$$

$$\text{Until} \quad Z_k \geq (\bar{Z} - \delta), \quad \forall k \in [R, N] \qquad (11)$$

Right Edge = R

where n is the number of the profile data, N is the number of the breakpoints, z_k: z coordinate of profile data, Z_k is the z coordinate of breakpoints on profile and δ is half of the tolerance band which is based on the size of joints and the resolution of laser camera.

It is shown that starting from the breakpoint which exceeds the average value of total profile data, the first breakpoint can be found in the tolerance band which is shaped from the centers of flat surfaces, S1 and S4.

In case of the root pass, P_4 can be found between left edge P_2 and rootpoint P_4 as follows.

Root:

$$If(Z_{k+1} > Z_k), \quad k = L \dots R \qquad (12)$$

Then, Root = k

This is eventually on the extreme point between two edge points. An example of root pass of V groove on the developed program is shown in Fig.7.

After the root pass, as root point, P_4 do not appear, an algorithm for finding wetting points, P_3 and P_5, made by weld bead is required. Of the breakpoints obtained from segmentation processing, the left wetting point, P_3 is defined as the first break point that makes an abrupt angle change, along the joint surface (in this case, the oblique side of the groove) which starts from the left edge point, P_2. A schematic algorithm shown in Fig. 8 is as follows:

Left Wet:

$$If(\tan \theta_2 = \frac{Z_{k+1} - Z_k}{Y_{k+1} - Y_k} < \tan \theta_1), \quad k = L, \cdots, R \qquad (13)$$

Then, Left wet = k

where, (Y_k, Z_k) are the coordinates of breakpoint on profile.

Fig.9 shows an example of determining the wetting point as a tracking position at the second pass on developed program. The right wetting point, P_5, can be determined by using a symmetrical form of the process.

5. DEVELOPMENT OF SOFTWARE FOR JOINT TRACKING USING GUI

In this study, Borland C++, a compiler, was used to develop the program. And by using the library for Korean language, user can perform the joint tracking with far less effort to understand English. Every operation can be made through menus which were constructed by GUI(Graphic User Interface) and the modification of parameters which are needed to joint tracking can be easily made. In welding operation, this program made it possible to monitor each parameter and profile on line while joint tracking. Fig.10 shows an example of the 3rd pass in tracking mode of program.

The features of weld joints and weld beads have been extracted through vision processing, , and by using the information about these features, the tracking point is determined through the layering process.

5.1 Flow chart for joint tracking in tracking mode

Flow chart for joint tracking is shown in Fig.11.

5.2 Multipass Layering Process

Most of the past joint tracking systems using vision sensor were restricted to single pass welding. In this study, an algorithm is proposed to apply not only to root pass but also the second pass and the third one as well.

In general, in case of the root pass, from the left edge point, P_2, and the right edge point, P_6, the root point, P_4 which is located in the middle of them can be found as a tracking point.
To apply joint tracking system to Vee groove welding requiring multi-pass welding, a few additional routines other than root pass are necessary. In case of the 2nd pass, if tracking position is hypothesized to be wetting point, P_3, the routines for P_3 and multi-pass procedure as that of Fig.12 are necessary. As shown in Fig. 13, tracking position can be calculated by

deciding multi-pass procedure, offset(m) and reference point obtained from extraction and recognition of weld beads. Fig.14 shows an example of finding the wetting point as a tracking position of the third pass on Vee groove, and here, the offset is zero.

6. JOINT TRACKING USING 3-AXIS CARTESIAN ROBOT

Tracking position (y, z) of welding torch is calculated on camera coordinate frame. And because the torch direction in relation to base metals can be determined, full 3 dimensional joint tracking is possible. After that, as the torch position can change according to past corrections or past tracking position sent to the motion controller, tracking position must be transformed to base coordinate frame. With the camera located in the look-ahead distance from the welding torch, the information of joint positions is used to find the weld path or trajectory.

To remove unexpected out-of-range corrections and undesirable abrupt changes, the trajectory found should be filtered.

6.1 Coordinate Transformation

For welding torch to follow the tracking positions measured by camera, the tracking positions should be represented. in base coordinate frame. Coordinate frames are defined as shown in Fig.15.

{B} : base coordinate frame
{C} : camera coordinate frame
{1} : slide 1 coordinate frame
{2} : slide 2 coordinate frame
{J} : jig slide coordinate frame

x coordinate of the joint in terms of base coordinate frame at k-th step is represented as follows:

$$^{B}P(k)_{SORG} = \ ^{B}_{J}T(k) \ ^{J}P_{SORG} \qquad (14)$$

where $^{B}P(k)_{SORG}$ is the position vector of the joint in terms of jig slide coordinate frame at k th step and $^{B}_{J}T(k)$ is the transformation matrix of jig slide coordinate frame relative to base coordinate frame at k-th step

The profiles obtained from the vision system at k-th step can be rewritten with respect to the base coordinate frame by using compound transformation.

$$^{B}P(k)_{SORG} = \ ^{B}_{1}T(k) \ ^{1}_{2}T(k) \ ^{2}_{C}T(k) \ ^{C}P_{SORG}$$
$$(15)$$

where the vector, $^{C}P_{SORG}$ has 3 components, i.e., x coordinate of zero, and y, z coordinate found from the profile at k-th step, respectively.

The coordinates of torch position relative to the base coordinate frame at k-th step are represented by the compound transformation of slide 1 and slide 2.

$$^{B}P(k)_{TORG} = \ ^{B}_{1}T(k) \ ^{1}_{2}T(k) \ ^{2}P_{TORG} \qquad (16)$$

Torch position at k-th step is a position of tracking position which is sensed at (k-n)th step. Where the value, n which is the number of steps between the camera and the torch, is fixed as constant. This can be found from Eq.'s (15), (16) as follows:

$$^{B}P(k)_{TORG} = \ ^{C}P(k-n)_{SORG} \qquad (17)$$

From Eq.'s (15), (16), (17), the sliding distance of slide 1 and 2, $d_1(k)$, $d_2(k)$ respectively, can be obtained.

6.2 Tracking Error Handling

Trajectory is based on image analysis depending mainly on the quality of profiles. Because low quality profile may cause tracking errors, modification is necessary before moving the torch to the tracking position. Under assumption that joint does not change abruptly, to validate tracking position and get rid of sudden unexpected movement, filter is used.

Moving average filter which gives a weighting factor on present data was used so that joint information could be obtained, according to the tendency of the past tracking positions.

7. CONCLUSION

As a part of welding automation, through observation and experiments of vision processing and joint tracking using machine vision, the following results were acquired.

(1) An effective vision processing technique is suggested for 3 steps of welding processes: preparation, welding process execution and post-weld inspection.

(2) By using the above technique, the characteristic features of the weld joints and the weld beads can be exactly extracted in this study.

(3) A software using the graphic user interface in Korean language is developed for monitoring each parameter and profile, and is applied to joint tracking.

(4) It is shown that, by specifying the layer procedure, the information of joint geometry of the weld joints and the weld beads can be used effectively not only in single pass but also in multi-pass welding.

REFERENCE

Pugh, A. (1986). Robot Sensors Vol. 1-Vision, Springer-Verlag

Agapakis, J. E. et al (1990). Approaches for Recognition and Interpretation of Workpiece Surface Features using Structured Lighting. In: *The International Journal of Robotics Research*, **Vol. 9, No. 5**, pp.3-16.

Agapakis, J. E. et al (1986). Joint Tracking and Adaptive Robotic Welding using Vision Sensing of the Weld Joint Geometry. In: *Welding Journal*, **Vol. 65, No. 11**, pp. 33-41.

Rioux, M. et al (1987). Design of a Laser Depth of View Three-Dimensional camera for Robot Vision. In: *Optical Engineering*, **Vol. 26, No. 12**, pp. 1245-1250.

Nayak, N. and Ray, A.(1993). Intelligent Seam Tracking for Robotic Welding, Springer-Verlag.

Nakata, S. and Jie, H.(1989a). Construction of Visual Sensing System for In-process Control of Arc Welding Process and Application in Automatic Weld line Tracking. In: *Japanese Welding Journal*, **Vol. 7, No. 4**, pp. 467-472.

Nakata, S. et al (1989b). Determination on Geometrical Arrangement of Optical Equipments and Photographic Parameters for Construction of Visual Sensing System. In: *Japanese Welding Society*, **Vol. 7, No. 3**, pp. 358-362..

Clocksin, W.F. et al (1985). An Implementation of Model-Based Visual Feedback for Robotic Welding of Thin Sheet Steel. In: *The International Journal of Robotics Research*, **Vol. 4, No. 1**, pp.13-26.

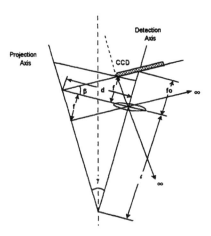

Fig. 2 Description of Scheimpflug Geometry

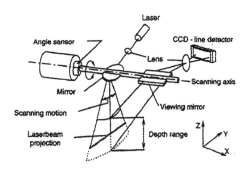

Fig. 3. Principle of Operation for the Autosinchronized Scanner

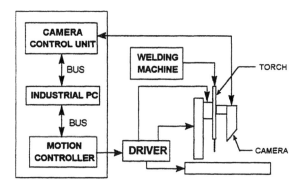

Fig. 4 System Configuration for Joint Tracking

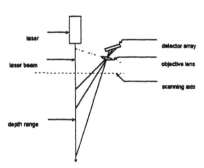

Fig. 1 Principle of Triangulation

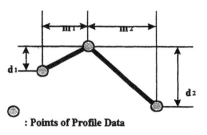

⬤ : Points of Profile Data

Fig. 5 Description of the Segmentation Processing

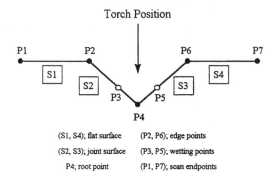

Fig. 6 Characteristic Features of a
Vee-grooved Weld Joint

(S1, S4); flat surface   (P2, P6); edge points
(S2, S3); joint surface   (P3, P5); wetting points
P4; root point   (P1, P7); scan endpoints

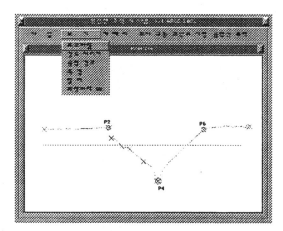

Fig. 7 Example for the Root Pass on Joint
Tracking Program

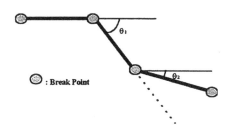

Fig. 8 Detection of the Wetting
Point

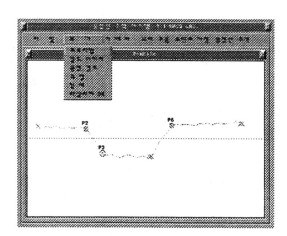

Fig. 9 Example for the 2nd Pass on Joint
Tracking Program

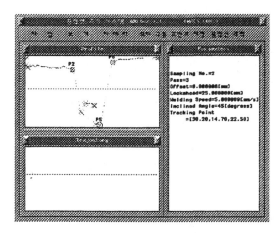

Fig. 10 Tracking mode for the 3rd Pass on
Joint Tracking Program

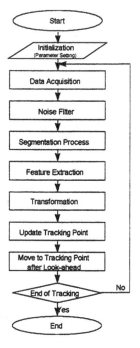

Fig. 11 Flow Chart of
Joint Tracking in
Tracking Mode

Fig. 12 Multipass Procedure

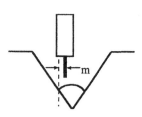

Fig. 13 Setting of Tracking Position

331

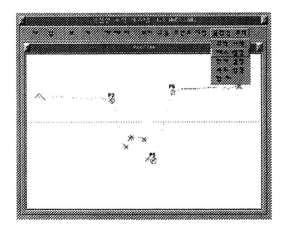

Fig. 14 Example for the 3rd Pass on Joint
        Tracking Program

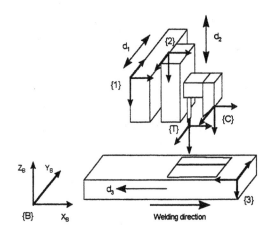

Fig. 15 Description of Coordinate Transformation

INTERACTIVE VISUALISATION OF SEQUENCE LOGIC AND PHYSICAL MACHINE COMPONENTS WITHIN AN INTEGRATED DESIGN AND CONTROL ENVIRONMENT

R. Harrison, A.A. West, P. Hopkinson and C.D. Wright

Manufacturing Systems Integration (MSI) Research Institute, Loughborough University, Loughborough, United Kingdom, LE11 3TU.

Abstract: An Integrated Machine Design and Control (IMDC) environment for the visual representation and integration of the physical machine components and control logic is discussed in this paper. The approach taken is unique in that (a) the control logic and physical models of the elements can be investigated individually for correctness and completeness, (b) the control logic can be easily integrated with the solid models to animate the model of the physical machine and (c) reconfiguration enables the same control logic to be applied to real world physical machine elements. At any stage during the machine design and implementation process, the user of the environment can pause and question the validity of certain operations and control system parameters.

Keywords: Machine, Control, Logic, Design, Distributed, Objects, Modelling, Petri-nets.

1. INTRODUCTION

The design and implementation of manufacturing machines is under increasing time and financial pressures as customers demand increased product variety and quality at reduced product cost (Young, 1995). Increased competition and governmental pressure to focus on environmental issues has forced modern machine builders and users to consider the requirements for the next generation of machines that allow the reconfiguration of both the control software and physical hardware (Rahkonen, 1995). Machines will be required to be developed in the minimum amount of time and comprise (a) vendor independent hardware components, (b) sophisticated control algorithms, (c) intelligent sensors and actuators and (d) user friendly interfaces. In addition, open systems issues concerning the ease of integration, interoperability of the software and hardware components and available standards must be addressed (Crowcroft, 1995) to ensure that reuse and reconfiguration can be achieved. The inherent complexity inevitably necessitates the increased application of machine modelling software toolkits (i.e. for control logic and physical machine element design and analysis) and advanced computer technology throughout the machine life cycle from requirements definition, through the design and build stages to maintenance and reconfiguration.

End users, machine designers and machine builders require technical and operational knowledge from disparate disciplines and specialised domain experts at various stages throughout the machine requirements, design, build, installation, set-up, maintenance and reconfiguration life cycle (Carrott, et al., 1997). Common frames of reference are vital to improve the communications between the above stakeholders in the machine design and build process. In particular it is important that discussions are focused around both the control system software and physical hardware to ensure that the required logical operation and physical functionality is achieved.

Visual representation of machine control software is currently limited. The majority of machine control system software is developed for implementation on programmable logic controllers (PLC's) (Michel, 1990). The basic representation of the logical mapping of sensor inputs to actuator outputs is inherently simple, but soon becomes complex as more function-

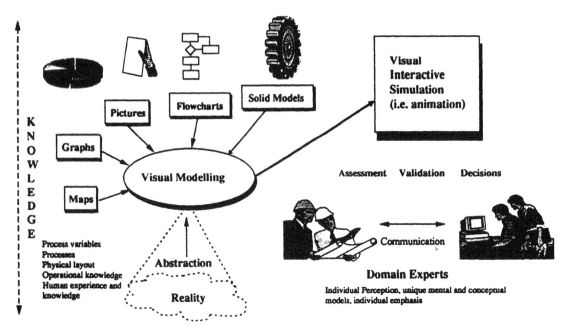

Fig. 1. Illustration of the Visual Interactive Simulation Process.

ality is included. The development of complex manufacturing logic for PLC's is a specialised activity, and it is difficult for a non expert to appreciate the operation of the complete system (Venkatesh, *et al.*1994). Ladder logic diagrams and sequential function charts (David and Alla, 1992) provide graphical representation of sequential operation but the design of flexible, reusable and maintainable software is nevertheless difficult to achieve.

There has been widespread use of visual representation of manufacturing products in terms of surface and solid models (Hoffmann, 1997). Computer aided design package usage e.g. AutoCAD and Unigraphics (Liang *et al.*,1996) enables design engineers to visualise physical components and layouts prior to commissioning. In certain cases, operational logic has been included into the solid model to enable a dynamic animation of the modelled system elements to be observed and optimised (Hoffmann, 1997). A major limitation with this approach has been the fact that in order to transfer the results of the modelling exercise to a real system, the logic must be reimplemented outside of the solid model and the opportunity for errors and sub optimal implementation behaviour proliferate (Wright and Case, 1995). There is a requirement for an environment in which (a) the control logic and physical models of the elements can be investigated individually for correctness and completeness, (b) the control logic can be easily integrated with the solid models to animate the machine solid model and (c) reconfiguration enables the same control logic to be applied to real world physical control elements.

A major research theme at the Manufacturing Systems Integration (MSI) Research Institute at Loughborough University, UK is the realisation of the next generation of machine systems (Carrott, 1996). An Integrated Machine Design and Control (IMDC) en-

vironment (an integrated software environment and toolset comprising third party and "in house" tools) has been developed that seeks to support the work of control system engineers and mechanical designers throughout the design and development life cycle of manufacturing machines. Visibility of the physical machine (using the IMDC Machine Modeller developed around the ACIS solid modelling kernel (Murry and Yue, 1993) and control logic software (using the Synect modelling tool produced by Hopkinson Computing Ltd. (Anon, 1995)) is a core requirement in the IMDC system and is discussed in this paper.

The environment is based around a distributed object oriented representation of manufacturing machines as aggregations of basic components (Joannis and Krieger, 1992): single axes, multi-axes, digital and analogue input / output and dumb and intelligent sensors (e.g. vision systems and robots). Distributed object technology (DOT) (Orfali, *et al.*, 1996) provides the core framework within which the tools and real system intercommunicate.

2. VISUAL INTERACTIVE SIMULATION.

The visual interactive simulation paradigm (also termed visual interactive modelling and visual interactive problem solving was originally applied to the discrete event simulation of job shop scheduling problems in manufacturing (Hurrion, 1980) and has mainly been utilised in Operational Research (Bell, 1995). Visual interactive simulation is particularly appropriate in the manufacturing machine domain (Sadashir, *et al.*, 1989) and can provide a common frame of reference to facilitate human communication of ideas. In the manufacturing machine domain it is vital that both the physical machine and control logic are available for scrutiny. Fig. 1 illustrates the process. A visual model (i.e. a model based upon non textural and non verbal elements to communicate the state of a system) is developed that provides an abstraction of reality.

The application of visual interactive simulation to industrial machine design and control projects using the IMDC environment has resulted in a number of observations and benefits:

- It is important to ensure general interaction and early involvement by developing an animated picture as soon as possible.
- Interaction allows the end users and machine builders to make complex decisions with increased confidence due to their increased understanding of the machine operation and interaction.
- The visual image is widely accepted and unexpected situations can be envisaged via what if? scenarios.
- Of vital importance is the integration of the control logic with the visual simulation in the IMDC environment. This enables the verification (by the developer) and validation (by the user) of the physical, functional and logical performance of the machine. In addition, realistic scenarios and results can be replayed to managers to ensure their participation and ownership of the project.

3. VISUALISATION OF SEQUENCE LOGIC

Manufacturing and process industry control system applications invariably include sequence logic. The complexity of the application logic typically involves the need to manage several concurrent activities, co-ordinating their behaviour to achieve the desired application goal.

In addition to the standard software engineering problems of defining how the proposed system is to work, and expressing the design in a form which is readily understood, control systems designers must ensure that the system's dynamics do not contain design errors, such as deadlock, livelock and undesirable modes of operation. Traditional approaches have tended to be ad-hoc or have inadequately addressed the designer's needs, leaving such errors to be identified late in the life-cycle, resulting in costly modifications.

Synect™ provides a set of software tools (application editor, compiler, logic engine, logic monitor and code generator) which are integrated into the IMDC platform (Anon, 1995). Applications are described by means of a graphical editor. Synect is based on the generation of a Petri-Net model from the application description (Peterson, 1981). The evolving design can then be examined analytically for structural and behavioural deficiencies e.g. checking for errors such as deadlocks and unwanted state combinations. An executable logic model is created which can then drive a visual solid model of the physical machine

(see section 4). The automatic code generation tool produces code which is then compiled with IMDC runtime libraries enabling the application logic to communicate with the distributed machine components.

3.1 Logic Visualisation / Representation

The designer of a sequential control application can visualise the problem and present the solution in a number of ways. Three possible approaches are:

- consider the application as consisting of a set of interlocks. If the target environment is hard-wired logic or, as in the majority of current industrial automation projects, a PLC programmed in ladder logic, the design solution maps easily on to the implementation (Michel, 1990).
- take a functional view. One of the more popular structured methods used for real-time applications has been the Ward-Mellor variant of the Yourdon method (Ward and Mellor, 1985). Control transforms are triggered by events and then react by sending signals to other transforms.
- take an object oriented view. Object orientation has grown from a modelling paradigm and, it is claimed, leads to more intuitive solutions which are more maintainable and more amenable to reuse.

The combinational logic approach is potentially the preferred option if the application logic is very simple or there is a strong need for the solution to require the minimum of memory or execute as fast as possible. Otherwise, the solution can be more difficult to verify, is not conducive to diagnosing operational mis-behaviour and is more difficult to modify without causing unwanted side-effects.

Functional approaches have the benefit of supporting the concept of sequences, typically by a form of state diagram. The functional approach helps the designer to consider how the different components need to be co-ordinated to carry out a particular operation.

Object-oriented approaches encapsulate the sequence logic into the objects which model the allowable behaviour of real-world components. The overall system logic is however now fragmented across different objects and this can lead to unanticipated modes of operation.

Synect provides a methodology which combines the co-ordination of the functional approach with the encapsulation of object orientation.

IMDC integrates the Synect simulator with the 3D solid modeller (see fig. 2). As described in section 4, the modeller incorporates concepts such as timing information and sensor emulation into the solid modelling software so that the solid model shows a real-

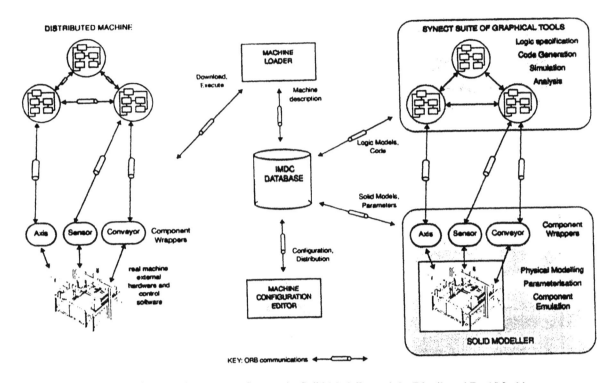

Fig. 2. Illustration of IMDC Interaction between Synect, the Solid Modeller and the Distributed Real Machine.

time emulation of how the implemented system will perform.

4. VISUALISATION OF MACHINE ELEMENTS

A machine designer can utilise solid models in order to visualise the physical machine elements and their interaction (Hoffmann, 1997). The conventional approach to the solid modelling of machines involves:

1. initial static visualisation of the machine elements and
2. determination of the system dynamics and machine element interaction by embedding operational logic within the modelling tool (Yong et al., 1985).

The limitation of this approach is that the coding for the real system operation is normally undertaken after the simulation phase and the software frequently bears little relation to the mechanisms used to drive the simulation. In addition this approach to the modelling of the physical machine elements cannot be used throughout the life cycle and cannot be used interactively with other design tools.

The approach adopted within the IMDC environment is fundamentally different and offers a means to overcome these difficulties. Graphical support is provided for design evaluation which enables designers to rapidly investigate what if scenarios, using the same control logic throughout the life cycle.

Models of physical machines can be graphically constructed within the machine modeller from component building blocks such as actuators, sensors, conveyors, alarms and structural elements. The model components incorporate both geometric and behav-

ioural models of the external systems being represented. The tool can be used to specify component states, operational parameters, motion parameters and locations. Components, parameters and complete sub-assemblies can be stored in the IMDC database.

The interface to components of each specific type is exported via a wrapper object using distributed object technology. External objects residing in remote processes can connect to components using the wrapper and either drive them (e.g. actuators, conveyors) or be notified as events occur inside them (e.g. sensors). Interaction via a GUI displaying the 3D model workspace allows the user to move around the model and examine elements in detail as the machine models are being driven.

5. ENVIRONMENT TO SUPPORT VISUAL INTERACTIVE SIMULATION: IMDC

The IMDC environment comprises four distinct areas of functionality:

- user tools to enable, for example, logic simulation, motion design, machine modelling and tools associated with the control and monitoring of the runtime system. The user tools can be progressively changed or extended via a set of generic interfaces.

- systems tools to enable system administration, access to information and security. Typical system configuration information includes the logical and physical system layout (machines and networks), user names, passwords, access rights, project and user environment information.

- the IMDC object oriented database (POET a product from the POET Software Corporation (Anon, 1994) which provides persistent storage for the outputs from life cycle activities, system configuration, traceability and version control and holds the generated elements of specific target control solutions

- the distributed runtime machine composed of communicating software components, a subset of which provide interfaces to external devices and third party control software for the monitoring and control of the physical machine.

These functional elements all inter-communicate via the underlying Object Request Broker (ORB) architecture as described in the following section.

5.1 IMDC System Architecture

The adoption of an object-oriented approach, particularly the use of distributed object technology, has been the key to providing flexibility in the choice of implementation technologies. Fig. 3 illustrates the principle of object distribution across heterogeneous host and network architectures. Identical client software located on different host platforms 1 and 3 communicate with the server object on host 2. Hosts 1 and 3 may be physically linked to host 2 by different network types.

A multi-schema architecture has been implemented to separate client (typically software tools) and server (such as data repository) applications via an underlying integration infrastructure. The integration infrastructure has been built using distributed object technology based upon the Common Object Request Broker Architecture (CORBA) specification from the OMG (Anon, 1991). The infrastructure acts as a system-wide broker for object services and provides abstraction from low level device specific problems. Object services are dynamically registered and de-registered with the infrastructure by processes which implement the services (object servers). Client processes query the infrastructure for available services and are given the necessary connection information to access services.

6. INTERACTIVE USE OF PHYSICAL AND CONTROL LOGIC MODELLING TOOLS

Application logic is not embedded in the modelling tool but is generated by use of the Synect logic toolset. By functioning as a server, the modeller can be controlled by remote processes for example, (a) the Synect logic simulator tool, or (b) task control software running in the target control system. This highlights an important feature of the run-time architecture, namely that the IMDC defined interfaces to the controlled elements of these solid models are identical to the IMDC interfaces to the real control system. Hence, the control logic processes can drive either the modelled hardware elements (thus animating the model), the real world hardware elements, or a mixture of the two. This permits incremental proving of the control logic in a hardware independent manner

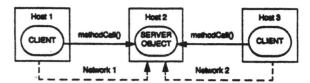

Fig. 3. The Principle of Object Distribution Across Heterogeneous Host and Network Architectures.

and also allows hardware to be included into the system as and when it becomes available.

6.1 Application Example

The proof of concept and effectiveness of the IMDC approach is being evaluated using representative industrial problems in packaging, assembly, transfer-line and PCB handling applications.

Within Synect, applications are described in terms of a hierarchy of communicating objects. Part of the object hierarchy for an example PCB handling machine is illustrated in the right hand window of fig. 4. The internal logic for each object is expressed in the form of state transitions diagrams (STDs). A fragment of a typical STD can be seen in the left hand window of fig. 4.

The PCB handling machine consists of a framework which supports a board carriage system, a movable gantry and a robot arm which are used to populate PCBs with components. To enable visual interactive simulation the solid model of each of these component can be associated with real world input and output contact points from the control logic (see fig. 2).

7. CONCLUSIONS

The novelty of the IMDC approach is that the control system software and physical machine components can be individually conceptualised and interactively tested. Furthermore IMDC enables modelled components to be progressively replaced with real system components until the final solution is attained. A particularly attractive feature is that the same sequence logic is used to control both the modelled and real world components.

The commercial benefits of IMDC relate to both its impact through improved manufacturing efficiency for the end user and its potential to provide a stimulus to machine and control system vendors by:

- reducing the development cost and time of highly automated applications, and encouraging reuse of software/hardware building blocks.

- reducing the cost of eliminating design faults (fast identification) and ease of service and maintenance.

- allowing more effective adaptation and alteration of control systems, without resorting to the expensive services of a specific system supplier.

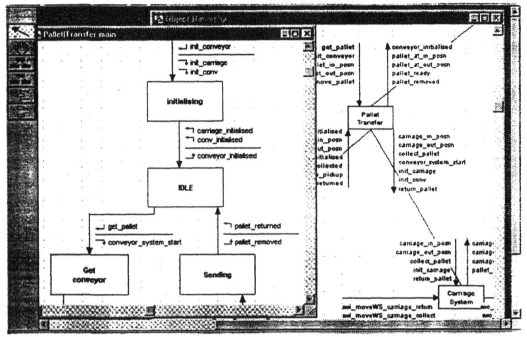

Fig. 4. Typical User Interface Screen for the Synect Editor showing an Object Hierarchy and STD.

ACKNOWLEDGEMENTS

The authors gratefully acknowledge the EPSRC for the provision of research funding and Hopkinson Computing Ltd. for their collaborative input.

REFERENCES

Anon, (1991). *The Common Object Request Broker: Architecture and Specification*, Object Management Group (OMG) Document, No. 91.12.1.

Anon, (1994). *POET Reference Manual, Version 2.1*, POET Software Corporation.

Anon, *Synect User Guide, (1995). Hopkinson Computing, Ltd.*, 29 Deepdale, Guisborough, UK.

Bell, P.C. and O'Keefe, R. (1995), Visual Interactive Simulation - History, Recent Developments and Major Issues, *Simulation*, 49, No 3, pp. 109-116.

Carrott, A.J., Moore, P.R., Weston, R.H. and Harrison, R., (1996). The UMC Software Environment for Machine Control System Integration, Configuration and Programming, *IEEE Trans. on Industrial Electronics*, 43, No 1, pp. 88-97.

Carrott, A.J., Wright, C.D., West, A.A., Harrison, R. and Weston, R.H. (1997). A Toolset for Distributed Real Time Machine Control, *SPIE Photonics East Procs.*, Ma. USA, 2913, pp.2-12.

Crowcroft, J.(1995). *Open Distributed Systems*, Boston; London, Artech House.

David, R. and Alla, H. (1992). *Petri Nets and Grafcet*, Englewood Cliffs, NJ, Prentice Hall.

Hoffmann C, Rossignac J (1997). Special issue: Solid modelling, *Comp.-Aided Des.*, 29, No.2, p.87.

Hurrion, R.D., (1980). An Interactive Visual Simulation System for industrial Management, *European journal of Operational Res.*, 5, pp. 86-93.

Joannis, R. and Krieger M., (1992). Object Oriented Approach to the Specification of Manufacturing Systems, *Computer Integrated Manufacturing Systems*, 5, No.2, pp. 133-145.

Liang M., Ahamed S. and vandenBerg B. (1996). A STEP Based Tool Path Generation System for Rough Machining of Planar Surfaces, *Computers in Industry*, 1996, 32, No.2, pp.219-231.

Michel, G. (1990). *Programmable Logic Controllers: Architectures and Applications*, Wiley, UK.

Murray J.L. and Yue Y. (1993). Automatic Machining of 2.5D Components with the ACIS Modeller, *Int. Journal of Computer Integrated Manufacturing*, 6, No.1-2, pp.94-104.

Orfali, R., Harkey, D. and Edwards, J., *(1996). The Essential Distributed Objects Survival Guide*, Wiley and Sons, Chichester.

Peterson J.L., (1981). *Petri Net Theory and Modelling of Systems*, Prentice Hall.

Rahkonen, T.(1995). Distributed Industrial Control Systems - A Critical Review Regarding Openness, *Ctrl. Eng. Pract.*, 3, No. 8, pp.1155-1162.

Sadashir, A., (1989) Software Modelling of Manufacturing Systems: The Case for an Object Oriented Programming Approach, *Annals of Operational Research*, 17, pp. 363-378.

Venkatesh, K., Zhou, M., and Caudill, R.J. (1994). Comparing Ladder Logic Diagrams and Perti Nets for Sequence Controller Design Through a Discrete Manufacturing System, *IEEE Trans. on Ind. Electronics*, 1994, 41, No 6, pp. 611-619.

Ward, T.W. and Mellor, S.J. *(1995). Structured Development for Real-Time Systems*, Vol. 1, Yourdon Press, Prentice-Hall.

Wright, C.D. and Case K. (1996) Emulation of Modular Manufacturing Machines using CAD Modelling, *Mechatronics*, 4, No. 7, pp. 713-735.

Young, S.L.(1995). Technology: The Enabler for Tomorrows Agile Enterprise, *ISA Trans.*, No 4, pp. 335-341.

Yong, Y.F. et al *(1985). Off-Line Programming of Robots, Handbook of Industrial Robotics*, John Wiley, New York, pp 366-386.

INTERFERENCE-FREE TOOL PATH PLANNING FOR FLANK MILLING OF TWISTED RULED SURFACE

Jung-Jae Lee and Suk-Hwan Suh

Computer Automated Manufacturing Lab
Department of Industrial Engineering
POSTECH, San 31 Hyoja-dong, Pohang, Korea, 790-784
E-mail: shs@vision.postech.ac.kr, http://camlab.postech.ac.kr

Abstract: Flank milling is the crucial feature that the five-axis NC machine offers. Compared with bottom-edge based machining (e.g., flat/fillet endmilling), the machinability can be greatly enhanced by the flank milling where the side cutting edge is mainly used. As far as tool path planning is concerned, conventional method taking the cutter-axis parallel to the ruling lines of the generator curves does impose interference including overcut/undercut problem. In this paper, a new interference-free tool path planning method for the *twisted ruled surface* is presented. Our strategy is to change the tool orientation and offset distance such that the undercut volume is minimized without global tool interference. Compared with previous method relying on the constant tool orientation or offset distance within the ruling vector, the undercut/overcut volume can be minimized without causing global tool interference. The validity and effectiveness of the presented algorithm is demonstrated via several illustrative examples.

Keywords: Tool path planning, flank milling, twisted ruled surface, five-axis machining, CAM, Automated Machining

1. INTRODUCTION

Compared with bottom-edge (with ballnose or flat endmill) based machining, the flank milling of ruled surface enables tremendous increase in productivity as well as improvement of the surface finish (Vicker and Quan, 1989). In spite of the strong potentials, the flank-milling method has not been widely used in practice mainly due to large investment for the five-axis CNC machine tools capable of controlling the five axes simultaneously. (Note that the typical five-axis machining center costs over 1 million US $ (Cover story, 1991). To this problem, additionally-five-axis machining method where the rotary/tilt table is attached to the three-axis CNC machine tool may be employed (Suh and Lee, 1996).

For flank-milling, only a few results were reported (Stute, *et al.*, 1979; Reisteiner, 1993). Tool path planning problems for flank milling is much more complicated than others, as the use of side cutting

edge often limits the accessibility (due to tool body interference) of the tool to the part surface. Further, the cutter deflection should be taken into consideration in tool path planning. Thus, the crucial issue in tool path planning for the flank milling is how to efficiently make use of the orientability of the toolto avoid the tool deflection problem and tool-interference problem.

Stute (Stute, *et al.*, 1979) presented an interference-free path planning algorithm for the flank milling of the ruled surface. His algorithm is able to avoid only "local" tool interference by changing the tool orientation. Further, as will be clear, this method may leave unnecessarily large undercut or overcut volume. From the point of differential geometry, Reihsteiner (Reisteiner, 1993) analyzed ruled surface for flank milling without presenting tool path planning algorithm. In this paper, we present a robust and optimal tool path planning algorithm for flank milling of twist ruled surface such that the tool

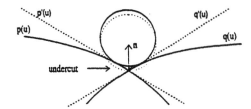

Fig. 1. Interference avoidance by changing offset distance

interference problem can be "globally" solved while "minimizing the undercut/overcut volume".

2. THEORETICAL BACKGROUNDS

Assuming the angular difference is proportional to $v \in [0,1]$, the maximum error ε is

$$\varepsilon = \begin{cases} R(1-\cos2\alpha) & \text{for } n = n(u,0) \text{ or } n(u,1) \\ R(1-\cos\alpha) & \text{for } n = n(u,0.5), \end{cases} \quad (1)$$

where, $n(u,v)$ is the unit surface norml vector at ruled surface. It indicates the minimum error is expected with the tool-axis vector taken around midpoint of the ruling line. (This is used as a heuristic in determining tool orientation vector in our solution algorithm to be given in 3) In fact, the maximum overcut error with the tool orientation of the surface normal at either of the ruling line is 3.5 - 4 times larger ($= (1-\cos2\alpha)/(1-\cos\alpha)$) than that of tool orientation of midpoint ($n(u,0.5)$).

To remove overcut, two approaches can be thought of. The first is to change the offset distance d, such that the tool does not overcut as shown in Figure 1. While this method can eliminate the overcut problem, it may increase undercut volume. If smaller tool is used to reduce the undercut volume, it raises another problems, such as increasing tool deflection, decreasing machining speed, deteriorating machining accuracy and efficiency.

The second approach is to adjust the tool orientation without changing offset distance proposed by Stute(Stute, et al., 1979). This method keeps the tool being closer to the part surface than the straightforward offsetting method, and hence it increases accuracy of machined surface. Note, however, that this method does not guarantee the *global* interference problem as the adjustment is made purely based on the local information.

3. PROBLEM AND SOLUTION PROCEDURE

Consider the ruling line defined by $p(u)$ and $q(u)$. From the midpoint of the ruling line $S(u,0.5)$, the tool center line (or tool orientation vector) is offset by d, then rotated by theta with respect to the surface normal vector $n(u,0.5)$. In general, the tool center

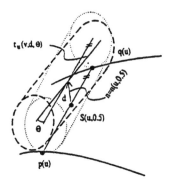

Fig. 2. Description of $t_u(v,d,\theta)$

line corresponding to the ruling line $\overline{(p(u)q(u))}$ can be represented as follows (See Figure 6):

$$t_u(v,d,\theta) = Rot[(u,0.5),\theta]Trans[n(u,0.5)]S(u,v), \quad (2)$$
$$v \in [0,1]$$

Let b_i be a point on $t_u(v,d,\theta)$ when $v = v_i$, and $H_u(v_i,d,\theta)$ be the minimum distance between b_i and the part surface measured on the plane containing b_i and $\overline{t_u}$ (v,d,θ), the normal vector of $t_u(v,d,\theta)$ (See Figure 3). Then, the performance of the tool path (position and orientation) for the u-th ruling line represented by $H_u(v,d,\theta)$ can be defined as follows:

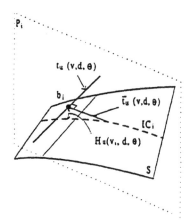

Fig. 3. Description of H_u

$$I_u = \sum_{i=0}^{m} H_u(v_i,d,q) \quad (3)$$

where $v_i = i/m$, and m is the number of digitized points on the ruling line. Note that $| H_u(v_i,d,\theta) - R| \le \varepsilon$, where ε is the tolerance. For the ruling line $g(u)$, the problem of optimal tool path planning is to find the offset distance d and the rotating angle θ such that I_u is minimized without global tool interference.

3.1 Determining rotational direction

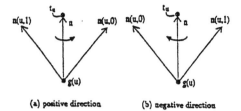

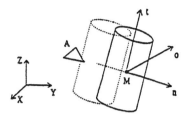

(a) positive direction   (b) negative direction

Fig. 4. Relationship between the rotating direction and $n(u,0)$, $n(u,1)$

For the sake of efficiency in the search procedure for finding the optimal rotation angle, it is worth determining direction of the rotation. Suppose the sign of θ is determined by the right-hand rule with the thumb pointing along the positive direction of $n(u,0.5)$. The rotating direction should be positive (CCW) or negative (CW) determined based on the relationship between $n(u,0)$ and $n(u,1)$.

If $(n(u,0) \times n(u,1)) \cdot t_n < 0$ positive direction
otherwise   negative direction   (4)

Figure 4 shows two cases of selecting the rotating direction. By the right-hand rule, the rotating direction is positive in Figure 4(a), and negative in Figure 4(b).

3.2 Determining rotational range

The tool axis vector should be included between n_p and n_q, where n_p and n_q are the lines starting from $p(u)$ and $q(u)$ along $n(u,0)$ and $n(u,1)$, respectively (Figure 5). Otherwise, the machining accuracy will be decreased dramatically due to increase of the unremoved volume. Therefore, it is desirable to determine the feasible range of theta such that the tool axis exists between n_p and n_q. From the trigonometric relationship, the maximum rotating angle θ_h (Figure 5) is given below:

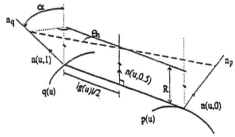

Fig. 5. Determination of θ_h

$$\theta_h = \tan^{-1}\left(\frac{R\tan\alpha}{|g(u)|/2}\right) \qquad (5)$$

Once the sign and the range of theta is determined, the problem of finding OTPO (Optimal Tool Position and Orientation) at ruling vector u can be

Fig. 6. Tool local coordinate [*not*]

formulated as follows.

$$\min I_u$$
$$S.T.\ I_u = H_u(v_i, d, \theta),\ v_i = i/m$$
$$H_u(v, d, \theta) \geq R - \varepsilon,\ \forall v \qquad (6)$$
$$0 \leq \theta \leq \theta_h\ \text{if}\ ((n(u,0) \times n(u,1))\cdot t_n \leq 0$$
$$-\theta_h \leq \theta \leq 0\ \text{if}\ ((n(u,0) \times n(u,1))\cdot t_n > 0$$

4. COMPUTING THE MINIMUM OFFSET AMOUNT

Suppose a triangle A and a cylindrical tool are defined in $[X,Y,Z]$ (Figure 6). And, R is the tool radius, t is tool axis vector, and M is the offset point of $S(u,0.5)$ distanced by R along $n(u,0.5)$. Assuming the tool is allowed to move along $n(u, 0.5)$ direction passing through M, we want to find tool position contacting triangle A.

The above problem is 3D problem and hence direct solution method is rather complicated. Without loss of generality, the 3D problem is cast into 2D domain as follows. First, set a local coordinate $[n,o,t]$, where $o = t \times n$. The transformation matrix T between $[X,Y,Z]$ and $[n,o,t]$ is given by

$$T = \begin{bmatrix} n^T & -n^T \cdot M \\ o^T & -o^T \cdot M \\ t^T & -t^T \cdot M \\ 0 & I \end{bmatrix}, \qquad (7)$$

where n^T is transpose of n. Using the relationship, triangle A can be transformed from $[X,Y,Z]$ coordinate frame to $[n,o,t]$ coordinate frame by

$$A^I = TA, \qquad (8)$$

and projected into the '*no*' plane (*XY*-plane in local coordinate $[n,o,t]$) by

$$V_i^z = 0,\ i \in [1,3], \qquad (9)$$

where V_i^z is the z value of ith vertex of A^I.

Now, the 3D problem can be restated as follows: In the *XY*-plane, consider a triangle A^I and a circle C radius of R centered at origin $c = [0,0]$. Assuming that the circle can move only along X direction, find the circle center position when the circle touches A^l. This is 2D problem which can be solved analytically.

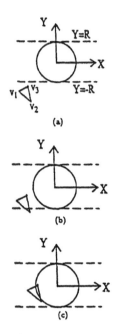

Fig. 7. Three types relationship between A^l and C

If y values of all the vertices A^l are greater (resp. less) than or equal to R (resp. $-R$), the circle does not interfere the triangle along X axis (Figure 7(a)). Thus, we do not have to consider the triangle when checking the local interference as follows: i.e.,

$$\text{if} \quad |V_i.y| \ge R \; \forall i, \text{ then no interference.} \quad (10)$$

In the case that at least one of vertices is between $X=-R$ and $X=R$, there is no interference, if the minimum distance between A^l and the center point (c) of C is greater than or equal to R (Figure 7(b)). Otherwise, interference occurs as shown in Figure 7(c). Thus, the following conditions are added when checking interference.

$$\text{if } distance(A^l, c) \ge R, \text{ no interference} \quad (11)$$
$$\text{otherwise,} \qquad \text{interference is detected.}$$

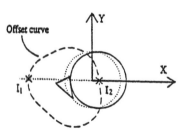

Fig. 8. Calculation of the minimum moving distance

If interference is detected, the circle C should be moved along X axis. The distance becomes minimum when the circle C contacts A^l (Figure 8), which can be

computed as follows. a) generate offset curve O of the triangle with offset distance R, b) find intersection points between O and X axis, and finally c) determine the intersection point having the largest x value I_2 in Figure 8). To avoid interference with A^l, the center of C should move back to the intersection point along X axis. The same procedure can be applied to the polyhedron surface model decomposed by a finite number of triangles.

5. COMPUTING $H_u(v, d, \theta)$

Let b_i be a point on $t_u(v, d, \theta)$ corresponding to $v = v_i$ on the ruling line. Then, $H_u(v, d, \theta)$ is the minimum distance between b_i and S on the plane P_i containing b_i and $\overline{t_u}$ (the normal vector of $t_u(v, d, \theta)$ as shown in Figure 3. Alternatively, $H_u(v, d, \theta)$ is the minimum distance between b_i and the intersection curve IC_i. Note that IC_i is defined on plane P_i as illustrated in Figure 3. From the relationships, $H_u(v, d, \theta)$ can be computed as follows.

Let t be the unit axis vector of $t_u(v, d, \theta)$. For the given t and b_i, P_i is defined as

$$P_i = \{w | (w_i - b_i) \cdot t = 0 \}. \quad (12)$$

Let $S(u', v')$ be a point on IC_i, then the following relationship holds:

$$(S(u', v') - b_i) \cdot t = (p(u') + v'g(u') - b_i) \cdot t = 0. \quad (13)$$

Rearranging Eq. (15):

$$v' = \frac{(b_i - p(u')) \cdot t}{g(u') \cdot t} \quad (14)$$

For the given t and b_i, the intersection curve(IC) satisfies the following;

$$IC(t, b_i, u') = (u', \frac{(b_i - p(u')) \cdot t}{g(u') \cdot t}), \; u' \in [0, 1].$$
$$(15)$$

Finally, $H_u(v, d, \theta)$ is determined as follows.

$$H_u(v_i, d, q) = \min_{u' \in [0,1]} |b_i - IC(t, b_i, u')|. \quad (16)$$

6. ILLUSTRATIVE EXAMPLE

To verify the effectiveness of the proposed algorithm for finding OTPO, the algorithm was applied for a ruled surface (Figure 9), where two boundary curves are given by

$$p(u) = (x, (2uy - y)\cos \alpha, (2uy - y)\sin \alpha) \quad (17)$$
$$q(u) = (x, (2uy - y)\cos \alpha, -(2uy - y)\sin \alpha),$$

where $x=5$, $y=5$, and alpha is the half twist angle.

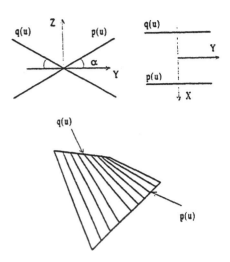

Fig. 9. Ruled surface example 1.

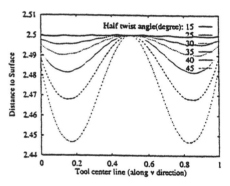

Fig. 10. Change of overcut with Stute's method for various twist angles (R=2.5).

Overcut volume for the various alpha and tool radius was analyzed. Recall that as illustrated in Figure 10, overcut volume of Stute proportionally increases as alpha increases. For instance, when $\alpha \geq 25$ the result is not satisfactory at all due to large overcut. For $\alpha = 30$, the cutting profile around $u = 0.5$ ruling line was compared for various methods in Figure 17, including the presented method OTPO (solid curve),

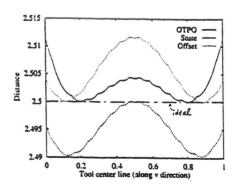

Fig. 11. Cutting profiles (R=2.5, $\alpha=30°$) of the proposed method and Stute's.

Stute (dashed curve), and "offset" (dotted curve). The offset curve was obtained by pulling down the Stute's tool curve by a constant amount such that the peak overcut is controlled to the ideal value (2.5 mm). The result clearly shows that the OTPO yields: a) small overcut in most of region, b) under cut volume less than the overcut volume of Stute, showing the superiority of the presented method.

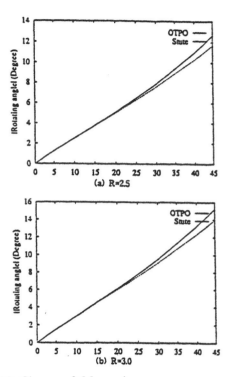

Fig. 12. Change of θ for various α

The effectiveness of our algorithm is due to the capability of optimizing the rotating angle and offset distance according to twist angles. From Figure 12,

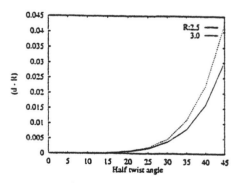

Fig. 13. Change of d for various α

rotating angle is not directly proportional to the twist angle, while Stute does. The angular difference between two methods is clear as the twist angle increases (Figure 12(a)), and as the tool radius increases (Figure 12(b)). Figure 13 shows the adjusting capability for the offset distance in the presented method. Note that Stute method does have the constant offset distance regardless of the twist angle.

Finally, with the developed method we generated tool path for flank milling of ruled surface included in the impeller shape shown in Figure 14. The impeller has nine blades, each modeled by twisted ruled surface. Specifically, left side surface of the blade is the target surface to be machined by flank milling. (Note that the right side surface cannot be machined by flank milling due to tool interference.) The wire form of the target surface is shown in Figure 14(a). By applying the developed algorithm, we obtained tool path shown in Figure 14(b).

7. CONCLUSION

In this paper, we addressed the powerfulness of the flank milling method for free surface machining, and developed tool path planning algorithm for the twisted ruled surfaces. Compared with previous method for flank milling, the developed algorithm is comprehensive covering twisted ruled surface, and can generate efficient tool path and orientation such that the undercut/overcut volume is minimized with global tool interference checking capability. Our method is characterized by the capability of adjusting the rotation angle together with the offset distance. Through various examples, we showed that the developed algorithm can produce effective tool path and orientation compared with previous methods.

REFERENCES

Vicker, G. and K. Quan (1989), "Ball-mills versus end-mills for curved surface machining," *ASME J. of Engineering for Industry*, vol. 111, pp. 22-26.

Cover Story(1991): "Machining Centers," *American Machinist*, **Sept.**

Suh, S., and J. Lee (1996), "Flank-milling of ruled surface with additionally-five-axis CNC machine," *Proc. 1996 IDMME Conf., Nnates, France*, **April**, pp. 375-384.

Stute, G., A. Storr, and W. Sielaff(1979), "NC programming of ruled surfaces for five-axis mahcining," *CIRP*, **vol. 28**, pp. 267-271.

Reisteiner, F. (1993), "Collision-free five-axis milling of twisted ruled surfaces," *CIRP*, **vol. 42**, pp. 457-461.

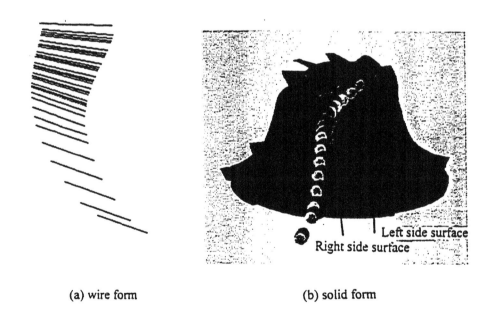

(a) wire form   (b) solid form

Fig. 14. The generated toolpath for matching an impeller

INTERACTION WITH THE RELATIONAL STRUCTURES
OF COMPUTER AIDED DESIGN MODELS

Franco Folini and Isabella Bozza

Dipartimento di Ingegneria Industriale
Università degli Studi di Parma
Parma Italy
e-mail: folini@ied.unipr.it

Abstract: The current evolution of the software systems supporting the mechanical design processes is gradually extending the "traditional" CAD systems domain to integrate CAM, CAE and PDM modules. Consequently, the CAD model is changing its geometry-oriented nature, integrating information and data coming from the whole design and production processes. A growing network of relations characterizes this new CAD model. At present, some relations are represented explicitly in the integrated CAD model while others are completely implicit; for some types of relations, CAD systems offer only a limited set of operation; finally, the relations are usually presented to the user though heterogeneous interaction modalities. This paper proposes a classification of the typical CAD relations and suggests a basic set of operations on them; finally, it discusses the emerging interaction modalities and delineates their possible application in the CAD field.

Keywords: CAD/CAM models, constraints, relational databases, design systems.

1. INTRODUCTION

The software systems currently used by engineers for designing the product and its production process are the result of a long evolution. Up till now, the product design was supported by a set of different systems, such as CAD, CAM, CAE, etc., typically connected by means of explicit data conversion and exchange. Nowadays, the market provides new integrated systems, able to cover almost all the phases of the design process: conceptual design, modeling, structural analysis, assembly, simulation, etc. Besides, new powerful integration technologies and standards, such as OLE, CORBA, etc., begin to appear. A pool of strongly integrated software modules, supplied by one or more vendors, constitutes a typical advanced configuration. Each module is designed to support a specific activity and relies on a shared database, usually called "master model". The advantages of this approach are evident: no data conversion between modules, reduced

consistency problems, coexistence of specific models of the product, uniform access to the data, better support for workgroups, etc. For example, engineers performing structural analysis can perform dimensional optimizations operating on the geometry of the part directly in the FEM/FEA module.

The use of a shared database allows a better parallelization of the activities, as indicated by the "concurrent engineering" principles. The master model is, therefore, a big container that collects and organizes heterogeneous information about the product, its functional and physical properties, its production process and its life cycle. Within the master model, a network of links exists to represent dependencies and relations between models, versions, components, etc.; these links are only a fraction of the huge set of relations involved in a real-world design and production process. The word "relation" is here used with the most general meaning and refers to all the kind of links between pieces of homogeneous or heterogeneous information, even at different levels of detail. For example, a relation can exists between two planes defined to be parallel by the geometric modeling module; a relation can connect a component in a library with an assembly that uses that component; a relation can link two versions of the same technical report. Anyhow, many relations, defined or generated during the design process, are not explicitly represented in the master model; in particular, relations between information from different tasks are frequently left to the user responsibility. For example, no relation is usually created to describe the link between the graphical results coming from a finite elements analysis post-processing and a thickness value that has been modified on the base of such results. Other relations, even if important, are hidden, or difficult to access by the user. For example, in many CAD modules, it is often difficult, or impossible at all, to explore the dependencies structure of a complex parametric model. Finally, basic operations on relations are presented, in each module of the integrated software system, with different interaction modalities. Such differences do not always depend only on the context lexicon and can confuse the user, increasing the design time.

2. THE PROBLEM

The geometry has always been the main topic of interest for researchers and developers in the mechanical CAD area. From the designer point of view, geometry is only one of the many aspects involved in the design of a mechanical part or product. During the different design tasks, designers and draftsman are required to take into account also many non-geometric information and features. In order to satisfy these needs, computer aided design systems are gradually evolving toward a more complete and general representation of parts and products. These systems are able to model new data types and offer new commands to manage non-geometric features of the product and information on the design and production processes.

In order to offer the mentioned level of functionality, CAD systems use either multiples heterogeneous data structures, integrated and centralized on a workstation, or a single model distributed on the network. In both the cases, the model is called master model. This model represents the product data and is strongly populated by relations at different levels of detail. Within the master model, relations can connect information items locally to a single data structure, such as in a parametric representation, or globally between different data structures (e.g. the relation between two different versions of the same project).

Today design systems provide a wide and effective

user support for almost all the user operations on the geometry; just about all CAD systems have a consistent and effective user interface and a well defined set of operations for the definition, editing and inquire of geometric data. The increasing variety and quantity of the data managed by a typical CAD system underlies an exponential number of relations to be managed. Furthermore, many relations previously manually managed in the design process and implicitly represented by identification code, versions numbering or explicitly described in technical documentation and reports, should now be automatically managed by means of the CAD and PDM systems. Despite the new evolutions and enhancements, important features of the CAD system, such as the user interaction modalities, remain mainly geometry-oriented and therefore inadequate for handling the new types of data. The more evident anomalies of the user-interface are in the management and interaction with the relations of the model. The user interface aspects and behavior and the set of operations proposed for the interaction with relations are less mature and homogeneous than the corresponding interface elements for the interaction with the geometry. Often, similar relations require the use of different interaction tools, depending on the current design phase, on the geometric context or on the software module used. In order to perform the same operation on different data, the user should learn and use different commands and interaction modalities, while his expectations are for software systems in which all the information is accessible end editable with homogeneous, simple and consistent tools. The following section propose a uniform and coherent approach to the interaction with the relational structures of computer aided design models.

3. RELATIONS CLASSIFICATION

In order to describe and analyze the state of the art of the representation, modelization and interaction with relations a minimal glossary is required.

Software system. A program able to manage, represent and store information. The software systems of interest for the mechanical design area are DBMS, CAD, CAM, CAE, PDM systems, etc. In the following, the symbol S is used to indicate a generic software system.

Node. A basic information element that can be referred by the software system. Nodes can be either atomic, no more splittable (e.g. a point in a 3D space), or composed of sub nodes (e.g. an assembly composed of parts and sub assemblies). In the following, nodes are indicated by lower case letters a, b, etc.

Relation. A link between two nodes, representing some kind of connection existing, at a semantic level, between the two information pieces. Relations can be explicitly represented (e.g. hyperlinks in an HTML page), or implicitly represented (e.g. for two different orthographic projections in a 2D CAD model, connected only by relative planar positions). In the following, relations are indicated by upper case letters. For example, $R_S(a,b)$ indicate a binary relation R, defined in the system S, between the nodes a and b.

Software systems, utilized in the design process, use and model many kinds of relations. It is possible to classify these links on the base of many criteria. In the following, it is proposed a new classification aimed at providing a lexical base for discussions on computer aided design tools and models. The classification is based on the *meaning* of the relations and focuses the typical requirements and problems of a general CAD/CAM system and data structure. The relations categories individuated are:

One-way constraint relation. It is a direct binary relation $R_S(a,b)$ imposing a constraint on a node b on the base of a node a, where a must be always determined before b (e.g., the relation between an offset surface and its master surface in a parametric system). This kind of relation is mainly used in parametric systems based on a procedural approach.

Two-ways constraints relation. It is a binary relation $R_S(a,b)$ imposing a constraint between the nodes a and b (e.g. the parallelism relation between lines, in a variational drafting system), where a can be used to determine b or, indifferently, b can be used to determine a. These relations are typically used in parametric systems based on a variational approach.

Attachment relation. It is a hierarchical binary relation $R_S(a,b)$ linking the entities a and b, where a is the master node and b is the slave or dependent node (e.g. the relation between a part and the document with its textual description, in a general CAD system). b provides a additional information, without imposing it any constraint. A classical example of attachment relation is the link between a textual document and a chart placed in the document as *OLE object*.

Sequence relation. It is a direct binary relation $R_S(a,b)$ describing a dependence, temporal, algorithmic or historical between a node a and a node b (e.g. one of the relations between the form feature HOLE and form feature EXTRUDE in a form-features system, where the hole is placed on the extruded volume). a always precedes b. Sequence relations usually connect a set of nodes to constitute lists.

Part-of relation. It is a hierarchical binary relation $R_S(a,b)$ expressing that b is one of the parts constituting b (e.g. the relation between a SHAFT model and the ENGINE assembly model where the SHAFT is used).

Reference relation. It is a direct binary relation $R_S(a,b)$ describing the link between a node a and a node b, where a is the source of the information or data and b is an instance or a copy of a (e.g. the relation between the instance of a standard SCREW, in a mechanical design model, and the original SCREW model, in a catalog where it is defined and described with attributes, price, availability, etc.).

Relations are generated and used in each phase of the design process to connect models or single numerical data. Analyzing a software system user-interface, it is possible to verify that the different kind of relations are referred with different names: some names are chosen according with the designers technical language or within the lexicon specific of the application context; other names are mathematical terms or come from implementation details. These discrepancies on naming can confuse the user and make the software system hard to learn and to use.

4. OPERATIONS ON RELATIONS

During the design process, the designer produce, correlate, organize and collect a huge quantity of information. All the information is stored in a form that can vary from a strongly structured set of data, managed by a well integrated set of software systems, to a group of heterogeneous data, produced by different software systems and located on different computers. In both the cases, many relations exist over the data produced and used by the designers and such relations should be visible, accessible and manageable, according to the model consistency and the local policies rules. It is important the designer to maintain the control over all relations connecting geometric data, part models, part libraries, symbol catalogues, design versions, etc. Therefore, the software system should provide a minimal set of basic operations on relations, covering almost all the designers needs. These basic operations should be chosen so that a complex operation can be expressed as a sequence of simpler operations. The basic operations individuated are:

Editing. This operation groups the basic activities on relations, such as creation, modification and deletion of relations. As for any other type of information stored in computer system, these operations should be controlled by a set of context dependent rules, in order to preserve the integrity and the meaning of the overall structure and to respect the local security policies. Under the "editing" label, we include also the management of the relations attributes and states. In many contexts and applications, the state of a relation and its attributes are as important as the relation itself (e.g. the relation between a parametric solid model of a part and the corresponding FEM model: this relation can have an attribute "aligned" with value true or false).

Browsing. This operation allows to navigate through relations following references (like in a hypertext) and inquiring the relations attributes and linked nodes. In an advanced use, the browsing activities require the capability of evaluating and displaying the exact or estimated *sensitivity area* for a given node or relation. The sensitivity area of a given node a is defined as the set of nodes and relations that are affected by a change made to the node a. In many design activities, it is important to have an exact or estimated map for the sensitivity area in order to evaluate the impact of a modification to a node or to a relation, before performing the change (e.g. before upgrading a component description in a shared symbols library, it is important to know which drawings will be affected by the change). Therefore, the software system should be able to generate such maps starting from a node, following outgoing relations and individuating the entities reached. The results should then be reported to the user in a graphical or textual description. Another operation included under the "browsing" label, is the measure of the path length, defined as the collection of relations connecting two nodes.

Viewing and zooming. The viewing operation translates the relations internal data structures in a textual, hyper-textual or graphical form. In order to be useful, the simple viewing of a relational structure has to be coupled with zooming operations. The zooming operations allow to view the set of nodes and relations at different levels of detail. The zooming capabilities involve activities like detailing, abstracting and filtering. Also in this case, the operation

behavior should be driven by a set of rules defined for the specific design context and aligned with the existing security policies.

Validating. This operation applies a set of general and/or context-specific rules for asserting and certificating the integrity and validity of a set of relations. It is important not only to check for dangling links, but also to apply syntactic and semantic rules in order to verify the type and the state of nodes referred by the relations (e.g. in feature based modeling system, the designer could ask for checking that all the "passing throw holes" features remain "passing throw", after non trivial modeling operations). In many contexts it is also important to monitor vital or strategic relations in order to ensure the validity of the general structure. Furthermore, for some application, it can be of interest to audit and log the use of some relation, in order to identify and document "who", "when" and "how" data in the model is used.

Publishing. This operation exports a set of relations and nodes, mainly for documentation purpose. Publishing CAD data is becoming increasingly important in order to make information visible and accessible within the enterprise Intranets or, in some case, within the whole Internet. The traditional publishing tools, designed to produce technical documentation and reports to distribute in a digital or printed format, are now being replaced by a new generation of software tools, able to extract the required information and publish it in real-time. These tools are mainly oriented to the digital formats and are intended to support the network access. This approach definitively solves the problem of consistency between the CAD model and the data available to the reader of the publication. In this operation, the information, relations and nodes, are converted, statically or dynamically, on a textual or graphical format. The emerging standards are the HTML language and the standard 3D description language VRML. The publishing operations must be performed preserving the existing policies rules defined in the publishing organization.

5. INTERACTION WITH RELATIONS

As highlighted previously, the software systems currently available to the designers offer partial or inappropriate tools for the interaction and the management of the nodes and relations. In general, these interaction tools are heterogeneous and insufficient for satisfying the users needs.

While the complexity level of the new integrated design systems is continuously growing, it is important to preserve and enhance the usability and learnability characteristics: a mature, coordinated and uniform approach to the interaction with relations is a must.

Current design systems use a mix of interaction techniques. The most frequently used are:

Properties box. It is a simple window describing a node or a relation (Fig. 1). No other operation is usually supported. Even if it can be considered the simplest interaction technique on data, it is frequently implemented by hidden or complex commands. The emerging standard is the "Properties tabbed window", proposed by Microsoft Windows. When it is properly implemented, this presentation technique allows a simple and direct access to all the information describing a node or a relation. In many cases it support also some modification on the node and/or on the relations connecting that node to others.

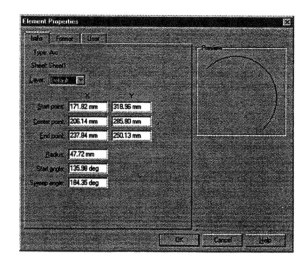

Fig. 1. An example of "Properties box", captured from the drawing module of Intergraph/SolidEdge 3.0. This tabbed window allows the modification of the node attributes and the view of the relation connecting it with other nodes.

Tabular description. This representation displays a list of nodes and relations in a table similar to the tables used in the databases systems (Fig. 2). This technique is used in many systems and is usually preferred for the visualization and interaction with homogeneous and well structured data, even

numerous. The B.O.M., Bill of Materials, is a typical example of a tabular description.

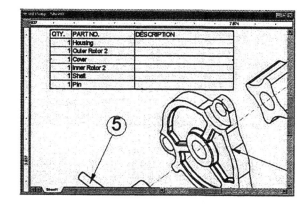

Fig. 2. A simple example of a "Tabular Description", automatically generated by the CAD system and placed within a drawing. The image comes from SolidWorks 97.

Hyper-textual description. Traditionally limited to the on-line helps, the hyper-textual descriptions are now used also for presenting also other types of information. The diffusion of Internet and of its standards, such as the HTML and VRML languages, is suggesting new and more powerful ways to use these techniques (Roller and Bihler, 1997). Furthermore, the recent diffusion of Intranet and Intranet servers, easily connectable to a database, is now supporting a fast and effective diffusion of these techniques. All the main CAD systems on the market include now a converter to the VRML format.

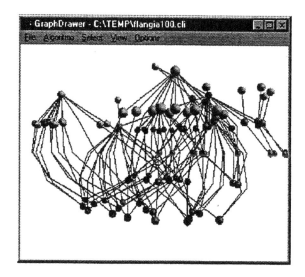

Fig. 3. A 3D view of the relations defined in a simple 2D parametric model of a FLANGE. This image demonstrates how the graphical representation of relational structures can easily become useless,

even using advanced 3D representations.

Graph-based descriptions. It is a graphical representation, usually 2-dimensional, of a set of nodes and relations. Many research activities have been done in recent years on graphs-visualization (Di Battista, *et al.*, 1994) but, from the CAD perspective, the problem is still open. When the complexity of the structure is more than trivial, the graphical description looses its readability characteristics and become unusable. An effective graphical representation can be obtained only for hierarchical structures (trees) or for graphs strongly structured in sub-graphs. The graph-based description is usually proposed as the representation method closer to the designer language and knowledge. Furthermore, the graph-based representation constitutes, usually, the preferred context where to implement an almost complete set of operation on relations and nodes.

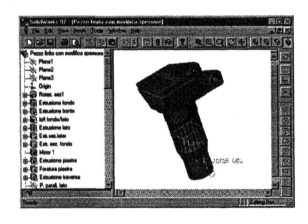

Fig. 4. A snapshot of a complex hierarchical structure. The image comes from the Features Modeling module of SolidWorks 97.

Hierarchical graphical description. This is a mixed textual and graphical description of hierarchical structures. These descriptions, proposed by the Macintosh and Windows operating systems for the interaction with the file-system, are becoming used also in the design systems. The hierarchical representation of the tree structure is simple to use and easy to read. The "expand" and "collapse" operations provide to the user the required level of flexibility and constitute a good example of the zoom operation.

6. CONCLUSIONS

The current evolution of the software systems supporting the mechanical design is gradually extending the "traditional" CAD systems domain to integrate CAM, CAE and PDM modules. Consequently, the CAD model is changing its geometry-oriented nature, integrating information and data coming from the whole design and production processes. A growing network of relations characterizes this new CAD model. At present, some relations are represented explicitly in the integrated CAD model while others are completely implicit; for some types of relations, CAD systems offer only a limited set of operation; finally, the relations are usually presented to the user though heterogeneous interaction modalities. The paper proposed a classification of the typical CAD relations and suggested a basic set of operations on them; finally, it discussed the emerging interaction modalities and delineated their possible application in the CAD field.

7. REFERENCES

Harary, F. (1969). *Graph theory*, Addison Wesley, Reading, MA.

Di Battista, G., P. Eades, R. Tamassia and Tollis I.G. (1994). Algorithms for drawing graphs: an annotated bibliography. *Computational Geometry*, **Volume 4**, pp. 235-282.

Gansner, E.R. (1993), *A Technique for Drawing Direct Graphs*, IEEE Transactions on Software Engineering, Volume 19 No. 3, pp. 214-230.

Roller D. and M. Bihler (1997), *Electronic Business: optimizing communication and information infrastructures using Web technology*. In: 30th ISATA, Mechatronics (D. Roller, (Ed.)), pp. 269-278, Automotive Automation Ltd., Croydon UK.

EXPERT DRAWBEAD MODEL
FOR FINITE ELEMENT ANALYSIS OF SHEET METAL
FORMING OPERATIONS

Y.T. Keum, J.W. Lee, and B.Y. Ghoo

Han-Yang University, Division of Mechanical Engineering
17, Haengdang-dong, Sungdong-ku, Seoul, 133-791, Korea

Abstract : The drawbead expert model which embodies the drawing characteristics of the drawbead suitable for a forming analysis is investigated. For introducing the expert model, we first present the equations calculating the drawbead restraining force and pre-strain of basic drawbeads with a multiple regression method. Then, the drawing characteristics of the combined drawbeads are described. Finally, to confirm the application of the expert model to a die design analysis, the forming process for a section of automotive inner hood panel is simulated.

Keywords: Modeling and Simulation of Sheet Metal Forming, Expert Model for Drawbeads, Virtual Manufacturing, Computer Aided Engineering

1. INTRODUCTION

The considerable trial-and-errors in stamping die design are required, in general, because the stamping variables such as tool geometry, friction, binder/blank holding forces etc. are various in tools. The drawbeads are usually used for the control of stamping variables. Therefore, the designed drawbeads would be modified in tryout processes for the proper stamping dies. The precise description of the drawbead used in stamping dies has been a long-cherished desire to many forming analysts. The reason is that the drawbead restraining force in sheet metal forming analysis is boundary condition and has an important effect on the accuracy of the finite element solution.

Nine (1982) investigated stamping variables affecting on the drawbead restraining force and proposed the drawbead restraining force and

strain when a circular drawbead is employed in a binder-holding process. Also, Wang (1982) formulated a mathematical model of circular drawbeads and evaluated the drawbead restraining force considering the bending, sliding, and unbending effects. Using the virtual work principle, Levy (1982) predicted the drawbead restraining force introducing the anisotropy and strain rate parameters of the sheet. Weinmann and Sanchez (1988) analyzed the drawbead restraining force of circular drawbeads in which they are assumed as rollers. Stoughton (1988) presented a mathematical model of drawbead in assumption that the work required to bend and straighten the sheet is identical to the work required to overcome the frictional forces.

In this research, the drawbead expert model which provides the drawbead restraining force and pre-strain for a forming processes is introduced.

2. BASIC DRAWBEADS

2.1 Circular drawbead

The geometric model of a circular drawbead and a deformed sheet are shown in Fig.1. The elastic deformation is due to the initial elastic displacement of the sheet when the drawbead closes and it can be determined from the maximum displacement of a fixed beam. The elastic force, F_e, and the displacement, δ, are respectively derived as follows :

$$F_e = \frac{16\ E\ w\ \delta\ t^3}{(2\ R_s + W + 2\ cl)^3} \quad (1)$$

$$\delta = min(\ h,\ 2(2R+t)\frac{RY_p}{t\ E}\ ) \quad (2)$$

The deformed sheet metal by a circular drawbead is shown in Fig.2. Because the drawbead restraining force results from the accumulation of the bending and/or unbending, tension, and frictional force, it is calculated by using the forces occurred when the sheet passes through the drawbead. To keep the state the upper and lower drawbeads fully closed, the exerted blank holding force is usually greater than the upper pressure. The discrepancy between the blank holding force and the upper pressure contributes to the frictional forces existing on both flanges.

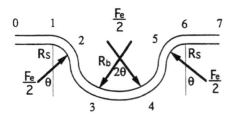

Fig. 1. Sectional view of the blank sheet formed by a single circular drawbead.

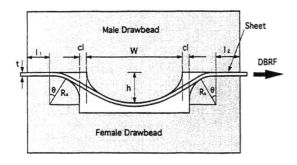

Fig. 2. Detail description of the sheet formed by a single circular drawbead.

Therefore, if the lengths of the flanges are respectively defined as l_1 and l_2, then the frictional forces, F_{c1} and F_{c2}, are derived as follows :

$$F_{c1} = 2\mu\ (BHF - F_R)\frac{l_1}{l_1 + l_2} \quad (3)$$

$$F_{c2} = 2\mu\ (BHF - F_R)\frac{l_2}{l_1 + l_2} \quad (4)$$

The force at point 1, F_1, is given to the bending force when the thickness is equal to the initial one and the strain of the drawing direction is zero. In the region from point 1 to point 2, the force is calculated by multiplying $e^{\mu\theta}$ to the force before the bending. The successive unbending force acts on the material at point 2. The frictional force, $\mu(F_e/2)$, also acts on the sheet. At point 3, the bending force, F_3, as well as the frictional force, $\mu(F_e/2)$ acts on the sheet by the elastic force. In the region from point 3 to point 4, the force after the sliding is obtained by multiplying $e^{\mu\theta}$ to the force before the bending. Because the unbending force at point 4, F_4, the frictional force, μF_e from point 4 to point 5, the bending force, F_5, at point 5, the frictional force, μF_e, from point 5 to point 6, the unbending force, F_6, at point 6, and the frictional force existing on the flange, F_{c2}, act on the sheet metal in due sequence, the drawbead restraining force, DBRF, is,

$$DBRF = [\{(F_{c1} + F_1)e^{\mu\theta} + F_2 + \mu F_e + F_3\}e^{2\mu\theta} \\ + F_4 + \mu F_e + F_5]e^{\mu\theta} + F_6 + F_{c2} \quad (5)$$

where F_i is the bending or unbending force. When the sheet metal is drawn out through a circular drawbead, the free body diagram of the upper male drawbead are depicted as Fig.3. The upper pressure normal components of the elastic force and frictional of the circular drawbead are the summation of force. The normal components of the force generated by the pressure distribution

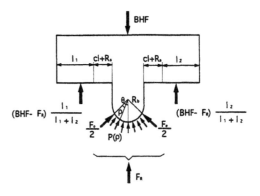

Fig. 3. Free body diagram for a single circular drawbead.

acting on the drawbead surface, B, can be computed as follows :

$$B = \int_0^{2\theta} F_I \ e^{\mu\rho} \cos(\theta - \rho) \ d\rho = F_I \ G(\mu, \theta) \qquad (6)$$

Therefore, using Eq.(1) and Eq.(6), the upper pressure, F_R is written as follows :

$$F_R = F_I \ G(\mu, \theta) + F_e \cos\theta \qquad (7)$$

In Eq.(7), F_I is expressed as follows :

$$F_I = [(F_{cl} + F_1)e^{\mu\theta} + \mu F_e + F_2 + F_3] \qquad (8)$$

When the sheet metal is drawn out through a drawbead, the bending and the tension affect the strains. The strain by the tension is equal to that by the drawbead restraining force after the material is drawn out. When the material is drawn out, assuming the plane strain and the plane stress state and using the Levy-Mises equation, the effective stress, $\bar{\sigma}$, and the effective strain, $\bar{\varepsilon}$, are respectively represented as follows :

$$\bar{\sigma} \fallingdotseq \frac{\sqrt{3}}{2}\sigma_\theta, \qquad \bar{\varepsilon} = \frac{2}{\sqrt{3}}\varepsilon_\theta \qquad (9)$$

Ignoring the strain-rate effects, the strain of the drawing direction, ε , is given by,

$$\varepsilon = \varepsilon_b + [\frac{\sigma_p}{K(2/\sqrt{3})^{n+1}}]^{1/n} \qquad (10)$$

where σ_p is the stress at the drawing out and ε_b is the strain by the bending and/or unbending.

2.2 Stepped drawbead

The geometric model of a stepped drawbead and a deformed sheet are shown in Fig.4. The ways of the equation derivation of stepped drawbeads

are similar to those of the circular drawbeads. The elastic force, F_e, is expressed as follows by the fixed beam theory :

$$F_e = \frac{16 \ E \ w \ \delta \ t^3}{(4R + 2cl)^3} \qquad (11)$$

where the elastically displaced distance, δ, is the same as Eq.(2).

The deformed sheet metal by a stepped drawbead is shown in Fig.5. It is assumed that the frictional forces in the contact area are derived from the discrepancy between the blank holding force and the upper pressure. If the lengths of the flanges are respectively defined as l_1 and l_2, the frictional force is given by Eq.(3) and Eq.(4). Following the way as in the circular drawbead, the drawbead restraining force of the stepped drawbead, DBRF, can be obtained as follows :

$$DBRF = [(F_{cl} + F_1)e^{\mu\theta} + F_2 + \mu F_e + F_3]e^{\mu\theta} + F_4 + F_{c2} \qquad (12)$$

7The free body diagram of the upper male drawbead is seen in Fig.6. The upper pressure, F_R, is written as follows :

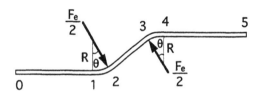

Fig. 5. Detail description of the sheet formed by a stepped drawbead.

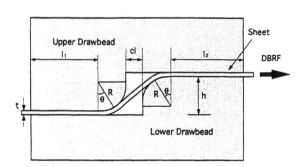

Fig. 4. Sectional view of the blank sheet formed by a single stepped drawbead.

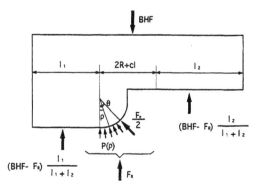

Fig. 6. Free body diagram for a single stepped drawbead.

$$F_R = B + (F_e/2)\cos\theta \qquad (13)$$

In Eq.(13), the normal force by the pressure, B, can be computed as follows :

$$B = \int_0^\theta F_I \, e^{\mu\rho} \cos\rho \; d\rho = F_I \, H(\mu,\theta) \qquad (14)$$

Ignoring the strain-rate effects, furthermore, the strain of the drawing direction may be calculated by Eq.(10) as in the circular drawbead.

2.3 Squared drawbead

In general, stamping die engineers design a squared drawbead for keeping the sheet along the boundary of the die. So, the squared drawbead has a very big drawbead restraining force so that it is defined with a considerably big value. In addition, the pre-strain does not come into existence because the sheet metal is not drawn out.

3. EXPERT MODELS

3.1 Circular drawbead

To obtain the precise predictions for the circular drawbead, the correction of the theoretical values introduced in the previous chapter is induced by comparing with the experimental measurements. If the deviations between theory and experiment were made linear multiple regression with input variables, they are known to be affected mainly by the friction. Therefore, the friction coefficient is employed to correct the theory for the precise drawbead restraining force of the circular drawbead. To consider the friction effects, the correction factor, a, is introduced as follows :

$$\mu' = a\,\mu \qquad (15)$$

where the a is obtained by substituting Eq.(15) into Eq.(5). If the correction factor, a, is assumed the function of input variables, it is expressed as follows :

$$a = \alpha - \beta\mu - \gamma h - x R_s + \lambda t - \chi BHF \qquad (16)$$

where μ is the friction coefficient, R_s is the radius of curvature of bead flange, t is the thickness of the sheet, and BHF is the blank holding force. In Eq.(16), α, β, γ, x, λ and χ are the constants. Therefore, the drawbead restraining force of the circular drawbead would be predicted more precisely by using the corrected friction coefficient, μ', expressed in Eq.(15).

3.2 Stepped drawbead

Similarly as the corrected theoretical values in the circular drawbead, the friction coefficient, μ', is corrected as follows :

$$\mu' = b\,\mu \qquad (17)$$

where the correction factor, b, can be obtained as follows :

$$b = \alpha - \beta\mu - \gamma BHF - xt + \lambda R \qquad (18)$$

where μ is the friction coefficient, BHF is the blank holding force, t is the thickness of the sheet, and R is the radius of drawbead shoulder. In Eq.(18), α, β, γ, x and λ are the constants.

3.3 Combined drawbead

The drawbead restraining force of the combined drawbead is assumed to be expressed as a summation of these of the basic drawbeads. That is, the drawbead restraining force, $DBRF_C$, of a combined drawbead, C, which consists of two basic drawbeads, A and B, can be given as follows :

$$DBRF_C = DBRF_A + DBRF_B \qquad (19)$$

where $DBRF_A$ and $DBRF_B$ are the drawbead restraining forces of drawbeads A and B, respectively, which are calculated from the equations mentioned in the previous chapter.

The strain of the combined drawbead is evaluated by summing that of the bending and that of the tension. The strain due to the bending, ε_C, can be obtained by the following equation :

$$\varepsilon_C = \varepsilon_A + \varepsilon_B \qquad (20)$$

where ε_A and ε_B are the strains of drawbeads A and B, respectively. Thus, the strain of the combined drawbead, ε, when the sheet metal is drawn out through the C, is given by

$$\varepsilon = \varepsilon_C + \left[\frac{\sigma_p}{K(2/\sqrt{3})^{n+1}}\right]^{1/n} \qquad (21)$$

where σ_p is the tensile stress due to the drawbead restraining force.

Fig.7 shows the double circular drawbead separated into two single circular drawbeads. Here, the assumption is made that the blank holding forces acting on both sides of the circular drawbead, BHF_1 and BHF_2, can be represented as the blank holding force, BHF

354

divided by the length ratios, as shown below :

$$BHF_1 = BHF \frac{L_1}{L_0} \qquad (22)$$

$$BHF_2 = BHF \frac{L_2}{L_0} \qquad (23)$$

where L_0 is the total length of the double circular drawbead, L_1 is that of one circular drawbead, and L_2 is that of the other drawbead. The thickness of the sheet entering the second circular drawbead is equal to that drawing out the first drawbead.

Fig.8 shows that a circular-and-stepped drawbead can be separated into a single circular drawbead and a stepped drawbead. In the same manner as mentioned in the double circular drawbead, the blank holding forces acting on a stepped drawbead and a circular drawbead, BHF_1 and BHF_2, can be calculated with the total blank holding force, BHF, multiplied by each length ratio, as given in Eq.(22) and Eq.(23). L_0 is the total length of the drawbead, L_1 is that of the circular drawbead, and L_2 is that of the stepped drawbead.

4. APPLICATION

To verify the reliability of the drawbead expert model developed, the section of a inner hood panel with a circular-and-stepped drawbead seen in Fig.9 is simulated. The tooling geometry for the analysis is illustrated in Fig.10. The point A shown in Fig.10 is the position of a circular-and-stepped drawbead.

The dimensions of a circular-and-stepped drawbead are like follows : circular drawbead height, HSC = 5mm, radius of circular drawbead shoulder, R_s = 8mm, stepped drawbead height, HS = 4mm, radius of male stepped drawbead shoulder, R_1 = 7.5mm, and radius of female stepped drawbead shoulder, R_2 = 5mm. The blank holding force, BHF, is 115N/mm. The terminal punch stroke is 43.7mm. The material properties of the sheet are listed in Table 1. The circular-and-stepped drawbead model provides the drawbead restraining force, DBRF = 67.7N/mm and pre-strain, ε = 10.7%, respectively.

Fig.11 shows thickness strain distribution. The left boundary of the sheet metal is drawn into the die cavity about 23mm and the right boundary is drawn into the die cavity about 1mm. The pre-strain is predicted to 10.4% after the sheet metal is drawn out. Without expert model, however, the pre-strain is predicted to 1.4%. This simulation reveals that

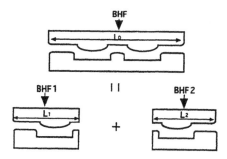

Fig. 7. Partition of a double circular drawbead into two single circular drawbeads.

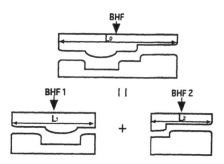

Fig. 8. Partition of a circular-stepped drawbead into a single circular bead and a stepped bead.

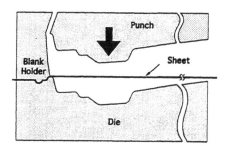

Fig. 9. Sectional view of tooling of a inner hood panel.

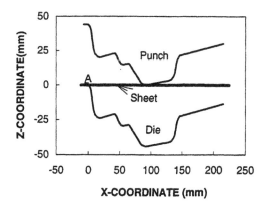

Fig. 10. Tooling geometry and modeling of a section of a inner hood panel, shown in Fig.14.

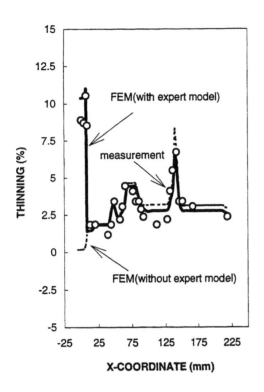

Fig. 11. Comparison of the strain distribution for the section, shown in Fig.15, among FEM simulations and measurement.

Table 1 Material properties of the blank sheet

Properties	Values
Plastic anisotropy parameter	1.72
Hill's yield function parameter	2.0
Strength coefficient [MPa]	489
Work-hardening exponent	0.228
Yield stress [MPa]	169
Young's modulus [GPa]	1.39
Thickness [mm]	0.6
Coulomb friction coefficient	0.18

the circular-and-stepped drawbead model developed describes the deformed state of the sheet metal around the edge of the blank holder more precisely than the usual model which requires the long working time for setting the boundary condition.

5. CONCLUSIONS

The expert drawbead models have been proposed which calculate the drawing characteristics of the sheet. The drawbead restraining force and pre-strain of basic drawbeads are formulated with a correction of the theoretical value by comparing with the measurements, and those of the combined drawbeads are computed from the basic drawbeads'. The following conclusions are induced in the research :

(1) Since the drawbead restraining force and pre-strain of basic drawbeads are mainly affected by the friction, they are evaluated employing the friction coefficient modified, based on the experimental measurements.

(2) The drawbead restraining force and pre-strain of combined drawbeads generated by the expert model are in good agreement with experimental measurements.

(3) The proposed models are more accurate in solution and more efficient in computation time than usual models.

ACKNOWLEDGEMENTS

This study is supported by Korean Ministry of Education through Research Fund.

REFERENCES

Levy, B.S. (1982). Development of a Predictive Model for Draw Bead Restraining Force Utilizing Work of Nine and Wang, *J. Applied Metal Working*, Vol. 3, No. 1

Nine, H.D. (1978). Drawbead Forces in Sheet Metal Forming. In : *Mechanics of Sheet Metal Forming*, pp. 179-211.

Nine, H.D. (1982). New Drawbead Concepts for Sheet Metal Forming. *J. Applied Metal Working*, Vol. 2, No. 3, pp. 185-192.

Stoughton, T.B. (1988). Model of Drawbead Forces in Sheet Metal Forming. *15th IDDRG*, pp. 205~215.

Wang, N.M. (1982). A Mathematical Model of Drawbead Forces in Sheet Metal Forming. *J. of Applied Metal Working*, Vol. 2, No. 3, pp. 193-199.

Weinmann, K.J. and L.R. Sanchez (1988). A General Computer Model for Plane Strain Sheet Flow and its Application to Flow between Circular Drawbeads. *15th IDDRG*, pp. 217~226.

VR BASED CAD SYSTEM FOR PROTOTYPING ELECTRIC MACHINES THAT ENABLE TESTING THEIR ELECTRO-MAGNETIC INFLUENCES TO HUMANS

Loskovska S., Kacarska M., Grcev L.

University "Sts. Kiril and Metodij"
Faculty of Electrical Engineering
Karpos II, bb
Skopje, Macedonija

Abstract: The increased number of electromagnetic fields sources and the growing use of electronics equipment in almost all human's working areas, makes the human life dependent on the quality of products that produce or/and use electromagnetic energy. Therefore, the important parameter of equipment quality becomes capability of its efficient utilization without destroying quality of the living environment and/or causing health problems. CAD system for prototyping electric equipment, currently underdevelopment, enables design of electric devices and testing electromagnetic fields influences to humans who are nearby or/and use designed devices. Requirements of the system that arise from the problems connected with design of electric machines with lowered emission of fields are described. System structure is presented. Examples of visualized electromagnetic fields effects generated with the prototype version of the systems monitoring part are selected to present application areas of the system

Keywords: device design, electromagnetic fields, virtual reality

1. INTRODUCTION

For the past twenty years there has been apprehension that the electric and magnetic fields (EMF) arising from the supply and use of electricity may have adverse effects on health. However, in spite of great number of large-scale studies, many of which are expected to continue for many years to come, the scientific evidence to date is not conclusive in assessment of the risk associated with low-frequency EMF exposure (-,1995). In the mean time, a necessity has been imposed on the in-dividuals and organizations concerned with electricity to manage the EMF issue (-,1996). In the situation of uncertain risk it was necessary to develop a strategy of caution. Such policy, often referred as "prudent avoidance", appears to be the approach most widely accepted (Perry, 1994). It is based on the fact that evidence exists suggesting enough of a potential problem to warrant further research, which consequently means that doing nothing may not be the best policy. On the other side, there are enough unknowns about the magnitude of the risk and which aspects of the fields cause the risk. The steps that constitute prudent avoidance are simple: increasing of the distance from a source of high fields, reduce time spent in high fields, and reduce emission of high fields, provided the cost is reasonable.

As a result manufacturers and suppliers of the products and systems seen as sources of high fields are redesigning or implementing systems that produce lower fields. In some cases, companies have been mandated to do so by state commissions (Perry, 1994), but usually the main driving force are recommendations issued by world renown institutions such as the Swedish Radiation Protection Institute (SSI). Today situation concerning the low frequency magnetic field health risk is similar to the situation concerning the health effects of work with computer displays (VDTs) escalated during the first part of the 80s and culminated in 1986-1988. SSI took the position that voluntary actions employing a principle of caution were highly desirable. Today, about ten years afterwards we can face the fact that nearly all VDTs sold all over the world comply with the recommendations issued in the voluntary SSI standards. SSI has recently issued recommendations on the low frequency magnetic fields (Kivisäkk, *et al.* 1993), which may be expected to be followed by many manufacturers all over the world.

Of special concern is the low frequency magnetic field health risk of some industry workers, which are exposed to fields of extremely high intensity. For example, the workers in 400 kV substations are exposed to magnetic fields between 10 and 100 μT, but welders are exposed to fields between few 1000 and 100000 μT (Weman, 1994). Such high fields are of special concern since the only scientific evidence so far that strongly indicates that EMF may cause biological effects is obtained in laboratory experiments utilizing relatively high fields.

Development of a system for prototyping electric devices with reduced emission of fields is the main interest of the group. It is supposed that system will offer to the developers to create and redesign products that fulfill standards for reasonable costs. The system which is based on the approach of virtual reality (VR), like the others similar systems, offers possibility to developers to create, investigate, interact and make judgments about the functional characteristics of the prototype. It differs from the others, because it considers the influ-

ences of electromagnetic processes to humans, synthesizing performances of CAD systems, 3D graphics interfaces with methods for determination of human's exposure parameters to electromagnetic fields.

The paper contains the description of the problem, requirements and objectives for such system. The system architecture and components are presents together with discussion about the first results.

2. PROTOTYPING PRODUCTS WITH REDUCED EMISSION OF FIELDS

Fig. 1 shows steps that correspond to tasks and operations performed when prototyping an electric device with lowered emission of fields. Entities shown on the figure in fact present parts that should be implemented in VR based prototyping system.

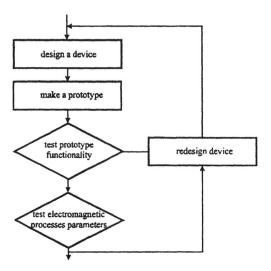

Fig. 1 Design of an electric device

There have already been developed many CAD systems. They are used for designing products, cars, crafts and microchips, to name a few application areas. At present, many of CAD interfaces, that are rooted in the modeling techniques used in physical construction, are based on using standard interface tool kits for interaction (2D mouse, tablets, ...) and construction techniques such as 2D projections for representing three dimensions. There is a fundamental problem here, which is that humans have difficulties in visualizing three-dimensional situations using their two-dimensional description. Some improvements are offered by the use of VR technology for prototyping (Göbel, 1996; Vijay, *et al,* 1996). One of the major potential benefits of VR in this field is opening new ways to build and interact with designs that are not limited on two-dimensional display surfaces.

So far, the phase of making a prototype and testing its functional characteristics consider one of the two approaches: development of a real model and performing several tests on such obtained model; or performing extensive numerical calculations for some functional aspects before producing the product. VR technology can be applied to improve this situation. In virtual prototyping, a product data model is used to build a virtual model, which can be handled like its physical representation. For each model, a functional simulation that represent its complex behavior is performed. This simulation is further used to evaluate various aspects of the construction (Göbel, 1996).

Testing electromagnetic influences of designed products on humans during prototyping phase is rarely considered. Almost all studies are performed for already developed systems and products. The problems that can appear if testing electromagnetic fields influences is applied for already developed products are

→ inability to reduce influences of other electromagnetic sources,
→ repletion of the measurements can not be performed under same conditions,
→ necessity of robust measurement devices which will cover all points of interests,
→ performed measurements generally produce enormous amount of data, and
→ redesign of the model requires additional costs for the new prototype.

The application of VR technology to introduce testing process in the prototyping system has the same advantages as development of VR based system for testing products functionality. Even more, several different mathematical methods can be implemented to calculate various aspects of electromagnetic processes. However, the main advantage would be visualization of the parameters directly on the human' models. The advantage of the approach is an intuitively better understanding of complex situation, besides the presentation of complex numerical calculations within virtual systems implies a loss of details and may result in incorrect data (as a result of reduced amount of data).

3. REQUIREMENTS

For users the system is to provide a powerful application-independent method for:
→ easy description, modification and extension of geometrical properties of prototypes;
→ easy construction and on-line presentation of scenes and testing scenarios;
→ selection of different mathematical methods for electromagnetic parameters, and on-line calculation and visualization of calculated parameters.

These requirements mean that run-time interface of the system must be simple and flexible (Green, et al., 1996). The user can add and delete parts of prototype, add and modify its functional and electric characteristics and change testing environments, modifying only parameters of definitions concerned with the change. In order to reduce extensive effort for interpretation of electromagnetic influences parameters, visualization interface must be expressive and simple.

As addition to previous requirements simplicity of database is important for the system design, too. The database stores definitions for objects, their functional characteristics, testing environments, and scenarios; therefore its organization is essential for systems users and developers. Simplicity of the database will be considerable if it enables fast modeling and prototyping
→ by using the programming approach or text editors to create object; and
→ by allowing the use of already designed tools to create objects.

4. SYSTEM STRUCTURE

Fig. 2 shows system structure and logical flow of the information inside the system and between the system and users. The system currently consists of three parts:
→ 3D based CAD system;
→ Scenario creating system; and
→ Monitoring system.

The CAD system enables design of 3D models with specific geometry, functionality and electric parameters. To specify the geometry of the model each user has to determine model dimensions, materials and colors. Device motion, visible position changes of its parts are

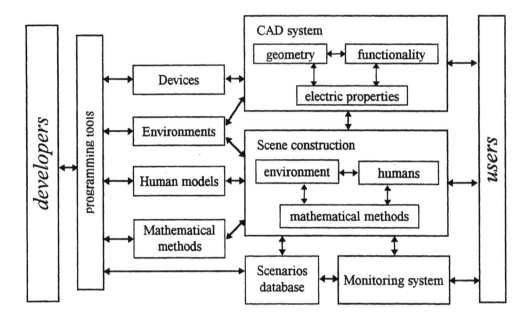

Fig. 2 System structure and information flow

required for specification functionality of each model. Finally, this system provides determining mathematical models for devices' electric sources and their position.

Scenario creating system provides a mechanism for determination and setting testing conditions and parameters. Selection of testing conditions consists of selection virtual environment, the number, placement and activity timing of electric devices, mathematical models of possible additional electric sources, and humans' behavior during testing period. At this point the user selects a method and parameters for representation electromagnetic influences to humans included into the system.

The monitoring system uses the scripts for prepared scenes, performs a task of scenes construction and presents the process to one or more possible output devices. For each human inside the system, it calculates selected parameters and correspondingly visualizes obtained values.

The final version of the system is supposed to work on O2 desktop workstation, and to use cameras, head mounted display (HMD), data gloves, standard keyboard and mice as peripherals. The users can navigate through the system using the standard keyboard, 2D mouse, or HMD and gloves.

5. APPLICATION EXAMPLES

Three different examples are presented to show a few possible application fields of the system. They include:
→ visualization of EMF influences on humans that work with a mixer;
→ visualization of EMF influences on welders; and
→ monitoring EMF effects in virtual kitchen.

Each example is prepared with the prototype version of monitoring system. For presentation are used measured values in real environments. The measurements were done for human model, that is divided on cubic regions with dimensions 5x5x5 cm^3. For each cubic part of the model flux density is measured for all positions of the human when performing particular task. Obtained values are recorded into datafiles, keeping four values for each cubic part (three for the cubic center coordinates and one for the flux density value). The system reads each value when rendering the picture converts it to correspondent color and shows visible sides in calculated color.

Example 1: Visualization of EMF influences on humans that work with a mixer is chosen for presentation because this devices belong to kitchen machines that are relatively often used, and measured values for flux density show bigger values than recommended. Fig. 3 shows visualized flux density on upper part of human body. Different levels of gray scale correspond to values of flux density. Lighter colors (white) is selected to present lower flux density levels, while black points represent values bigger than 20 μT.

Fig. 3 Visualized flux density during work with mixer

Example 2: Another area where the system can be applicated is the area of development equipment and protective clothes for welders. Welders belong to one of the worker groups that are exposed to the highest intensities of EMF. Problems arise because arc welding requires high currents, equipment is close to the welders, and welding cables are often in direct contact with the body. For example, during resistance welding with alternate current (50/60 Hz) of 10kA, environmental magnetic fields with more than 1.000 μT are created.

Fig. 4 Visualized flux density during an arc welding

The system is applicable for the manufacturers of welding equipment for improving the designs in this areas. For example, the system can be used to determine position of cables and supply and return conductors. Another possible application is development of protective clothes for welders. Fig. 4 presents effect of EMF for an arc welding with AC current (50 HZ and

200 A). Flux density values are in range from 54 μT to 226 μT.

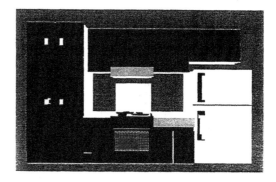

Fig. 5 A model of virtual kitchen (the left side from the door)

Example 3: This example is selected to show more general application of the system (Loskovska, et al 1997). Development of a virtual kitchen and investigating EMF influences to humans inside the kitchen has several reasons. When humans are at home they spend the most of their time in the kitchen or living room and most of instruments in the kitchen are sources of electromagnetic fields. The use of the system for monitoring EMF effects in virtual environments where more electric devices are present, can help when performing extensive studies. Advantages are the same testing condition for several successive measurements, setting different methods for EMF methods calculation, possibility of kitchen elements rearrangements, ...

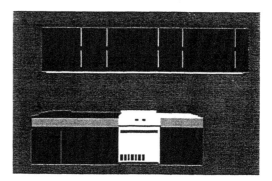

Fig. 6 A model of virtual kitchen (the right side from the door)

Fig. 5 and Fig. 6 show parts of the experimental kitchen. Electric sources considered are a refrigerator, cooker, mixer, coffee machine and dish washing machine. Electric installations are placed on both walls on approximately 1,90 cm from the floor. The other parts are not electric sources.

Fig. 7 and Fig. 8 present visualized flux density values on a standing human in front of dish washing machine (approximately 10 cm from it) when all devices in the kitchen are working.

Fig. 7 Visualized flux density on human in a kitchen

Fig. 8 Visualized flux density on human

6. CONCLUSION

Although the work is still in the research stages and the system has not yet been evaluated, there are many potential benefits of it. Advantages offered by the system are:
→ electromagnetic influences are calculated directly by monitoring human's position during the work;
→ the same pattern of human behavior can be used in several environment prototypes;
→ the same pattern of human behavior can be used for different calculation methods; and
→ having a pattern of the behavior, standards will be set according to duration of the operation held in such environment.

Monitoring system and implementation of several mathematical methods for influences calculation are considered as the next step of the project.

REFERENCES

-, (1995) *Electric and magnetic fields and cancer. An update*, CIGRE Working group 36.06, *Electra*, No. 161, pp. 131-141.

-, (1996) *EMF Issue Management*, Panel 4, 1996 CIGRE Session, Paris, France.

Göbel, M., (1996) Industrial applications of VEs, *IEEE Computer Graphics and Applications*, January, pp. 10 - 13

Green, M., Halliday, S., (1996) A Geometric Modeling and Animation System for Virtual Reality, *Communications of the ACM*, Vo.39, No.5, pp.46 - 53

Kivisäkk, E., Moberg, L., (1993) SSI Policy: Health Risks form Electromagnetic Fields, *SSI News*, No. 2.

Loskovska, S., Ololoska-Gagovska L, Kacarska, M., Grcev, L., (1997) Visualization on Electromagnetic Fields Influences on Humans in Virtual Kitchen, *CSSIT'97*, Las Vegas 30. June - 3 July, (to appear)

Perry, T.S. (1994) Toady's View of Magnetic Fields, *IEEE Spectrum*, Vol.31, No. 12, November, pp. 14-23.

Vijay, M. Swami, C. Edmond Prakash, (1995) Voxel-Based Modeling for Layered Manufacturing, *IEEE Computer Graphics and Applications*, Vol.15, No.6, pp.42-47

Weman, K., (1994) Health Hazards Caused by Electromagnetic Fields During Welding, *Svetsaren (Sweden)*, Vol. 48, No. 1, pp. 14-16.

PATH COMPENSATION WITH RESPECT TO MANUFACTURING TOLERANCES

Tae-Il Seo   Philippe Dépincé   Jean-Yves Hascoët

Institut de Recherche en Cybernétique de Nantes
UMR CNRS 6597 - Ecole Centrale de Nantes
1, rue de la Noë, BP 92101, 44321 Nantes Cedex 3, FRANCE
Tel. : (33) 2.40.37.16.71 Fax : (33) 2.40.74.74.06
e-mail: Jean-Yves.Hascoet@lan.ec-nantes.fr

Abstract: In this paper, we present our research dealing with the choice of the reference of compensation with respect to a given tolerance and the compensation tool trajectory in the milling process. The general purpose of our research is to compensate the tool deflection that occurs during the milling process. We consider it under two aspects : surface prediction generated by the tool deflection and surface prediction generated by the contact points between the cutting tool flute and the workpiece. These two aspects imply two different approaches to carry out the compensation method, called *mirror method*, which can generate a new tool trajectory. In order to compare these two approaches, we present some practical examples.

Keywords: CAD/CAM, Machine tool, Path Compensation, End Milling, Manufacturing Tolerance

1. INTRODUCTION

The end milling operation is very useful in the fields of industry. Despite the multiplication of NC machines-tools and the increase of CAD/CAM software performances, the dispersions due to the tool deflection make an inaccuracy of milling operation. Therefore, we cannot produce what we hope, in spite of the generation of a nominal tool trajectory by the CAD/CAM system. In order to increase the accuracy of milling operation, we have to take into account an approach to determine a new tool trajectory, which allows to compensate the errors of the tool deflection. Our ultimate purpose is to develop an approach to compensate the tool deflection by the generation of a new tool trajectory without reducing the productivity.

To accomplish our purpose, firstly, we consider a general milling process shown in figure 1. This process schematizes a cutting process with an off-line compensation, which can determine a com-

pensated tool trajectory. This path generator requires three steps, cutting force model, tool deflection model and path compensation.

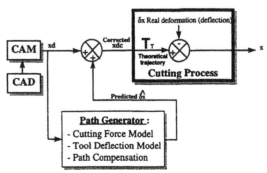

Fig. 1. Milling process with compensation

We have used a cutting force model (Kline *et al.*, 1982), a tool deflection model (Armarego and Deshpande, 1994; Suh *et al.*, 1995) and a path compensation method, called mirror method (Hascoët *et al.*, 1997). For the path compensation, the mirror method requires a reference to search a compensated tool trajectory. This reference of compensation can be considered according to two different aspects of the milled surface generation. We have firstly considered the surface generation by the average tool deflection (Suh et al., 1995; Hascoët et al., 1997). This approach allows us to simplify the process of the milled surface prediction. We have also considered the surface generation by the contact points between the tool flute and the workpiece (Fujii et al., 1979; Tlusty et al., 1991). This approach allows us to predict more exactly the milled surface. For these two different approaches, we present our considerations of the tool path compensation with respect to a given manufacturing tolerance.

2. CUTTING FORCE MODEL

Among the different cutting force models, we have chosen the model of Kline and DeVor, which allows to determine not only cutting forces but also force centers (Kline et al., 1982). The force model is based on the determination of two specific coefficients K_T and K_R which can make the proportional relation between cutting forces and an unit chip section area. From experimental results, we characterize a tool-matter couple by these coefficients K_T and K_R, modeled by a function of the three cutting parameters, radial depth of cut, axial depth of cut and feed rate per tooth. From this model, not only the cutting forces along any direction (distributed along the tool flute) can be computed but also a cutting force center is applied to simplify the tool deflection model. The cutting force center is a point where a concentrated cutting force applied for the cantilever beam model. So, the bending moment produced by a concentrated cutting force and his center is equivalent to that produced by distributed cutting forces.

We have characterized a tool-matter couple. We have chosen an end milling tool, 4 flutes, diameter 6 mm, 30°-helix angle, 30 mm-used length and a steel workpiece (middle carbon steel). The spindle speed is fixed by 1250 RPM. A set of experiments has been realized in order to determine the models of K_T and K_R. All simulation and experimental results presented in this paper are carried out with this specification of cutting tool and workpiece.

3. TOOL DEFLECTION MODEL

To avoid the deflection model becoming sophisticated, the calculation of the deflection is then based on a cantilever beam model (Armarego and Deshpande, 1994; Suh et al., 1995). A concentrated cutting force is loaded at a force center instead of considering the distributed cutting

forces along the tool flute. With a view to improving inaccuracy of the computed tool inertia, we have applied an equivalent tool diameter D_E of a cantilever beam, which is determined in order to make its inertia the same as that of the real tool diameter D (Kops and Vo, 1990). We have calculated the tool deflection as a function of the tool angular position because the cutting force evolves according to the tool orientation in the workpiece. We have also calculated the average tool deflection for several tool angular positions under the same cutting conditions, radial depth, axial depth and feed rate per tooth. In fact, these two types of tool deflection are required to predict the milled surface.

4. COMPENSATION METHOD

The path compensation consists in correcting a theoretical path by integrating errors due to the deflection.

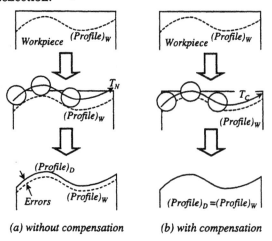

Fig. 2. Path compensation

Figure 2 shows two cases of the milling operations : with and without compensation. In the case of the figure 2-(a), $(profile)_W$ is the nominal profile (wished profile) which has to be obtained, and T_N is the nominal trajectory computed to obtain $(profile)_W$. In fact, T_N is provided by a CAM system without taking into account the effects of the tool deflection. We obtain the deflected profile $(profile)_D$ which is deviated as compared to $(profile)_W$ because of the tool deflection. In the case of the figure 2-(b), our aim is to determine a new trajectory T_C, that associated to the defects of the tool deflection will allow to generate the profile P_C equal to the wished profile $(profile)_W$. To obtain the compensated trajectory T_C, we use a method, called *mirror method*. The mirror method is an iterative process that allows to compute a new path which compensates the tool deflection (Hascoët et al., 1997).

The mirror method is an iterative algorithm to generate a new tool trajectory which allows to compensate tool deflection errors. In each step of iteration, we need a reference to generate a temporary compensated trajectory which compensates previous tool deflection errors. With a view to determine this reference of compensation, we have considered the process of milled surface prediction.

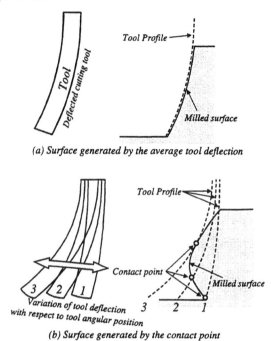

(a) Surface generated by the average tool deflection

(b) Surface generated by the contact point

Fig. 3. Surface generation

5. REFERENCE FOR THE COMPENSATION

In order to apply the mirror method, the question of how to predict the milled surface takes precedence over all others. So, we have taken into account the two following aspects : (1) prediction of surface generated by the profile of deflected tool, and (2) prediction of surface generated by the trace of the contact point between the tool flute and the workpiece. In the first case, we have made an assumption that the milled surface corresponds to the deflected tool profile. We have taken into account a deflected tool profile to generate a milled surface. In the view of calculation time, this assumption gives us some advantages to accomplish the tool trajectory compensation. But, if we want to predict a precise milled surface, we must consider the variation of the tool deflection with respect to the tool angular position. In the second case, we have made a different assumption that the milled surface is generated by contact points between a cutting tool and a workpiece. In this case, we must take into account the variation of the cutting force as a function of the tool angular position. Figure 3 shows the process of the surface generation by the profile of deflected tool and by the trace of the contact points. We have two different ways to predict the milled surface, as a

consequence, we have shown that the two milled surfaces are different (Dépincé et al., 1997).

We present the consideration of a reference of compensation to compare these two methods of surface generations. When a certain tolerance is given, we must respect this tolerance to manufacture a workpiece. So, we also consider the choice of the reference of compensation regarding to the given tolerance.

5.1 Compensation when milled surface is generated by tool profile

In this section, we present a compensation when milled surface is generated by tool profile. The profile of the deformed tool is a curve and its deflection is variable.

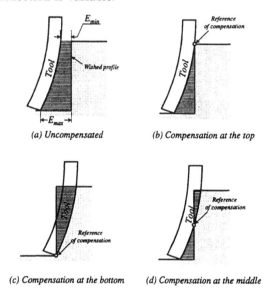

(a) Uncompensated   (b) Compensation at the top

(c) Compensation at the bottom   (d) Compensation at the middle

Fig. 4. Levels of compensation

The error due to the tool deflection differs along the tool axis, when cutting tool mills a workpiece by a nominal trajectory. Figure 4-(a) shows a case of milling process without compensation. In this case, we consider maximum error E_{max} and minimum error E_{min}. The errors E_{max} and E_{min} are distances measured from the wished profile. If E_{max} (or E_{min}) produces an overcut error (or undercut error), the value of E_{max} (or E_{min}) is negative (or positive). For the calculation of the compensation, we have to choose a reference along the cutting tool. Three examples of compensation are presented on figure 4 : (b) compensation with respect to the top of the part, which entails an undercut error compared to the theoretical profile, (c) compensation with respect to the bottom of the part, which entails an overcut error, (d) some intermediate situation between (b) and (c) (Suh et al., 1995).

In the fields of industry, the dimension to obtain is toleranced : one has to produce a surface in a valid domain. In our case, the errors due to the deflection have to belong to this area. We propose to determine the level of compensation (see figure 4 (b), (c) and (d)) according to the toleranced dimension.

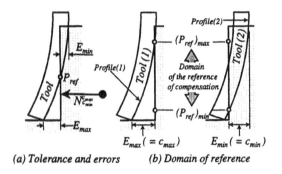

(a) Tolerance and errors   (b) Domain of reference

Fig. 5. Reference of compensation taking into account a tolerance

The first step consists in verifying the possibility to respect a given tolerance $N_{c\min}^{c\max}$. When we manufacture a workpiece by the nominal tool trajectory, if a tolerance $N_{c\min}^{c\max}$ is given, we can define a following discriminant Δ_{tol} to distinguish between a possibility and an impossibility for respecting the tolerance.

$$\Delta_{tol} \equiv (c_{\max} - E_{\max}) \cdot (c_{\min} - E_{\min}) \quad (1)$$

We can verify possibility to respect $N_{c\min}^{c\max}$ by following definitions.

- If $\Delta_{tol} \geq 0$, impossible to respect $N_{c\min}^{c\max}$. The compensation is necessary.
- If $\Delta_{tol} < 0$, possible to respect $N_{c\min}^{c\max}$. The compensation is not necessary.

If we require the compensation, we try to determine a reference of compensation P_{ref} (see figure 5-(a)). After the milling process with the compensated trajectory for the reference P_{ref}, if $\Delta_{tol} < 0$, the compensation is successful. In order to consider a domain of the reference of compensation, we take into account two limits of the reference. In figure 5-(b), two different profiles of the deformed tool are shown. Profile (1) is determined so that the maximum error arrives at the point $N + c_{\max}$. In this case, $(P_{ref})_{\max}$ is the reference of compensation. So, in order to determine $(P_{ref})_{\max}$, we continue to carry out the mirror method until the maximum error will be equal to $N + c_{\max}$. By the same way, Profile (2) is determined so that the minimum error arrives at the point $N + c_{\min}$. In this case, $(P_{ref})_{\min}$ is the reference of compensation. So, in order to determine $(P_{ref})_{\min}$, we continue to carry out the mirror method until the minimum error is equal to $N + c_{\min}$. Therefore, if $(P_{ref})_{\max} > (P_{ref})_{\min}$, the two limits of reference $(P_{ref})_{\max}$ and $(P_{ref})_{\min}$ make a domain of reference of compensation with respect to $N_{c\min}^{c\max}$. All the references of compensation taken inside this

domain $[(P_{ref})_{\max} : (P_{ref})_{\min}]$ assure the respect of the dimension $N_{c\min}^{c\max}$.

We have tested this choice of compensation reference on an example of cylindrical workpiece (Dépincé et al., 1997). In this case, we have succeeded in reducing the maximum error by approximately 80%.

5.2 *Compensation when milled surface is generated by the only one moving contact point*

As mentioned in the previous section, this case is different, but can be considered in the same way. In this case, we choose three points for the reference of compensation.

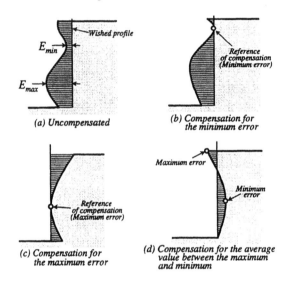

(a) Uncompensated

(b) Compensation for the minimum error

(c) Compensation for the maximum error

(d) Compensation for the average value between the maximum and minimum

Fig. 6. References of compensation

Figure 6 shows uncompensated profile and three compensated profiles. Figure 6-(a) shows the milled profile without compensating. Here, $E_{\max}$ and $E_{\min}$ are the minimum and maximum errors which move as function of the cutting conditions. So, we cannot easily predetermine positions of these errors. But, we can compensate with respect to three points as the previous case. Three examples of compensation are presented in figure 6 : (b) compensation for the minimum error, (c) compensation for the maximum error and (d) compensation for the average value between maximum and minimum errors. Firstly, we can verify a possibility to respect the given tolerance $N_{c\min}^{c\max}$. In figure 7-(a), the milled profile generated by contact points and a given reference of compensation are shown. This milled profile is obtained after compensation with respect to a given reference. As compared to a tolerance $N_{c\min}^{c\max}$, if both $E_{\max}$ and $E_{\min}$ are in the domain of tolerance $N_{c\min}^{c\max}$ (or if $\Delta_{tol} \geq 0$), the compensation is not necessary. But, if not, the compensation must be carried out to respect the given tolerance.

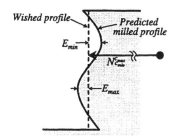

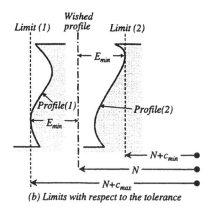

(b) Limits with respect to the tolerance

Fig. 7. Limits of compensation with respect to the tolerance

We can consider the domain of reference of compensation in the way as in the previous section. In figure 7, two profiles generated by contact points are shown. They are different pattern profiles due to the variation of the tool deflection as a function of the tool angular position. To obtain profile (1), we continue to carry out the iteration of the mirror method until the maximum error $E_{\max}$ is tangent to the limit (1). After that we have determined the profile (1) by iterations of the mirror method, we can know a compensated position indicated by $(P_C)_{\max}$. By the same way, for a limit (2), we can determine a profile (2) and a compensated position indicated by $(P_C)_{\min}$. As a matter of fact, $(P_C)_{\max}$ and $(P_C)_{\min}$ are two limits of the compensated position for an arbitrary nominal position among all nominal positions decomposed on a nominal trajectory. If we consider a set of $(P_C)_{\max}$ and $(P_C)_{\min}$ for whole nominal positions of a nominal trajectory, we can obtain a bandwidth which allows us to assure the respect of a given tolerance $N_{c_{\min}}^{c_{\max}}$. Figure 8 schematizes this bandwidth.

In figure 8, as the nominal trajectory T_N is decomposed in N nominal tool positions, $(P_C)_{\max}^i$ is a compensated tool position which can compensate the tool deflection so that a maximum error $E_{\max}$ is equal to $c_{\max}$ for a ith nominal position, and $(P_C)_{\min}^i$ is also a compensated tool position which can compensate the tool deflection so that a minimum error $E_{\min}$ is equal to $c_{\min}$ for a ith nominal position. $(T_C)_{\max}$ is a limit consisting of a set of $(P_C)_{\max}^i$ $(\forall i = 1, 2, 3, \cdots, N)$, $(T_C)_{\min}$ is a limit consisting of a set of $(P_C)_{\min}^i$ $(\forall i = 1, 2, 3, \cdots, N)$.

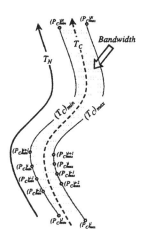

Fig. 8. Bandwidth allowing to guarantee the tolerance

Both $(T_C)_{\max}$ and $(T_C)_{\min}$ constitute a bandwidth. So, we can make the following conclusion :

- If an arbitrary compensated tool trajectory T_C exists in a bandwidth consisting of two limits compensated trajectories $(T_C)_{\max}$ and $(T_C)_{\min}$, T_C can compensate the errors of the tool deflection to guarantee a given tolerance $N_{c_{\min}}^{c_{\max}}$.

For this conclusion to be reasonable, the following condition must be satisfactory :

- The minimum error $E_{\min}$ of the profile (1) must be less than $C_{\min}$, and the maximum error $E_{\max}$ of the profile (2) must be less than $C_{\max}$. $(\Delta_{tol} < 0)$

Unless this condition, the intersection (bandwidth) of the two sets produced by $(T_C)_{\max}$ and $(T_C)_{\min}$ cannot exist.

We have treated a practical example of simulation. Figure 9 shows two cases of milling operations without and with compensation. Firstly, there is a plan to manufacture a workpiece. The purpose of this plan is to cut a surface indicated by a toleranced dimension $17^{+0.15}_{-0.15}$. In case (1), a milling process without compensation is presented. After this milling process, the milled surface stays outside the criteria of the tolerance. So, we cannot obtain the milled surface respecting the tolerance. On the contrary, in case (2), a milling process with compensation is presented. Before the milling process, we have obtained a compensated trajectory by the mirror method. After the milling process with compensated radial depth of cut, we have obtained a milled surface which stays inside the criteria of the tolerance. So, we have succeeded in obtaining the milled surface respecting the tolerance.

We have carried out another simulation for the compensation of the milling operation of a cylindrical workpiece (Dépincé et al., 1997). We have

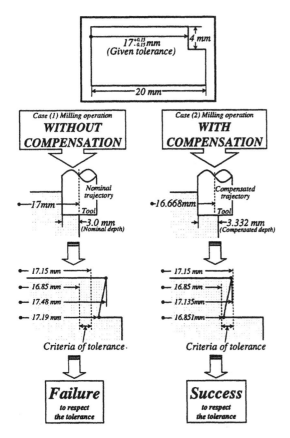

Fig. 9. Example of compensation to respect the given tolerance

fore, we choose the second method to carry out the compensation.

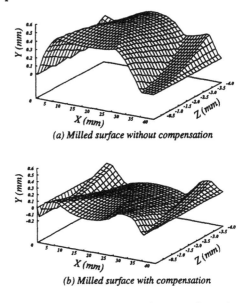

(a) Milled surface without compensation

(b) Milled surface with compensation

Fig. 10. Predicted milled surfaces with and without compensation

chosen the average value of the maximum and minimum errors for the reference of compensation. Two simulation results are shown in figure 10. After compensation, we have globally reduced the errors as compared to the figure 10, but some errors cannot be reduced. We could also take a form tolerance as criteria, instead of a dimension tolerance.

6. CONCLUSION

We have presented the tool path compensation with respect to a given tolerance. Firstly, we have summarized the modeling and methodology to approach reasonable considerations for the path compensation. Then, we have taken into account the determination of reference of compensation. We have described the choice of the reference of compensation for the two different methods to predict the milled surface. The possibility to respect a given tolerance was mathematically verified in the two different cases of the surface prediction. In the impossible case of respecting the tolerance, we have taken into account references of the compensation which allow to guarantee this tolerance. We have presented some examples for the two cases of compensation. We have obtained reasonable results for the two cases. But, the second method allows to more precisely compensate the errors and to predict the milled surface. There-

7. REFERENCES

Armarego, E.J.A. and N.P. Deshpande (1994). Force prediction models and CAD/CAM software for helical tooth milling processes : 3. end milling and cutting analyses. *International Journal of Production Research.*

Dépincé, Ph., J.Y. Hascoët and T.I. Seo (1997). Surface prediction in end milling. *2rd International ICSC Symposia on IIA97 in Nîmes, FRANCE.*

Fujii, Y., H. Iwabe and M. Suzuki (1979). Effect of dynamic behaviour of end mill in machining on work accuracy. *Bulletin of the Japan Society of Precision Engineering* **13**(1), 20–26.

Hascoët, J.Y., T.I. Seo and Ph. Dépincé (1997). Compensation des déformations d'outils pour la génération de trajectoires d'usinage. *16th Canadian Congress of Applied Mechanics CANCAM97 in Québec, CANADA.*

Kline, W.A., R.E. Devor and I.A. Shareef (1982). The prediction of surface accuracy in end milling. *Transactions of the ASME* **104**, 272–278.

Kops, L. and D.T. Vo (1990). Determination of the equivalent diameter of an end mill based on its compliance. *Annals of CIRP* **39**/1, 93–96.

Suh, S.H., J.H. Cho and J.Y. Hascoët (1995). Incorporation of tool deflection in tool path computation. *International Journal of Manufacturing System* **15**(3), 190–199.

Tlusty, J., S. Smith and C. Zamudio (1991). Evaluation of cutting performance of machining centers. *Annals of CIRP* **40**/1, 405–410.

DATABASE DESIGN FOR FLEXIBLE MANUFACTURING CELLS

Per Gullander[†], Sven-Arne Andréasson[‡], Anders Adlemo[‡]

[†] *Department of Production Engineering,*   [‡] *Department of Computing Science*
Chalmers University of Technology,   SE-412 96 GÖTEBORG, SWEDEN

Abstract: One very important aspect when designing modern manufacturing systems is flexibility, such as product flexibility, expansion flexibility and routing flexibility. This paper presents a reference architecture for control systems which has a structure that promotes flexibility. Since a control system is a highly information-intensive system it is very important that the information is stored in a suitable manner. One viable approach is to rely on a database system for the information storage. A database system has inherently a good separation between the structure of a design and the related data. This paper presents the conceptual information design of our generic control system and it is compared with a case study of an existing machining cell.

Keywords: Computer Integrated Manufacturing, Flexible Manufacturing System, Modeling.

1. INTRODUCTION

Flexibility is one of the key issues to achieve manufacturing systems that can adapt to the ever-changing customer needs and incorporate new production technology without extensive investment in time and money. Because of the complexity to design and implement integrated shop-floor control systems, the resulting controllers often become, not only expensive, but also inflexible as well, despite the fact that their components are inherently flexible. For instance, robots and machine tools can be used in a flexible manner in a large variety of application, while the control software that synchronizes these lower-level automation components most often cannot function with other machine tools or products than those it was originally designed for. Each time a new product or a new resource is introduced, the control system must be more or less re-implemented.

The aim of the research reported in this paper, is to support the development of control systems that are truly flexible and, consequently, less expensive since they can be used by many installations and during a long period in time, adopting to changes in the production environment. Based on this, a generic reference architecture for cell control systems has been developed at Chalmers University of Technology, Sweden. The architecture describes how the control system of a generic manufacturing cell should be designed and implemented to be truly flexible.

Since cell-controllers are information-intensive systems, a database system is normally needed to store the required information. A good database design is essential for successful controllers, e.g. see Xiang *et al.*, 1994. Most of the research related to the design of cell control databases has provided specific designs which are only applicable to particular cell types. In contrast to such specific designs, the scope of this paper is to present a conceptual database design which can be used for many cell types.

Next chapter briefly presents the reference architecture for a flexible manufacturing system, a FMS. In chapter 3, a case study is described and the experiences drawn from it concerning the design and implementation of an adequate database are outlined. Then, in chapter 4, information that is normally required by a cell controller regarding products and resources is presented. In chapter 5, the conceptual design of this information is described. Finally, the obtained results are discussed and some conclusions are given.

2. ARCHITECTURE FOR A FMS

The architecture that is presented in this section describes a generic cell-control system; a more detailed description can be found in Fabian *et al.*, 1997.

The main features of the architecture are: (i) physical separation of generic functions from the functions specific to the products and the resources currently in use; (ii) separation of product and resource related information; (iii) storage of the frequently changing data on current products and resources in a database; and (iv) a modular structure of the control system.

To avoid the inflexibility of control systems designed for specific applications, the control-system structure must be separated from the run-time information. The generic structure of the reference architecture is designed and implemented once. The required static and dynamic information is stored in a database, and can be accessed and easily modified as production proceeds at run-time. In this way, only the data in the database has to be changed when the control system is subject to alterations caused by the environment.

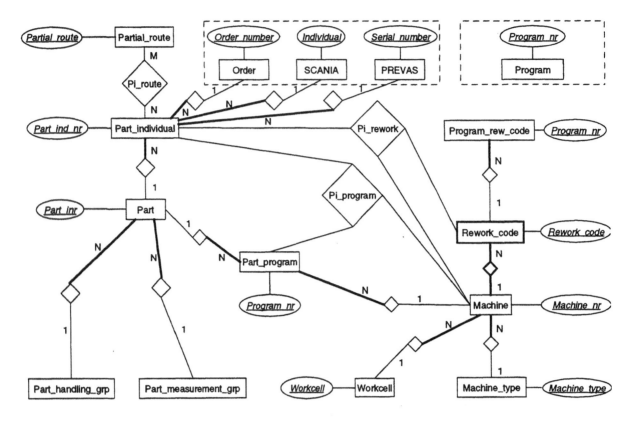

Fig. 1: The conceptual model of the database in the machining cell, modelled as an Entity-Relationship scheme. Only the key attributes are shown.

Based on the information stored in the database, a supervisor can automatically be synthesized. The supervisor is a description (represented by an automaton) that includes all possible routes that the products can pass through a manufacturing cell in order to fulfill the production requirements, see Fabian, 1995. The scheduler chooses one of these possible routes specified in the supervisor. The result from the scheduler is then converted by the dispatcher into commands to the sub-ordinate control system modules that controls the resources.

The controller comprises several co-operating modules, each responsible for specific functions. There are modules that execute the scheduling, dispatching, and monitoring functions, and for each physical resource in the cell, there is a control module that executes all resource-specific control tasks. To increase flexibility, the generic and the specific aspects of this resource control, are separated. These generic resource functions are described in models called Generic Resource Models (GeRMs), see Gullander *et al.*, 1995, which is used for implementation of the resource-specific control-modules. The controller's modules have a client/server relationship and communicate using messages as specified in a scheme called GeMPS (Generic Message-Passing Structure).

Products and resources are described using data models that are implemented in the database and accessible from all the modules in the control system. The resource data model describes capabilities and constraints of the manufacturing resources. The product model includes information on the required operations, gripping information, and error handling, see Andréasson *et al.*, 1995.

3. CASE STUDY OF MACHINING CELL

A cell for production of rear axles for heavy vehicles has been analyzed. The cell consists of a lathe, a milling device, a quality control station, a gantry crane, two output and one input buffer.

Rear axles are manually positioned by the operator at the input station where a bar-code reader registers the incoming parts. When the gantry crane has fetched a new part from the input port, it waits for the milling machine to finish its current task. When the mill has been unloaded and loaded, the gantry waits for the lathe to finish its current job. Finally, when the lathe has been unloaded and loaded, the processed part is left at the output buffer, whereupon the gantry fetches a new part from the input station. The operator can, at any time, request a specific part to the quality control station.

The control system was designed using an object-oriented method which resulted in a number of control modules according to Shlaer and Mellor, 1992. Some of these modules, represents the physical devices, while some other control modules implement functions for scheduling, dispatching and monitoring. The objects communicate with each other through messages synchronizing and requesting operations to be carried out.

3.1 Information model

The database is an essential part of the cell controller in the case study and used extensively to store control information and system status. Fig. 1 depicts the design as an E-R-scheme using the symbols of Elmasri and Navathe, 1994.

The entities Workcell, contain information about the cell, and the template entities from which new entities are dynamically created, e.g. when a product arrives to the cell. There are two template entities: Machine_type and Part. The handling and measurement of the parts depend on the part's geometry, but many part types requires similar handling and measurement. Therefore, each part belongs to a Part_handling_grp and to a Part_measurement_grp.

When a part is identified at the entry conveyor, it is related to one of the Part_individual entities. Each Part individual is related to Partial_route which defines a number of routes through the cell.

All programs needed for the normal operations are listed in Part_Program. The machine specified in the product's route is then used to identify the specific program to be used (relation Pi_program). There is also another type of program called Rework that is used when a product is re-routed to a machine because of poor quality. The correct rework program numbers are identified by the use of the product's rework code and the entities Program_rew_code and Rework_code.

3.2 Evaluation of the case study

The cell control system has in many ways unusually high flexibility. It is, for example, possible to concurrently produce any mix of product types, introduced to the cell in any sequence (i.e. high mix flexibility). Furthermore, the implementation is carried out using a very modular software package and a standardized communication protocol that promote flexibility and openess.

However, the cell control programs are heavily influenced by the product routes and cell configuration relevant at the time of implementation. Therefore, introduction of products that require completely different part flows, or introduction of new machine tools, means that the control programs are affected. Furthermore, by specifying the routes for the products, individual machine tools are identified. If a new machine tool is added, it also affects the product description, and vice versa. And, if there are multiple machines capable of performing a certain operation, one of the machines must be specified in advance, not taking into account the run-time status.

Finally, the routes can only describe sequential routes and not complex lists including alternative operations and parallel operations.

4. DATA DEFINITION

The traditional database design process involves four major steps: (i) *data definition*, i.e. collection of user requirements and data; (ii) *conceptual design*, i.e. implementation-independent data model, both static (e.g. by an Entity-Relationship scheme) and dynamic aspects (data access and manipulation); (iii) *logical design*, i.e. translation of the conceptual schema into a logical schema using concepts of the chosen implementation technology (e.g. relational databases or object-oriented databases); and (iv) *physical design*, i.e. creation of the physical schema taking into account the physical constraints of the operational environment.

A clear separation of these steps simplifies the development of information systems. In this paper, the focus is on database data definition and conceptual design (Chapters 4 and 5 respectively). The logical and physical design are briefly discussed in Chapter 6.

4.1 Object orientation

When designing the data models of the manufacturing system, we use an object-oriented methodology to get a natural, understandable model of the system that is straight-forward to change and maintain, see Jacobsson *et al.*, 1992. It is important to first define the control structure and then the object-oriented information model which will provide the design of the database needed by the system (Vernadat, 1994; DiLeva *et al.*, 1993).

True flexibility can only be achieved when different non-related aspects of a system are kept apart from each other as much as possible. A change should not mean that there has to be changes in more than one part of the manufacturing system. Therefore, the information of the *resources* that are present in the cell and the information of the *products* to be produced should be kept apart. The linking between these is represented by the matching of product operation requirements with resource operation capabilities.

4.2 Product data

In this subsection, the product-related data is presented. A more detailed description can be found in Andréasson *et al.*, 1995. The product data, both dynamic and static, contains the information described in a tuple, $<O, G, E, V>$, that formally models the product information: O describes all the possible combinations of process operations necessary to manufacture each product, G describes how the product should be transported by the mover at different stages of production, E describes the operations to be performed on the product in each error situation (called exceptions), V describes information variables specific for the product individual as well as variables common for all products of the specific type, e.g. flags for identification of error exceptions.

When a product enters a cell, a description of the individual products, e.g. the operation lists, will be calculated from a product template. The operations and error handling of the products are described in more detail in the following sub-sections.

Operations on the products. Information about the set of operations that each product requires must be made accessible to the control system. To be able to easily introduce new products (product flexibility), these operation lists are preferably stored in a database instead of hard-coded into the control programs. It is important to note that these operation lists do not identify which specific resources that should be utilized since this is determined at run-time to take full advantage of the current production status.

Apart from these process operations, a number of operations must also be performed on the product to transport it within the cell. However, these *transport operations* are not really required by the product, but are rather requirements put on production by the manufacturing system's layout. Which transport operation to perform depends on the resources involved in the transportation. Since these resources are not identified in the operation lists, the products' transport operations cannot be included in the operation lists, but are stored separately in the database.

In order to fully utilize the separation of product operations from resource capabilities and to achieve a high degree of flexibility, the products' operation lists should not specify a strictly sequential list of operations. Instead, the process planning should leave as much freedom as possible to the control system regarding the sequences of operation steps. In this way, the routing flexibility of the system will increase. Therefore, the operation lists can describe, not only simple sequences of operations, but also complex combinations including conditional, alternative, and parallel (synchronized and asynchronous) operations. A formal language has been defined to specify operations lists, see Andréasson *et al.*, 1995.

Error exceptions. Error exceptions, i.e. descriptions of the actions that should be performed to recover from an error, are defined and stored in the database. These exceptions are triggered by flags set by the controller module that performs the error diagnosis, or manually by an operator.

4.3 Resource data

A manufacturing system can be viewed as a set of resources needed to manufacture products. The resources that are present in manufacturing cells can be divided into three groups according to their functionality:

– *Producers* that make the necessary physical or logical changes in product properties, e.g. CNC machines (physical changes) and measuring stations (logical changes).

– *Locations* that only store products and cannot change any properties of the product, e.g. local buffers and storage systems.

– *Movers* that transport the products between producers and locations, e.g. AGVs, robots and conveyors.

The similarities between the different resource types makes it advantageous to apply object-oriented principles when modeling the resources. Attributes common for all resources can be described in a resource template model and inherited to its sub-class resources.

When a resource is added to a cell, a description of the individual resource with all its specific characteristics is calculated from a resource template (instanciation), in analogy with what happens when a product arrives to a cell.

In the following, generic resource attributes are defined. Some of these attributes are common for all resource types, while others are specific for the classes of generic movers or generic producers. Locations are not treated specifically but are regarded as a kind of producer that does not change any properties of the product. Hence, both producers and locations are treated as subclasses to a common generic producer entity.

Generic resource attributes. For each manufacturing resource, the following information is normally needed by the control system:

– Resource-descriptive information, i.e. information that defines the individual resource's properties. This information is static or close to static.

– Resource status information, i.e. dynamic information that describes the current status of the resource. This information is retrieved and updated frequently by the control system's modules.

– Resource error handling information, i.e. information that describes the actions to be performed in a given error situation.

– Resource capability information, i.e. information that describes which operations a specific resource is capable of performing. This information is divided into a (nearly) static part defining the programmed capabilities, and a dynamically updated part describing the current capabilities (which is a subset of the programmed capabilities).

– Resource constraints information, i.e. the constraints enforced by a resource, e.g. if products only can flow in one direction between two machine tools.

Resource-descriptive information. Resource-descriptive information defines the individual resource's properties and is either static or only rarely updated. This type of information describes several characteristics of the resource that must be available to the control system, including:

– Data on the resource manufacturer, model, control system version, the resource's location in the factory layout, etc.

– The hierarchical decomposition of the resource. A resource can generally be regarded as an aggregate of hierarchically subordinate resources, e.g. tools and fixtures.

– The number of products that can be handled concurrently in the specific resource.

Some pieces of this information are specific for the resource class, while other pieces are common for a group of resource types and can be inherited from the super-class.

Resource status information. Information describing the current status of the resources is retrieved and updated frequently by the control system's modules. The status information includes:

– Current state of the resource, e.g. working, finished, empty, loaded, etc.

– Current fixture, current tools available, current product or products, currently loaded program, etc.

– Error code defining the specific error in erroneous situations. This code is stored when error diagnosis has been carried out.

– Operation mode of the resource, e.g. if it is manually or automatically controlled.

The status information can be stored with different levels of detail. The more system status details available, the easier it is to evaluate error situations and to diagnose errors correctly. Hence, a quick restart of the system after failures is easier, leading to an increased availability of the system. The disadvantage of storing detailed information about the system status is that the frequent updates require the database management system and the communication link to be very fast. As a consequence, it is difficult to give general guidelines to the appropriate level of detail of the status information. The experiences drawn from the case study, however, show that an extensive use of a database to store detailed system status is both feasible and advantageous.

Resource error-handling information. Error exceptions, i.e. descriptions of the actions that should be performed to recover from an error, are defined and stored in the database. For the resources involved in an error situation the proper actions to carry out depend on: (i) the error type, e.g. tool or fixture breakage, program fault, or product fault, (ii) the number and types of resources involved, (iii) which specific resources that are involved, and (iv) if a product is involved in the error, and in that case, which product.

Resource capability information. Resource capability information describes which operations a specific resource is capable of performing. Both the producers and the movers have capabilities which can be treated in a similar manner. A separation into two parts is advocated:

– a (nearly) static part defining the *programmed capabilities,*

372

– a dynamically updated part describing the *current capabilities* (which is a subset of the programmed capabilities).

For each such operation, the actual programs to be used in the robots, machine tools etc. must be identified. The reason for this separation is that the capability of a resource changes dynamically. In general, the availability of an operation depends on the availability of the resources utilized in this operation. In a situation where two producers share an expensive tool, both are capable of performing the operation (i.e. a programmed capability) but only the producer that currently holds the tool has the operation available (current capability).

Based on a product design and information about a specific producer and available tools, a program for the producer can easily be developed that execute the operation. If programs corresponding to a certain operation are developed for multiple producers in a system, the system's flexibility is increased.

In analogy with the producers, the movers must also be programmed in advance to be able to perform the transportation operations. In the general case when there are more than one mover present, the flexibility and utilization rate of the system can be increased if multiple movers are programmed to execute each transport operation, whenever possible. For the movers, not only information about which transport operations they are capable of performing should be stored, but also which handshake operations they are capable of performing.

Resource constraints information. The products usually restrict the sequence in which the operations can be performed by defining the allowed operation se-

quences. In analogy to this, the resources, in the general case, constrain the ways production can be carried out. First, a resource cannot perform an operation on any number of products concurrently. The number of products that can be loaded to a particular producer or the number of products that a mover can transport limit the usage of the resource. Secondly, the sequence in which a resource can perform the operations might be restricted, e.g. a certain operation might not be allowed directly after another. These constraints can be described using a language similar to the one used to define the products operation lists.

5. CONCEPTUAL DATABASE DESIGN

The part of the database that is used by the supervisor calculator (mapping) is illustrated in Fig. 2. A conceptual design of the products database is described in Andréasson *et al.*, 1995. Only entities, relationships and key attributes are given while most of the information attributes have been left out. A step of the calculation starts with a part in a given resource when an operation has just been completed.

The calculation must then identify all possibilities to perform the next operation and the actions to be taken to achieve this. In other words, the producers that can perform the operations have to be identified and the corresponding movement operations have to be calculated.

Consider an example when a producer M_1 has finished operation number 1 on part P_1. The next step is to calculate all possibilities how the next operation, e.g. operation number 2 (which is given in the operation list in the product template), could be performed in the cell. The calculator can from the entity Prod_ind_Operations get the corresponding global opera-

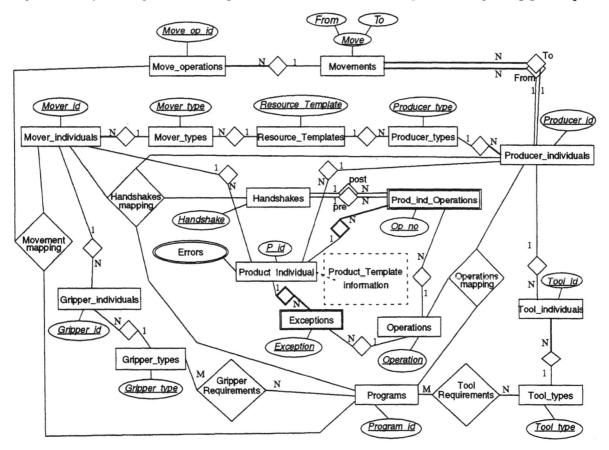

Fig. 2: The Entity-Relationship (ER) schema of the conceptual database design. Only entities, relationships and their key attributes are given.

tion identity from the entity Operations, say Op_n, which is then used for the mapping. The relationship Operations_mapping provides information on producers that include a corresponding program for the given operation. For each program there is a relationship Tool_Requirements that gives the required tools needed by the program. This information should be compared to the tools in Tool_Individuals that are coupled to the corresponding producer in Producer_Individuals. Only producers with the requirerd tools should be considered. Since this information is updated dynamically by the system, only producers that have all the capabilities at the calculation moment will be considered.

The next step is to calculate all possible ways to move P_1 from its present location to each of the possible producers. By doing this, the dispatcher has complete freedom to choose among the producers at run-time. Let us consider how to get P_1 from M_1 to M_2. First it must be decided how to move the product out of its present location, M_1. This is described in the post-operation handshake information given for operation number 1 in the entity Prod_Ind_Operations. This handshake is then mapped on the possible movers in the same way that the operations were mapped on the producers. The differences are: (i) that the producer, in this case M_1, also is considered since it, just like the mover, must have a handshake program, and (ii) that the availability of grippers instead of tools must be considered.

In the same way, the handshake for the receiving producer, M_2, has to be mapped. The type of handshake is given as a pre-operation handshake for operation number 2 in Prod_Ind_Operations. Finally, a mover program must be mapped for each destination and each possible mover. In this way the supervisor calculator can build the supervisor structure that will become the input to the dispatcher. During this calculation an optimization should be performed to mitigate the state explosion.

The run-time system should continuously update the database in order to get schedules that most of the time provide valid suggestions to the dispatcher. There also exist a possibility for the dispatcher to test a suggested alternative producer in the supervisor structure to verify if the considered producer fulfills all the requirements at the dispatching moment. The control system also indicates in the Prod_Ind_Operations entity when an operation has been fulfilled. Then, if a rescheduling occurs, the supervisor calculator only considers unfulfilled operations.

6. DISCUSSION

A considerable amount of research has been carried out that aims at increasing the flexibility of manufacturing cells in order to reduce costs and increase quality and availability. The aim is, however, usually not to achieve a generic controller that fulfills these requirements, as is the case in our research, but instead to find a generic development methodology or a control system applicable to a specific cell. The database design presented is based on both the experiences from the development of a cell controller and from the analysis of a case study presented in Chapter 3.

The scope of this paper is on the data requirement definition and the conceptual design of the database. However, the implementation of this design (i.e. which tools, standards, etc. that are used) is of major importance for the characteristics of the resulting system. The design can, for example, be implemented in a relational database (Vernadat, 1994) or an object-

oriented database (Jørgensen, 1994). The choice of implementation technology affects the system flexibility, and must therefore not be forgotten.

ACKNOWLEDGMENTS

This article was partially supported by the Swedish National Board for Industrial and Technical Development (NUTEK) under grant number 9304792-2.

REFERENCES

Andréasson, S.-A., A. Adlemo, Martin Fabian, Per Gullander and Bengt Lennartson (1995), "Database Design for Machining Cell Level Product Specification", *21st IEEE International Conference on Industrial Electronics Control and Instrumentation, IECON'95*, Orlando, U.S.A., pp. 121-126.

DiLeva, A., P. Giolito and F. B. Vernadat (1993), "M*:OBJECT — An Object-Oriented Database Design Methodology for CIM Information Systems", *Control Engineering Practice*, **1** no. 1, pp. 183-187.

Elmasri, R. and S. B. Navathe (1994), *Fundamentals of Database Systems*, (second edition). Benjamin/ Cummings Publishing Company.

Fabian, M. (1995), *On Object Oriented Non-Deterministic Supervisory Control*, Technical report no. 282, Ph. D. thesis, Control Engineering Laboratory, Chalmers University of Technology, Göteborg, Sweden.

Fabian, M., B. Lennartsson, P. Gullander, S.-A. Andréasson and A. Adlemo (1997), Integrating Process Planning and Control of Flexible Production Systems", Will appear at ECC'97, Brussels, Belgium, July 1997.

Gullander, P., M. Fabian, S.-A. Andréasson, B. Lennartson and A. Adlemo (1995), Generic Resource Models and a Message-passing Structure in an FMS Controller, *1995 International Conference on Robotics and Automation, ICRA'95*, Nagoya, Japan, pp. 1447-1454.

Jacobsson, I, M. Christersson, P. Jonsson and G. Övergaard (1992), *Object-Oriented Software Engineering - A Use Case Driven Approach*, Addisson-Wesley.

Jørgensen, K. A. (1994), Object-Oriented Information Modelling, *Modern Manufacturing: Information Control and Technology*, M. B. Zaremba and B. Prasad (eds.), Springer Verlag Advanced Manufacturing Series, pp. 47-84.

Shlaer, S. and S. J. Mellor (1992), *Object Lifecycles — Modeling the World in States*. Yourdon Press Computing Series, Prentice-Hall.

Vernadat, F. B. (1994), Databases for CIMS and IMS, *Modern Manufacturing: Information Control and Technology*, M. B. Zaremba and B. Prasad (eds.), Springer Verlag Advanced Manufacturing Series, pp. 85-114.

Xiang, D., C. O'Brien and C. S. Syan (1994), Information Models for Control of a Flexible Manufacturing Cell, *Proceedings of Fourth International FAIM Conference*, Virginia, USA, pp. 79-88.

TOOL MACHINE ALLOCATION PROBLEM IN A CELLULAR MANUFACTURING SYSTEM

Braglia M.*, Zanibelli A.**, and Gentili E*.

* *Dipartimento di Ingegneria Meccanica, Università degli Studi di Brescia*
Via Branze, 38 - 25123 Brescia, Italy
** *Dipartimento di Meccanica, Politecnico di Milano*
Piazza Leonardo Da Vinci, 32 - 20133 Milano, Italy

Abstract: In this paper the tool allocation problem of a flexible manufacturing system is discussed. Two algorithms are developed and compared with one another in the presence of the "part move" strategy which moves the item to the workstation having the necessary tooling. The tools can not be shared between the machines. The aim is to find the appropriate cutting tool allocation which minimises the part flow between the CNC centers, i.e. the total distance travelled by the material handling system.

Keywords: management systems, optimisation, heuristics, flexible manufacturing systems

1. INTRODUCTION

It is known that several tool management issues affect the productivity of a flexible manufacturing system (FMS). In fact, as indicated in Gray *et al.* (1993), a lack of attention to such issues (e.g., tool allocation, tool inventory control, tool replacement, etc.) is a primary reason for the poor performance of many FMSs.

As described in Hahn and Sanders (1994), the design problem of the tool management system for an FMS starts with the selection of a material flow strategy: *to move parts, to move tools* or a mix of the two policies. The *move part* strategy corresponds to moving the workpiece to the machine having the necessary tooling. On the contrary, the *move tool* strategy moves the cutting tools to the machine where they are needed.

The advantages of the *move part* strategy are mainly to avoid: (i) the cost due to a tool handling system, and (ii) the complexity of the tool flow control. The advantages of the *move tool* strategy are mainly: (i) to avoid the reposition of the part and/or the recalibration of the position of the tool, and (ii) to reduce the work-in-process (a workpiece is released into the shop only when a CNC center becomes

available, Hahn and Sanders (1994)). In conclusion, the *move part* strategy reduces the flexibility and the productivity of the cell. But, on the contrary, it is less costly and easier to manage. Generally, when considering the low performances of the actual commercial tool handling systems, the *move part* strategy dominates the *move tool* strategy in the FMS design, even when taking account of the many potential advantages of the tool sharing philosophy.

With respect to the *move part* strategy, the tool management can be viewed as the policy used to solve the tool allocation problem, that is the optimal distribution of the cutting tools in the tool magazines of the workstations (Veeramani *et al.* (1991)). In this case, one speaks of *static approach* to the tool allocation problem, in contrast with the *dynamic approach* characteristic of the *move tool* policy. With the static approach, given a specific part mix to be processed, tools are irrevocably assigned to a particular machine for a frozen production cycle (or "production window"). The cutting tools are re-allocated only at the end of each cycle. Several static policies have been studied and compared, mainly by simulation (Amoako-Gyampah *et al.* (1992), Amoako-Gyampah and Meredith (1996), Co *et al.* (1990), Hedin *et al.* (1997), Ventura *et al.* (1990)):

resident tools, bulk exchange, sharing tools, tool clustering, tool migration, etc.. It is worth noting that in a static approach, even if a tool is removed from a CNC center, when a part type is completed (see, for example, the tool migration policy), this tool is not put into another machine during the same production cycle.

The aim of this study is to present two different heuristic procedures for the optimal placement problem of tools in the CNC *on-board* magazines (Tool Machine Allocation Problem (TMAP)).

2. THE MODEL

The automatic manufacturing cell considered in this paper can be described as follows:

1. four (similar) CNC centers with an automated tool magazine are placed along a linear rail served by a single conveyor. (See Figure 1: Figures and Tables are reported at the end of the paper.) The distance between two consecutive machines is $d = 5$m;

2. the capacity of each tool magazine is of 6 tool slots;

3. the manufacturing cell can process 50 different parts;

4. the 24 tools, placed in the 4 tool holders, cover all types of production. The problem of loading duplicate copies of certain tools into the CNC *on-board* magazines is not considered it in this paper.

5. the process of each item requires from 5 to 20 tools;

6. each item is characterised by a daily demand volume;

7. tool reliability problem (i.e., the tool replacement strategy due to the wear of the cutting tools) is not considered;

8. the cell does not use an automatic tool handling system. For this reason, if the necessary tools for the process of the item are not present in the magazine of the current CNC machine, the part will be worked on another workstation.

3. THE TOOL ALLOCATION ALGORITHMS

When considering the process requests of the $k = 50$ different workpieces, the final objective is to place the $n = 24$ cutting tools on the $m = 4$ machines minimising the expected (average) distance travelled by the material handling device (MHD), i.e., the movement of the parts from a (CNC) workstation to another one. Two heuristic procedures, based on different approaches, are proposed to solve this particular TMAP.

3.1 An algorithm based on clustering methods

As commonly reported (see, for example, Amoako-Gyampah *et al.* (1992)), to treat the TMAP (which is a combinatorial *cluster* problem) one can adopt procedures similar to the various techniques developed for the *machine cell and part family design problem*. In fact, it is often sufficient to substitute *the magazines of the CNC machines* to *the cells* and *the tools* to *the machines*, respectively, in the two different cluster problems. Of course, when the TMAP is considered, the number and the size of the "cells" (i.e., number of tool slots in the magazines) are known *a priori* (in the present case, six elements/tools for each one of the four groups/machines). According to this philosophy, the algorithm proposes an approach based on a (known) *similarity coefficient*, as in Seifoddini and Wolfe (1987) (HEURISTIC I). The *kxn part-type-tool-type* component matrix $\mathbf{M} = \left[m_{ij} \right]$, where $m_{ij} = 1$ if part i requires tool j, otherwise $m_{ij} = 0$ (see, for example, Table 1), is subjected to a cluster analysis based on the *Jaccard similarity coefficient* method: given two tools i and j, the similarity coefficient is defined as:

$$ s_{ij} = \begin{cases} \dfrac{n_{ij}}{\left(n_i + n_j - n_{ij} \right)} \,, & \text{if } i \neq j \\ 0 \,, & \text{if } i = j \end{cases} $$

where n_{ij} is the number of parts requiring both tool i and tool j, n_i the number of parts requiring a processing on tool i and, analogously, n_j for the tool j. The nxn matrix containing pairwise similariry measures between the tools is called *similarity matrix* S. S is an input of the clustering procedure which groups the tools in an iterative process:

Step 1) group the most similar pair of tools (clusters);

Step 2) revise S by calculating the similarity measures between newly formed clusters (treated as new fictious tools);

Step 3) repeat Steps 1 and 2 until all tools are grouped into the m clusters (i.e., machines).

The similarity level at which two or more tools/clusters are joined together is called *threshold value* (see, Seifoddini and Wolfe (1987)). The results of the clustering procedure can be best illustrated using a particular tree diagram called *dendogram* (see Figure 2). A similarity scale is adopted to show the similarity level at which differents tools/clusters are grouped.

Defined the clusters, i.e. the allocation of the different tools into the m workstations, it is also important to calculate the respective disposition of the machines in the manufacturing cell. In this case it is then necessary to introduce another (final) Step in the procedure:

Step 4) select the optimal physical placement of the m workstations (clusters) along the linear MHD, taking into account all the possible $m!$ combinations (note that m is in general a small value).

3.2. An algorithm based on "local search" methods

A second heuristic procedure (HEURISTIC II) is developed by a new approach to TMAP. As the aim is to reduce the distance travelled by the conveyor, the TMAP is treated as a *facility layout problem* (FLP), i.e. the physical allocation of machines in a layout.

Briefly, the object of an optimal layout is to reduce the *expected* movement of the MHD between the machines of the cell. Therefore, the fundamental value to be determined is the *frequency* f_{ij} of part moves between machine i (MI) and machine j (MJ) in the cell. Moves into and out of the cell are not considered. Generally, in an FLP one defines the following quantities:

n = number of machines;

v_{ij}^k = expected volume of items of type k to be carried from MI to MJ in a given time horizon;

n_{ij} = expected number of different items to be carried from MI to MJ in a given time horizon;

u^k = number of items of type k to be carried in a single trip of the MHD;

d_{ij} = distance (or time) required by the MHD to travel from MI to MJ when the two machines are adjacent to each other;

$$f_{ij} = \sum_{k=1}^{n_{ij}} \left\lceil \frac{v_{ij}^k}{u^k} \right\rceil = frequency \text{ of part moves between}$$

MI and MJ in the cell (where $\lceil x \rceil$ is the smallest integer greater than or equal to x);

$\tilde{f}_{ij} = f_{ij} \cdot d_{ij} = adjusted\ frequency.$

Frequencies, travel distances (times) and adjusted frequencies are given by three matrices, respectively: (i) *flow matrix* (also referred to as *travel chart, cross chart,* or *from-to chart*), (ii) *distance (time) matrix,* and (iii) *adjusted flow matrix.* The *distance matrix* can be inferred from a *machine size table* reporting the machine dimensions. The clearance between each pair of machines is assumed to be constant and included in the machine dimensions. Then, the FLP cost function is given by

$$F(S) = \sum_{i=1}^{n} \sum_{j=i+1}^{n} f_{ij} d'_{ij}$$

where S is a layout solution (i.e., sequence of machines) and d'_{ij} is the travel distance required from MI to MJ when the layout is S. Notice that, for each new layout, it is necessary to create a new $\mathbf{D'} = \begin{bmatrix} d'_{ij} \end{bmatrix}$ distance matrix.

It is possible to treat the TMAP as an FLP if one "replaces" the machines of the FLP with the tools of the TMAP. The generic *frequency* f_{ij} of parts displacements between tool i and tool j is easily determined when considering the route and operation sheets of the n parts and the (daily) demand volumes. With respect to the distances, when two tools, in a given solution S, are assigned to the same workstation, a distance d'_{ij} is imposed between the two tools equal to 0 (in fact, the use of the MHD to move the workpiece is not required). On the contrary, if the two tools are allocated in two different machines, the distance between the two tools is equal to the distance between the two machines (i.e., $d'_{ij} = d$ or a multiple of d).

Moreover, even a generic tool cluster solution S of the TMAP, can be seen as an ordered array of size n: the first c tools are clustered on machine 1, the second c tools in machine two, and so on (see, for instance, Figure 3).

The procedure proposed in this case is derived from a modified version of the heuristic procedure developed by Sarker *et al.* (1994) to solve the SRLP minimising the backtracking cost. The steps of the algorithm proposed here are the following:

Step 1) in correspondence of a given *random* initial tool allocation solution, S*, calculate the *n*x*n* *adjusted flow matrix*;

Step 2) for each tool calculate the sum of the elements of the corresponding row in this matrix;

Step 3) take the tool associated with the largest row sum and test all the neighbourhood solutions obtained by exchanging the tool with all the others, grouped in different clusters (i.e., machines) in S*. Let S the best neighbourhood solution that is obtained;

Step 5) if S is better than S* then S* = S and goto Step 2. Else, take the tool associated with the second largest row sum and test all the neighbourhood solutions as in Step 3. The procedure stops when all tools are tested and no improvement with respect to the starting solution S* is obtained.

This technique may be considered an "intelligent" local search, where the element that must be moved is selected on the basis of the characteristics of the current solution.

In this algorithm the neighbourhood solutions are obtained by *exchanging* the position of two tools in the solution S. In the original procedure of Sarker *et al.* (1994), the neighbourhood solutions are obtained with an *insertion* of the selected "element" in the different positions of solution S. In fact, it is evident that the conventional *insertion scheme* can not be used in this cluster application.

4. EXPERIMENTAL RESULTS

The results of Table 2 it follows that it is possible to obtain better results with the second technique (which does not treat the TMAP as a cluster problem) than those provided by the (conventional) first procedure, despite the little increse of the computation times (from 1 to 3 seconds with a PC Pentium 100 Mhz). The values of the expected distances traveled by the MHD reported in Table 2 are obtained as average on 20 random problems generated in the following way:

1. the number h of tools (processes) required by the route and operation sheet for each of the 50 items to be processed is generated according to a Uniform distribution U(5,20);
2. the type of each of the h tools (ordered with respect to the route sheet) is selected by a distribution U(1,24);
3. the daily demand volume for each items is selected by a distribution U(1,20).

Given these data, it is easy to calculate the corresponding *flow matrix* **F** (see HEURISTIC I), or the *part-type-tool-type* component matrix **M** (see HEURISTIC II).

The results of Table 2 imply a *Relative Percentage Deviation* (RPD)

$$RPD = \frac{28006 - 22806}{28006} \cdot 100 = 18.57 \ \%.$$

This result is probably due to the fact that the second approach (i.e., the FLP) on which HEURISTIC II is based, permits to consider not only the type of tools which are present in the *operation sheet* of each item, but also the order of utilisation (i.e., the *route sheet*). It can be noted that it is difficult to use this important, supplementary information about the workpiece in the TMAP, if one develops a heuristic based on a cluster approach, currently considered in the literature.

5. CONCLUSIONS AND REMARKS

This paper presents two heuristic procedures developed to solve the tool allocation problem in a FMS. It is shown that a technique derived from another combinatorial problem, i.e. the facility layout problem, outperforms an algorithm based on the concept of clustering methods, generally used in the literature. This is probably due to the fact that, with the first approach, it is possible to consider the important information relevant to the order of utilisation of the tools required by a generic item (i.e., the route sheets). Possible future works may concern an extension of this technique, avoiding some of the hypothesises introduced in this paper.

REFERENCES

Amoako-Gyampah, K., J.R. Meredith and A. Raturi, (1992). A comparison of tool management strategies and part selection rules for a flexible manufacturing system. *International Journal of Production Research*, **30**, 733-748.

Amoako-Gyampah, K. and J.R. Meredith, (1996). A simulation study of FMS tool allocation procedures. *Journal of Manufacturing Systems*, **15**, 419-431.

Co, H.C., J.S. Biermann and S.K. Chen, (1990). A methodical approach to flexible-manufacturing-system batching, loading and tool configuration problems. *International Journal of Production Research*, **28**, 2171-2186.

Gray, A.E., A. Seidmann and K.E. Stecke, (1993). A synthesis of decision models for tool management in automated manufacturing. *Management Science*, **39**, 549-567.

Hahn, H.S. and J.L. Sanders, (1994). Performance analysis of a LIM-based high-speed tool delivery system for machining. *International Journal of Production Research*, **32**, 179-207.

Hedin, S.R., Philipoom, P.R. and Malhotra, M.K., 1997, A comparison of static and dynamic tooling policies in a general flexible manufacturing system, *IIE Transactions*, **29**, 69-80.

Sarker, B.R., W.E. Wilhelm and G.L. Hogg, (1994). Backtracking and its amoebic properties in one-dimentional machine location problems. *Journal of the Operational Research Society*, **45**, 1024-1039.

Seifoddini, H. and P.M. Wolfe, (1987). Selection of a Threshold Value Based on Material Handling Cost in Machine Component Grouping. *AIEE Transaction*, **19**, 266-270.

Veeramani, D., D.M. Upton and M.M. Barash, (1992). Cutting-tool management in computer-integrated manufacturing. *International Journal of Flexible Manufacturing Systems*, **3/4**, 237-265.

Ventura, J.A., F.F. Chen and C.-H. Wu, (1990). Grouping parts and tools in flexible manufacturing systems production planning. *International Journal of Production Research*, **28**, 1039-1056.

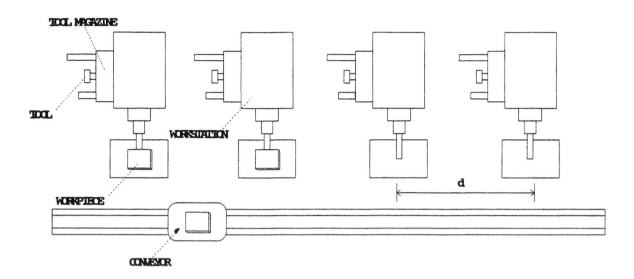

Figure 1. The layout scheme of the manufacturing cell

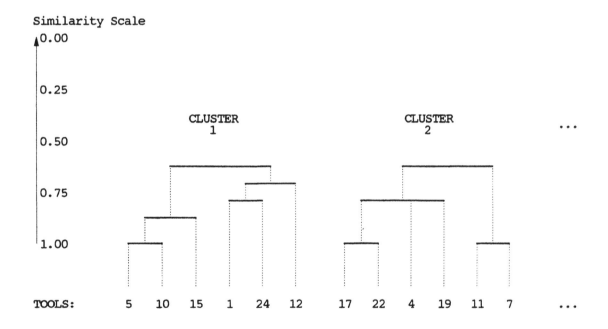

Figure 2. A dendogram example

TOOLS	3	18	15	1	2	8	20	24	11	4	17	9	13	12	21	7	23	6	14	22	10	16	5	19
GROUPS	Machine 1						Machine 2						Machine 3						Machine 4					

Figure 3. An example of tool cluster solution S

Table 1. An example of *part-type-tool-type* component matrix

											T	O	O	L											
		1	2	3	4	5	6	7	8	9	10	11	12	13	14	15	16	17	18	19	20	21	22	23	24
	1	1		1					1				1			1	1		1				1		
P	2			1		1	1					1								1		1			
A	3				1			1	1		1								1				1	1	
R	4		1			1									1			1			1				1
T	...																								
	50		1		1		1							1		1			1			1			

Table 2. Experimental results (on 20 random problems)

	HEURISTIC I	HEURISTIC II
AVERAGE VALUE	28006	22806
AVERAGE CPU TIME	0.6	3.38

OPEN CELL CONTROL SYSTEM FOR COMPUTER-INTEGRATED MANUFACTURING

Sung-Chung Kim
School of Mechanical Engineering
Chungbuk National University
48 Gaeshin-Dong, Cheongju, Chungbuk, 360-763, Korea

Kyung-Hyun Choi
School of Mechanical Engineering
Pusan National University
San 30, Jangjeon-Dong, Kumjung-Ku, Pusan, 609-735, Korea

Abstract: A Flexible manufacturing Cells(FMC) considered as a best production tool for Computer-Integrated Manufacturing (CIM) usually consists of a number of NC machines, sensor systems for the cell monitoring, robots for all par loading and unloading, and a cell controller. Most of the machines comprising an FMC are typically from different machine tool manufacturers with different communication hardware standards, protocols, and control languages. These problems lead to difficulties in integrating cell constituents and employing FMCs in a real manufacturing environment. In this study a cell controller is being developed to deal with above problems. It is designed based upon object-oriented principles, and has tow major modules: configuration and multi-layered control. Since the cell controller developed is independent of specific machine tool requirement, it allows not only operators to choose suitable machine tools for their manufacturing systems but also the machine tools innovation.

Keywords: Computer-Integrated manufacturing(CIM), Cell control, Open System, Object-Oriented Principle

1. INTRODUCTION

As product life cycles reduce, modern manufacturing systems are required to have sufficient responsiveness to adapt their behaviours efficiently to a wide range of circumstances. Responses to these demands include progress in the automation of flexible manufacturing systems, the use of manufacturing knowledge bases, and shorter programming times. The efforts to achieve advanced automated factory bring into focus the development of manufacturing cells with high levels of flexibility, Flexible Manufacturing Cells (FMCs). An FMC may consist of NC machines, robots, conveyors, and sensor devices (i.e., simple sensors and vision systems), and a cell controller which coordinates these

constituents to perform given tasks in an FMC. Most of the machines comprising the cell are typically from different machine tool manufacturers. Each manufacturer of machine tools or machine control units designs a control language which highlights the features of the particular line of machinery they produce. Currently the machine tool producers do not design their machinery with integration to other manufacturer's machinery in mind. A major roadblock to the true integration of these components into FMCs is the abundance of machine programming languages and communication protocols.

In order to deal with these problems, there have been several attempts to produce standards from which all controller manufacturers could draw upon to develop their system(Modern machine shop 1990). However, none has enough support to date to be widely accepted.

In this study a cell controller is being developed to deal with above integration and operation problems problems. It is designed based upon object-oriented principles, and has tow major modules, i.e., configuration module and multi-layered control module. In this way one system is used to setup, program and control a machine tool or set of machine tools independent of the type or vendor of machines. Therefore, a company could purchase the best type of machinery suited for a particular application and not have the control system as important criteria.

2. FLEXIBLE MACHINES CELL

As illustrated in Fig. 1, an FMC may consist of NC machines, robots, conveyors, and sensory devices (i.e., simple sensors and vision systems). In order to perform given tasks in an FMC, these constituents have to operate harmoniously under the overall command of a cell controller.

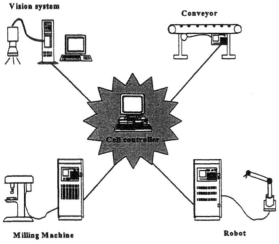

Fig. 1. Typical components and communication in an FMC

The material processing operations in an FMC must be automated, programmable, and easily alterable. From this perspective, CNC turning centres and CNC machining centres comprise the majority of the processing equipment in a machining FMC. These machines are not only capable of being easily reprogrammed, but are also capable of accommodating a variety of tools via tool changer and tool storage system. It is common for a CNC machining centre to contain 60 or more tools (mills, drills, boring, etc.), and for CNC turning centres to contain 12 or more tools (right-hand turning tools, left-hand turning tools, boring bars, drills, etc.) (Chang, et al. 1991).

The use of industrial robots is becoming an established practice in manufacturing. Although the possible tasks performed by robots are varied, their major function in an FMC is material and tool handling and transportation. Since robots can be reprogrammed and retooled relatively quickly and easily, they are the most attractive material handling equipment in FMCs dealing with small batch jobs.

A gripper or an end effector is defined as the special device that attaches to the manipulator's wrist to enable the robot to accomplish a specific task. Because of the wide variations in tasks that are performed by robots, the gripper must usually be customized for a specified job. In some cases a gripper changer may be provided to allow the robot system to adapt to different gripping requirements. A conveyor system is used when materials must be moved in relatively large quantities between specific locations over a fixed path. Most conveyor systems are powered to move the materials along the pathways, other use gravity to cause the load to travel from one elevation in the system to a lower one. In an FMC, conveyors are usually used in conjunction with some other material handling devices, such as robots or automatic loaders, and are generally short in length. A component is placed onto the conveyor at one end and is routed directly to its destination. In an FMC the conveyor must be controlled by the cell controller.

Sensing is one of the most essential aspects of an FMC. The main reason of using sensors in a manufacturing cell is to monitor the system to detect errors, and to obtain data such as part location, orientation, etc., which can be used by the cell controller, or a cell constituent. Currently used sensors in manufacturing cells can be categorized as simple sensors (i.e., limit switches, etc.) and programmable sensors (i.e., vision systems). Limit switches can be mechanical or optical based. They have proven to be reliable and easy to interface in a manufacturing cell. Their purpose includes the detection of the movement of parts, part carriers, or to report on cell status. These sensors can be interfaced to either a specific cell constituent and report to its controller, or directly to the cell controller. A vision system on the other hand is typically a programmable sensor. Vision systems are currently used in cells to perform, among other things, the important tasks of part identification, part location and orientation, and part inspection.

3. OPEN CELL CONTROLLER STRUCTURE

The Cell Controller is designed based upon object-oriented principles, and has tow major modules: configuration and multi-layered control. The graphical configuration editor is used to define the configuration in a manner like simulation systems. The cell configuration contains a list,layout, and characteristics of machine tools, which are installed in a specific flexible manufacturing cess environmnet. The Object-

Oriented nature including encapsulation and multi-instanciation allows to create machine object(MO) from the configured machine tools. These MOs are used as the lowest layer in the multi-layed control module. The multi-layered control module consists of scheduler layer and MO layer, and functions autonomously in coordinating the operatins of cell constituents by sending eents and reports between layers. Since the cell controller developed is independent of specific machine tool requirement, it allows not only operators to choose suitable machine tools for their manufacturing systems but also the machine tools innovation.

3.1 Cell Configuration Module

The cell configuration module plays an important role in building an object data model by interacting with operators. This model includes the characteristics of the equipment forming the cell as well as the relationships they can have with other equipment. The information to be kept can be divided into two categories: static and dynamic information. Static information does not change as a result of cell activities, and is organized based upon a component's general feature, geometric, communication, and sensory characteristics.

The general characteristics include such salient features as the number of axes, programming language, etc. The geometric components of the equipment model contain the geometry of the parts that the equipment is composed of. All equipment has a physical manifestation. Therefore, all equipment requires a geometric description, such as volume, location, and rest position, etc. The geometric characteristics would enable the environment to produce a visual schematic representation of equipment's structure, the detection of collisions with other objects in the cell, and determine the spacial relationship between that piece of equipment and others. A solid geometry model is particularly useful for these purposes.

The communication characteristics describe the means that a specific equipment controller has for exchanging information with other equipment. These vary form digital or analog input/output lines, or more sophisticated protocols such as RS232, IEEE4888, and Ether Net, etc. The communication component of the equipment model captures the set of communication channels present in the equipment, and their parameters, such as voltage levels, limits, rates, and addresses. The sensory components of the model contain information about the equipment's ability to perceive various attributes of its environment using sensors such as pressure, temperature, force, and vision, etc. To fully describe the real-world objects in classes, these characteristics are represented by instance variables within each class.

The dynamic information includes a machine's current status and the time history of the state variables. These are used to coordinate the activities of the components in the cell. It is anticipated that the time history will support the development of error recovering mechanisms for the cell.

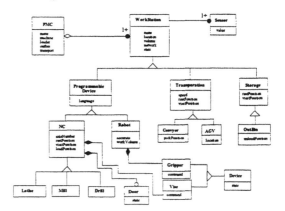

Fig.2. Architecture of the Cell Facility Model abstract

To represent this real-world semantics using an object-oriented model, the generalization, the aggregation, and the composition abstractions are used. Several object classes are defined in the system according to a desired manufacturing cell. The architecture of classes corresponding to the objects is shown in Fig. 2. With reference to Fig.2, an instance of the class **FMC** contains many instances of the class **WorkStation** or the subclasses of the class **WorkStation**. Each instance of the class **WorkStation** uses many instances of the class **Sensor** or the subclasses of the class **Sensor**. The relationship between the instance of the class **Robot** and the instance of the class **Gripper** may be an *association* or a *contain* relationship depended on the robot configuration. The filled diamond symbol represents this relationship. The other two instances of the class **Vise** and the class **Door**, may have the same relationships to an instance of class **NC** as an instance of class **Gripper** to an instance of class **Robot**.

Details of the important classes of the cell configuration model are described as follows:

Class FMC, is the abstraction of common characteristics of flexible manufacturing cells. This class describes the relevant properties of a cell through seven instance variables. These are *name, specification, machines, robots, outbins, transports*, and *controller*. The value of the instance variable *name* is a character string that is identified to the user. The value of *specification* represents the size of FMC object. Other values of instance variables are cell constituent objects.

Class WorkStation is an abstract class. All equipment classes constituting an FMC, such as NC, Robot, AGV, etc., are subclasses of class WorkStation. This class includes geometric, communication, and sensory characteristics of its sub classes as instance variables. These instance variables and method related to them are inherited by all the **WorkStation** subclasses.

Class Programmable Device is used to represent machines which can interpret and execute locally stored programs. Each object of the class **ProgrammableDevice** has its own programming language.

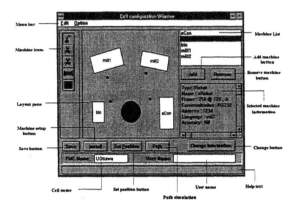

Fig.3. User Interface for the Graphic Editor

As shown in Fig. 3, the graphic editor is responsible for configuring object **aFMC**, that is, building the Cell Object Model. This editor lets a user manipulate objects (instances of the class or subclasses of **Workstation**) in the model by editing their visual representations in a syntactical and semantical consistent fashion. The user is provided with icon-based graphical modelling capabilities, as shown in that Fig. This allows the user to generate two representations of the system at the same time: (1) external (visual) representation, as a real-like icon-based picture on the screen, and (2) internal (logical) representation, as a set of objects representing corresponding system components, which comprise the Cell Configuration Model.

The modelling process starts with the defining general characteristics of the cell. A particular component is modelled by activating the corresponding interactive button. The user is asked to define the characteristics of the component. This editor places an icon and symbol, representing that component, on the screen in the proper location. The symbols and their spacing are scaled according to the physical dimensions. To facilitate the editing process, various panes and buttons are designed and arranged on the graphic editor.

3.2 Cell control module

Cell activity representation

For representation of cell activities, the methodology developed here regards the tasks to be performed by the cell or any of its constituent machines for being primal. Sensory signals indicating the change of state of machines are used to trigger or initiate tasks. A task may be simple and require a relatively short time to execute, or may be complex and lengthy. The methodology developed here may be depicted by a set of diagrams called "Task Initiation Diagrams" and their accompanying rules

The task initiation diagrams are multi-layered. The topmost layer shows the dynamic behaviour of the task-level program of the cell. Fig. 4 shows an example of a Task Initiation Diagram and corresponding cell program. For conciseness, states and tasks are cryptic instead of verbose. The task initiation diagram consists of two components: tasks and states. Tasks in the task initiation diagram would generally consist of a concerted group of subtasks or operations involving more than one constituent of the cell, and in such a case are termed composite tasks. These are shown by the framed boxes in the diagram. As an example **TE1** (load part1 to Mill1) involves concerted operations involving the Machine and the Robot. Tasks involving a single machine are called simple tasks and are shown by a box, e.g., **TE2** (machine part).

The cell state is given at any instant by the collection of states of its constituents. For initiation of a given task, a composite state consisting of a subset of the cell state needs to exist. These composite states are shown in the Task Initiation Diagram by ellipses, e.g., *R11/3* or *M13/4*. The last number of the symbols indicates how many individual states are required to determine this composite state.

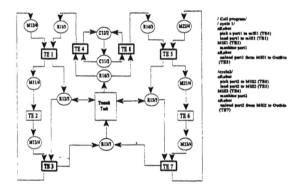

Fig. 4. Task Initiation Diagram example

Cell Controller Operation

Cell operation involves tasks that are performed on single machines independent of others, and tasks that require the cooperation of two or more machines. In cases where a task calls for the coordination of two or more machines, the cell controller has to be involved to ensure proper execution of that task. For tasks involving a single machine, the controllers primary function is to schedule the start of the task, and waits for its completion in order to command the nest task. In order to accomplish these functions, the Holonic controller is designed as a hybrid structure of both hierarchal controller and decentralised controllers as shown in Fig. 5. Holon is an autonomous and cooperative building block of a manufacturing system for transforming, transporting, storing and/or validating information and physical objects. The controller consists of three different layers. The Scheduler, the Decentralized Control layer, and the Virtual Device layer. In the figure, dark lines indicate physical connections and light lines indicate logical

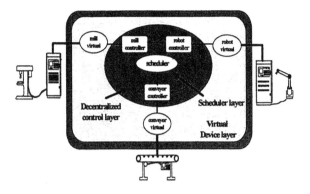

Fig. 5. The structure of the Open Cell Controller

connections. Information and message passing are indicated by arrows.

The Scheduler is a core component which receives the states of all the machines in the cell from the Decentralized Control layer, and decides the next task. It then dispatches the next task to be executed to the Decentralized Control layer. It uses the process knowledge bases that contain the routine cell task rules which are generated from the TID.

The Decentralized Control layer consists of Virtual Holons for the physical machines. Their main role is to perform the harmonization and the cooperation between the cell components in order to carry out the task called for by the Scheduler layer. They provide a device independent interface to the actual cell components by translating the generic commands and error messages of the corresponding machine. The

Virtual holons in the layer communicate and pass messages with each others. Each is correlated to a dedicated Virtual Device and has no connections with other Virtual Devices. A Virtual holons send commands to the corresponding physical machine, and receives the state of that machine, through that Virtual Device in the Virtual Device layer.

The lowermost layer of the controller consists of the Virtual Devices which monitor and continuously mirror, in real time, the state of the physical machine they represent. Each machine state is analyzed by its Virtual Device and reported to the corresponding Virtual holons as required. The Virtual Devices also act as conduits for commands from the Virtual holons to the physical machines.

4. CONCLUSIONS

A open cell controller is developed to deal with problems associated integrating multi-vendor machines into an Flexible Manufacturing Cell. This controller is designed based on object oriented and artificial intelligence principles. The task given to the controller is carried out by the including two modules, that is configuration and multi-layered control modules. As results of this study, it allows not only operators to choose suitable machine tools for their manufacturing system but also the machine tool innovation.

REFERENCES

Chang, T.C., Wysk, R.A., Wang, H.P., (1991) Computer- Aided Manufacturing, Prentice-Hall, Enllewood Cliffs, New Jersey.
Modern Machine Shop (1990), NC/CIM1990 Guidebook, Gardner Publication, Vol.62, No.9A.

AUTHOR INDEX